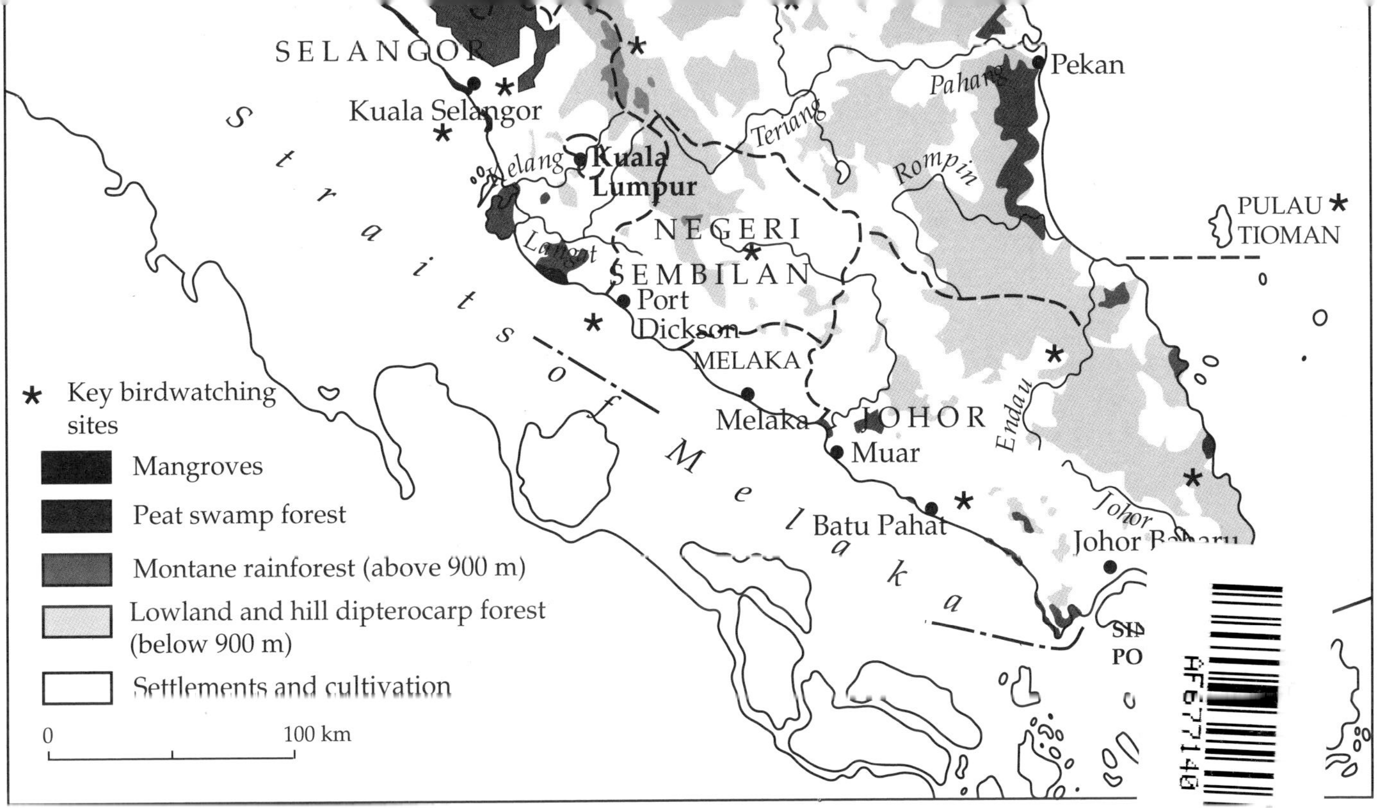
SELANGOR
Kuala Selangor
Kuala Lumpur
Kelang
Langat
NEGERI SEMBILAN
Port Dickson
MELAKA
Melaka
JOHOR
Muar
Batu Pahat
Johor Baharu
Johor
Endau
Rompin
Teriang
Pahang
Pekan
PULAU TIOMAN
Straits of Melaka
Key birdwatching sites
Mangroves
Peat swamp forest
Montane rainforest (above 900 m)
Lowland and hill dipterocarp forest (below 900 m)
Settlements and cultivation
0
100 km

A FIELD GUIDE TO
THE BIRDS OF PENINSULAR MALAYSIA AND SINGAPORE

A FIELD GUIDE TO

The Birds of Peninsular Malaysia and Singapore

Second Edition

Allen Jeyarajasingam

ILLUSTRATIONS BY

Alan Pearson

OXFORD
UNIVERSITY PRESS

OXFORD
UNIVERSITY PRESS

Great Clarendon Street, Oxford OX2 6DP

Oxford University Press is a department of the University of Oxford.
It furthers the University's objective of excellence in research, scholarship,
and education by publishing worldwide in

Oxford New York

Auckland Cape Town Dar es Salaam Hong Kong Karachi
Kuala Lumpur Madrid Melbourne Mexico City Nairobi
New Delhi Shanghai Taipei Toronto

With offices in

Argentina Austria Brazil Chile Czech Republic France Greece
Guatemala Hungary Italy Japan Poland Portugal Singapore
South Korea Switzerland Thailand Turkey Ukraine Vietnam

Oxford is a registered trade mark of Oxford University Press
in the UK and in certain other countries

Published in the United States
by Oxford University Press Inc., New York

First published 1999
This edition published 2012

British Library Cataloguing in Publication Data

Data available

Library of Congress Cataloging in Publication Data

Data available

Plates 40, 41, 43, 68, 69, 70, and 71 were kindly sponsored by Schmidt Marketing
(Malaysia) Sdn. Bhd and Minox.

Typeset by SPI Publisher Services, Pondicherry, India
Printed in China
on acid-free paper by CC Offset

ISBN 978–0–19–963942–7 (Hbk)
978–0–19–963943–4 (Pbk)

1 3 5 7 9 10 8 6 4 2

This book is dedicated to the memory of Alan Pearson's late father Bill Pearson, who gave Alan so much encouragement as a naturalist and an artist. This edition is specially dedicated to the memory of the late Laurence Poh Soon Pin, whose pioneering work in digiscopy to capture digital images of wild birds in their natural habitats has greatly bridged the information gap of the biology of many species of birds in both Peninsular Malaysia and Singapore.

Acknowledgements to first edition

In preparing the text for this guide, I am indebted to the large number of people who pooled in their expertise and efforts, from providing material and information on the various subject matters to the compilation of data as well as reading and reviewing the text.

David R. Wells, who advised on the boundaries as well as scope of materials for this guide, made a detailed check and commented on the checklist on which this guide is based. Christopher J. Hails went through the initial proposal for this guide and provided numerous comments and suggestions. Geoffrey W.H. Davison and Subaraj Rajathurai read the whole text, giving numerous comments and suggestions besides providing additional information. Taej Mundkur read the texts on the waterbird plates and gave numerous comments and useful pointers. Others who reviewed sections of the text and colour plates, allowed me access to their unpublished field notes, and gave valuable comments were Ooi Chin Hock, Glenda Noramly, and Mike H.N. Chong. I would also like to thank others from whose numerous discussions I benefited and who also provided me with their unpublished field notes, reference material, and sounds records: Dennis Yong, Mano Tharmalingam, Noramly Muslim, James Wong, Kanda Kumar Doraisamy, S.A. Hussain, Nather Khan, Yong Hoi Sen, Anthony Walker, Anthony C. Sebastian, Louis Ratnam, John R. Howes, Morten Strange, Sutari Supari, Ho Hua Chew, and Daphne and David N. Bakewell.

I am also grateful to Halimah Haji Ashari, who suggested the importance of including the vernacular names for the birds so that Malaysians could easily relate to them. In preparing the specific vernacular names, I called upon the generous expertise of Hairul Badarul Shah Mohd Hashim and Amir Noor Rashid Osman, who both commented at length on aspects of terminology. For aspects of geography on both West Malaysia and Singapore, I am grateful to Mahendan Tholasy and Tan Mee Guat, who both allowed me access to their material.

I would also like to thank Harry Toolseram, Taej Mundkur, Patrick Lee, Mohd Basil Haji Damanhuri, and Lim Bak Tiang for giving me useful tips and helping me with the finer aspects of Microsoft Word, which I used to write the entire text.

Particular thanks are due to the staff and personnel of the Department of Wildlife and National Parks, Peninsular Malaysia, who gave me permission to examine the bird skins in the Department's museum and go on their field research trips throughout the years, which enabled me to gather and collate information on the birds of West Malaysia: Mohamed Khan Momin Khan, Musa Nordin, Jasmi Abdul, Ebil Yusuf, Saharuddin Anan, Shariff Daim, Abdul Rahman Ismail, Sallehuddin Awang Long, Zainuddin Awang Lela, Khairuddin Abdul Rani, the late Ibrahim Haji Omar, Mohd Noordin Haji Harun, Mior Jamaluddin Mior Hamzah, Mohd Noordin Kassim, and Abu Bakar Ahmad.

Others who helped in some way or other include Syed Ali Abdul Kader, Zainab Hamzah, Kung Wing Chuang, Bakri Tumiran, Vejayan Rajaretnam, Mohd Hafizul Ong Abdullah, Md Rizal Liamen, Azizah Haji Hussain, Adi Ridzwan Abdullah, Nachatrasingam Raju, and Andrew Ponnampalam.

Allen Jeyarajasingam
Kuala Lumpur
March 1998

The artwork for this book would not have been possible without a great deal of assistance, encouragement and inspiration from others. I would like to thank Jon Curson, David Beadle, and Pat Morris for their valuable advice while the project was in its formative stages. Peter Colston, Robert Prys-Jones, and Don Smith provided help during my visits to the Natural History Museum (BMNH) collection at Tring, and James A. Dick similarly helped at the Royal Ontario Museum, Toronto. Paul Sweet at the American Museum of Natural History, New York, gave me valuable information. I would also like to thank the staff of the Jurong Bird Park, Singapore, and those of Melaka Zoo and Kuala Lumpur Bird Park for their assistance during my visits.

Reference material was loaned, in some cases for lengthy periods of time, by John Cox, Richard Lansdown, Sandra McGannity, and Nick Pope. Tim Pankhurst gave a lot of useful advice on lowland forest birds, based on his own extensive field observations. Birding World helped by publishing my request for information on shortwings. David Wells has been extremely helpful throughout this project and I am deeply indebted to him for all his time and advice. Geoffrey Davison has helped and encouraged this project from the beginning and has given many hours of his time. Valuable and constructive comments on the plates were made by Jon Curson, Jonathan East, John Shaughnessy, Ooi Chin Hock, Richard Lansdown, Rasainthiran Menayah, and Sanadure Kartar Singh.

The many hours I have spent in the field in pursuit of Malaysian birds have been made all the more enjoyable by the companionship of others. I have learnt a great deal from many people, but particular mention should be made of Phil Heath, Richard Lansdown, Andrew Lawson, Rasainthiran Menayah, Nick Pope, Crawford Prentice, Nachatrasingam Raju, Sanadure Kartar Singh, and Dennis Yong.

Jonathan East has been a wonderful companion with whom to share the experiences of travelling and bird-watching in Malaysia, bringing with him a fresh eye and a great sense of humour. He has also given me an enormous amount of help during my work back in England.

My wife's family have been extremely welcoming throughout my visits to Malaysia, tolerating my obsession with birds and associated eccentric behaviour. They have been equally hospitable to my travelling companions and remained unruffled by arrivals and departures at all hours of the day and night. Particular thanks go to Christina Yap and Cecilia Scott for making me feel so at home and helping with travel arrangements, permits, and bookings for my field trips. Martin Scott has also been a close friend and put in countless hours of driving for my benefit; it was largely due to his encouragement that I started this project in the first place. I should also like to thank Thomas and Julian Yap, the Ponnampalam family, the Ponniah family, the Jayasuria and Majalap families in Kota Kinabalu, and the Kingham family.

The following people have contributed in some way or another to my work on this book: Charles Chan, Uma Chandran, Martin Hale, Rob Hodgkison, Tigga Kingston, Mary Lacey, Shahroze Merali, Christopher Perrins, David Shepherd, Ivor Sweetnam, and the late Ken Wood. I would also like to thank all my other friends and colleagues, too numerous to name individually, for all their encouragement and support. Towards the end of the project I was given a term's sabbatical by my employers, Saint Felix School, Southwold, to help me complete my illustrations. In particular I would like to thank Susan Campion and the school's board of governors for granting my request for leave of absence.

I am grateful to the staff at Oxford University Press for their considerable help.

I would like to thank my friend and author Allen Jeyarajasingam for all his enthusiasm and commitment throughout this project. His expertise was apparent right from the start and I have enjoyed the many hours we have spent together, either in the field or indoors, discussing aspects of bird identification and of this book. He has also lent me large amounts of reference material which have been invaluable for the illustrations.

My close friend Ben, Benedict Lourdesamy, deserves a special mention for all his companionship and help during my travels. His deep love of the forests, great sense of humour, endless patience, and sharp powers of observation have been invaluable assets, and even our necessary excursions back to 'civilization' together have been really enjoyable.

My family have given me an enormous amount of encouragement and support with my work, making the long, solitary hours at the drawing board less arduous. My late father, Bill, to whom this book is dedicated, would always greet each new plate with particular enthusiasm and interest. Sadly he did not live to see the completed book.

Last, and certainly not least, I thank my wife Anne, who introduced me to the beauty of Malaysia and has helped in so many ways during this project. In return she has tolerated three years of neglect, my possession of a cast-iron excuse to get out of anything I didn't want to do, and the promise that I would do it 'once the book is finished'.

Alan Pearson
Suffolk
May 1998

Acknowledgements to second edition

Revising the text for this guide throughout 2010 and early 2011 proved to be just as meticulous, painstaking, and challenging as writing it from 1994–1998. So many changes had taken place within 11 years since the publication of the first edition in March 1999. Besides changes in the taxonomical sequence, the information gaps of many species had already been filled and these covered voice, behaviour, status, and distribution. In addition, 25 new species had occurred in the review area (including four feral species in Singapore) and this involved writing up new texts and illustrating the additional figures. In revising the text, I'm indebted to the various people who pooled their expertise and efforts, from providing material and information on the various subject matters to the compilation of data as well as the reading and reviewing of the revised text.

Mohamed Rafi Abdul Kudus and Angela Rafi-Francombe were instrumental in getting me to take the first step. They would constantly remind me of the need to have a new and updated edition in view of the wealth of new information as well as the changes that had taken place. They allowed me access to their reference material and I also benefited from the numerous discussions I had with them.

David R. Wells who was our advisor when the project first started in 1994, advised on the boundaries as well as scope of materials for this edition. Having made a detailed study of the draft proposal for the second edition, he gave us the latest taxonomic sequences available, from which we chose *The Howard and Moore Complete Checklist of the Birds of the World – 3rd Edition*. His two landmark volumes, *The Birds of the Thai-Malay Peninsula – Volume One* and *Volume Two* are based largely on this checklist and were the main sources of reference in updating this guide. Taej Mundkur, Paul T. Green, and Richard Noske also reviewed the draft proposal and provided numerous comments and suggestions. Geoffrey W.H. Davison read through substantial portions of the revised text, giving numerous comments and suggestions besides providing additional information on Singapore. John R. Howes read the texts on the waterbirds and wetland habitats, giving numerous comments and useful pointers. Yang Chong made available reference material as well as field notes besides being a good companion in the field. Chiu Sein Chiong supplied a series of sound recordings of the calls of the resident Oriental Honey-buzzard, which I was able to compare with my personal recordings of the vocalizations of this species. Ooi Beng Yean, Hum Kim Choy, Lim Kim Chye, and Lim Swee Yian kindly allowed me access to their numerous digital images for my perusal. Nina Cheung and Eileen Chiang, both Bird i-Witness (BiW) administrators, gleaned information for me from BiW. Bakri Tumiran kindly allowed his illustration of the profile of the equatorial rainforest to be used once again in this edition.

I would also like to thank others from whose numerous discussions I benefited and who also provided me with unpublished field notes and sound recordings: Yeap Chin Aik, Khoo Swee Seng, Caroline Ho, Lim Aun Tiah, Ooi Chin Hock, Wilbur Goh, David Lai, Andrew Sebastian, Choo Jin Wah, Theresa Ong, Irshad Mubarak, Ravinder Kaur, and Diya Rashidah Abu Rahman. David Bakewell and James Eaton provided published papers on latest taxonomical revisions for a number of species and discussed at length splits within a number of species. Being a member of the Malaysian Nature Society Bird Conservation Council Records Committee, I had access to unpublished submissions of rare records, and with the permission of the committee, was able to use them.

In revising the vernacular names for the birds, I called on the expertise of Mike H. N. Chong, who was instrumental in coining suitable descriptive names of birds in *Bahasa Melayu*. His publication *Khazanah Burung Bukit Fraser* (The Ornithological Heritage of Fraser's Hill) co-authored with Sutari Supari, gave suitable descriptive names to most montane resident birds in Peninsular Malaysia. This publication provided the basis for revising the vernacular names of most birds in the review area, recommending descriptive names that were not only species specific but which were also practical to use.

I am especially grateful to my good friend and artist Alan Pearson for his immensely important contributions from revising nine plates and painting two completely new plates as well as illustrating the figures of the new species to correcting other figures in the existing plates. Our discussions were endless from telephone conversations and emails to text messages, the results of which you see in his fine artwork in this volume.

I would also like to record a note of thanks to my immediate superiors at school, principal Haji Mohd Idrus Abd. Hamid and vice-principal Razmi Abdul Razak for relieving me of examination invigilation duties during the school break of November and December 2010, to allow me to proceed with my work uninterrupted. My sincere thanks also go to Nor Kamaliah Hasan, who offered to replace me as chief invigilator during that period.

I am also grateful to Schmidt Marketing (Malaysia) Sdn Bhd for kindly sponsoring the revision of seven plates by Alan Pearson for this second edition. My sincere thanks to Mr Vincent Chow, the general manager, and Mr Woo Kien Ping, the senior product specialist for making this possible.

Finally I would especially like to thank my family: my mother Gladys, my late father P. Jeyarajasingam, and my sister Annabelle, who gave me a lot of encouragement and support when I first took up birdwatching as a hobby 43 years ago; my wife Yatie and daughter Sarah, who had to endure my busy schedule when the revision was in progress; and my father-in-law, Haji Abdullah Mamat, together with my brothers-in-law, Ahmad Ridhuan Abdullah and Ahmad Hafizzi Abdullah, who took me around north-east Kelantan in my relentless pursuit of birds and their related habitats to compile new data.

Allen Jeyarajasingam
Kuala Lumpur
April 2011

I would like to thank Mohamed Rafi Abdul Kudus, together with Angela Rafi, for their enthusiasm and for lighting a bonfire in the necessary place to get this, the second edition, moving.

The provision of additional artwork for this book was made possible by funding from Schmidt Marketing (Malaysia) Sdn Bhd, and I would like to thank Woo Kien Ping (Senior Product Specialist) and Vincent Chow (General Manager) in particular. Their generosity has allowed us to improve some of the plates dramatically.

My friend Lesley Roberts also made suggestions that have had a deep impact on the visual style of the plates.

It has been a pleasure working with Allen J. once more, and again I have learnt a huge amount from him and been constantly encouraged along the way.

The following people have also helped in different ways: Kenneth Wong, Joo Ean Tan, Joo Lett Tan, the Murtaza family, Jennifer Owen, William Wood, and Brian Small. I would like to thank the staff at Oxford University Press for all their work and guidance in bringing this project through to fruition.

It was not possible to thank my friend Jonathan East adequately in the first edition, but I would like to do so again here for his continued humour, encouragement, and support.

My wife's family in Malaysia have continued to look after me with a generosity and kindness well beyond the call of duty and I am deeply indebted to them.

Finally, I thank my wife Anne for allowing the book to take over our lives a second time and for her endless support and patience.

Alan Pearson
Suffolk
April 2011

Contents

APPENDICES

List of plates

The plates are to be found directly after page xxxii.

Anatomy and plumage of a bird

PASSERINE: soft parts and body plumage tracts

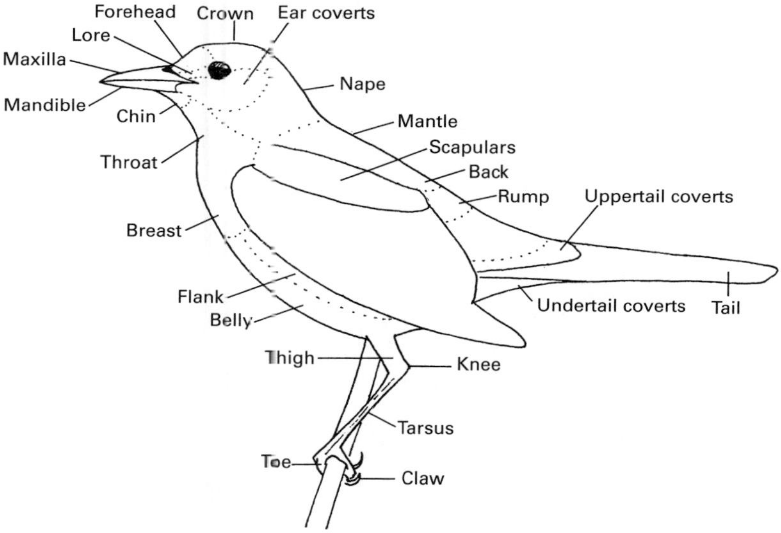

HEAD: soft parts and plumage markings

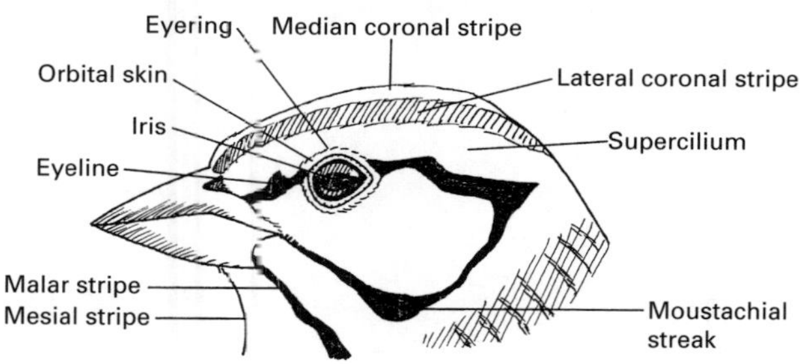

PASSERINE: wing feather tracts

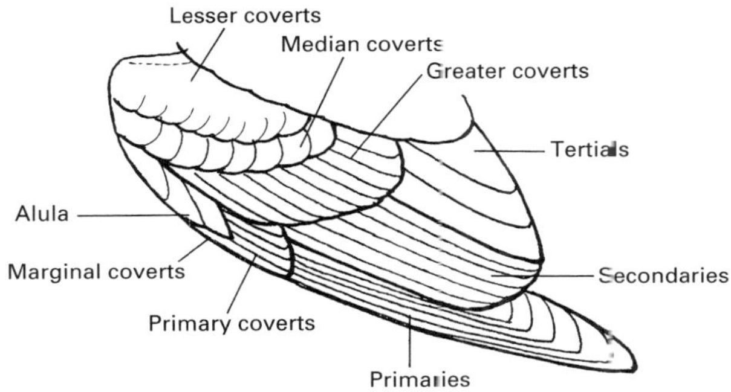

UPPERWING

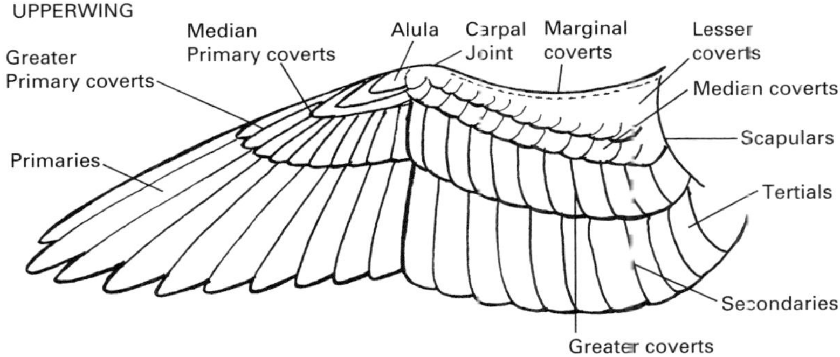

UNDERWING

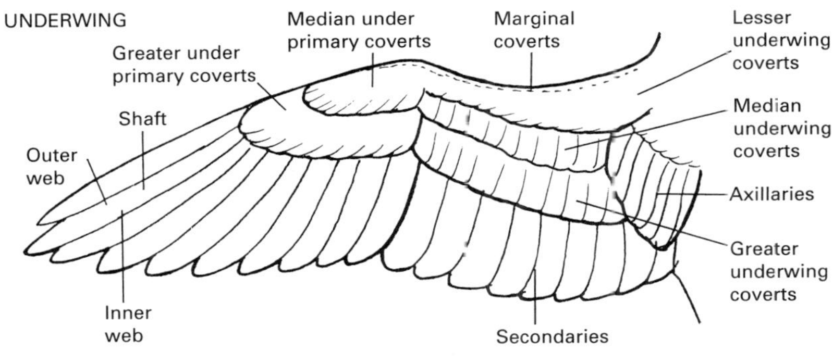

SHOREBIRD: soft parts, wing feather tracts, and body plumage tract

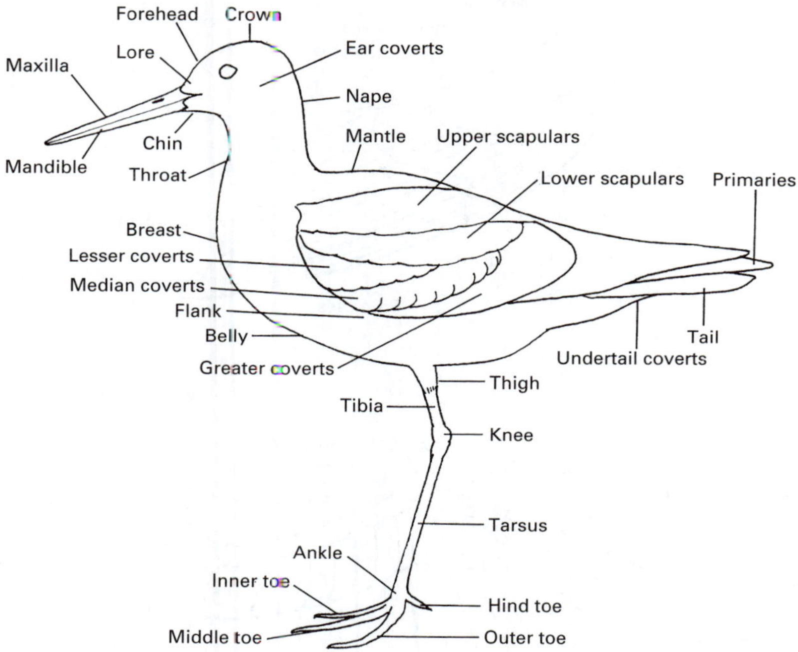

Glossary and abbreviations

Glossary

accipiter loose term to describe goshawks and sparrowhawks of the genus *Accipiter*
aerial pertaining to the air as in aerial feeding by swifts and swiftlets
allopreening the mutual preening of feathers by bonded pair of birds of the same species
antiphonally in duets (see duetting)
arboreal tree dwelling
aquatic living in or near water
Australasia zoogeographical region which includes Indonesia east of Wallace's line, New Guinea, Australia, and New Zealand
autumn passage the movement of migratory birds from northern temperate breeding grounds south to tropical wintering quarters
axillaries the feathers of the axil, i.e. the 'armpit'
bangau Bahasa Melayu term for egret
basal pertaining to base
belatuk Bahasa Melayu term for woodpecker
belly portion of the underparts between the breast and undertail coverts
bill-clattering the clashing of the maxilla and mandible to produce sound as in storks
breast band a suffusion of contrasting colour across the breast
breeding plumage a distinct and usually bright coloured plumage or soft part of the male, acquired during breeding
brood host bird in whose nest has been deposited the eggs of brood parasites
brood parasite bird which lays its eggs in the nest of other birds, leaving them to incubate the eggs and raise the young, e.g. parasitic cuckoos
bukit Bahasa Melayu term for hill
burung Bahasa Melayu term for bird
cap the top of the head which incorporates the forehead, crown, upper nape, and supercilium
carpal joint bend of wing
casque enlargement on the upper part of the bill, as in hornbill
cere bare fleshy structure at base of maxilla containing the nostrils
cheek area refers to both malar area and ear coverts combined
clutch number of eggs
collar distinct band or bar passing around either the front or back of neck
colony a bird species breeding in large numbers at a particular location, e.g. boobies, terns, swiftlets

comb fleshy protuberance on the head as in Red Junglefowl
congener a member of the same genus; the Green Imperial Pigeon *Ducula aenea* and the Mountain Imperial Pigeon *Ducula badia* are congeners because they belong to a common genus, *Ducula*
coronal pertaining to crown
coronal stripe streak on the crown, running from front to back
crepuscular active at dawn and dusk
crest tuft of feathers on the head, which may be raised or lowered
crop a pouch-like enlargement of the oesophagus of many birds, in which food undergoes partial preparation for digestion before passing on to the stomach
crown used in two senses. Either the top of the head of a bird or the topmost branches and foliage of a tree
cryptic dull protective camouflage pattern to blend with the surroundings
diagnostic distinct features in plumage or sound of sufficient character to allow for positive identification
dipterocarp tree belonging to the family Dipterocarpaceae
diurnal active by day
dimorphic existing in two well-defined genetically determined plumage types as in male and female
distal terminal or outer portion
drumming the rapid pecking on wood by woodpeckers to produce sound; associated with territoriality
duetting male and female singing to each other in response, usually during courtship (see antiphonally)
ear coverts feathers overlying the position of the ears
ear tufts erect tufts of feathers on sides of head as in eared-nightjars and some owls
echolocation the use of high-frequency sounds to navigate in dark
eclipse plumage post-breeding plumage on which distinct breeding colours are obscured, such as in male ducks and some male sunbirds
endemic indigenous to a particular geographical location and not found elsewhere
enggang Bahasa Melayu term specifically for Rhinoceros Hornbill but also applied to some of the larger hornbills
epiphytes plants growing on trees, e.g. orchids, mistletoes, and some ferns
evergreen refers to forests that retain their leaves throughout the year as in equatorial rainforests
eyebrow dark or light contrasting colour above eye. For the purposes of this book, this term is specifically for owls which have eyes facing forward (see supercilium)
eyering feathered ring surrounding orbital skin
eyeline thin line of contrasting colour from base of bill through eye (see mask)
face loose term to describe the sides of head, forehead and throat
facial disc the flat, rounded appearance of an owl's face
facial skin an extensive area of bare skin around sides of head, as in some pheasants
feral non-indigenous species that have established breeding populations
first winter the distinct plumage type of a bird in the winter of its first calendar year

fledging period time from when a young bird hatches till it leaves the nest
flight feathers the primaries and secondaries which give buoyancy in flight
frontal shield the bare, horny protuberance or fleshy skin on forehead which extends down to base of maxilla
frugivorous fruit eating
gape the fleshy interior of a bird's bill
glean to search leaves, branches, or bark for food or prey
gliding straight and level flight with wings outstretched or slightly swept back with no flapping
gorget a distinctively coloured patch on the throat or upper breast
Greater Sundas Sumatra, Borneo, Java, and Bali. Together with the Malay Peninsula, collectively referred to as the Sunda subregion
gregarious in large groups
gular pouch loose patch of elastic skin on a bird's throat (or mandible as in pelicans)
gunung Bahasa Melayu term for mountain
hackles long, pointed neck feathers as in Red Junglefowl and Nicobar Pigeon
hepatic the brown colour morph found in some cuckoos
hood dark coloured head, including throat
incubation period time taken for an embryo in an egg to develop before hatching
insectivorous insect eating
jizz shape and general characteristics
jowl heavily rounded appearance of cheek and throat as in Cattle Egret
juvenile the distinctive first plumage of a young bird once it has fledged
kelong fish trap consisting of tall poles, standing in the sea
kleptoparasitic harassing feeding birds and stealing their food as in frigatebirds and jaegers
lallang tall scrubland grass *Imperata cylindrica*
lanceolate slender and pointed like a lance
lappet fold of skin hanging from head
leading edge the forward edge of the wing
lobe an area of skin extending outward from a toe as in grebes and coots
malar stripe stripe running from base of bill to sides of cheek
mantle back, upperwing coverts and scapulars
mandible lower part of bill
mask very broad eyeline which includes the ear coverts
maxilla upper part of bill. The ridge of the maxilla is called the culmen
median pertaining to middle
median stripe streak running down the centre of mid-section
mesial stripe streak running down centre of throat and mid section
mist net (noun) fine silken mesh used to trap birds and bats
mist-net (verb) the act of trapping using mist nets
monotypic genus a genus with only one species, for example, the genus *Eupetes*, has only one species, the Rail-babbler *Eupetes macrocerus*
montane pertaining to mountains; for the purposes of this book, used for areas above 900 m

morph a distinct plumage type that is genetically determined and which may exist in the same breeding population as one or more distinct forms
moult the process of shedding worn-out feathers and growing new ones in place
nape the back of a bird's neck
nasal pertaining to the nose, as in sounds uttered through the nose
nectarine pertaining to the family Nectarinidae (sunbirds and spiderhunters)
nidicolous naked and helpless on hatching and requiring a period of time to attain growth before being able to move around
nidifugous have covering of down or feathers on hatching and are able to move about almost immediately
nocturnal active at night
nomadic the erratic movement from one place to another
non-breeding plumage the usually dull coloured plumage of the male outside the breeding season
non-passerine all orders of birds other than Passeriformes
nuchal pertaining to nape (e.g. nuchal crest, i.e. crest on the back of neck)
nuptial pertaining to courtship display and breeding
ocelli eye like iridescent spots on the plumage of some pheasants
omnivorous feeding on both animal and vegetable matter
onomatopoeic based on sound
orbital skin bare skin surrounding the eye
Oriental Zoogeographical region which includes the Indian subcontinent, southern China, South-East Asia, and the Greater Sundas
Palaearctic Zoogeographical region which includes Europe, North Africa, the Middle-East, northern temperate Asia east to Japan; divided into eastern and western Palaearctic
patagial bar a conspicuous fold along the angle between the upper arm and forearm of birds, especially in some vultures
pan-tropical distributed around the world's tropics
passage migrant migratory bird that stops briefly over at a place to refuel before moving on to wintering quarters
passerine bird in the order Passeriformes (perching birds)
pelagic frequenting the ocean
pied patterned in black and white
pishing making squeaky noises to imitate the alarm calls of birds to entice them to come closer for better views
plumage the feathers covering a bird
plume a greatly elongate feather, usually associated with courtship behaviour as in back and breast plumes as well as nuchal plumes in egrets
polyandrous reversal of sex roles where the female has more than one male mate, leaving them to incubate the eggs and care for the young, e.g. buttonquails, paintedsnipe, and jacanas
polymorphic occurring in many different plumage varieties as in the Oriental Honey-buzzard
primary forest original undisturbed forest, usually void of excessive undergrowth. Also called primeval forest
pulau Bahasa Melayu term for island

racquet elongate, racquet-shaped tail feathers as in some drongos
raptor loose term to describe diurnal birds of prey (Falconidae and Accipitridae)
resident a bird which breeds and remains in the same region throughout the year
rictal bristles tufts of bare feather shafts around base of bill
ringing the process where a numbered metal ring is placed on the leg of a bird
roost resting or sleeping place for birds
scapulars feathers of the shoulder
secondary forest new forest regenerated after logging of primary forest. Usually has luxuriant undergrowth
shaft streaks contrasting line of colour along the middle of feather which forms either a pale or dark streak
shorebird loose term applied to birds of the families Rostratulidae, Haematopodidae, Charadriidae, Scolopacidae, Recurvirostridae, Dromadidae, and Glareolidae
skin bird specimen prepared for scientific study
skulk moving close to the ground in an unobtrusive manner, usually amongst thick vegetation
soaring rising flight on rising thermals with fully outstretched wings without flapping, utilized by storks and large raptors
solar navigation navigation during the day, using the sun as a cue
soft parts iris, bill, eyering, orbital skin, cere, legs, and claws of a bird. Also called bare parts
spangles iridescent spots on a bird's plumage
speculum brightly coloured iridescent squarish patch on a duck's secondaries
split the scientific process of describing the differentiating genetic characteristics of the different subspecies of a single species to reclassify it (them) as a separate species—a very complex process most accurately and validly achieved by DNA analysis rather than analysis and comparison of morphological and vocal differences
spring passage movement of migratory birds from southern tropical wintering quarters north to temperate breeding grounds
spur used in two senses. Either the sharp projection on the hind leg of some pheasants or the pale crescentic markings on the underwing of female and juvenile frigatebirds
stellar navigation navigation at night, using the stars as a cue
storey a level of the forest
streamers elongate tail feathers
subadult a later stage in juvenile plumage prior to acquiring full adult plumage
subsong a subdued or abbreviated version of the usual territorial song
subspecies a specific population of a species in one geographical area which is morphologically distinct from other populations of the same species in different geographical areas throughout its range
subterminal near the end
sundaic pertaining to the Sunda subregion (see Greater Sundas)
supercilium dark or light contrasting line above eye. For the purposes of this book, specifically refers to birds with eyes set on the sides of head as opposed to owls (see eyebrow)
talons the feet and claws of birds of prey

terminal the end
terrestrial ground dwelling
thermalling soaring on rising thermals
thermals hot rising air
tiong Bahasa Melayu term specifically for Hill-Myna but also applied to other mynas, rollers, and Common Moorhen because of white patches on wing
trailing edge the hind edge of the wing
underparts throat, breast, belly, flanks and undertail coverts
underwing the entire under surface of the wing
upperparts entire crown, nape, mantle and uppertail coverts
upperwing the entire upper surface of the wing
vagrant rare and irregular in occurrence
vermiculations very fine wavy barrings on plumage
vent specifically refers to the cloaca but also applied to the feathers of the lower belly and undertail coverts
ventral pertaining to the underside
wattle fleshy lobe of skin on face or base of bill
web used in two senses. Either a fold of skin stretched between toes as in ducks or a vane of a feather
wingbar line of contrasting colour across wing
wing coverts the lesser, median and greater coverts of both the upperwing and underwing
wing lining loose term to describe underwing coverts
wing panel contrasting area on wing, formed usually by pale or brightly coloured edges to folded secondaries and primaries
winter visitor migratory bird that spends the entire winter in a particular tropical location
zygodactylous two toes pointing forward and two backwards as in parrots, barbets, honeyguides, and most woodpeckers

Abbreviations

Directions

NNE North-North-East
NNW North-North-West
SSE South-South-East
SSW South-South-West

Authoritative texts and sources

BAC Bucknill, J.A.S.. and Chasen, F.N. (1927)
BiW Bird I Witness, Malaysia
L and G Lim, K.S. and Gardner, D. (1997)
L and R Lekagul, B. and Round, P.D. (1991)

M and W	Medway, Lord and Wells, D.R. (1976)
P and K	Pizzey, G. and Knight, F. (1997)
RCS	Unpublished MNS-BCC Records Committee submissions
SHY	Siti Hawa Yatim (1988)
W and H	Wang, L.K. and Hails, C.J (2007)
W1	Wells, D.R. (1999)
W2	Wells, D.R. (2007)

Plates

1. **Long-billed Partridge** *Rhizothera longirostris* 36 cm (14 in)
   Large and robust with long, slightly down-curved bill. **p. 74**
   (a) Male: Greyish breast and collar.
   (b) Female: Rufous breast and collar.

2. **Black Partridge** *Melanoperdix niger* 25 cm (9.5 in)
   (a) Male: Glossy black with chestnut brown wings. **p. 74**
   (b) Female: Dark chestnut with whitish throat and belly and black bars on scapulars.

3. **Blue-breasted Quail** *Coturnix chinensis* 15 cm (6 in)
   Yellow legs with hind toe. **p. 74**
   (a) Male: White and black throat markings, slaty blue head and flanks and bright chestnut flanks and belly.
   (b) Female: Buff brown plumage marked with fine blackish bars across breast and flanks.

4. **Malaysian Partridge** *Arborophila campbelli* 25 cm (10 in)
   Black bill, red legs and orbital skin, grey underparts with black and rufous bars on flanks. **p. 75**

5. **Chestnut-necklaced Partridge** *Arborophila charltonii* 28 cm (11 in)
   Dark chestnut breast band, greenish yellow bill with red base and yellow legs. **p. 75**

6. **Ferruginous Partridge** *Caloperdix oculeus* 26 cm (10.5 in)
   Bright rusty rufescent underparts, black eyeline, black and white scales on back and flanks, blackish spots on brown wings. Female lacks spur above hind toe. **p. 75**

7. **Crested Partridge** *Rollulus rouloul* 26 cm (10.5 in) **p. 76**
   (a) Male: Dark glossy blue plumage, red orbital skin and broad maroon crest.
   (b) Female: Green plumage with chestnut wings.

133. **Small Buttonquail** *Turnix sylvaticus* 14 cm (5.5 in)
   Smaller than 134 with pink legs distinctive. Yellow-buff underparts with blackish spots from foreneck to bend of wing. **p. 144**

134. **Barred Buttonquail** *Turnix suscitator* 17 cm (6.5 in)
   Greyish legs with no hind toe. Bright rufous buff flanks, belly and undertail coverts. **p. 144**
   (a) Male: Lacks black throat and upper breast.
   (b) Female: Black throat and upper breast.

133

1(a)
1(b)
2(b)
2(a)
4
5
6
7(a)
7(b)
3(a)
3(b)
134(b)
134(a)

8. **Red Junglefowl** *Gallus gallus* 42–75 cm (17–30 in)
Slender streamlined body, white ear patch and dark greyish legs distinctive. **p. 76**
(a) Male: Bright orange-yellow neck hackles. Long blackish tail with white patch at base.
(b) Female: Reddish chestnut head and neck streaked brown and buff.

9. **Crestless Fireback** *Lophura erythrophthalma* 49 cm (20 in) **p. 78**
(a) Male: Greyish tinge on mantle, flame orange patch on rump, rufous buff tail.
(b) Female: Dark glossy blue with brownish head and buff brown throat.

10. **Crested Fireback** *Lophura ignita* 67 cm (27 in) **p. 78**
(a) Male: Glossy dark blue with flame orange back and white central tail feathers.
(b) Female: Chestnut brown with white streaks on breast and white scales on blackish flanks.

11. **Mountain Peacock-pheasant** *Polyplectron inopinatum* 46–66 cm (18–26 in) **p. 78**
(a) Male: Chestnut brown mantle with green ocelli on wings and tail.
(b) Female: Shorter tail and smaller ocelli.

12. **Malaysian Peacock-pheasant** *Polyplectron malacense* 40–53 cm (16–21 in) **p. 79**
(a) Male: Dark buff brown upperparts marked with blue-green ocelli.
(b) Female: Duller with shorter tail with smaller and black ocelli on mantle.

13. **Crested Argus** *Rheinardia ocellata* 76–238 cm (30–94 in)
Long bushy crest, pale buff supercilium and greyish legs distinctive. **p. 79**
(a) Male: Blackish brown plumage and long, broad tail marked with white spots.
(b) Female: Shorter crest and tail, dark brown plumage.

14. **Great Argus** *Argusianus argus* 75–200 cm (30–80 in) **p. 80**
(a) Male: Lengthened secondaries and long central tail feathers marked with buff brown ocelli.
(b) Female: Short tail, lacks both lengthened secondaries and ocelli.

9(b)
10(b)
9(a)
10(a)
8(a)
8(b)
11(b)
12(b)
11(a)
12(a)
14(b)
14(a)

PLATE 3 **DUCKS AND GREBE**

16. Wandering Whistling-duck *Dendrocygna arcuata* 45 cm (18 in) **p. 82**
(a) Blackish spots on breast, creamy white flank plumes with black streaks.
(b) Flight: White flank plumes separate dark underwing from chestnut flanks

17. Lesser Whistling-duck *Dendrocygna javanica* 41 cm (16 in) **p. 82**
(a) Greyish buff neck and face, tawny rufous belly, dark grey mantle with rufous bars.
(b) Flight: Neck held lower than body with legs trailing tail.

18. White-winged Duck *Cairina scutulata* 76 cm (30 in) **p. 83**
(a) White head with heavy black spotting on crown and nape.
(b) Flight: Grey speculum, black flight feathers and white wing coverts.

19. Cotton Pygmy-goose *Nettapus coromandelianus* 33 cm (13 in) **p. 83**
(a) Male breeding: Glossy dark green upperparts and band around breast.
(b) Female: Dull brown upperparts, band around neck and breast scaled blackish.

20. Gadwall *Anas strepera* 50 cm (20 in) **p. 84**
(a) Male breeding: Dark greyish plumage, black bill.
(b) Female: Black bill with orange-yellow sides.

21. Eurasian Wigeon *Anas penelope* 48 cm (19 in) **p. 84**
(a) Male breeding: Chestnut head and neck, yellow crown; white wing coverts.
(b) Female: Deep brown, greyish brown wing coverts.

22. Northern Shoveler *Anas clypeata* 50 cm (20 in) **p. 85**
(a) Male breeding: Black bill, green head, chestnut belly and flanks.
(b) Female: Brown with dark brown markings, black bill with orange sides.

23. Northern Pintail *Anas acuta* 53 cm (21 in) **p. 85**
(a) Male breeding: Long black tail, brown head, narrow white head stripe.
(b) Female: Brown with dark brown markings on back.

24. Garganey *Anas querquedula* 41 cm (16 in) **p. 85**
(a) Male breeding: Broad white supercilium over eye to nape.
(b) Female: Mottled brown, black eyeline, pale supercilium.

25. Eurasian Teal *Anas crecca* 37 cm (15 in) **p. 86**
(a) Male breeding: Rusty red head with broad curving dark green patch.
(b) Female: Mottled brown with dark eyeline.

26. Tufted Duck *Aythya fuligula* 40 cm (16 in) **p. 86**
(a) Male breeding: Glossy black head, black breast, white belly and flanks.
(b) Female: Dull brown head and upperparts, pale brown belly and flanks.

33. Little Grebe *Tachybaptus ruficollis* 25 cm (10 in) **p. 90**
(a) Breeding: Bright chestnut sides of head and neck.
(b) Non-breeding: Sandy buff sides of head and flanks.
(c) Juvenile: White stripes on head and neck.

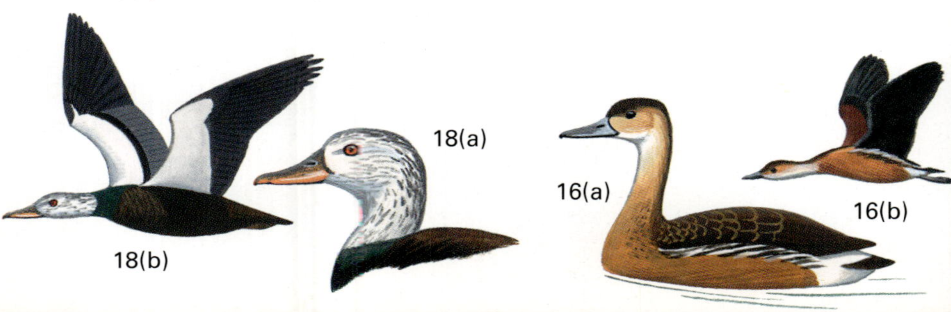

17(b)
17(a)
23(b)
23(a)
25(b)
25(a)
20(b)
20(a)
21(b)
21(a)
24(a)
24(b)
26(b)
26(a)
22(b)
22(a)
19(b)
19(a)
33(c)
33(a)
33(b)

PLATE 4 **SHEARWATERS, PETRELS, STORM-PETRELS, PELICANS, CORMORANTS, AND DARTER**

27. Streaked Shearwater *Calonectris leucomelas* 48 cm (19 in)
Pale head, horn-coloured bill, dark brownish streaks on hindcrown and nape. **p. 87**

28. Wedge-tailed Shearwater *Puffinus pacificus* 43 cm (17 in)
Dark morph: Overall uniform sooty-brown, long wedge-shaped tail. **p. 88**

29. Short-tailed Shearwater *Puffinus tenuirostris* 41 cm (16 in) **p. 88**
(a) Short, square rounded tail with toes projecting beyond.
(b) Conspicuous white markings along centre of underwing.

30. Bulwer's Petrel *Bulweria bulwerii* 27 cm (10.5 in) **p. 88**
(a) Sooty brown with whitish band on secondaries and long wedge-shaped tail.
(b) Shows all dark underwing.

31. Wilson's Storm-petrel *Oceanites oceanicus* 17 cm (7 in) **p. 89**
(a) White rump and black square-cut tail with legs extending beyond tail.
(b) Patters along water surface

32. Swinhoe's Storm-petrel *Oceanodroma monorhis* 20 cm (8 in) **p. 89**
(a) Sooty-brown plumage, dark rump and moderately forked tail.
(b) Does not patter along water surface.

65. Great White Pelican *Pelecanus onocrotalus* 157 cm (60 in)
White plumage, black primaries and secondaries, bright yellow gular pouch. **p. 108**

66. Spot-billed Pelican *Pelecanus philippensis* 140 cm (55 in)
White with greyish tinge, flesh-coloured gular pouch. dark spots on maxilla. **p. 109**

70. Little Cormorant *Phalacrocorax niger* 52 cm (20 in)
(a) Slender build with proportionately smaller and more slender bill than 71.
(b) Swims with body partly submerged. **p. 111**

71. Great Cormorant *Phalacrocorax carbo* 82 cm (32 in)
Large size with black plumage with dark bronze sheen and white throat. **p. 111**

72. Oriental Darter *Anhinga melanogaster* 91 cm (36 in)
(a) Long snake-like neck and long pointed bill.
(b) Swims with body submerged and neck fully extended and exposed. **p. 112**

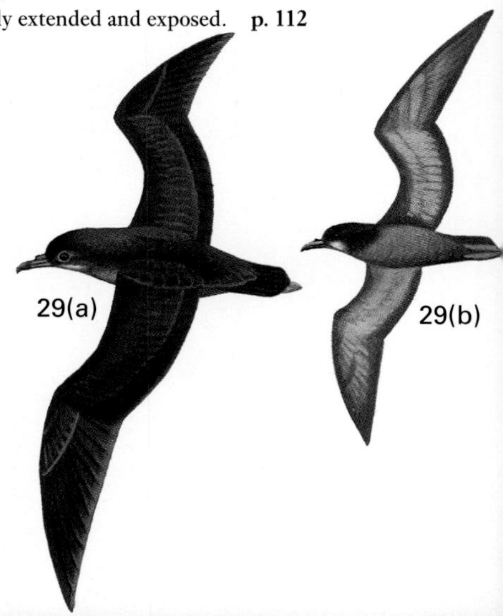

65
66
27
30(a)
30(b)
28
32(a)
31(a)
72(b)
32(b)
71
72(a)
31(b)
70(a)
70(b)

34. Milky Stork *Mycteria cinerea* 97 cm (38 in) **p. 91**
(a) White plumage with black flight feathers and tail.
(b) In flight, black flight feathers contrast sharply with white wing coverts.

35. Painted Stork *Mycteria leucocephala* 102 cm (40 in) **p. 92**
(a) White plumage with closely barred blackish breast band and wing coverts.
(b) Black flight feathers show little contrast with blackish grey wing coverts.

36. Asian Openbill *Anastomus oscitans* 81 cm (32 in) **p. 93**
(a) Feathered face and dull greenish horn pincer-shaped bill diagnostic.
(b) Black flight feathers and tail contrast sharply with white wing coverts.

37. Woolly-necked Stork *Ciconia episcopus* 91 cm (36 in) **p. 93**
(a) White neck and breast, blackish bill, black belly and upperparts.
(b) White neck and breast contrast with black belly and underwing.

38. Sunda Stork *Ciconia stormi* 89 cm (34 in) **p. 94**
(a) Orange-red bill with small knob at base, yellow orbital skin.
(b) In flight, shows all black underwing, breast, and belly.

39. Lesser Adjutant *Leptoptilos javanicus* 114 cm (45 in) **p. 94**
(a) Yellowish orange unfeathered head and neck.
(b) In flight, neck tucked back in an 'S' shape.

40. Black-headed Ibis *Threskiornis melanocephalus* 76 cm (30 in) **p. 95**
(a) White plumage, long black down-curved bill and featherless neck.
(b) In flight, shows patch of red skin along leading edge of underwing.

41. Glossy Ibis *Plegadis falcinellus* 64 cm (25 in) **p. 96**
(a) Rich dark chestnut plumage heavily glossed with green and purple.
(b) In flight, appears all dark.

132. Sarus Crane *Grus antigone* 152 cm (60 in) **p. 143**
(a) Greyish plumage with bare red head and upperneck.
(b) In flight, blackish flight feathers distinctive.

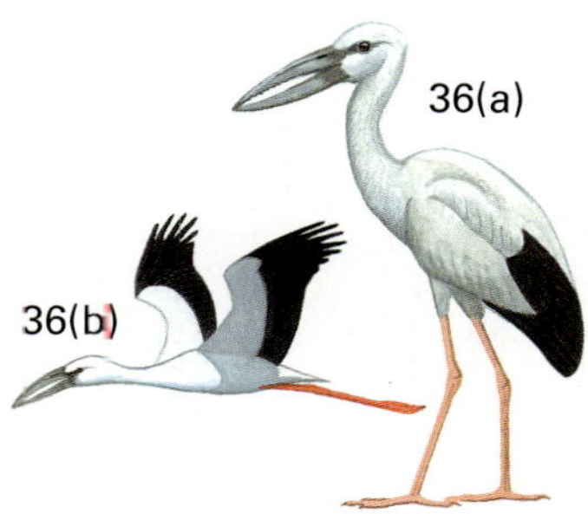

34(a)
34(b)
35(a)
35(b)
37(a)
37(b)
38(b)
38(a)
39(a)
39(b)
41(a)
41(b)
132(a)
132(b)
40(a)
40(b)

PLATE 6 **BITTERNS, NIGHT-HERONS, LITTLE HERON, AND POND-HERONS**

43. Yellow Bittern *Ixobrychus sinensis* 38 cm (15 in) **p. 97**
(a) Male: Pale brown mantle and pale yellowish buff wing coverts.
(b) Juvenile: Bold streaks on upperparts and underparts.

44. Chestnut Bittern *Ixobrychus eurhythmus* 38 cm (15 in) **p. 98**
(a) Male: Deep chestnut head, neck and mantle.
(b) Female: Dark crown and heavily streaked plumage.

45. Cinnamon Bittern *Ixobrychus cinnamomeus* 38 cm (15 in) **p. 98**
(a) Male: Uniform buff upperparts and wings.
(b) Juvenile: Mottled brown upperparts and streaked plumage.

46. Black Bittern *Dupetor flavicollis* 54 cm (21 in) **p. 98**
(a) Male: Black plumage with buff yellow stripe down side of neck.
(b) Female: Black parts replaced with blackish brown.

47. Rufous-necked Night-heron *Gorsachius melanolophus* 48 cm (19 in) **p. 99**
(a) Adult: Bright rufous neck and breast.
(b) Juvenile: Black crown finely streaked white.

48. Black-crowned Night-heron *Nycticorax nycticorax* 61 cm (24 in) **p. 99**
(a) Adult: Black crown, mantle, grey wings, white underparts.
(b) Juvenile: Whitish buff spots above, whitish buff streaks below.

49. Little Heron *Butorides striata* 46 cm (18 in) **p. 100**
(a) Adult: Grey-blue plumage, brown wing coverts with yellowish streaks.
(b) Juvenile: Buff streaks on underparts and buff spots on wing coverts.

51. Chinese Pond-heron *Ardeola bacchus* 46 cm (18 in) **p. 101**
(a) Breeding: Dark chestnut head and neck, rich brownish maroon breast.

52. Javan Pond-heron *Ardeola speciosa* 46 cm (18 in) **p. 101**
(a) Breeding: Pale beige head and nape and rich buff breast.

50/51/52. Indian/Chinese/Javan Pond-heron
Non-breeding: Whitish buff head, neck and breast streaked bold brown. **p. 100, 101, 101**

51(a)

50/51/52

52(a)

48(b)
47(a)
47(b)
48(a)
49(a)
49(b)
46(b)
46(a)
43(a)
43(b)
44(b)
44(a)
45(a)
45(b)

PLATE 7 **BITTERNS, NIGHT-HERONS, LITTLE HERON, AND POND-HERONS, MOSTLY IN FLIGHT**

43. Yellow Bittern *Ixobrychus sinensis* **p. 97**
(c) Male: Pale brown mantle, yellowish buff wing coverts, blackish flight feathers.
(d) Juvenile: Streaked upperparts, blackish flight feathers.

44. Chestnut Bittern *Ixobrychus eurhythmus* **p. 98**
(c) Male: Deep chestnut mantle, greyish black flight feathers.
(d) Juvenile: Streaked mantle and wing coverts.

45. Cinnamon Bittern *Ixobrychus cinnamomeus*
Shows no black in wings. **p. 98**
(c) Male: All cinnamon-rufous above.
(d) Juvenile: All mottled brown upperparts.

46. Black Bittern *Dupetor flavicollis* **p. 98**
(c) All dark upperparts contrasting with buff-yellow stripe on neck.

47. Rufous-necked Night-heron *Gorsachius melanolophus* **p. 99**
(c) Blackish flight feathers with white tips to primaries, chestnut upperparts.

48. Black-crowned Night-heron *Nycticorax nycticorax* **p. 99**
(c) Adult: Black crown and back contrasting with pale grey wings and tail.
(d) Juvenile: Dark brown upperparts with whitish buff spots.

49. Little Heron *Butorides striata* **p. 100**
(c) Adult: Blue-grey mantle contrasting with dark crown and wings.
(d) Juvenile: Dark brown upperparts contrasting with buff streaked underparts.

50. Indian Pond-heron *Ardeola grayii* 46 cm (18 in) **p. 100**
(a) Breeding: Brownish-buff head, neck, and breast, dark-brownish-maroon mantle.
(b) Breeding: Maroon mantle and scapulars contrast with all white wings.

51. Chinese Pond-heron *Ardeola bacchus* **p. 101**
(b) Breeding: Chestnut head and black back contrasting with white wings.

52. Javan Pond-heron *Ardeola speciosa* **p. 101**
(b) Breeding: Pale beige head and black back contrasting with white wings.

50/51/52. Indian/ Chinese/Javan Pond-heron
Non-breeding: Brownish back, streaked head and breast. **p. 100, 101, 101**

51(b)
52(b)
50/51/52
49(c)
49(d)
47(c)
48(c)
48(d)
44(c)
44(d)
43(d)
43(c)
45(c)
45(d)
46(c)

PLATE 8 **LARGE HERONS, EGRETS, AND GREAT BITTERN**

42. Great Bittern *Botaurus stellaris* 76 cm (30 in) **p. 97**
(a) Large and stocky with black malar stripe and mottled sandy buff plumage. (b) Upperwing flight: Sandy buff wing coverts blackish-barred flight feathers.

53. Cattle Egret *Bubulcus ibis* 51 cm (20 in) **p. 102**
(a) Breeding: Orange-yellow plumes on head, breast and back. (b) Non-breeding: All white. (c) Flight.

54. Grey Heron *Ardea cinerea* 102 cm (40 in) **p. 102**
(a) Pale grey upperparts, white head and neck with black line above eye. (b) Upperwing flight: Grey wing coverts and upperparts, black primaries and secondaries.

55. Great-billed Heron *Ardea sumatrana* 114 cm (45 in)
Large size, long stout bill, dusky grey plumage and blackish legs. **p. 103**

56. Purple Heron *Ardea purpurea* 97 cm (38 in) **p. 103**
(a) Adult: Rufous neck and upperparts, greyish wings. (b) Juvenile: Sandy rufous head and neck, brownish wing coverts. (c) Adult (upperwing flight): Grey wing coverts, blackish flight feathers. (d) Juvenile (upperwing flight): Brownish wing coverts, blackish flight feathers.

57. Great Egret *Casmerodius albus* 90 cm (35 in) **p. 104**
(a) Breeding: Black bill, blue facial skin, reddish tarsi, plumes on lower back. (b) Non-breeding: Yellow bill and facial skin. (c) Flight. Neck bulge prominent.

58. Intermediate Egret *Mesophoyx intermedia* 71 cm (28 in) **p. 104**
(a) Breeding: Black bill with yellowish base, plumes on breast and lower back. (b) Non-breeding: Yellow bill with dark tip. (c) Flight.

59. Little Egret *Egretta garzetta* 61 cm (24 in) **p. 104**
(a) Breeding: Long nuchal plumes and plumes on breast and back. (b) Non-breeding: Lacks nuchal plumes. (c) Flight

60. Pacific Reef-egret *Egretta sacra* 58 cm (23 in)
Grey facial skin and greenish legs, feathered up to knee. **p. 105**
(a) Pale morph: All white with yellow mandible, dark maxilla and tip. (b) Dark morph: Slaty grey with black bill with some yellow on mandible.

61. Chinese Egret *Egretta eulophotes* 69 cm (27 in) **p. 105**
(a) Breeding: Orange-yellow bill, shaggy crest, plumes on breast and back. (b) Non-breeding: Lacks crest, dark maxilla, pinkish mandible.

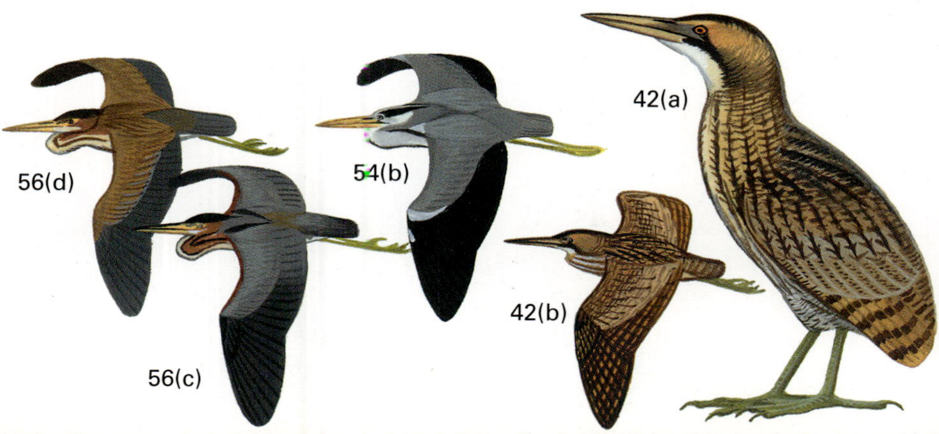

55
54(a)
56(a)
56(b)
53(b)
53(a)
53(c)
60(a)
60(b)
61(a)
61(b)
57(b)
58(a)
58(b)
57(a)
59(a)
59(b)
59(c)
58(c)
57(c)

62. Christmas Island Frigatebird *Fregata andrewsi* 95 cm (37 in) **p. 107**
- (a) Male: All black with white belly.
- (b) Female: White breast and belly, white spurs on base of underwing.
- (c) Juvenile: Tawny buff head, blackish breast band.

63. Great Frigatebird *Fregata minor* 95 cm (37 in) **p. 107**
- (a) Male: All black plumage.
- (b) Female: White breast, whitish throat.
- (c) Juvenile: Tawny head, complete or incomplete breast band.

64. Lesser Frigatebird *Fregata ariel* 78 cm (31 in) **p. 108**
- (a) Male: All black with white patch on underwing.
- (b) Female: Concave hind border of white breast, white spurs on underwing.
- (c) Juvenile: Rusty buff head, complete or incomplete breast band.

67. Masked Booby *Sula dactylatra* 86 cm (34 in) **p. 110**
- (a) Adult: White plumage, black mask and flight feathers, yellowish bill.
- (b) Juvenile: Pale brown with white collar.

68. Red-footed Booby *Sula sula* 70 cm (28 in) **p. 110**
- (a) Pale morph: White plumage, blue facial skin and bill, red feet.
- (b) Dark morph: All brown with red feet.
- (c) Juvenile: All brown with pale neck.

69. Brown Booby *Sula leucogaster* 74 cm (29 in) **p. 111**
- (a) Adult: Dark brown head and upperparts, white underparts, whitish bill.
- (b) Juvenile: All brown, paler below.

67(b)
67(a)
68(b)
68(a)
68(c)
69(a)
69(b)
63(b)
63(a)
63(c)
62(a)
62(b)
62(c)
64(a)
64(b)
64(c)

PLATE 10 **FALCONS AND BAT HAWK**

73. Black-thighed Falconet *Microhierax fringillarius* 16 cm (6 in) **p. 113**

(a) Tiny size with blackish thighs and upperparts with white narrow down-curving supercilium. White underparts with rufous undertail coverts and buff suffusion on lower breast and flanks.

(b) Underwing flight pattern: White wing coverts and underparts, buff lower belly.

75. Common Kestrel *Falco tinnunculus* 31–35 cm (12–14 in)
Diffuse blackish malar stripe in both sexes. **p. 114**

(a) Male: Greyish head and tail, rufous upperparts with black spots.

(b) Female: Brown upperparts with blackish bars.

(c) Upperwing flight pattern (male): Rufous mantle and wing coverts contrast with blackish flight feathers.

(d) Upperwing flight pattern (female): Entirely barred upperparts.

(e) Underwing flight pattern: Buff underparts and wing coverts with blackish streaks, black subterminal tail band with white tip.

(f) Male hovering: Tail broadly fanned.

77. Eurasian Hobby *Falco subbuteo* 31–35 cm (12–14 in) **p. 115**

(a) Adult: Blackish upperparts, broad black malar stripe, streaked underparts and chestnut thighs and undertail coverts.

(b) Juvenile: Blackish brown upperparts with paler fringes, buff thighs and undertail coverts.

(c) Underwing flight pattern (adult): Streaked underparts and chestnut undertail coverts distinctive.

79. Peregrine Falcon *Falco peregrinus* 38–48 cm (15–19 in)
Large size with broad and pointed wings distinctive. **p. 116**

(a) Adult: *F. p. ernesti*. Black hood and yellow cere.

(b) Juvenile: *F. p. ernesti*. Blackish bars on underparts, blue-grey cere.

(c) Underwing flight pattern (adult): Distinctive flight silhouette with lavender grey underparts with bold dark bars contrasting with white throat.

(d) Adult making steep dive.

84. Bat Hawk *Macheiramphus alcinus* 46 cm (18 in) **p. 121**

(a) Wispy crest, black plumage, whitish throat with dark mesial stripe, yellow iris.

(b) Underwing flight pattern: Large, broad and pointed wings with whitish throat contrasting with dark mesial stripe and dark underparts.

75(c)
75(d)
75(f)
75(a)
75(b)
75(e)
77(c)
77(a)
77(b)
79(c)
79(b
84(b)
79(a)
84(a)
79(d)

PLATE 11 **NEW MIGRANT RAPTORS—FALCONS AND SPARROWHAWK**

74. Lesser Kestrel *Falco naumanni* 29–32 cm (11.5–13 in)
Slimmer and proportionately shorter tailed than 75. **p. 113**
(a) Male: Unmarked bluish-grey crown, nape, and sides of head, unmarked rufous-chestnut mantle with bluish-grey tertiaries and greater coverts.
(b) Crown more finely streaked than female 75, more V-shaped dark markings on upperparts.
(c) Upperwing flight (male): Rufous chestnut upperwing coverts, bluish-grey tertiaries.
(d) Upperwing flight (female): Barred uppertail coverts.
(e) Underwing flight pattern (male): Buff breast and upper belly, white throat, underwing, lower belly and undertail coverts.

76. Amur Falcon *Falco amurensis* 28–31 cm (11–12.2 in)
Red eyering, cere, and legs distinctive in adult. **p. 114**
(a) Male: slate-grey head and upperparts pale grey underparts, rufous-chestnut thighs and undertail coverts.
(b) Female: Similar to adult 77 but soft part colouration different (red versus yellow) dark grey upperparts marked with dark bars, thighs and undertail coverts white, and malar stripe narrower
(c) Juvenile: Yellowish cere and eyering. From juvenile 77 by reddish legs and more prominent and broader buff fringes above.
(d) Underwing flight (male): White wing coverts, black flight feathers and pale grey underparts.
(e) Underwing flight (female): Like adult 77 but base of coverts whiter with dark bars on undertail.
(f) Underwing flight (first summer male): Greyish breast and belly, rufous-chestnut undertail coverts.

78. Oriental Hobby *Falco severus* 28–33 cm (11–13 in) **p. 115**
(a) Adult: Black hood, buff-white throat, unmarked reddish-rufous underparts.
(b) Juvenile: Breast and belly marked with dark streaks.
(c) Underwing flight (adult): Reddish-chestnut underparts and wing coverts distinctive.
(d) Underwing flight (juvenile): Streaked underparts and wing coverts.

106. Eurasian Sparrowhawk *Accipiter nisus* 28–38 cm (11–15in)
Long-tailed appearance and the absence of mesial stripe distinctive. **p. 131**
(a) Male: Slaty-grey upperparts with orange-rufous wash on sides of head and faint orange-rufous bars on breast and flanks.
(b) Female: Brown tinged upperparts with distinct pale buff supercilium.
(c) Juvenile: Heavy rufous-chestnut streaks on upper breast, barred underparts.
(d) Upperwing flight (male) Distinct slaty-grey upperparts.
(e) Underwing flight pattern (male): Rufous-orange markings on underparts and wing coverts.
(f) Underwing flight pattern (female): Darker and more pronounced markings on underparts and wing coverts.
(g) Underwing flight pattern (juvenile): Distinct streaks on upper breast.

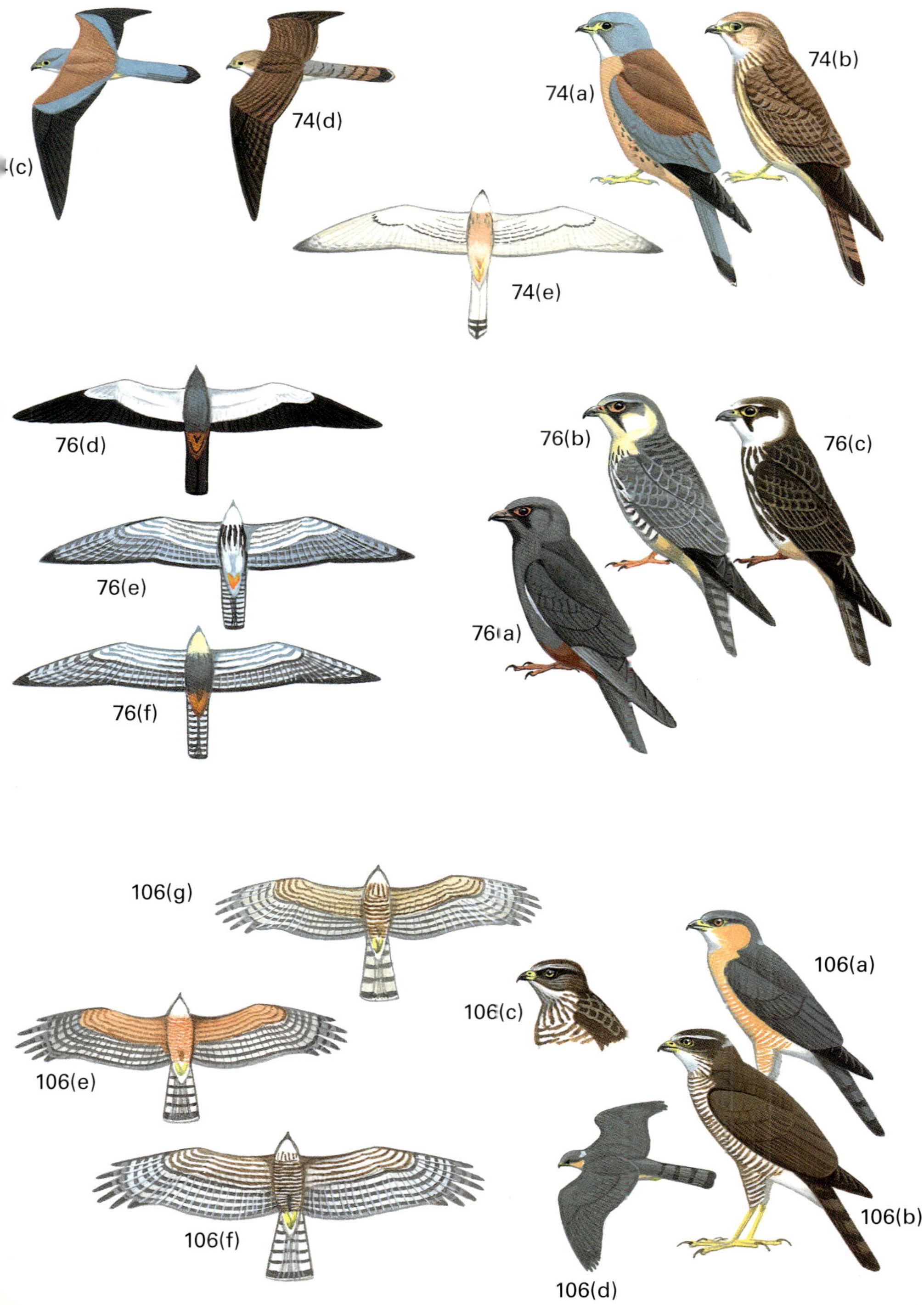
(c)
74(d)
74(a)
74(b)
74(e)
76(d)
76(b)
76(c)
76(e)
76(a)
76(f)
106(g)
106(a)
106(c)
106(e)
106(b)
106(f)
106(d)

80. Osprey *Pandion haliaetus* 55–61 cm (20–23 in) **p. 118**

(a) Blackish brown upperparts, white head and neck with superficial nuchal crest and black eyeline.
(b) Upperwing flight pattern: All dark upperparts and banded tail.
(c) Underwing flight pattern: Whitish with black patch at carpals.

81. Jerdon's Baza *Aviceda jerdoni* 46 cm (18 in) **p. 119**

(a) Long erectile white-tipped crest, rufous bars on lower breast, belly and thighs.
(b) Underwing flight pattern: Rufous and whitish wing coverts, black and white wingtips, alternate dark and pale tail bands.

82. Black Baza *Aviceda leuphotes* 33 cm (13 in) **p. 119**

(a) Long erectile black crest, white breast band and narrow rufous barrings on underparts.
(b) Underwing flight pattern: Grey flight feathers and tail contrast with black wing coverts and wingtips.

85. Black-winged Kite *Elanus caeruleus* 28–33 cm (11–13 in) **p. 121**

(a) Adult: Pale grey upperparts, white underparts, black shoulder patch, red iris.
(b) Juvenile: Upperparts tinged brown with whitish scales, yellow iris.
(c) Adult hovering.
(d) Upperwing flight pattern (adult): Pale grey upperparts contrasting with black shoulder patch.
(e) Underwing flight pattern (adult): White underparts contrasting with blackish flight feathers and wingtips.

86. Black Kite *Milvus migrans* 61–66 cm (24.4–26 in) **p. 121**

(a) Dark brown plumage, long wings and forked tail. *M. m. lineatus* (illustrated): Bluish-grey cere and legs.
(b) Upperwing flight pattern: Dark upperparts with long angular wings drooped at carpal joint with broadly fanned tail.
(c) Underwing flight pattern: Distinct flight silhouette with whitish patch at base of primaries.

87. Brahminy Kite *Haliastur indus* 43–51 cm (17.2–20 in) **p. 122**

(a) Adult: Bright chestnut upperparts, white head, neck and breast with blackish brown streaks.
(b) Juvenile: Dull brown with buff spots and streaks.
(c) Underwing flight pattern (adult): Chestnut wing coverts, rufous flight feathers, and blackish wingtips.
(d) Underwing flight pattern (juvenile): Pale patch at base of primaries.

80(c)
80(b)
80(a)
81(b)
82(b)
82(a)
81(a)
85(e)
86(c)
85(d)
85(c)
87(c)
85(a)
85(b)
87(d)
86(a)
86(b)
87(a)
87(b)

PLATE 13 **ORIENTAL HONEY-BUZZARD**

83. Oriental Honey-buzzard *Pernis ptilorhyncus* 51–61 cm (20.4–24.4 in)
In underwing flight pattern, both subspecies show a dark trailing edge to wing. **p. 119**

(a) *P. p. orientalis* (adult male pale morph): Very short crest, greyish sides of head, pale throat, brownish upperparts, pale underparts (highly variable) with rufous bars, dark brown iris, grey bill and cere. Tail has two complete, well-spaced bold dark bands with a broad white central band.

(b) Underwing flight pattern (male pale morph): Pale underparts and wing coverts with rufous bars (clearly visible at low altitudes).

(c) Underwing flight pattern (female pale morph): Pale underparts and wing coverts like male but tail shows narrower alternate dark and pale bands

(d) Underwing flight pattern (male dark morph): All dark underparts and wing coverts and central white tail band.

(e) Underwing flight pattern (female dark variant): All dark underparts (different shade) and wing coverts with narrow alternate dark and pale tail bands.

(f) Upperwing flight pattern (female). Note tail pattern and characteristic longish neck with small pigeon-like head.

(g) Flocks riding thermals during migration.

(h) *P. p. torquatus* (adult male typical morph): Crest tall and wispy, sides of face grey, upper breast all brown with fine barrings on lower breast, belly, and flanks, yellow iris.

(i) *P. p. torquatus* (adult male tweeddale morph): Black head and upperparts, white underparts with blackish streaks on upper breast and blackish bars on lower breast, belly, flanks, thighs, and undertail coverts.

(j) Underwing flight pattern (male typical morph): Brown underparts and wing coverts with whitish bars, male tail pattern.

(k) Underwing flight pattern (male tweedale morph): White underparts with blackish streaks on upper breast and bars across lower breast, belly and undertail coverts.

(l) Underwing flight pattern (female typical morph): Brown underparts and wing coverts with whitish bars, female tail pattern.

83(j)
83(h)
83(k)
83(l)
83(i)
83(b)
83(f)
83(g)
83(c)
83(d)
83(a)
83(e)

PLATE 14 **SEA-EAGLE, FISH-EAGLES, AND HIMALAYAN GRIFFON**

88. White-bellied Sea-eagle *Haliaeetus leucogaster* 60–69 cm (24–28 in) **p. 122**
- (a) Adult: White head and underparts contrasting with grey upperparts.
- (b) Juvenile: Pale brownish plumage.
- (c) Underwing flight pattern (adult): Black flight feathers contrast with white wing coverts and underparts.
- (d) Underwing flight pattern (juvenile): Creamy wing coverts, blackish wingtips and flight feathers with pale patch at base of primaries.

89. Lesser Fish-eagle *Ichthyophaga humilis* 51–64 cm (20.4–26 in) **p. 123**
- (a) Adult: Like adult 90 but smaller with all dark tail.
- (b) Juvenile: Like juvenile 90 but lacks whitish streaks on head and breast.
- (c) Underwing flight pattern (adult): Like adult 90 but tail all dark with whitish mottling at base.
- (d) Underwing flight pattern (juvenile): Like juvenile 90 but shows darker brownish mottles on wing coverts.

90. Grey-headed Fish-eagle *Ichthyophaga ichthyaetus* 61–74 cm (24.4–30 in) **p. 123**
- (a) Adult: Pale grey head, brownish breast and upper belly, white lower belly and undertail coverts. Shows white tail with black terminal band.
- (b) Juvenile: Pale brown head and breast with whitish streaks on breast.
- (c) Underwing flight pattern (adult): White tail contrasts with black terminal band with wings and breast contrasting with white lower belly and undertail coverts.
- (d) Underwing flight pattern (juvenile): Pale brownish wing with long white bar, dark terminal tail band shows less contrast with rest of brownish tail.

93. Himalayan Griffon *Gyps himalayensis* 115–125 cm (46–50 in) **p. 124**
- (a) Juvenile: Pale head and underparts, dark brownish upperparts with long narrow whitish streaks, blackish bill and cere.
- (b) Underwing flight pattern (juvenile): Pale wing coverts and underparts contrasting with dark flight feathers and short wedge-shaped tail.

Note: White-rumped (91), Indian (92), Red-headed (94), and Cinereous Vultures (95) are not illustrated in colour. See text for details.

93(a)

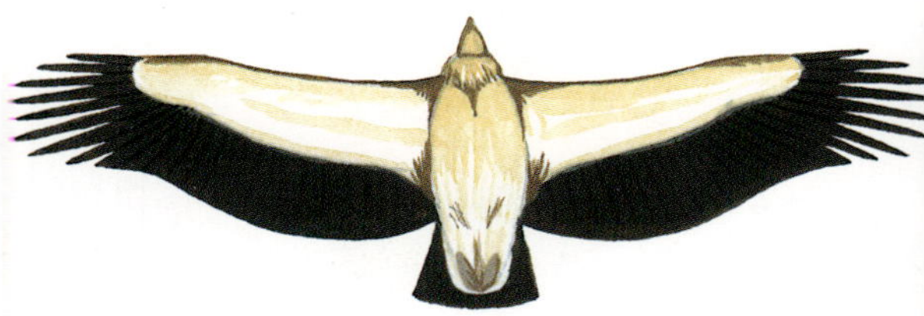

93(b)

88(c)
88(a)
88(d)
89(c)
88(b)
89(d)
89(b)
90(c)
89(a)
90(d)
90(a)
90(b)

96. Short-toed Snake-eagle *Circaetus gallicus* 60–81 cm (24–32 in) **p. 127**
(a) Adult: Rounded head, dark brown lores and ear coverts. (b) Upperwing flight: Greyish brown upperparts and wing coverts, blackish flight feathers and wingtips. (c) Underwing flight: Pale underparts with faint dark streaks.

97. Crested Serpent-eagle *Spilornis cheela* 53 cm (21 in) **p. 127**
(a) Adult: Stocky and short-tailed with dark brown plumage with blackish nuchal crest with white spots. (b) Juvenile: Dull brownish upperparts with pale fringes. (c) Underwing flight (adult): Shallow 'V' distinctive with broad white central tail band and white band bordering narrow black trailing edge of wing. (d) Underwing flight (juvenile): Whitish underparts with dark streaks, tail marked with white and dark bands.

115. Changeable Hawk-eagle *Nisaetus cirrhatus* 56–75 cm (22–30 in) **p. 136**
(a) Dark morph: Blackish brown with dark terminal tail band. (b) Pale morph: Dark brown above and whitish underparts with bold streaks below. (c) Juvenile: White head and underparts. (d) Underwing flight (dark morph): Dark underparts with pale band on flight feathers. From 109 by narrower wings and broader tail. (e) Underwing flight (juvenile): Pale underparts, blackish wingtips, barred tail.

116. Mountain Hawk-eagle *Nisaetus nipalensis* 66–75 cm (26–30 in)
Large size and tall erectile crest distinctive. In flight, wings broader than other Hawk-eagles, showing more prominent 'finger' projection. **p. 136**
(a) Adult: Streaked upper breast and barred underparts. (b) Juvenile: Like adult but tawny buff head and underparts unmarked. (c) Underwing flight (adult): Dark mesial stripe distinctive, breast and underwing coverts tawny with blackish streaks on upper breast.
(d) Underwing flight (juvenile): All white throat, unmarked buff wing coverts and underparts

117. Blyth's Hawk-eagle *Nisaetus alboniger* 51–58 cm (20–23 in) **p. 137**
(a) Adult: Long erectile crest, whitish underparts with dark mesial stripe. (b) Underwing flight pattern (adult): Broad central white tail band.

118. Wallace's Hawk-eagle *Nisaetus nanus* 46 cm (18 in) **p. 137**
(a) Adult: Smaller than 117, tail has alternate darker and paler bands. (b) Juvenile: Brownish upperparts, unmarked buff underparts and pale bands on tail. In field plumage pattern similar to juvenile 117. (c) Underwing flight (adult): Rufous wing coverts, dark trailing edge of wings.

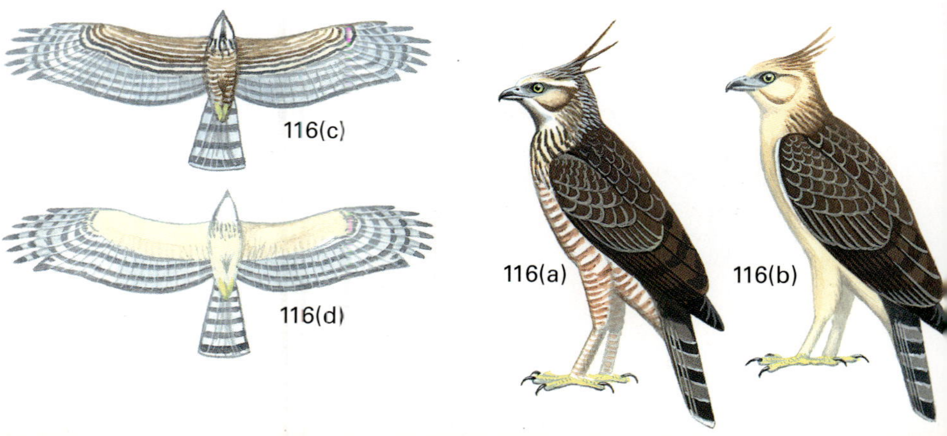

96(c)
96(a)
96(b)
97(a)
97(c)
97(b)
97(d)
115(b)
115(c)
115(d)
115(a)
115(e)
117(a)
117(b)
118(a)
118(b)
118(c)

98. Eastern Marsh-harrier *Circus spilonotus* 48–56 cm (19–22 in) **p. 127**

(a) Male: White head, neck and breast streaked black, broad black mask, white underparts and grey tail.
(b) Female: Pale buff head and bright rufous underparts.
(c) Upperwing flight (male): Blackish mantle and wingtips, white rump.
(d) Upperwing flight (subadult male): Brown wing coverts, banded tail.
(e) Upperwing flight (female): Pale buff crown, dark rufous upperparts with pale markings.

99. Hen Harrier *Circus cyaneus* 43–51 cm (17–20 in) **p. 128**

(a) Male: Pale grey head and breast, white underparts and black wingtips.
(b) Female: Heavily barred underwing.
(c) Upperwing flight (male): Pale grey head, upperparts and tail, blackish wingtips and white rump.
(d) Upperwing flight (female): Dark brown mantle and wings, white rump, banded tail.

100. Pied Harrier *Circus melanoleucos* 43–46 cm (17–18 in) **p. 128**

(a) Male: Black head and upper breast, black wingtips, white underparts.
(b) Female: White underwing with dark barrings.
(c) Upperwing flight (male): Black head and mantle, white bar on wing coverts, narrow black bar on wing, blackish wingtips, white tail.
(d) Upperwing flight (female): Dark brown mantle and wing coverts, pale bluish grey flight feathers with black bands and narrow 'V'-shaped rump patch.

107. Grey-faced Buzzard *Butastur indicus* 41–46 cm (16–18 in) **p. 131**

(a) Adult: Grey ear coverts, white throat and dark mesial stripe.
(b) Juvenile: White underparts with narrow brown streaks.
(c) Upperwing flight (adult): Brown plumage with blackish streaks and bars.
(d) Underwing flight (adult): Rufous-brown breast and belly barred whitish.

108. Common Buzzard *Buteo buteo* 50–56 cm (20–22 in) **p. 132**

(a) Adult: Stocky build, dark brown upperparts and whitish underparts with dark streaked breast.
(b) Upperwing flight: Broad, rounded wings and tail, dark upperparts.
(c) Underwing flight: Black patches at carpal joint, pale wings with black wingtips and trailing edge.

108.1. Long-legged Buzzard x Upland Buzzard **p. 132**

(a) Upright stance, largish white head, barred tail, dark rufous 'trousers'.
(b) Upperwing flight: Brownish wings with pale patch on blackish primaries.
(c) Underwing flight: White underwing with blackish slightly rufous median and greater wing coverts, with whitish secondaries marked with faint barrings.

108(c)

108(a)

108(b)

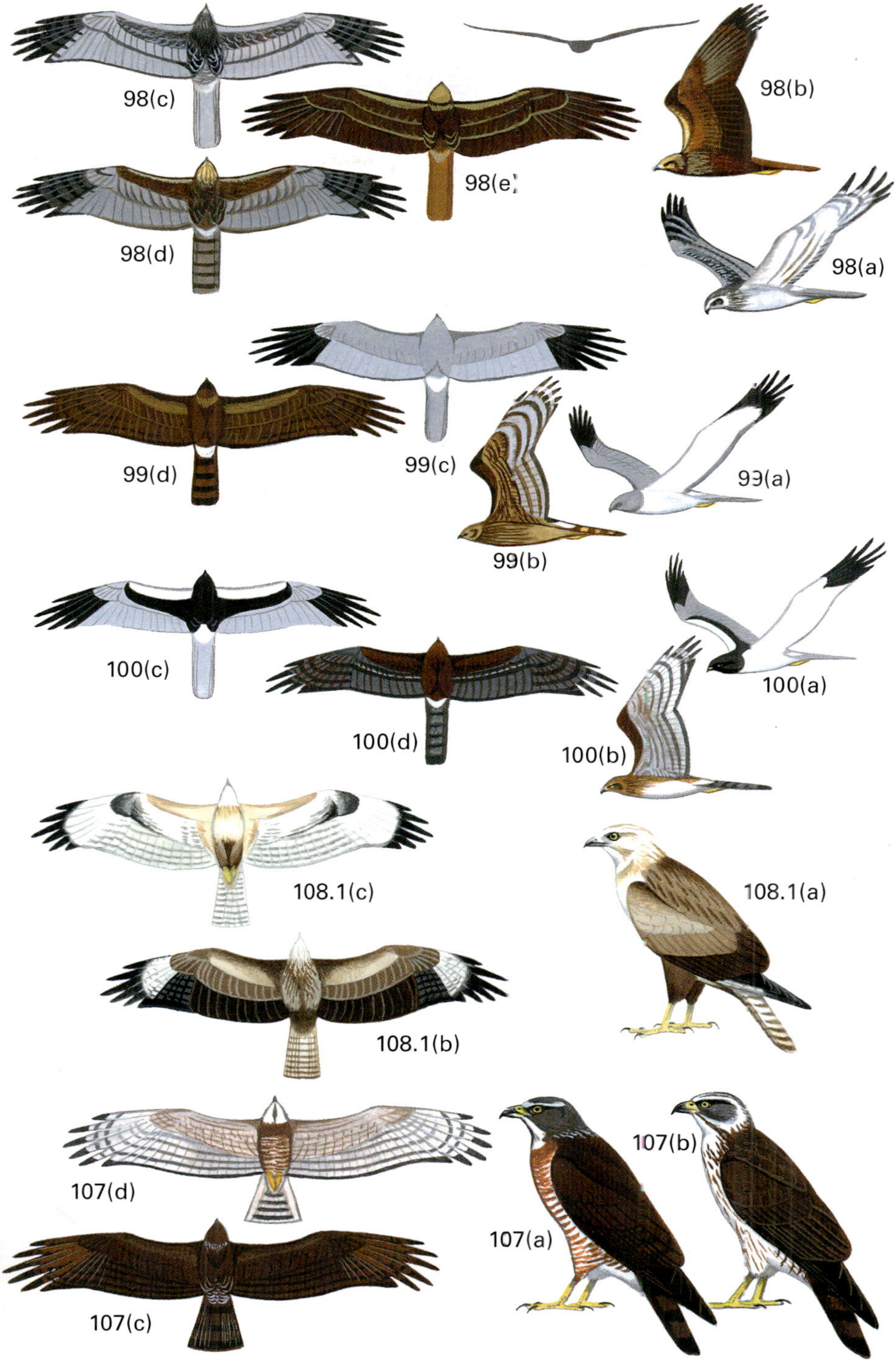
98(c)
98(b)
98(e)
98(d)
98(a)
99(c)
99(d)
99(a)
99(b)
100(c)
100(a)
100(d)
100(b)
108.1(c)
108.1(a)
108.1(b)
107(d)
107(b)
107(a)
107(c)

101. Crested Goshawk *Accipiter trivirgatus* 40–46 cm (16–18 in) **p. 129**
- (a) Adult: Slaty grey head, white throat and dark mesial stripe, rufous streaks on breast and rufous bars on rest of underparts.
- (b) Juvenile: Brown head and spotted belly.
- (c) Underwing flight (adult): White underparts with sparse dark markings, dark wingtips and barred tail.
- (d) Underwing flight (juvenile): Streaked breast and spotted belly.

102. Shikra *Accipier badius* 30–36 cm (12–14 in) **p. 129**
- (a) Male: Bluish grey upperparts and whitish underparts with pale rufous bars.
- (b) Female: Upperparts darker with brownish tinge.
- (c) Juvenile: White supercilium, dark mesial stripe and streaked breast.
- (d) Underwing flight (male): Pale underparts dark wingtips barrings on central tail feathers.
- (e) Underwing flight (female): Wings marked with dark barrings, underparts and wing coverts marked with more rufous bars.

103. Chinese Sparrowhawk *Accipiter soloensis* 30–36 cm (12–14 in) **p. 130**
- (a) Adult: Dark slaty grey above, whitish below, orange suffusion on breast.
- (b) Juvenile: Streaked breast and lacking dark mesial stripe.
- (c) Underwing flight (adult): White underparts, dark wingtips.
- (d) Underwing flight (juvenile); Pale underparts, streaked breast.

104. Japanese Sparrowhawk *Accipiter gularis* 25–30 cm (10–12 in)
White throat with faint mesial stripe. **p. 130**
- (a) Male: Pale rufous breast and belly with faint whitish bars.
- (b) Female: Narrow dark brown bars on breast and belly.
- (c) Juvenile: Bold dark streaks on breast.
- (d) Underwing flight (male): Pale rufous breast, belly and wing coverts.
- (e) Underwing flight (female): Dark brown bars on breast, belly and wing coverts.

105. Besra *Accipiter virgatus* 30–36 cm (12–14 in)
Dark mesial stripe distinctive at rest and in low flight. **p. 131**
- (a) Male: Dark slaty grey above, white throat and rufous and white below, streaks on upper breast.
- (b) Female: Dark brown upperparts with underparts marked with brownish bars.
- (c) Juvenile: Dark brown above with bold dark streaks on breast.
- (d) Underwing flight (male): Streaked breast, rufous and white below.
- (e) Underwing flight (female): Streaked breast, brownish bars below.

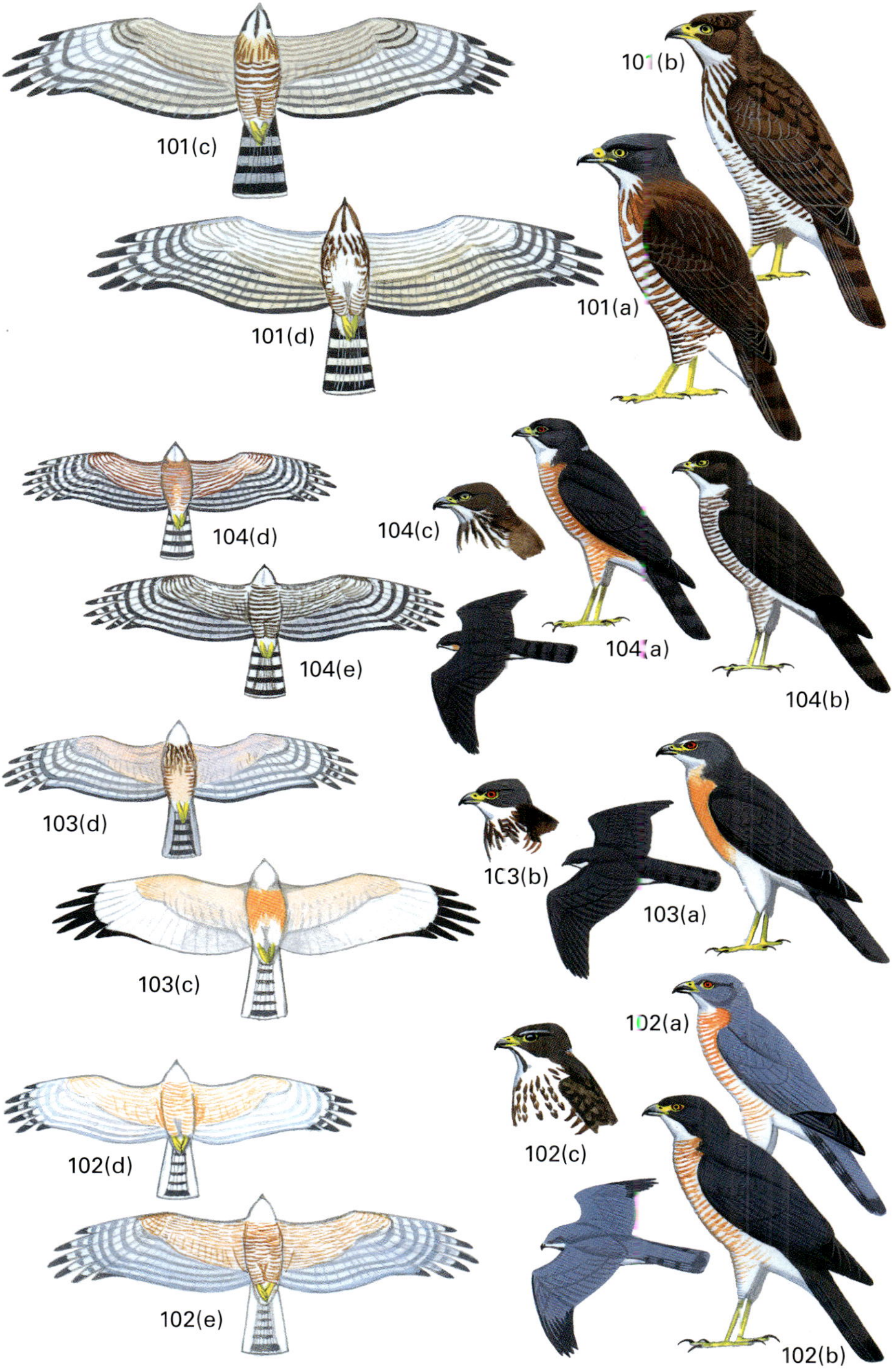
101(b)
101(c)
101(a)
101(d)
104(d)
104(c)
104(e)
104(a)
104(b)
103(d)
103(b)
103(a)
103(c)
102(a)
102(c)
102(d)
102(e)
102(b)

PLATE 18 **LARGE DARK EAGLES, BOOTED EAGLE, AND RUFOUS-BELLIED EAGLE**

109. Black Eagle *Ictinaetus malayensis* 69–78 cm (28–31 in) **p. 133**
(a) Adult: Large size with blackish plumage. (b) Juvenile: Browner plumage with buff streaks and banded tail. (c) Underwing flight (adult): Large, broad wings and closely held tail.

110. Greater Spotted Eagle *Aquila clanga* 66–74 cm (26–30 in) **p. 133**
(a) Adult: Blackish brown plumage, gape does not extend behind eye. (b) Upperwing flight (adult): Broad wings, pale crescentic rump patch. (c) Underwing flight (adult): Wing coverts darker than primaries.
(d) Upperwing flight (juvenile): Whitish spots on wings, broader rump patch.

111. Steppe Eagle *Aquila nipalensis* 76–86 cm (30–34 in) **p. 134**
(a) Adult: Larger than 110 with gape extending behind eye. (b) Upperwing flight (juvenile): Pale patch at base of primaries. (c) Underwing flight (adult): Wing coverts slightly darker than primaries.

112. Eastern Imperial Eagle *Aquila heliaca* 79–84 cm (32–34 in) **p. 134**
(a) Adult: Diagnostic tawny buff crown and hindneck. (b) Upperwing flight (juvenile): Whitish panel on inner primaries. (c) Underwing flight (adult): Silvery grey flight feathers, basal part of tail, darker wing coverts and distal part of tail.

113. Booted Eagle *Hieraaetus pennatus* 48–56 cm (19–22 in) **p. 135**
(a) Pale morph: White underparts, dark ear coverts. (b) Upperwing flight: Pale brown wing coverts, pale rump. (c) Underwing flight (pale morph): White wing coverts and underparts grey tail. (d) Underwing flight (dark morph): Dark underparts, cinnamon buff tail.

114. Rufous-bellied Eagle *Lophptrorchis kienerii* 51–61 cm (20–24 in) **p. 135**
(a) Adult: Black head, crest and upperparts, rufous belly, flanks and thighs. (b) Juvenile: Blackish brown upperparts, dark eye patch. (c) Underwing flight (adult): Rufous wing coverts and underparts. (d) Underwing flight (juvenile): White underparts.

113(b)

113(c)

113(d)

113(a)

114(a)

114(b)

114(c)

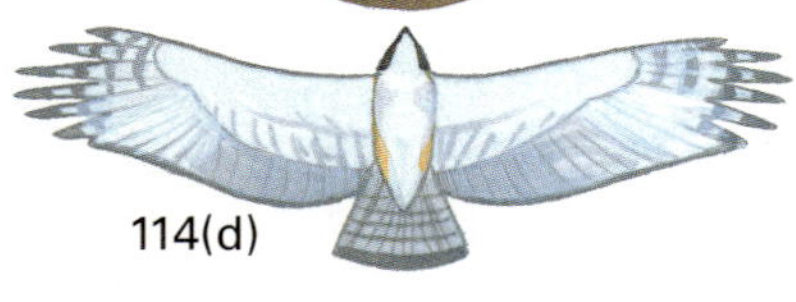
114(d)

109(c)
109(b)
109(a)
110(c)
110(b)
110(d)
110(a)
111(c)
111(b)
111(a)
112(c)
112(b)
112(a)

PLATE 19 **RAILS AND CRAKES**

119. Red-legged Crake *Rallina fasciata* 25 cm (9 in)
Reddish brown head, nape and breast, red legs, white markings on primaries. **p. 138**

120. Slaty-legged Crake *Rallina eurizonoides* 25 cm (9 in)
Dark greenish grey legs, narrow white barrings on flanks. **p. 138**

121. Slaty-breasted Rail *Gallirallus striatus* 30 cm (12 in)
Chestnut crown and nape, slaty grey breast, underparts finely barred white. **p. 139**

122. White-breasted Waterhen *Amaurornis phoenicurus* 33 cm (13 in)
Slaty grey upperparts, white underparts, cinnamon undertail coverts. **p. 139**

123. Baillon's Crake *Porzana pusilla* 19 cm (7.5 in)
Small size, brownish upperparts flecked white, black and white bars on flanks and undertail coverts. **p. 139**

124. Ruddy-breasted Crake *Porzana fusca* 22 cm (8.5 in)
Reddish chestnut underparts, dark olive brown upperparts, red legs. **p. 140**

125. Band-bellied Crake *Porzana paykullii* 23 cm (9 in)
Dark olive-brown crown, nape and mantle, whitish supercilium. **p. 140**

126. White-browed Crake *Porzana cinerea* 20 cm (8 in)
Greenish bill, black eyeline, narrow white supercilium and white stripe below eye. **p. 140**

127. Watercock *Gallicrex cinerea* 43 cm (17 in)
Large size. **p. 141**

(a) Breeding male: Blackish grey mottled buff brown on upperparts, red bill, frontal shield and legs.

(b) Female and non-breeding male: Overall buff brown streaked dark above, yellowish green bill and legs.

121
119
120
123
124
126
125
122
127(a)
127(b)

PLATE 20 **OTHER RAILS, FINFOOT, JACANAS, PAINTEDSNIPE, OYSTERCATCHER, CRAB PLOVER, AND THICK-KNEE**

128. Purple Swamphen *Porphyrio porphyrio* 43 cm (17 in)
Purplish blue plumage, pale grey head, red bill, frontal shield and legs. **p. 141**

129. Common Moorhen *Gallinula chlorophus* 33 cm (13 in)
(a) Adult: Blackish plumage, white line along flanks, red bill.
(b) Juvenile: Brownish, line along flank less distinct. **p. 141**

130. Common Coot *Fulica atra* 41 cm (16 in)
Glossy black plumage, white bill and frontal shield. **p. 142**

131. Masked Finfoot *Heliopais personatus* 53 cm (21 in)
Yellow bill, narrow white line running down sides of neck. **p. 142**
(a) Male: Black throat.
(b) Female: White throat.

135. Beach Thick-knee *Esacus neglectus* 53 cm (21 in) **p. 145**
(a) Massive black bill with yellow base, broad blackish eyeline, white supercilium.
(b) Flight: Greyish upperwing coverts and white underwing, black flight feathers.

136. Eurasian Oystercatcher *Haematopus ostralegus* 45 cm (18 in) **p. 146**
(a) Breeding: Black upperparts and throat, white underparts, red bill and legs.
(b) Flight: White rump, base of tail and wingbars.

137. Crab Plover *Dromas ardeola* 38 cm (15 in) **p. 146**
(a) Thickset with heavy black bill.
(b) Upperwing flight: White wing coverts, black mantle and flight feathers.

152. Greater Paintedsnipe *Rostratula benghalensis* 25–28 cm (10–11 in) **p. 155**
(a) Female: White eyering and stripe behind eye, rich chestnut sides of neck.
(b) Male: Yellowish white eyering and eyestripe, brownish above with buff spots.

153. Pheasant-tailed Jacana *Hydrophasianus chirurgus* 30 cm (12 in) **p. 155**
(a) Breeding adult: Long tail, dark underparts.
(b) Non-breeding male: Tail much shorter, white underparts.
(c) Flight: White upper and lower wings contrast with black wing tips.

154. Bronze-winged Jacana *Metopidius indicus* 28 cm (11 in) **p. 156**
(a) Dark bronzy plumage with diagnostic white supercilium and yellow bill.
(b) Flight: Blackish flight feathers, olive bronze upperwing coverts.

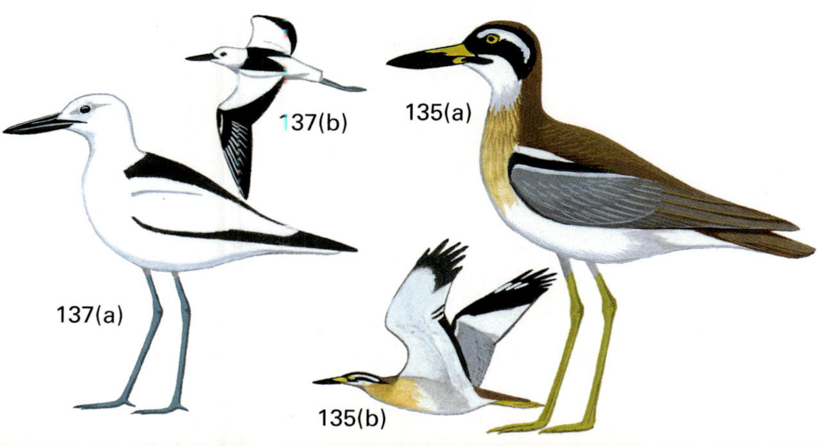

129(b)
128
129(a)
130
131(b)
131(a)
154(a)
153(b)
153(c)
154(b)
153(a)
136(b)
136(a)
152(b)
152(a)

PLATE 21 STILT, LAPWINGS, LARGE PLOVERS, PHALAROPE, AND PRATINCOLES

138. Black-winged Stilt *Himantopus himantopus* 38 cm (15 in)
White head, neck and underparts with long pink legs distinctive. **p. 147**
(a) Male: Black upperparts
(b) Female: Dark brown upperparts
(c) Upperwing flight:

139. Yellow-wattled Lapwing *Vanellus malabaricus* 27 cm (10 in) **p. 148**
(a) Long yellow wattle, narrow white stripe between black cap and brown mantle.
(b) Upperwing flight:

140. Grey-headed Lapwing *Vanellus cinereus* 35 cm (14 in)
(a) Grey head and breast, blackish breast band and yellow bill with dark tip.
(b) Upperwing flight: **p. 148**

141. Red-wattled Lapwing *Vanellus indicus* 33 cm (13 in) **p. 149**
(a) Black head and breast, white ear patch and red wattle in front of eye.
(b) Upperwing flight:

142. Pacific Golden Plover *Pluvialis fulva* 25 cm (10 in) **p. 149**
(a) Breeding: Upperparts mottled black and golden yellow, black face, throat.
(b) Non-breeding: Upperparts dark brown mottled golden yellow.
(c) Upperwing flight (non-breeding): Entire upperparts golden yellow.
(d) Underwing flight (non-breeding): All grey wings.

143. Grey Plover *Pluvialis squatarola* 28 cm (11 in) **p. 150**
(a) Breeding: Grey and white mottled upperparts, black face, throat, and belly.
(b) Non-breeding: Brownish grey upperparts, whitish underparts.
(c) Upperwing flight: Greyish mantle and wing coverts, white rump and barred tail.
(d) Underwing flight: Diagnostic black axillaries.

191. Red-necked Phalarope *Phalaropus lobatus* 21 cm (7.5 in)
Long, slender needle-like bill distinctive. **p. 174**
(a) Female breeding: Bright chestnut upper breast and sides of neck.
(b) Non-breeding: Black crown, white neck and black eyeline.
(c) Upperwing flight (non-breeding): White wingbars, broad white base of outer tail.

192. Oriental Pratincole *Glareola maldivarum* 25 cm (10 in) **p. 175**
(a) Breeding: Creamy throat, rufescent wash on breast and flanks.
(b) Upperwing flight: Dark upperparts, distal part of tail, white rump and basal part of tail.
(c) Underwing flight: Chestnut wing coverts, blackish flight feathers.

193. Small Pratincole *Glareola lactea* 18 cm (7 in) **p. 175**
(a) Pale sandy grey crown and mantle, pale buff grey breast, white underparts.
(b) Upperwing flight: Slightly forked tail, pale grey mantle and wing coverts.
(c) Underwing flight: Black wing coverts, white base of secondaries.

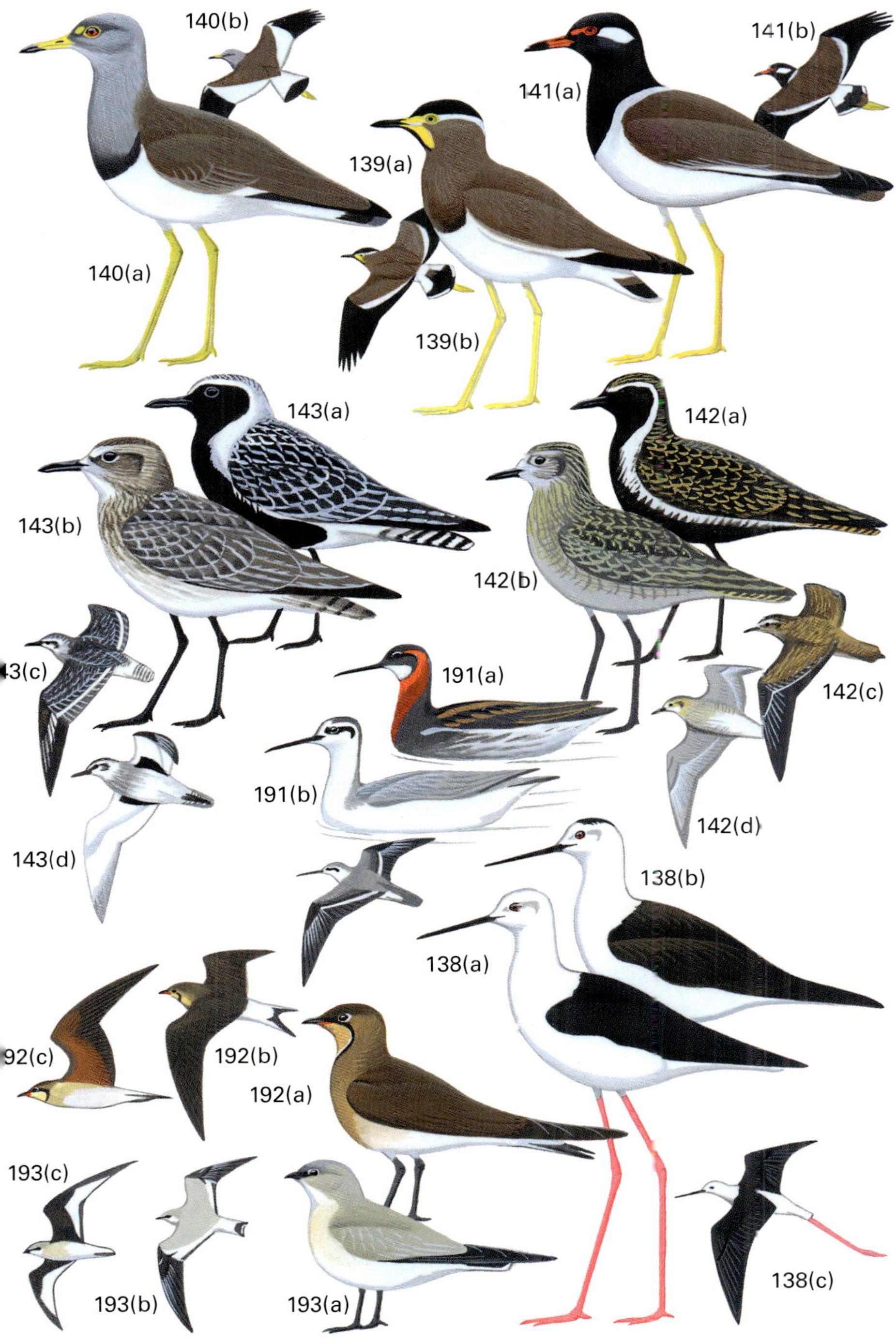

140(b)
141(b)
141(a)
139(a)
140(a)
139(b)
143(a)
142(a)
143(b)
142(b)
3(c)
191(a)
142(c)
191(b)
142(d)
143(d)
138(b)
138(a)
92(c)
192(b)
192(a)
193(c)
193(b)
193(a)
138(c)

144. Common Ringed Plover *Charadrius hiaticula* 19 cm (7.5 in) **p. 150**
(a) Breeding: White forehead, black midcrown, broad black breast band. (b) Non-breeding: Black parts dark brown, breast band broken in the middle.

145. Long-billed Plover *Charadrius placidus* 21 cm (8.5 in)
Long bill and long yellow legs distinctive. **p. 150**
(a) Breeding: Black eyeline, white forehead, black midcrown. (b) Non-breeding: Black facial markings brown, breast band much reduced.

146. Little Ringed Plover *Charadrius dubius* 18 cm (7 in)
Yellow orbital skin distinctive. **p. 151**
(a) Breeding: Black breast band and eyeline. (b) Non-breeding: Breast band less distinct.

147. Kentish Plover *Charadrius alexandrinus* 15 cm (6 in) **p. 151**
(a) Male breeding: Chestnut brown crown and hindneck, black midcrown, black eyeline and narrow black broken breast band. (b) Female and non-breeding male: Crown, eyeline and broken breast band greyish brown.

147.1. Kentish Plover (White-faced Plover) *Charadrius. a. dealbatus* 16 cm (6.25 in) **p. 152**
(a) White forehead, lores, supercilium, and lower ear coverts distinctive. (b) Male breeding: Bright orange crown paler and brighter than 147, narrower lateral breast patches than 147, black bar on mid crown broader than in 147. (c) Female breeding: Strong orange-tone to brown lores, crown, upper ear coverts, and whitish forehead and supercilium.

148. Malaysian Plover *Charadrius peronii* 15 cm (6 in)
Greyish brown upperparts with faint whitish scalings distinctive. **p. 153**
(a) Male: Black hindneck collar. (b) Female: Rufous crown and ear coverts. (c) Juvenile: Lacks rufous crown and ear coverts. (d) Upperwing flight (male): Black hindneck collar, white forecollar.

149. Lesser Sand Plover *Charadrius mongolus* 20 cm (8 in)
Greyish legs distinctive. **p. 153**
(a) *C. m. atrifrons* (breeding): Black lores, forehead and eyeline, white throat and orange rufous breast, crown and nape. (b) *C. m. mongolus* (non-breeding): Pale facial markings, pale buff supercilium.

150. Greater Sand Plover *Charadrius leschenaultii* 24 cm (9 in) **p. 154**
(a) Like breeding 148 but bill longer and thicker. Legs yellow. (b) Non-breeding: Brownish and white facial markings with pale supercilium.

151. Oriental Plover *Charadrius veredus* 24 cm (9 in) **p. 154**
(a) Male breeding: White head, rufous orange breast, narrow black breast band below.
(b) Female breeding: Rufous orange tinged head and breast.

144(a)
144(b)
146(a)
146(b)
147(a)
147(b)
148(a)
148(b)
148(c)
148(d)
145(a)
145(b)
149(a)
149(b)
150(a)
150(b)
151(a)
151(b)

PLATE 23 **WOODCOCK, SNIPE, TURNSTONE, KNOTS, AND RUFF**

155. Eurasian Woodcock *Scolopax rusticola* 36 cm (14 in) **p. 157**
- (a) Triangular shaped head, transverse black bars on hindcrown and nape.
- (b) Upperwing flight: Shows broad wings with rounded tips.

156. Pintail Snipe *Gallinago stenura* 25 cm (10 in) **p. 158**
- (a) Brown bill with dark tip and yellow legs.
- (b) Upperwing flight: Legs trail further beyond tail than 157 and 158.
- (c) Underwing flight: Greyish black flight feathers, heavily barred wing coverts.

157. Swinhoe's Snipe *Gallinago megala* 28 cm (11 in) **p. 158**
- (a) Bill proportionately longer than 156, legs greyish.
- (b) Upperwing flight: Legs trail just beyond tail.
- (c) Underwing flight: Greyish flight feathers, heavily barred wing coverts.

158. Common Snipe *Gallinago gallinago* 28 cm (11 in) **p. 158**
- (a) At rest like 156 and 157
- (b) Upperwing flight: Shows narrow white trailing edge to wing.
- (c) Underwing flight: White bars on wing coverts, greyish flight feathers.

176. Ruddy Turnstone *Arenaria interpres* 23 cm (9 in) **p. 167**
- (a) Male breeding: White head, black facial markings, bright chestnut above.
- (b) Non-breeding: Pale brown head, much duller chestnut upperparts.
- (c) Upperwing flight (non-breeding): White wingbars, back, rump and uppertail coverts.

177. Great Knot *Calidris tenuirostris* 30 cm (12 in)
Larger and longer-billed than 178. **p. 168**
- (a) Breeding: Brownish black above with rufous chestnut patches on scapulars.
- (b) Non-breeding: Greyish above with pale fringes, black spots on sides of breast and flanks.
- (c) Upperwing flight: Shows faint wingbars and all white rump.
- (d) Underwing flight: All white wings.

178. Red Knot *Calidris canutus* 25 cm (10 in) **p. 168**
- (a) Breeding: Bright rufous chestnut head, breast and upper belly.
- (b) Non-breeding: Pale greyish above with faint whitish scales, pale supercilium.
- (c) Upperwing flight: Narrow white wingbars and pale greyish rump with fine streaks.
- (d) Underwing flight: White markings on pale greyish wings.

190. Ruff *Philomachus pugnax* Male: 30 cm (12 in) Female: 23–28 cm (9–11 in) **p. 173**
- (a) Breeding males: Distinct and highly variable neck ruff and ear tufts.
- (b) Non-breeding male: Blackish brown upperparts with narrow pale fringes.
- (c) Female: Smaller and shorter billed than male with variable plumage.
- (d) Upperwing flight (non-breeding): white sides to rump, black tail tip.

176(a)
176(c)
156(b)
176(b)
156(c)
157(c)
156(a)
157(a)
157(b)
158(c)
158(a)
155(b)
155(a)
158(b)
177(c)
177(a)
178(a)
178(c)
177(b)
178(b)
177(d)
178(d)

PLATE 24 **DOWITCHER, GODWITS, CURLEWS, WHIMBREL, AND SANDPIPER**

159. Asian Dowitcher *Limnodromus semipalmatus* 35 cm (14 in) **p. 159**
- (a) Breeding: Rufous chestnut head, breast and flanks, whitish bars on sides of breast.
- (b) Non-breeding: Dark crown and eyeline, blackish grey above with white fringes.
- (c) Upperwing flight: Whitish triangular rump patch, dark bars near tail tip.

160. Black-tailed Godwit *Limosa limosa* 38 cm (15 in) **p. 159**
- (a) Breeding: Bright rufous chestnut head, breast, blackish bars on flanks.
- (b) Non-breeding: Grey-brown upperparts, whitish underparts.
- (c) Upperwing flight: White wingbars, black terminal tail band.

161. Bar-tailed Godwit *Limosa lapponica* 42 cm (16 in) **p. 160**
- (a) Breeding: Bright rufous chestnut head, breast and flanks.
- (b) Non-breeding: Upperparts heavily marked with whitish scales.
- (c) Upperwing flight: *L. l. lapponica*. Shows white triangular rump patch.

162. Little Curlew *Numenius minutus* 30 cm (12 in) **p. 160**
- (a) Smaller than 163 with shorter and straighter bill.
- (b) Upperwing flight: Shows no pale rump.
- (c) Underwing flight pattern: Buff brown wing coverts, greyish flight feathers.

163. Whimbrel *Numenius phaeopus* 43 cm (17 in) **p. 161**
- (a) Shows black lateral coronal stripe and white median coronal stripe.
- (b) Upperwing flight: *N. p. phaeopus*. Shows pale triangular rump patch.

164. Eurasian Curlew *Numenius arquata* 58 cm (23 in) **p. 161**
- (a) Brownish buff above and whitish below, dark streaks on mantle, neck, breast and flanks.
- (b) Upperwing flight: Shows white triangular rump patch.
- (c) Underwing flight: Shows all white wings.

165. Far Eastern Curlew *Numenius madagascariensis* 60 cm (24 in) **p. 162**
- (a) Very long down-curved bill, overall brownish buff.
- (b) Upperwing flight: Rump concolorous with rest of upperparts.
- (c) Underwing flight: Wings barred blackish.

184. Pectoral Sandpiper *Calidris melanotos* 23 cm (9 in)
Forehead and breast heavily streaked, demarcated from white belly. **p. 171**
- (a) Breeding: Neck and breast washed buff.
- (b) Non-breeding: Duller without buff tones
- (c) Upperwing flight: Shows dark centre of rump and base of tail.

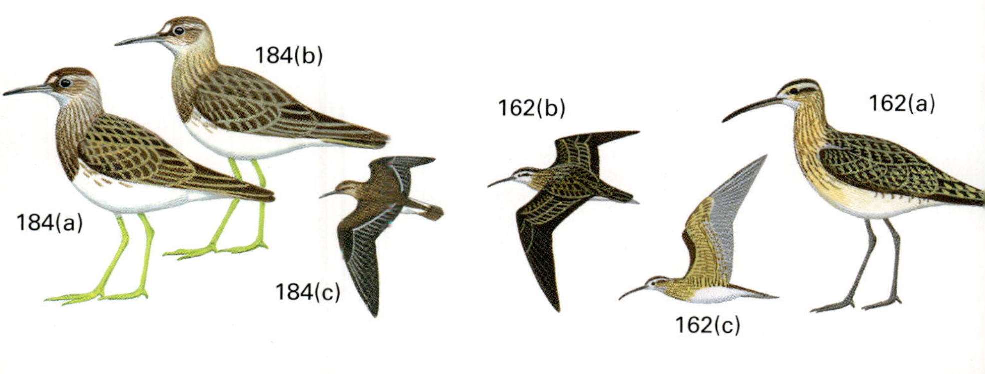

165(a)
165(b)
164(b)
164(a)
146(c)
165(c)
160(a)
160(b)
163(b)
160(c)
163(a)
161(c)
161(a)
159(c)
159(a)
161(b)
159(b)

PLATE 25 REDSHANKS, GREENSHANKS, SANDPIPERS, AND TATTLER

166. Spotted Redshank *Tringa erythropus* 30 cm (12 in) **p. 162**
(a) Breeding: Black plumage with white mottles and spots above.
(b) Non-breeding: Grey upperparts mottled white on mantle, white supercilium, red legs.
(c) Upperwing flight (non-breeding).

167. Common Redshank *Tringa totanus* 28 cm (11 in) **p. 163**
(a) Breeding: Greyish brown head and upperparts with blackish streaks, underparts white with bold brown streaks on breast.
(b) Non-breeding: Plumage lighter with less distinct streaking below.
(c) Upperwing flight (non-breeding).

168. Marsh Sandpiper *Tringa stagnatilis* 25 cm (10 in) **p. 163**
(a) Breeding: Indistinct supercilium, underparts white with blackish brown spots on breast.
(b) Non-breeding: Prominent white supercilium, entire underparts white.
(c) Upperwing flight (non-breeding).

169. Common Greenshank *Tringa nebularia* 35 cm (14 in) **p. 163**
(a) Breeding: Head, neck, breast and flanks boldly streaked black.
(b) Non-breeding: White underparts, fine streaks on sides of neck, breast and flanks.
(c) Upperwing flight (non-breeding).

170. Spotted Greenshank *Tringa guttifer* 32 cm (13 in)
Proportionately thicker billed than 169. **p. 164**
(a) Breeding: White underparts densely spotted on breast and flanks.
(b) Non-breeding: Pale grey upperparts, unmarked white underparts.
(c) Upperwing flight (breeding).

171. Green Sandpiper *Tringa ochropus* 24 cm (9.5 in) **p. 164**
(a) Non-breeding: Greenish brown smudges on sides of neck and breast.
(b) Upperwing flight: White square rump and tail with broad dark bars at terminal end.

172. Wood Sandpiper *Tringa glareola* 23 cm (9 in) **p. 165**
(a) Non-breeding: Brownish grey above, white spots, greyish brown streaks on breast.
(b) Upperwing flight: White square rump and tail marked by narrow dark bars at terminal end, legs trail further beyond tail than 171.

173. Grey-tailed Tattler *Tringa brevipes* 25 cm (10 in)
Pale grey upperparts and white underparts with short yellow legs. **p. 166**
(a) Breeding: Dark grey barrings on breast and flanks.
(b) Non-breeding: Lacks barrings on breast and flanks.
(c) Upperwing flight: All grey upperparts with unmarked tail.

174. Terek Sandpiper *Xenus cinereus* 23 cm (9 in) **p. 166**
(a) Breeding: Black scapular lines, blackish streaks on breast and neck.
(b) Non-breeding: Lacks scapular lines, indistinct streaks on breast.
(c) Upperwing flight.

175. Common Sandpiper *Actitis hypoleucos* 20 cm (8 in)
White curve up bend of wing. **p. 167**
(a) Non-breeding: Indistinct streaks on sides of neck and breast.
(b) Upperwing flight.

166(a)
166(b)
166(c)
167(a)
167(b)
167(c)
170(a)
170(c)
170(b)
169(a)
169(b)
169(c)
172(a)
171(b)
172(b)
171(a)
168(a)
168(c)
168(b)
174(a)
174(b)
174(c)
173(a)
173(b)
173(c)
175(b)
175(a)

179. Sanderling *Calidris alba* 20 cm (8 in) **p. 168**
(a) Breeding: Deep chestnut head, neck and breast, streaked dark brown. (b) Non-breeding: Black carpal patch. (c) Upperwing flight (non-breeding).

180. Red-necked Stint *Calidris ruficollis* 15 cm (6 in) **p. 169**
(a) Breeding: Rufous head and upper breast, chestnut upper scapulars, blackish streaks on white side of lower breast. (b) Non-breeding: Pale grey-brown upperparts with narrow dark shafts. (c) Upperwing flight (non-breeding).

181. Little Stint *Calidris minuta* 13 cm (5 in) **p. 169**
(a) Breeding: Bright orange-brown sides of face and breast. (b) Non-breeding: Pale brown breast, white throat and underparts. (c) Upperwing flight (non-breeding).

182. Temminck's Stint *Calidris temminckii* 15 cm (6 in) **p. 170**
(a) Breeding: Short white supercilium and grey-brown breast and sides of neck. (b) Non-breeding: Markings on breast and sides of neck less distinct. (c) Upperwing flight (non-breeding).

183. Long-toed Stint *Calidris subminuta* 16 cm (6.25 in) **p. 170**
(a) Breeding: Rufous crown and rufous fringes on upperparts. (b) Non-breeding: Lacks rufous crown and rufous fringes on upperparts. (c) Upperwing flight (non-breeding).

185. Sharp-tailed Sandpiper *Calidris acuminata* 22 cm (8.5 in) **p. 171**
(a) Breeding: Rufescent wash above, rufous buff tinge on neck and breast. (b) Non-breeding: Grey-brown above, faint streaks on sides of breast and flanks. (c) Upperwing flight (non-breeding).

186. Curlew Sandpiper *Calidris ferruginea* 22 cm (8.5 in) **p. 172**
(a) Breeding: Bright chestnut head, neck, breast and belly. (b) Non-breeding: Pale supercilium, streaks on crown and sides of neck, white underparts. (c) Upperwing flight (non-breeding): Note white rump.

187. Dunlin *Calidris alpina* 19 cm (7.5 in) **p. 172**
(a) Breeding: Black patch on centre of belly. (b) Non-breeding: Pale grey above, white below, greyish streaks on sides of breast. (c) Upperwing flight (non-breeding).

188. Spoon-billed Sandpiper *Eurynorhynchus pygmeus* 15 cm (6 in)
Bill distinctive in all plumages. **p. 172**
(a) Breeding. (b) Non-breeding. (c) Upperwing flight (non-breeding).

189. Broad-billed Sandpiper *Limicola falcinellus* 18 cm (7 in)
Longish slightly down-curved black bill and split whitish supercilium. **p. 173**
(a) Breeding. (b) Non-breeding. (c) Upperwing flight (non-breeding).

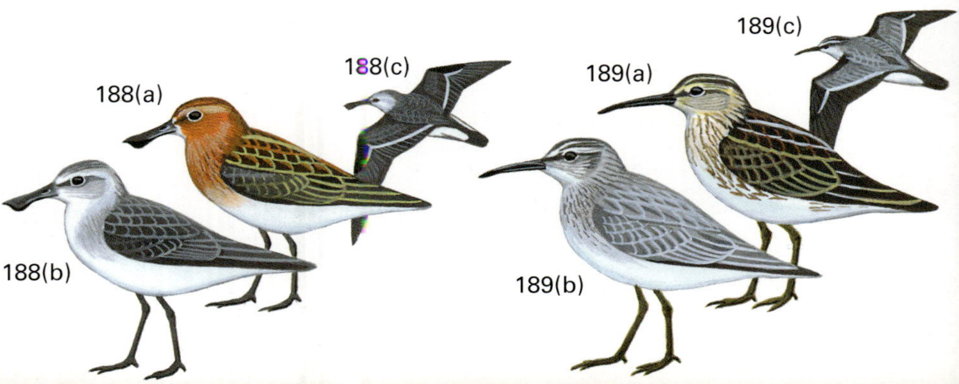

180(a)
181(a)
181(b)
180(b)
180(c)
181(c)
182(a)
183(c)
183(a)
182(b)
182(c)
183(b)
185(c)
185(a)
185(b)
187(c)
187(a)
187(b)
186(a)
186(c)
179(c)
179(a)
186(b)
179(b)

PLATE 27 **GULLS, TERNS, AND JAEGERS**

194. Heuglin's Gull *Larus heuglini* 65 cm (25.6 in) **p. 176**
(a) Breeding: All white head, dark grey upperparts, yellow bill and legs. (b) Juvenile:. Blackish bill, white head, underparts and flanks, streaked.

195. Brown-headed Gull *Larus brunnicephalus* 46 cm (18 in)
Black wingtips with white spots diagnostic. **p. 177**
(a) Breeding: Chocolate brown hood. (b) Non-breeding: White head with dark crescent on ear coverts.

196. Common Black-headed Gull *Larus ridibundus* 40 cm (15 in)
All white leading edge of upperwing, black edge to wing tips. **p. 177**
(a) Breeding: Chocolate brown hood, red bill. (b) Non-breeding: White head, dark crescent on ear coverts, bill dark tipped.

197. Laughing Gull *Larus atricilla* 41 cm (16.1 in) **p. 178**
(a) Breeding: Dark slate hood extends over nape to hindneck. (b) Non-breeding: Head white, dark greyish smudged ear coverts and hindcrown

198. Gull-billed Tern *Gelochelidon nilotica* 38 cm (15 in) **p. 178**
(a) Breeding: Black cap. (b) Non-breeding: White head with black spot on ear coverts.

199. Caspian Tern *Hydroprogne caspia* 53 cm (21 in) **p. 179**
(a) Breeding: All black cap. (b) Non-breeding: Black cap finely streaked white.

209. Whiskered Tern *Chlidonias hybridus* 26 cm (11 in) **p. 183**
(a) Breeding: Red bill, black cap, broad white cheek stripe. (b) Non-breeding: Black bill, fine black streaks on crown.

210. White-winged Tern *Chlidonias leucopterus* 23 cm (9 in) **p. 183**
(a) Breeding (underwing flight): Red bill, black head, breast, upper belly and wing coverts. (b) Breeding (upperwing flight): Black back, white wing coverts and rump. (c) Non-breeding: Black bill, black streaked crown and hindneck.

212. Pomarine Jaeger *Stercorarius pomarinus* 50 cm (20 in)
Twisted spoon-shaped elongate central tail feathers. **p. 184**
Pale morph: White throat and belly separated by broad blackish breast band.

213. Parasitic Jaeger *Stercorarius parasiticus* 45 cm (18 in) **p. 185**
Pale morph: Paler breast band showing less contrast against white throat and belly.

214. Long-tailed Jaeger *Stercorarius longicaudus* 50 cm (20 in) **p. 185**
Pale morph: White throat, breast, and belly.

212
213
214
195(a)
196(a)
196(b)
195(b)
209(b)
210(a)
210(b)
209(a)
210(c)
198(a)
199(a)
198(b)
199(b)

PLATE 28 **TERNS AND NODDY**

200. Lesser Crested Tern *Thalasseus bengalensis* 40 cm (15 in)
Slender orange bill and bluish grey upperparts. **p. 179**
Non-breeding: White forehead and crown, black nuchal crest.

201. Great Crested Tern *Thalasseus bergii* 46 cm (18 in)
Large size, yellow bill and dark grey upperparts. **p. 179**
Non-breeding: White forehead, streaked white forecrown, black nuchal crest.

202. Roseate Tern *Sterna dougallii* 39 cm (15 in) **p. 180**
(a) Breeding: Black cap, red bill with dark tip and red legs.
(b) Non-breeding: White forehead, black nape and stripe behind eye, yellowish legs.
(c) Juvenile: Black crown and pale grey back with blackish 'U' shaped marks

203. Black-naped Tern *Sterna sumatrana* 31 cm (12 in) **p. 180**
(a) Adult: All white plumage, deeply forked tail, black eyeline and nape band.
(b) Juvenile: Tail unforked and rounded, upperparts scaled blackish.

204. Common Tern *Sterna hirundo* 35 cm (14 in)
Slender build and deeply forked tail, black trailing to underwing. **p. 181**
(a) *S.h. longipennis*. All black bill and cap, greyish wash to white underparts..
(b) Non breeding: All black bill and legs, black markings extend around eye.
(c) *S.h. minussensis*. Red bill with dark tip, red legs.
(d) Juvenile: White forehead and underparts, black bill and legs. Upperwing should be scaled black and white.

205. Little Tern *Sterna albifrons* 23 cm (9 in) **p. 181**
(a) Breeding: Black crown, lores, eyeline, yellow bill with dark tip, yellow legs.
(b) Non-breeding: White crown and lores, black bill and legs.
(c) Juvenile: Upperparts grey with buff scalings; bill black with yellow base.

206. Aleutian Tern *Sterna aleutica* 33 cm (13 in)
Dark white-edged bar on secondaries in underwing flight pattern. **p. 182**
Breeding: White forehead, whitish grey flanks, breast and belly.

207. Bridled Tern *Sterna anaethetus* 37 cm (14.5 in)
Deeply forked tail, brownish black above, white of forehead extends above eye. **p. 182**

208. Sooty Tern *Sterna fuscata* 41 cm (17 in)
Deeply forked tail, sooty black above, white of forehead does not extend above eye. **p. 182**

211. Brown Noddy *Anous stolidus* 42 cm (17 in)
Wedge-shaped tail, dark chocolate brown, greyish white crown and nape. **p. 184**

206

202(c)

204(d)
204(b)
202(a)
204(a)
204(c)
202(b)
207(a)
203(a)
203(b)
205(a)
207(b)
208
205(c)
205(b)
200
211
201

PLATE 29 ROCK PIGEON, DOVES, CUCKOO-DOVES, NICOBAR PIGEON, AND IMPERIAL-PIGEONS

215. Rock Pigeon *Columba livia* 33 cm (13 in)
Blue-grey plumage with purplish and greenish gloss on head and neck, broad black wingbars and black terminal tail band. **p. 186**

216. Red Collared-dove *Streptopelia tranquebarica* 23 cm (9 in)
Narrow black collar around hindneck distinctive. **p. 187**
(a) Male: Blue-grey head and vinaceous red plumage.
(b) Female: Greyish brown head and dull brown plumage.

217. Spotted Dove *Streptopelia chinensis* 30 cm (12 in)
Long- tailed with brown upperparts with black streaks and broad black hindneck collar with fine white spots. **p. 187**

218. Barred Cuckoo-dove *Macropygia unchall* 38 cm (15 in)
Long graduated barred tail distinctive, upperparts marked with narrow and dark rufous bars. **p. 187**
(a) Male: Crown, nape and upper back washed iridescent green and purple, buff belly and undertail coverts.
(b) Female: Dark buff underparts, marked with narrow blackish bars, darker and bolder on breast.

219. Little Cuckoo-dove *Macropygia ruficeps* 30 cm (12 in)
Bright orange rufous head and pale orange rufous underparts distinctive. **p. 188**
(a) Male: Whitish scalings on upper breast.
(b) Female: Duller with blackish scalings on upper breast.

220. Emerald Dove *Chalcophaps indica* 25 cm (10 in)
Dark iridescent green mantle, vinaceous red sides of head and underparts and red bill diagnostic. **p. 188**
Male (illustrated): Blue-grey crown and nape, two whitish bars on lower back.

221. Zebra Dove *Geopelia striata* 21 cm (8.5 in)
Small size with proportionately long tail and grey head, blue-grey orbital skin, pale horn bill and pale brown upperparts with fine black and white barrings. **p. 189**

222. Nicobar Pigeon *Caloenas nicobarica* 41 cm (16 in)
Large and plump with short white tail, dark metallic green plumage with long iridescent bronze hackles around neck and back. **p. 189**

232. Green Imperial-pigeon *Ducula aenea* 43 cm (17 in)
Dull bronze green upperparts and dark brown undertail coverts. No tail band. **p. 193**

233. Mountain Imperial-pigeon *Ducula badia* 47 cm (18.5 in)
Brownish maroon upperparts, buff undertail coverts and broad dull greyish terminal band on blackish tail. **p. 193**

234. Pied Imperial-pigeon *Ducula bicolor* 40 cm (16 in)
Creamy white plumage with black flight feathers and broad black terminal tail band. **p. 194**

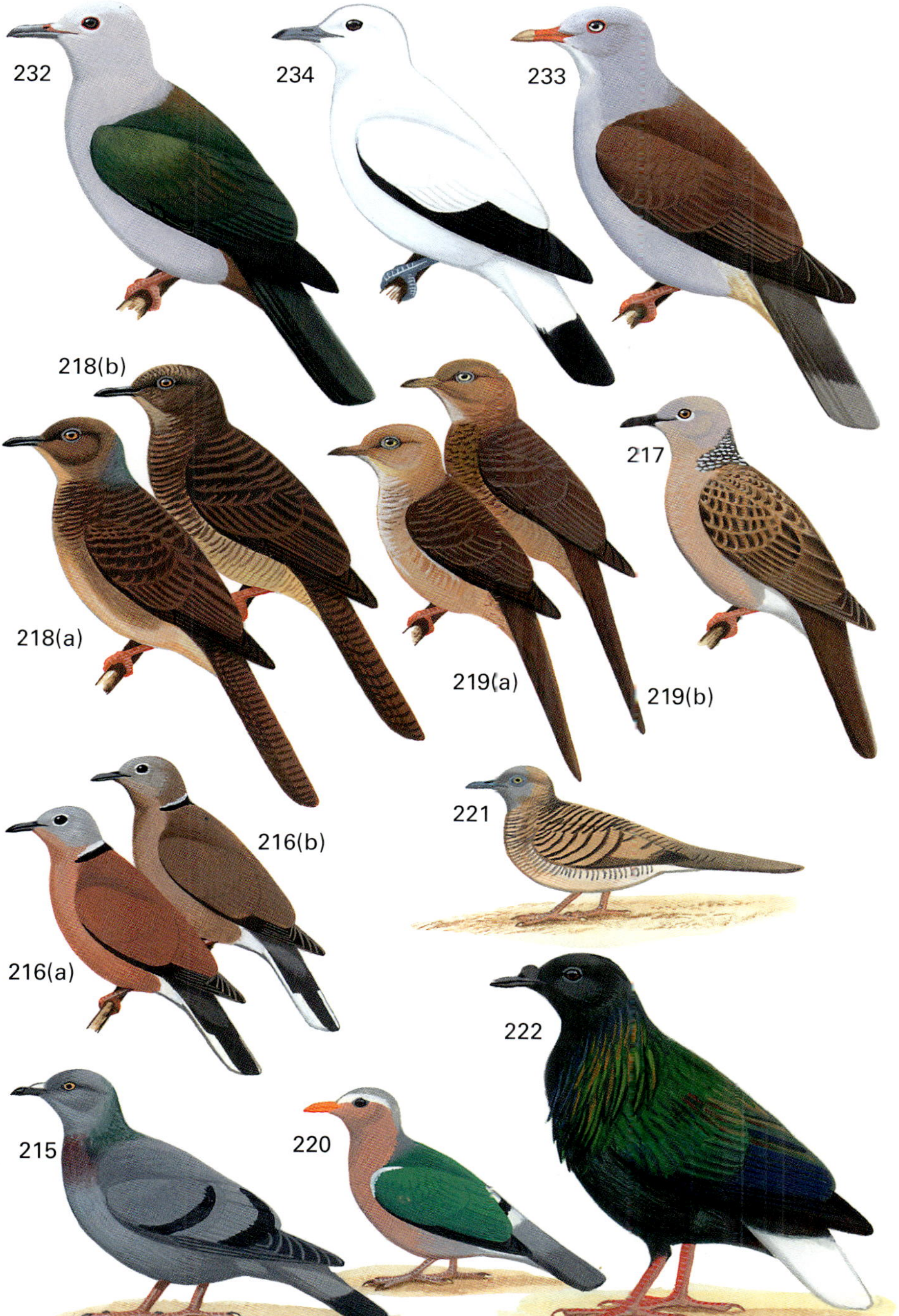
232
234
233
218(b)
217
218(a)
219(a)
219(b)
221
216(b)
216(a)
222
215
220

PLATE 30 **GREEN-PIGEONS AND FRUIT-DOVE**

223. Cinnamon-headed Green-pigeon *Treron fulvicollis* 26 cm (10 in) **p. 189**
- (a) Male: Rusty chestnut head, maroon mantle, ochraceous orange breast.
- (b) Female: Yellowish thighs.

224. Little Green-pigeon *Treron olax* 20 cm (8 in) **p. 190**
- (a) Male: Dark grey head, orange breast patch, maroon mantle.
- (b) Female: Whitish throat, greyish crown merging into greenish sides of head.

225. Pink-necked Green-pigeon *Treron vernans* 25 cm (10 in) **p. 190**
- (a) Male: Grey head, purplish pink neck and upper breast.
- (b) Female: Underparts mildly suffused yellow, buff undertail coverts.

226. Orange-breasted Green-pigeon *Treron bicinctus* 26 cm (10 in) **p. 191**
- (a) Male: Purplish pink upper breast, orange lower breast.
- (b) Female: Greyish suffusion on hindcrown and nape.

227. Thick-billed Green-pigeon *Treron curvirostra* 27 cm (11 in) **p. 191**
- (a) Male: Maroon mantle, cinnamon undertail coverts.
- (b) Female: Greenish undertail coverts with bold dark green streaks.

228. Large Green-pigeon *Treron capellei* 37 cm (14 in) **p. 191**
- (a) Male: Bright orange breast patch, dark brown undertail coverts.
- (b) Female: Duller orange breast patch, buff undertail coverts streaked green

229. Yellow-vented Green-pigeon *Treron seimundi* 32 cm (12 in) **p. 192**
- (a) Male: Maroon shoulder patch.
- (b) Female: Lacks maroon shoulder patch.

230. Wedge-tailed Green-pigeon *Treron sphenurus* 32 cm (12 in) **p. 192**
- (a) Male: Maroon shoulder patch, greyish tinge on mantle.
- (b) Female: Lacks both maroon shoulder patch and greyish tinge on mantle.

231. Jambu Fruit-dove *Ptilinopus jambu* 27 cm (11 in) **p. 193**
- (a) Male: Crimson face, black throat, white breast and belly, rosy breast patch.
- (b) Female: Purplish face and greenish grey breast.

227(b)
27(a)
223(a)
223(b)
225(b)
225(a)
226(b)
226(a)
224(b)
224(a)
228(b)
228(a)
231(b)
231(a)

PLATE 31 **PARROT, HANGING-PARROT, PARAKEETS, AND COCKATOOS**

235. Blue-crowned Hanging-parrot *Loriculus galgulus* 14 cm (5.5 in) **p. 195**
(a) Male: Blue crown patch, red throat and golden-yellow patch on mantle.
(b) Female: Lacks blue crown patch and red throat.
(c) Flight: Tiny size, short head and blunt head distinctive.

236. Tanimbar Corella *Cacatua goffini* 32 cm (12.5 in)
Crestless with greyish white bill and pinkish lores. **p. 195**

237. Yellow-crested Cockatoo *Cacatua sulphurea* 33 cm (13 in)
Large and robust, white plumage, black bill and long erectile yellow crest. **p. 195**

238. Rose-ringed Parakeet *Psittacula krameri* 36 cm (14 in) **p. 196**
(a) Male: Narrow pink collar on hindneck, black facial stripe from base of bill to pink collar.
(b) Female: Lacks pink collar and black facial stripe.

239. Red-breasted Parakeet *Psittacula alexandrii* 36 cm (14 in)
Blue head, pinkish breast and broad black malar stripe. **p. 196**
(a) Male: Pinkish breast tinged lilac, red maxilla, black mandible.
(b) Female: Pinkish breast tinged orange, all black bill.

240. Reddish-cheeked Parakeet *Psittacula longicauda* 41 cm (16 in) **p. 196**
(a) Male: Green crown, reddish sides of head and nape, broad black malar stripe, red maxilla, brown mandible.
(b) Female: Dark green malar stripe and all dull brown bill.

241. Blue-rumped Parrot *Psittinus cyanurus* 19 cm (7.5 in)
Stocky and short-tailed. **p. 197**
(a) Male: Blue-grey head and neck, blackish back, red maxilla and grey mandible.
(b) Female: Brown head and dull brown bill.
(c) Flight: Shows diagnostic red lining on underwing.

240(a)
239(b)
238(b)
240(b)
238(a)
239(a)
236
241(c)
241(b)
241(a)
237
235(c)
235(a)
235(b)

242. Pied Cuckoo *Clamator jacobinus* 33 cm (13 in)
Erect black crest, long broad-ended tail, black hindneck and wings white underparts and wing patch. **p. 198**

243. Chestnut-winged Cuckoo *Clamator coromandus* 46 cm (18 in)
Erect black crest, long broad ended tail, chestnut wings. **p. 199**

244. Large Hawk-cuckoo *Hierococcyx sparveriodes* 33–44 cm (13–16 in) **p. 199**
(a) Adult *H. s. sparveriodes*: Grey head, nape and cheeks, brownish grey upperparts, rufous breast with bold dark streaks, whitish belly with rufous brown bars.
(b) Juvenile: Broad rufous brown bars on upperparts, underparts whitish with brownish streaks.

245. Moustached Hawk-cuckoo *Hierococcyx vagans* 30 cm (12 in)
Broad slaty malar stripe pale greyish cheek and white throat. **p. 200**

246. Hodgson's Hawk-cuckoo *Hierococcyx fugax* 29 cm (11.5 in)
Narrow rufous tail tip diagnostic. **p. 200**
(a) *H. f. fugax*: Upperparts dark slaty grey, underparts whitish with dark streaks.
(b) *H. f. nisicolor*: Pale grey above, whitish below tinged rufous on breast.

247. Indian Cuckoo *Cuculus micropterus* 33 cm (13 in)
Dark brown upperparts, greyish green orbital skin and broad blackish subterminal band on white tipped brownish tail. **p. 201**
Female: Rufous tinge on breast.

248. Oriental Cuckoo *Cuculus saturatus* 33 cm (13 in) **p. 201**
(a) Typical morph: Grey upperparts, yellow orbital skin, blackish tail with white tip.
(b) Hepatic morph female: Rufous bay head, upper breast and entire upperparts marked with blackish bars.

257. Asian Koel *Eudynamys scolopaceus* 43 cm (17 in)
Long broad-ended tail, red iris and pale greenish bill diagnostic. **p. 205**
(a) Male: Glossy black plumage.
(b) Female: Dark brown with bold buff spots and streaks.

242

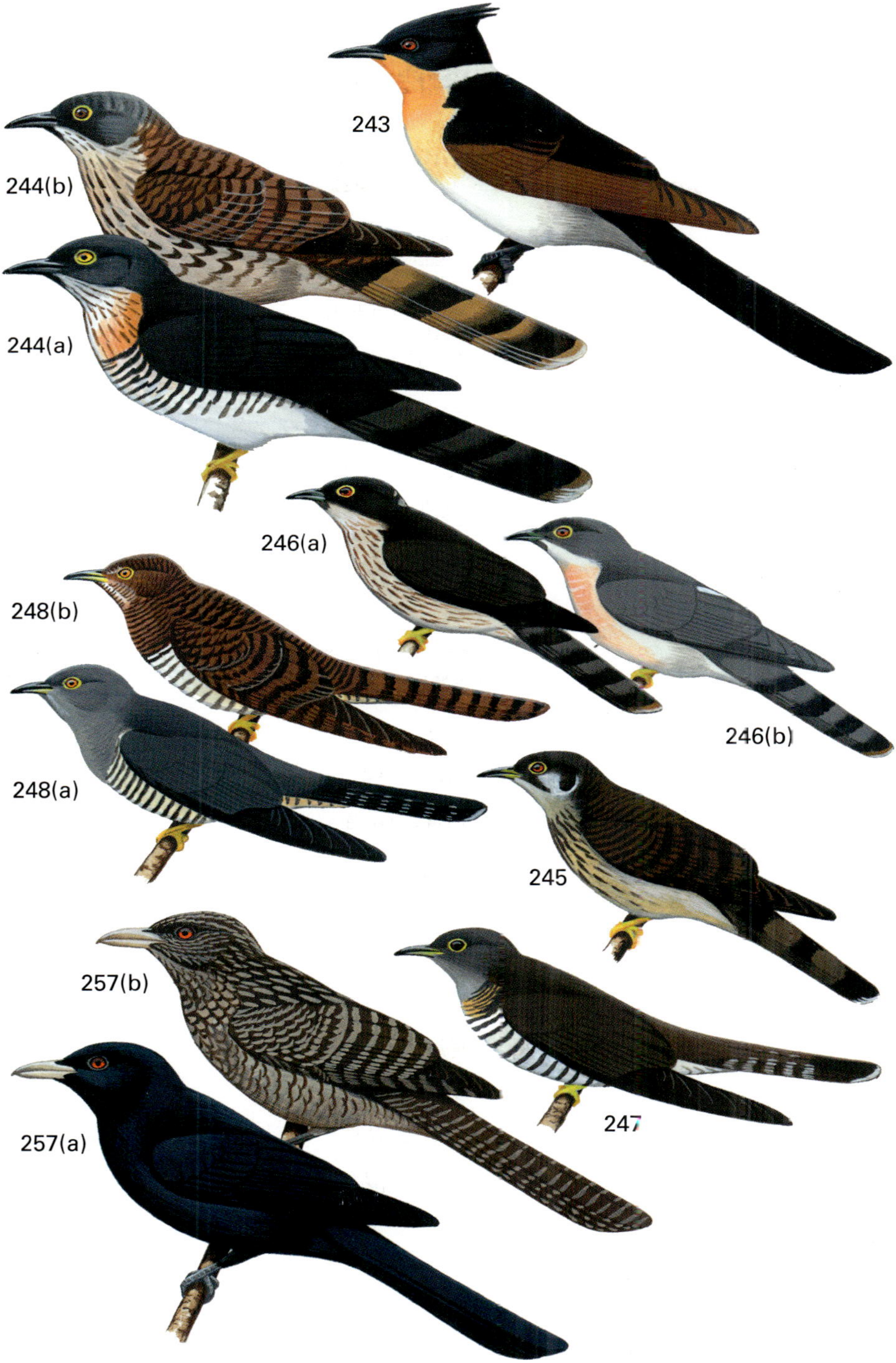
243
244(b)
244(a)
246(a)
248(b)
246(b)
248(a)
245
257(b)
247
257(a)

249. Banded Bay Cuckoo *Cacomantis sonnerattii* 23 cm (9 in)
Long slightly down-curved bill, dark rufous upperparts with fine blackish bars, broad dark patch behind eye, whitish supercilium, and whitish underparts with fine blackish bars. **p. 202**

250. Plaintive Cuckoo *Cacomantis merulinus* 22 cm (8.5 in) **p. 202**
(a) Adult: Pale grey head, throat and breast contrasting with rufous buff lower breast and belly.
(b) Juvenile: Rufous bay head and upperparts and whitish underparts marked with fine blackish barrings all over. Has faint whitish supercilium and small patch below eye.

251. Rusty-breasted Cuckoo *Cacomantis sepulcralis* 24 cm (9.5 in) **p. 203**
(a) Adult: Rufous buff underparts extends to breast and throat, merging with grey sides of head.
(b) Juvenile: Head and entire upperparts blackish with narrow rufous bars.

252. Little Bronze-cuckoo *Chrysococcyx minutillus* 16 cm (6.25 in)
Whitish forehead, brown upperparts tinged bronzy green, whitish underparts with fine blackish bars on throat and breast and broader bars on belly and flanks. **p. 203**
(a) Male: Narrow red orbital skin.
(b) Female: Narrow green orbital skin.

253. Horsfield's Bronze-cuckoo *Chrysococcyx basalis* 16 cm (6.25 in)
Dark brown ear coverts, pale supercilium, bold barrings on flanks and rufous base to sides of tail. **p. 204**

254. Asian Emerald Cuckoo *Chrysococcyx maculatus* 18 cm (7 in) **p. 204**
(a) Male: Iridescent dark green head, upperparts and upper breast with dark green barrings on white lower breast and belly.
(b) Female: Chestnut crown and nape contrasting with pale coppery green upperparts with entire underparts barred.

255. Violet Cuckoo *Chrysococcyx xanthorhynchus* 17 cm (6.5 in) **p. 204**
(a) Male: Iridescent dark purple head, upperparts and upper breast with dark purple barrings on white breast, orange bill with red base.
(b) Female: Reddish bronze upperparts, dull orange yellow bill and rufous outer tail feathers.

256. Drongo Cuckoo *Surniculus lugubris* 25 cm (9.25 in)
Glossy black plumage, slightly forked tail, white barrings on undertail coverts. **p. 205**

249
256
251(a)
251(b)
250(a)
250(b)
54(b)
255(b)
254(a)
255(a)
253
252(a)
252(b)

 MALKOHAS AND COUCALS

258. Black-bellied Malkoha *Rhopodytes diardii* 37 cm (14.5 in)
Small size with blackish plumage and red orbital skin. **p. 206**

259. Chestnut-bellied Malkoha *Rhopodytes sumatranus* 40 cm (16 in)
Chestnut belly and undertail coverts, pinkish orange orbital skin. **p. 206**

260. Green-billed Malkoha *Rhopodytes tristis* 56 cm (22 in)
Large and long-tailed with robust green bill and narrow white border surrounding red orbital skin. **p. 206**

261. Raffles's Malkoha *Rhinortha chlorophaea* 33 cm (13 in)
Small size with pale greenish orbital skin distinctive. **p. 207**
(a) Male: Buff rufous head, throat and breast and blackish tail marked with narrow dark rufous bars.
(b) Female: Pale grey head, throat and upper breast and all chestnut mantle and tail.

262. Red-billed Malkoha *Zanclostomus javanicus* 45 cm (18 in)
Red bill, blue orbital skin and pale cinnamon throat and upper breast. **p. 207**

263. Chestnut-breasted Malkoha *Zanclostomus curvirostris* 46 cm (18 in)
Broad red orbital skin, chestnut underparts and the lack of white tip to tail. **p. 207**

264. Short-toed Coucal *Centropus rectunguis* 37 cm (14.5 in) **p. 208**
(a) Adult: Miniature 265 with proportionately shorter tail.
(b) Juvenile: Rufous head and breast with pale buff streaks and pale buff supercilium.

265. Greater Coucal *Centropus sinensis* 53 cm (21 in)
Large size with long, broad-ended tail. **p. 208**
(a) Adult: Glossy black head, upperparts and bright chestnut mantle and wings.
(b) Juvenile: Black parts duller with faint greyish barrings or spots with chestnut parts barred blackish.

266. Lesser Coucal *Centropus bengalensis* 38 cm (15 in) **p. 208**
(a) Adult: Dull black head, underparts and tail with long buff streaks on head, neck, and mantle.
(b) Juvenile: Long white buff streaks on brown head and mantle with pale buff bill.

258
259
260
261(b)
262
263
264(b)
261(a)
264(a)
266(b)
265(b)
265(a)
266(a)

PLATE 35 **BARN OWL, BAY OWL, AND MEDIUM TO SMALLER TYPICAL OWLS**

267. Barn Owl *Tyto alba* 34 cm (13.5 in) **p. 209**
(a) White heart-shaped facial disc, long legs, golden buff and grey upperparts, white underparts.
(b) Shows all white underwing in flight.

268. Oriental Bay-owl *Phodilus badius* 29 cm (11.5 in)
Broad facial disc and forehead, white necklace, dark smudges below brown eyes. **p. 210**

269. White-fronted Scops-owl *Otus sagittatus* 27 cm (10.5 in)
Broad whitish forehead and eyebrows, whitish bill, rufescent plumage. **p. 211**

270. Reddish Scops-owl *Otus rufescens* 19 cm (7.5 in)
Rufescent plumage, underparts marked with fine blackish spots, brown iris. **p. 211**

271. Mountain Scops-owl *Otus spilocephalus* 20 cm (8 in)
Rufescent plumage, whitish spots on scapulars, fine whitish bars on underparts, yellow iris. **p. 211**

272. Collared Scops-owl *Otus bakkamoena* 23 cm (9 in)
Buff nuchal collar, brown iris, dark brown triangular patch on forehead contrasting with whitish buff eyebrows. **p. 212**

273. Oriental Scops-owl *Otus sunia* 19 cm (7.5 in)
Bold blackish streaks on underparts, yellow iris. **p. 212**
(a) Grey morph: Greyish plumage.
(b) Rufous morph: Brownish with strong rufescent tinge.

280. Collared Owlet *Glaucidium brodiei* 16 cm (6.5 in)
Tiny size, buff nuchal collar, whitish frontal collar, greyish head with fine spots, yellow iris. **p. 215**

281. Brown Boobook *Ninox scutulata* 30 cm (12 in)
Round head without facial disc, yellow iris, broad brown streaks on whitish underparts. **p. 215**

282. Short-eared Owl *Asio flammeus* 38 cm (15 in) **p. 216**
(a) Pale greyish white facial disc with white eyebrows and yellow eyes surrounded by blackish feathers.
(b) In flight shows black patch on carpals of upperwing and dark crescentic patch on carpals of underwing.

267(a)
267(b)
268
269
70
271
273(a)
273(b)
272
281
282(b)
282(a)
280

PLATE 36 **LARGER TYPICAL OWLS AND FROGMOUTHS**

274. Barred Eagle-owl *Bubo sumatranus* 48 cm (19 in)
Whitish underparts with fine brownish scaly bars, brown iris, pale yellowish bill, brownish eyebrows and ear-tufts finely barred whitish. **p. 212**

275. Dusky Eagle-owl *Bubo coromandus* 53 cm (21 in)
Pale greyish buff underparts with fine blackish streaks, yellow iris. **p. 213**

276. Brown Fish-owl *Ketupa zeylonensis* 53 cm (21 in)
Pale buff underparts with blackish vertical streaks and very fine indistinct horizontal bars, yellow iris. **p. 213**

277. Buffy Fish-owl *Ketupa ketupu* 50 cm (18 in)
Rich buff underparts with blackish streaks, whitish forehead, yellow iris. **p. 214**

278. Spotted Wood-owl *Strix seloputo* 48 cm (19 in)
White spots on dark brown upperparts, pale rufous facial disc, pale buff underparts with bold blackish brown horizontal bars. **p. 214**

279. Brown Wood-owl *Strix leptogrammica* 55 cm (22 in)
Broad rufous 'spectacles', dark smudges surrounding brown eyes, buff underparts with very fine horizontal bars. **p. 215**

283. Large Frogmouth *Batrachostomus auritus* 41 cm (16 in)
Large size, chestnut upperparts with white double collar, brown iris. **p. 217**

284. Gould's Frogmouth *Batrachostomus stellatus* 25 cm (9.75 in)
Creamy underparts with tawny rufous scales, brown and buff barred tail, yellow iris. **p. 217**

285. Javan Frogmouth *Batrachostomus javensis* 25 cm (9.75 in)
Heavy growth of bristles above base of bill and behind ears, white collar and yellow iris. **p. 217**

(a) Male: Sandy grey brown upperparts with white markings on wings, breast and belly.
(b) Female: Uniform rufous plumage with white markings on wings and breast. Tail marked with fine inconspicuous bars.

274

275

276

277

278

279

283

284

285(b)

285(a)
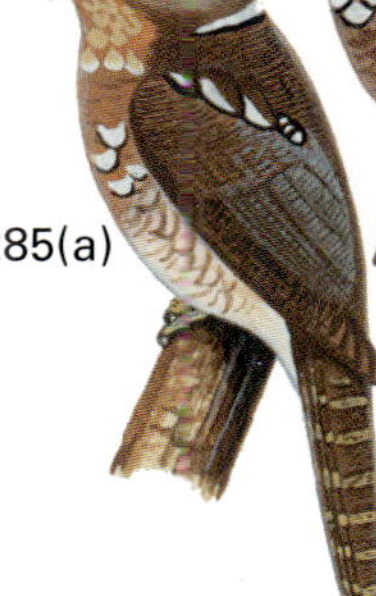

286. Malaysian Eared-nightjar *Eurostopodus temminckii* 27 cm (10.5 in) **p. 218**
(a) At rest: Medium size with erect ear tufts and conspicuous white collar.
(b) In flight, shows no white on wings or tail.

287. Great Eared-nightjar *Eurostopodus macrotis* 41 cm (16 in) **p. 219**
(a) At rest: Very large with erect ear tufts.
(b) In flight, resembles harriers, shows no white on wings or tail.

288. Grey Nightjar *Caprimulgus indicus* 28 cm (11 in) **p. 219**
(a) Male at rest: Dark grey plumage with white throat patch distinctive.
(b) Male in flight: Small white wing patch and white subterminal spots on tail.
(c) Female in flight: Shows buff wing patch, lacks white spots on tail.

289. Large-tailed Nightjar *Caprimulgus macrurus* 30 cm (12 in) **p. 220**
(a) Male at rest: Greyish brown plumage with bold black streaks on scapulars.
(b) Male in flight: Shows white wing patches and white corners to tail.
(c) Female in flight: Shows buff wing patches and buff corners to tail.

290. Savanna Nightjar *Caprimulgus affinis* 25 cm (10 in) **p. 220**
(a) Male at rest: Slim and short tailed. Greyish brown plumage mottled blackish on upperparts.
(b) Male in flight: Shows white wing patches and all white outer tail feathers.
(c) Female in flight: Wing patches buff grey, entire tail greyish brown.

286(b)
286(a)
287(a)
287(b)
288(a)
288(c)
288(b)
289(c)
289(b)
290(c)
290(b)
289(a)
290(a)

PLATE 38 **TREESWIFTS, SWIFTLETS, SPINETAIL, NEEDLETAILS, AND SWIFTS**

291. Grey-rumped Treeswift *Hemiprocne longipennis* 23 cm (9 in)
Prominent crest, dark blue-green crown and mantle greyish rump. **p. 221**
(a) Male: Chestnut ear coverts. (b) Female: Lacks chestnut ear coverts. (c) Upperwing flight: Shows pale grey patch on tertials.

292. Whiskered Treeswift *Hemiprocne comata* 15 cm (6 in)
Bronzy brown plumage with two horizontal white facial stripes. **p. 221**
(a) Male: Chestnut ear coverts. (b) Female: Lacks chestnut ear coverts. (c) Upperwing flight: Shows white patch on tertials.

293. Waterfall Swift *Hydrochous gigas* 16 cm (6.5 in)
All dark large swiftlet with notched tail. **p. 222**

294. Glossy Swiftlet *Collocalia esculenta* 10 cm (4 in) **p. 223**
(a) Upperwing flight: Glossy blue-black upperparts. (b) Underwing flight: Greyish throat and breast, white belly.

295. Himalayan Swiftlet *Aerodramus brevirostris* 13 cm (5 in)
Wings longer than 296 and 297, tail fork deeper. **p. 223**

296. Black-nest Swiftlet *Aerodramus maximus* 14 cm (5.5 in)
Slightly stockier than 297, rump greyish, slight notch in tail. **p. 223**

297. Edible-nest Swiftlet *Aerodramus fuciphaga* 14 cm (5.5 in)
Dark with whitish rump and slight notched in tail. **p. 224**

298. Silver-rumped Spinetail *Rhaphidura leucopygialis* 11 cm (4.5 in)
White rump and entire uppertail coverts, broad 'butter knife' wings. **p. 224**

299. White-throated Needletail *Hirundapus caudacutus* 20 cm (8 in)
White throat, silvery white patch on back and white patch on tertials. **p. 225**

300. Silver-backed Needletail *Hirundapus cochinchinensis* 20 cm (8 in)
Greyish throat, silvery white patch on back. **p. 225**

301. Brown-backed Needletail *Hirundapus giganteus* 25 cm (10 in)
Large and robust with bronzy brown patch on back. **p. 225**

302. Asian Palm-swift *Cypsiurus balasiensis* 13 cm (5 in)
Tail feathers held together like a spine when closed. **p. 225**

303. Fork-tailed Swift *Apus pacificus* 19 cm (7.5 in)
Deeply forked tail, narrow curved white band on rump. **p. 226**

304. House Swift *Apus affinis* 15 cm (6 in)
Slightly notched tail, white throat and broad white rump patch. **p. 226**

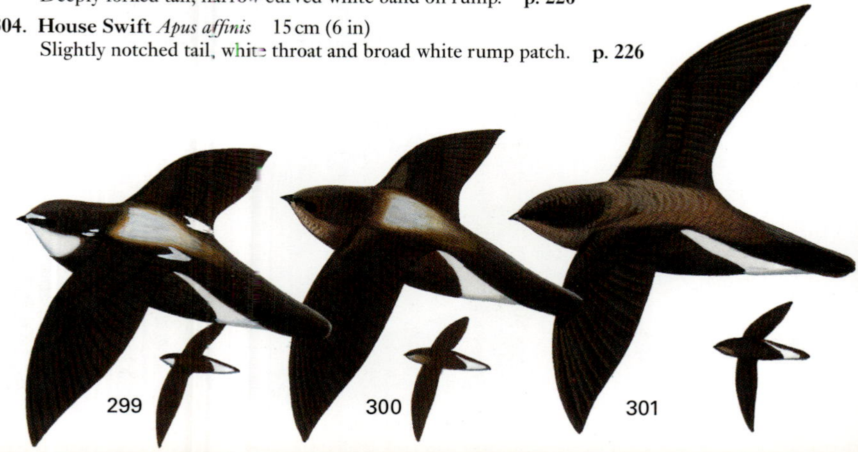

293
297
296
295
294(b)
303
294(a)
298
304
302
292(b)
291(c)
292(c)
291(b)
291(a)
292(a)

305. Red-naped Trogon *Harpactes kasumba* 32 cm (12.5 in)
Deep blue orbital skin **p. 227**
(a) Male: Red nuchal patch, white crescent across breast.
(b) Female: Dark brownish grey throat and breast contrasting with tawny brown underparts.

306. Diard's Trogon *Harpactes diardii* 30 cm (12 in)
Purple orbital skin, fine vermiculations on greyish white underside of retrices. **p. 227**
(a) Male: Pinkish crescent across breast, brownish maroon hindcrown, red nuchal patch.
(b) Female: Brownish head and breast, pinkish red underparts.

307. Cinnamon-rumped Trogon *Harpactes orrhophaeus* 25 cm (10 in) **p. 228**
(a) Male: Cinnamon rump concolorous with mantle and uppertail coverts.
(b) Female: Sides of head and underparts rusty buff.

308. Scarlet-rumped Trogon *Harpactes duvaucelii* 25 cm (10 in) **p. 228**
(a) Male: Bright red rump contrasting with mantle and uppertail coverts.
(b) Female: Pinkish red rump, reddish pink underparts.

309. Orange-breasted Trogon *Harpactes oreskios* 30 cm (12 in) **p. 228**
(a) Male: Olive green head, bright orange breast, orange-yellow underparts.
(b) Female: Olive green mantle, olive-green head and paler orange-yellow underparts.

310. Red-headed Trogon *Harpactes erythrocephalus* 34 cm (13.5 in)
Broad white crescent across breast. **p. 229**
(a) Male: Red head.
(b) Female: Brownish cinnamon head and breast.

305(a)
305(b)
307(a)
307(b)
308(a)
308(b)
306(a)
306(b)
309(a)
309(b)
310(a)
310(b)

313. Rufous-collared Kingfisher *Actenoides concretus* 24 cm (9.5 in)
Rufous collar and underparts, dark blue malar stripe, yellow bill. **p. 231**
(a) Male: Blue mantle.
(b) Female: Green mantle with buff spots.

314. Banded Kingfisher *Lacedo pulchella* 23 cm (9 in) **p. 231**
(a) Male: Upperparts black with blue and white bars.
(b) Female: Upperparts rufous with black bars.

315. Brown-winged Storkbill *Pelargopsis amauroptera* 36 cm (14 in)
Large red bill, chocolate brown wings and tail, bright blue rump. **p. 232**

316. Common Storkbill *Pelargopsis capensis* 37 cm (14.5 in)
Large red bill, pale brown head, blue upperparts, bright blue rump. **p. 232**

317. Ruddy Kingfisher *Halcyon coromanda* 25 cm (10 in)
Bright violet rufous upperparts, bright rufous buff underparts. **p. 233**

318. White-throated Kingfisher *Halcyon smyrnensis* 28 cm (11 in)
White throat and breast, chocolate brown head, bright blue wings. **p. 233**

319. Black-capped Kingfisher *Halcyon pileata* 30 cm (12 in)
Black head, white collar and breast, rufous belly and flanks. **p. 233**

320. Collared Kingfisher *Todiramphus chloris* 24 cm (9.5 in)
Blue-green upperparts, white collar and underparts. **p. 234**

321. Oriental Dwarf Kingfisher *Ceyx erithaca* 14 cm (5.5 in)
Tiny size. Coral red bill and feet, bright yellow underparts. **p. 234**
(a) Black-backed form: Blackish mantle tinged dark blue on scapulars, dark blue spot on forehead and sides of neck.
(b) Rufous-backed form: Dark rufous mantle tinged lilac.

322. Blue-banded Kingfisher *Alcedo euryzona* 20 cm (8 in)
Blackish blue upperparts with fine blue speckles on wing coverts. **p. 235**
(a) Male: Scaly blue breast band, white underparts, pale rufous flanks.
(b) Female: Pale rufous underparts.

323. Blue-eared Kingfisher *Alcedo meninting* 16 cm (6 in)
Dark blue mantle and ear coverts, dark rufous underparts. **p. 235**

324. Common Kingfisher *Alcedo atthis* 17 cm (7 in)
Blue-green mantle, rufous underparts and ear coverts. **p. 235**

322(b)
324
323
322(a)
321(b)
321(a)
317
319
318
320
313(b)
313(a)
314(a)
314(b)

ROLLERS, BEE-EATERS, AND HOOPOE

311. Indian Roller *Coracias benghalensis* 33 cm (13 in) **p. 229**
(a) Brown upperparts, throat and upper belly, pale blue belly and undertail coverts.
(b) Upperwing flight pattern: Pale blue rump conspicuous.

312. Dollarbird *Eurystomus orientalis* 30 cm (12 in) **p. 230**
(a) Glossy dark blue-green plumage, red bill.
(b) In flight, silvery white patch at base of primaries conspicuous.

325. Red-bearded Bee-eater *Nyctyornis amictus* 32 cm (13 in) **p. 236**
(a) Male: Crimson breast and throat, lilac forehead.
(b) Upperwing flight pattern: Bright green upperparts conspicuous.
(c) Juvenile: Entire head and underparts mostly green

326. Blue-tailed Bee-eater *Merops philippinus* 30 cm (12 in) **p. 237**
(a) Yellowish upper throat, brownish lower throat, elongate central tail feathers.
(b) Upperwing flight pattern: Blue rump and uppertail coverts.

327. Blue-throated Bee-eater *Merops viridis* 28 cm (11 in) **p. 237**
(a) Dark chestnut head and upper back, blue throat, elongate central tail feathers.
(b) Upperwing flight pattern: Iridescent pale blue rump conspicuous.

328. Chestnut-headed Bee-eater *Merops leschenaulti* 23 cm (9 in) **p. 237**
(a) Chestnut head and upper back, creamy yellow throat and sides of neck.
(b) Upperwing flight pattern: Pale iridescent blue rump conspicuous.

329. Eurasian Hoopoe *Upupa epops* 30 cm (12 in) **p. 238**
(a) Pinkish buff plumage, black wings with white bars, crest erect when alarmed.
(b) In flight, black wings and white bars very conspicuous, crest held flat on head.

328(b)
328(a)
326(a)
326(b)
327(a)
325(c)
325(a)
7(b)
325(b)
312(a)
312(b)
311(b)
329(b)
329(a)
311(a)

330. Bushy-crested Hornbill *Anorrhinus galeritus* 89 cm (35 in)
Blackish plumage and bill, whitish orbital skin. **p. 239**

331. Oriental Pied Hornbill *Anthracoceros albirostris* 70 cm (27 in) **p. 239**
(a) Male: Ivory-white bill and casque.
(b) Female: Bill and casque marked blackish.

332. Black Hornbill *Anthracoceros malayanus* 76 cm (30 in) **p. 240**
(a) Male: White casque and bill, (b) broad white supercilium in some individuals.

333. Great Hornbill *Buceros bicornis* 122 cm (48 in)
Black mask contrasting with pale yellow bill, casque and neck. **p. 240**
Male: Red iris.

334. Rhinoceros Hornbill *Buceros rhinoceros* 122 cm (48 in)
Ivory-white bill with upturned reddish and yellow casque. **p. 241**
Female: White iris.

335. Helmeted Hornbill *Rhinoplax vigil* 127 cm (50 in)
Short bill, deeply truncated casque and elongated central tail feathers. **p. 241**
Male: Bare and coarsely wrinkled red skin around neck.

336. White-crowned Hornbill *Berenicornis comatus* 90 cm (36 in) **p. 241**
(a) Male: White neck
(b) Female: Black neck

337. Wrinkled Hornbill *Aceros corrugatus* 81 cm (32 in) **p. 242**
(a) Male: Reddish casque, white neck and gular pouch.
(b) Female: Black head and neck, pale blue gular pouch.

338. Plain-pouched Hornbill *Rhyticeros subruficollis* 90 cm (36 in)
Both sexes resemble respective sexes of 339 but are smaller and lack corrugations on base of bill as well as dark bar on gular pouch. **p. 242**
(a) Male: Yellow gular pouch.
(b) Female: Blue gular pouch.

339. Wreathed Hornbill *Rhyticeros undulatus* 100 cm (42 in)
All white tail with black bar on gular pouch diagnostic. **p. 242**
(a) Male: White casque and bill.
(b) Female: Black head and neck, pale blue gular pouch.

338(a)

338(b)

336(a)
336(b)
330
337(a)
337(b)
339(a)
332(b)
332(a)
331(b)
331(a)
339(b)
335
333
334

340. Fire-tufted Barbet *Psilopogon pyrrolophus* 28 cm (11 in)
Yellow and black breast band, grey cheek patch, pale green bill with tuft of black and red bristles. **p. 244**

341. Lineated Barbet *Megalaima lineata* 29 cm (11.5 in)
Light brown head with buff streaks, pale straw bill and orbital skin. **p. 245**

342. Gold-whiskered Barbet *Megalaima chrysopogon* 30 cm (12 in)
Bright yellow oval cheek patch, greyish buff forecrown. **p. 245**

343. Red-crowned Barbet *Megalaima rafflesii* 27 cm (10 in)
Red crown, yellow cheek patch, blue throat and supercilium. **p. 245**

344. Red-throated Barbet *Megalaima mystacophanos* 22 cm (9 in) **p. 246**
(a) Male: Red throat, yellow forehead.
(b) Female: Pale greenish forehead, blue midcrown, red hindcrown.

345. Golden-throated Barbet *Megalaima franklinii* 22 cm (9 in)
Yellow throat and crown, silvery grey sides of face. **p. 246**

346. Black-browed Barbet *Megalaima oorti* 20 cm (8 in)
Black supercilium and malar stripe, yellow throat and crown. **p. 246**

347. Yellow-crowned Barbet *Megalaima henricii* 21 cm (8.5 in)
Yellow forehead and supercilium, blue throat. **p. 247**

348. Blue-eared Barbet *Megalaima australis* 17 cm (7 in)
M. a. duvaucelii: Blue throat broad red stripes above and below black ear coverts. **p. 247**

349. Coppersmith Barbet *Megalaima haemacephala* 15 cm (6 in)
Red legs, greenish white underparts with bold green streaks, red and yellow breast band. **p. 247**

350. Brown Barbet *Calorhamphus fuliginosus* 18 cm (7.5 in)
Brown upperparts, whitish underparts, orange legs. **p. 248**

352. Eurasian Wryneck *Jynx torquilla* 18 cm (7.5 in)
Distinctive cryptically patterned plumage, with short pointed bill, brown eyeline and indistinct whitish-grey supercilium. **p. 249**

352

340
341
342
343
344(b)
346
344(a)
345
347
348
349
350

PLATE 44 **HONEYGUIDE, PICULETS, AND WOODPECKERS**

351. Malaysian Honeyguide *Indicator archipelagicus* 18 cm (7 in)
Bulbul-shaped with dark olive brown above with dark streaks on flanks. **p. 248**
Male: Has narrow lemon yellow shoulder patch.

353. Speckled Piculet *Picumnus innominatus* 10 cm (4 in) **p. 250**
(a) Male: Small orange patch on forecrown. (b) Female: Entire crown olive green.

354. Rufous Piculet *Sasia abnormis* 9 cm (3.5 in) **p. 250**
(a) Male: Yellow patch on forehead. (b) Female: Rufous forehead.

355. Sunda Woodpecker *Dendrocopos moluccensis* 14 cm (5.5 in)
Brown cap, prominent dark malar stripe. **p. 250**
(a) Male: Red streak behind eye. (b) Female: Lacks red streak behind eye.

356. Grey-capped Woodpecker *Dendrocopos canicapillus* 15 cm (6 in)
Grey crown, indistinct dark malar stripe. **p. 250**
(a) Male: Red streak behind eye. (b) Female: Lacks red streak behind eye.

359. Banded Yellownape *Picus miniaceus* 25 cm (10 in)
Red crest with yellow tip, red wings, barrings above and below. **p. 251**
(a) Male: Red sides of head (b) Female: Brown sides of head with brown spots.

360. Crimson-winged Yellownape *Picus puniceus* 25 cm (10 in)
Red crown, yellow nuchal crest and red wings. **p. 252**
(a) Male: Red malar stripe. (b) Female: Malar area all green.

361. Lesser Yellownape *Picus chlorolophus* 27 cm (10.5 in)
Red supercilium, white facial bar and whitish bars below. **p. 252**
(a) Male: Red malar stripe. (b) Female: Lacks red malar stripe.

362. Chequer-throated Yellownape *Picus mentalis* 28 cm (11 in)
Chestnut sides of head and collar, throat streaked black and white. **p. 252**
(a) Male: Malar area marked with fine black streaks. (b) Female: Malar area chestnut and unstreaked.

363. Greater Yellownape *Picus flavinucha* 34 cm (13.5 in)
Dark brown crown and dark brown flight feathers with black bars. **p. 253**
(a) Male: Creamy yellow throat and malar area. (b) Female: Brown throat and malar area.

364. Laced Woodpecker *Picus vittatus* 30 cm (12 in)
Greyish sides of head, dark lacy streaks below. **p. 253**
(a) Male: Red crown. (b) Female: Black crown.

365. Streak-breasted Woodpecker *Picus viridanus* 30 cm (12 in)
Dark streaks on throat, whitish scalings on breast. **p. 253**

366. Grey-faced Woodpecker *Picus canus* 33 cm (13 in) **p. 254**
(a) Male: Red forecrown. (b) Female: Entire crown black.

376. Grey-and-Buff Woodpecker *Hemicircus concretus* 13 cm (5 in)
Small, crested, grey plumage with pale yellowish scales above. **p. 257**
(a) Male: Red crown. (b) Female: Grey crown.

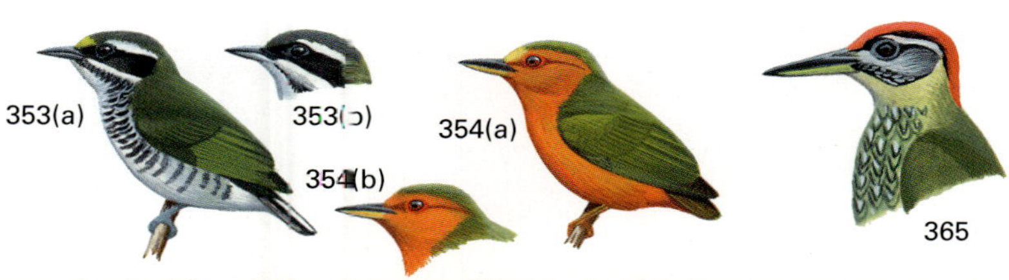

364(a)
366(a)
363(a)
366(b)
364(b)
363(b)
360(a)
361(a)
361(b)
360(b)
362(a)
356(a)
356(b)
362(b)
359(a)
355(a)
355(b)
359(b)
351
376(a)
376(b)

357. Rufous Woodpecker *Celeus brachyurus* 25 cm (10 in) **p. 251**
(a) Male: Red cheek patch. (b) Female: Lacks red cheek patch.

358. White-bellied Woodpecker *Dryocopus javensis* 43 cm (17 in) **p. 251**
(a) Male: Malar stripe and entire crown red. (b) Female: Lacks red malar stripe, forecrown black.

367. Olive-backed Woodpecker *Dinopium rafflesii* 28 cm (11 in) **p. 254**
(a) Male: Red crown and crest. (b) Female: Black crown and crest.

368. Common Flameback *Dinopium javenense* 30 cm (12 in)
White face with black eyeline and malar stripe, brown iris. **p. 254**
(a) Male: Red crown and crest. (b) Female: Black crown and crest with whitish spots.

369. Greater Flameback *Chrysocolaptes lucidus* 33 cm (13 in)
Larger and more robust than 368. White face with black eyeline, yellow iris. **p. 255**
(a) Male: Red crown and crest. (b) Female: Black crown and crest marked with white spots.

370. Bamboo Woodpecker *Gecinulus viridis* 28 cm (11 in) **p. 255**
(a) Male: Red crown. (b) Female: Brownish buff crown.

371. Maroon Woodpecker *Blythipicus rubiginosus* 23 cm (9 in) **p. 256**
(a) Male: Red nape and neck patch. (b) Female: Lacks red nape and neck patch.

372. Bay Woodpecker *Blythipicus pyrrhotis* 30 cm (12 in) **p. 256**
(a) Male: Red patch on neck. (b) Female: Lacks red neck patch.

373. Orange-backed Woodpecker *Reinwardtipicus validus* 30 cm (12 in) **p. 256**
(a) Male: Red crown and crest, red underparts and orange line down back. (b) Female: Dark brown crown and crest.

374. Buff-rumped Woodpecker *Meiglyptes tristis* 19 cm (7.5 in)
Small black bill and slight nuchal crest distinctive. Fulvous buff head, pale rump. **p. 256**
(a) Male: Red malar stripe. (b) Female: Lacks red malar stripe.

375. Buff-necked Woodpecker *Meiglyptes tukki* 22 cm (8.5 in)
Diagnostic buff neck patch. **p. 257**
(a) Male: Red malar stripe. (b) Female: Lacks red malar stripe.

377. Great Slaty Woodpecker *Mulleripicus pulverulentus* 50 cm (20 in) **p. 257**
(a) Male: Red malar stripe. (b) Female: Lacks red malar stripe.

357(a)
357(b)
372(a)
372(b)
371(a)
371(b)
367(a)
367(b)
368(a)
368(b)
369(a)
369(b)
370(a)
370(b)
377(a)
377(b)
358(a)
358(b)
373(a)
373(b)

378. Little Green-broadbill *Calyptomena viridis* 19 cm (7.5 in)
Chunky with bright grass green plumage. **p. 258**
(a) Male: Black neck spot and wingbars
(b) Female: Lacks black neck spot and wingbars

379. Black-and-Red Broadbill *Cymbirhynchus macrorhynchus* 25 cm (10 in)
White streak on wing, black breast band separating reddish maroon throat and underparts. **p. 259**

380. Long-tailed Broadbill *Psarisomus dalhousiae* 28 cm (11 in)
Long blue tail, green plumage, black and yellow head pattern **p. 259**

381. Silver-breasted Broadbill *Serilophus lunatus* 18 cm (7 in)
Broad black supercilium, yellow orbital skin, chestnut rump, greyish white underparts. **p. 259**
(a) Female: Narrow silvery white breast band
(b) Male: Lacks silvery white breast band

382. Banded Broadbill *Eurylaimus javanicus* 23 cm (9 in)
Red head, yellow streaks on upperparts and rump. **p. 260**
(a) Male: Narrow black upper breast band.
(b) Female: Lacks narrow upper breast band.

383. Black-and-yellow Broadbill *Eurylaimus ochromalus* 17 cm (6.5 in)
Black head, white collar, yellow streaks on wing, pink breast, yellow belly and undertail coverts. **p. 260**
(a) Male: Complete black breast band
(b) Female: Black breast band usually broken

384. Dusky Broadbill *Corydon sumatranus* 28 cm (11 in)
Large size, blackish plumage, pink bill and orbital skin. **p. 260**

384
379
382(a)
382(b)
383(b)
383(a)
381(a)
381(b)
380
378(b)
378(a)

PLATE 47 **PITTAS**

385. Rusty-naped Pitta *Pitta oatesi* 25 cm (10 in) **p. 261**
(a) Male: Bright rufescent head and underparts, saturated dark green upperparts.
(b) Female: Head and underparts duller, dark green upperparts tinged rusty.

386. Giant Pitta *Pitta caerulea* 29 cm (11.5 in)
Large size with thickset bill, blackish brown band around neck. **p. 261**
(a) Male: Bright pale blue upperparts, hindcrown and nape brownish black.
(b) Female: Rusty brown upperparts, hindcrown scaled brownish.

387. Banded Pitta *Pitta guajana* 23 cm (9 in)
Broad yellow and orange supercilium, golden brown mantle, white streaks on wing. **p. 262**
(a) Male: underparts dark blue with orange bands across breast.
(b) Female: underparts marked with narrow buff brown and black bands.

388. Garnet Pitta *Pitta granatina* 18 cm (7 in)
Dark electric violet- blue mantle and wings, bright red crown, nape and underparts. **p. 262**

389. Hooded Pitta *Pitta sordida* 19 cm (7.5 in)
P.s. cucullata: Black head, brown crown, bright green breast and upper belly, white wing patch. **p. 262**

390. Blue-winged Pitta *Pitta moluccensis* 20 cm (8 in)
Dark green mantle, bright blue wing coverts, bright rufous orange breast and belly, buff lateral coronal stripe, white wing patch. **p. 263**

391. Mangrove Pitta *Pitta megarhyncha* 20 cm (8 in)
Long and heavy bill, dark green mantle, bright blue wing coverts, bright rufous orange breast belly, white wing patch. **p. 263**

385(b)
390
385(a)
386(b)
391
386(a)
387(b)
387(a)
389
388

PLATE 48 **WOODSHRIKE, CUCKOOSHRIKES, TRILLER, MINIVETS, AND FLYCATCHER-SHRIKES**

394. Large Woodshrike *Tephrodornis virgatus* 23 cm (9 in) **p. 265**
Male: Black eyeline, greyish upperparts, white rump.

402. Javan Cuckooshrike *Coracina javensis* 30 cm (12 in)
Thickset black bill, diffuse black eyeline, black lores, brown iris. **p. 269**

403. Bar-bellied Cuckooshrike *Coracina striata* 30 cm (12 in)
White iris. **p. 270**
(a) Male: Diffused whitish barrings on lower underparts.
(b) Female: Grey and white barrings on underparts and rump.

404. Lesser Cuckooshrike *Coracina fimbriata* 23 cm (9 in) **p. 270**
(a) Male: Dark grey with black wings and white vent
(b) Female: Underparts whitish with narrow greyish bars.

405. Pied Triller *Lalage nigra* 18 cm (7 in)
Dark crown, white supercilium, staggered white wing bars. **p. 270**
(a) Male: Black crown, mantle, wings and tail, white underparts.
(b) Female: Black parts brown, with narrow brownish bars below.

406. Rosy Minivet *Pericrocotus roseus* 20 cm (8 in) **p. 271**
(a) Male: Rosy pink underparts, reddish rump and red uppertail coverts.
(b) Female: Pale yellow underparts.

407. Ashy Minivet *Pericrocotus divaricatus* 20 cm (8 in) **p. 271**
(a) Male: Black hindcrown and nape, white forecrown.
(b) Female: Crown and nape grey, flanks and sides of head suffused grey.

408. Fiery Minivet *Pericrocotus igneus* 15 cm (6 in) **p. 272**
(a) Male: Orange-yellow wingbar.
(b) Female: Flame orange rump.

409. Grey-chinned Minivet *Pericrocotus solaris* 18 cm (7 in) **p. 272**
(a) Male: Greyish suffusion on throat and sides of head.
(b) Female: Greyish crown and sides of head.

410. Scarlet Minivet *Pericrocotus flammeus* 18 cm (7.5 in) **p. 272**
(a) Male: Black head, throat, wings and upperparts.
(b) Female: Bright yellow forecrown, throat and sides of head.

411. Bar-winged Flycatcher-shrike *Hemipus picatus* 15 cm (6 in) **p. 273**
(a) Male: Black upperparts and wings, buff suffusion on breast.
(b) Female: Sooty brown upperparts and wings.

412. Black-winged Flycatcher-shrike
Hemipus hirundinaceus 15 cm (6 in) **p. 273**
(a) Male: All black upperparts and wings.
(b) Female: All sooty brown upperparts and wings.

411(b)
1(a)
412(b)
412(a)
394
402
403(a)
403(b)
04(b)
404(a)
407(b)
407(a)
405(b)
405(a)
408(a)
408(b)
409(a)
409(b)
410(a)
410(b)

PLATE 49 **WOODSWALLOWS, WHISTLER, SHRIKES, AND STARLINGS**

397. Ashy Woodswallow *Artamus fuscus* 16 cm (6.3 in) **p. 267**
(a) Brownish grey head, breast and upperparts, lighter on underparts.
(b) In flight, broad triangular wings distinctive.

398. White-breasted Woodswallow *Artamus leucorhynchus* 17 cm (6.5 in) **p. 267**
(a) Brownish grey head, breast and upperparts, white rump and underparts.
(b) In flight, broad triangular wings and white rump distinctive.

413. Mangrove Whistler *Pachycephala cinerea* 17 cm (6.5 in)
Greyish head, dull brown upperparts with pale ashy grey breast. **p. 274**

414. Tiger Shrike *Lanius tigrinus* 19 cm (7.5 in)
Bill thicker and tail proportionately shorter than 415. **p. 274**
(a) Male: Grey crown and nape, mantle and uppertail barred blackish.
(b) Juvenile and First Winter: Ear coverts barred blackish, bold barrings above and below.

415. Brown Shrike *Lanius cristatus* 20 cm (8 in)
Broad black eyeline and whitish supercilium distinctive. **p. 275**
(a) *L. c. cristatus*: Dull chestnut upperparts, greyish white forehead.
(b) *L. c. lucionensis*: Pale grey crown merging into ash brown upperparts.
(c) Juvenile and First Winter: Scaled above and below.

416. Long-tailed Shrike *Lanius schach* 25 cm (10 in)
Long black tail and white wing patch distinctive. **p. 276**
(a) Adult: Grey cap and back, black forehead, eyeline, wings, tawny flanks.
(b) Juvenile: Duller with faint barrings on flanks.

553. Asian Glossy Starling *Aplonis panayensis* 20 cm (8 in) **p. 329**
(a) Adult: Dark glossy green plumage and red iris distinctive.
(b) Juvenile: Blackish brown above whitish below, with bold streaks.

554. Purple-backed Starling *Sturnus sturninus* 19 cm (7.5 in) **p. 330**
(a) Adult: Metallic purple back and patch on nape, two narrow pale wingbars.
(b) Juvenile: Metallic purple areas replaced with brown.

556. White-shouldered Starling *Sturnus sinensis* 20 cm (8 in) **p. 330**
(a) Male: Pale greyish plumage, broad white shoulder patch.
(b) Female: Smaller white shoulder patch.

413
398(a)
398(b)
415(c)
415(a)
414(b)
414(a)
415(b)
416(b)
416(a)
553(b)
553(a)
556(b)
556(a)
554(a)
554(b)

PLATE 50 **ORIOLES AND DRONGOS**

420. Dark-throated Oriole *Oriolus xanthonotus* 20 cm. (7.5 in) **p. 278**
(a) Male: Black head, throat, upper breast, whitish lower breast and belly streaked blackish.
(b) Female: Olive brown above, streaked whitish lower breast and belly.

421. Eurasian Golden Oriole *Oriolus oriolus* 25 cm (10 in) **p. 278**
(a) Male: Bright yellow head and mantle, black wings with small yellow patch.
(b) Female: Yellow upperparts strongly tinged greenish, underparts streaked.

422. Black-naped Oriole *Oriolus chinensis* 27 cm (10.5 in)
O. c. maculatus is illustrated **p. 279**
(a) Male: Bright yellow plumage, broad black eyeline and nape.
(b) Juvenile: Dull olive yellow above, streaks on breast, bill should be blackish.

423. Black-hooded Oriole *Oriolus xanthornus* 25 cm (10 in) **p. 279**
(a) Adult: Black hood with bright yellow mantle and underparts
(b) Juvenile: Head dull blackish, throat, breast whitish with fine streaks.

424. Black-and-crimson Oriole *Oriolus cruentus* 23 cm (9 in) **p. 279**
(a) Male: Black plumage with dark crimson breast patch and wing patches.
(b) Female: Dull black plumage and bluish grey bill.

425. Black Drongo *Dicrurus macrocercus* 28 cm (11 in)
Deeply forked tail with slightly upturned tips. **p. 280**

426. Ashy Drongo *Dicrurus leucophaeus* 29 cm (11.5 in) **p. 280**
(a) *D. l. nigriesecens*. Blackish plumage with slight gloss
(b) *D. l. leucogenis.* Pale grey with white cheek patch.
(c) *D. l. salangensis.* Pale grey with white surrounding eye only.

427. Crow-billed Drongo *Dicrurus annectans* 28 cm (11 in)
Proportionately heavy bill, broad tail with widely splayed tips. **p. 281**

428. Bronzed Drongo *Dicrurus aeneus* 23 cm (9 in)
Black plumage heavily glossed blue, tail deeply forked. **p. 281**

429. Lesser Racquet-tailed Drongo *Dicrurus remifer* 27 cm (10.5 in)
Square cut tail with long ribbon- shaped racquets. **p. 282**

430. Greater Racquet-tailed Drongo *Dicrurus paradiseus* 32 cm (13 in)
Short crest, tail slightly forked with spoon-shaped racquets. **p. 282**

426(c)
426(a)
426(b)
430
425
427
429
428
423(a)
420(a)
422(b)
422(a)
420(b)
423(b)

PLATE 51 PHILENTOMAS, FANTAILS, MONARCH, AND PARADISE-FLYCATCHERS

395. Rufous-winged Philentoma *Philentoma pyrhoptera* 18 cm (7 in) **p. 266**
(a) Male common morph: Blue head and mantle, rufous wings and tail.
(b) Male blue morph: All blue plumage.
(c) Female: Greyish head, brownish mantle.

396. Maroon-breasted Philentoma *Philentoma velata* 20 cm (8 in) **p. 266**
(a) Male: Black throat, sides of head and dark maroon breast.
(b) Female: Lacks maroon breast.

431. White-throated Fantail *Rhipidura albicollis* 19 cm (7.5 in)
White throat, supercilium and tip of tail. **p. 283**

432. Pied Fantail *Rhipidura javanica* 18 cm (7 in)
Greyish brown breast band separating white throat and breast. **p. 283**

433. Spotted Fantail *Rhipidura perlata* 18 cm (7 in)
White spots on dark grey throat and breast. **p. 283**

434. Black-naped Monarch *Hypothymis azurea* 17 cm (6.5 in) **p. 284**
(a) Male: Bright blue plumage, black nape and gorget.
(b) Female: Dull blue head greyish brown upperparts.

435. Asian Paradise flycatcher *Terpsiphone paradisi* 21 cm (8.5 in) **p. 284**
(a) Male dark morph: Greyish throat and sides of head, rufous upperparts and tail.
(b) Male pale morph: White with glossy black head and throat.
(c) Female: Like dark morph male but lacking elongate tail.

436. Japanese Paradise-flycatcher *Terpsiphone atrocaudata* 20 cm (8 in) **p. 285**
(a) Male: Glossy black plumage and tail.
(b) Female: Blackish head, dark rufous mantle and tail.

435(a)
436(b)
435(c)
436(a)
435(b)
432
431
433
396(b)
395(c)
434(a)
395(b)
396(a)
395(a)
434(b)

PLATE 53 **SWALLOWS AND MARTINS**

446. Collared Sand Martin *Riparia riparia* 13 cm (5 in)
Slightly forked tail. **p. 289**
(a) Brown upperparts.
(b) White underparts separated by brown breast band.

447. Dusky Crag-martin *Ptynoprogne concolor* 14 cm (5.5 in)
(a) and (b) Dark plumage with slightly notched tail, marked with white subterminal spots. **p. 290**

448. Barn Swallow *Hirundo rustica* 15 cm (6 in)
Deeply forked tail with elongate outer tail feathers. **p. 290**
(a) Steel blue upperparts with white subterminal spots on tail.
(b) Blue breast band, chestnut throat.

449. Pacific Swallow *Hirundo tahitica* 14 cm (5.5 in)
Slightly forked tail. **p. 291**
(a) Steel blue upperparts with white subterminal spots on tail.
(b) Smoky grey underparts and chestnut throat.

450. Red-rumped Swallow *Cecropis daurica* 18 cm (7 in)
Dirty white underparts and underwing coverts marked with dark vertical streaks distinctive. **p. 291**

451. Striated Swallow *Cecropis striolata* 19 cm (7.5 in) **p. 291**
(a) Deeply forked tail and chestnut rump distinctive.
(b) Dark chestnut face, underparts and underwing coverts.

452. Asian House-martin *Delichon dasypus* 13 cm (5 in)
Slightly notched tail. **p. 292**
(a) White rump contrasting with steel blue upperparts.
(b) Pale underparts, greyish flight feathers, blackish wing coverts.

446(a)
447(a)
452(a)
446(b)
447(b)
452(b)
448(a)
449(a)
448(b)
449(b)
451(a)
451(b)
450

453. Straw-headed Bulbul *Pycnonotus zeylanicus* 29 cm (11.5 in)
Yellowish orange crown, white throat, black eyeline and malar stripe. **p. 293**

454. Black-and-white Bulbul *Pycnonotus melanoleucos* 18 cm (7 in)
Black plumage, white wing coverts and lining. **p. 293**

455. Black-headed Bulbul *Pycnonotus atriceps* 18 cm (7 in) **p. 294**
(a) Common morph: Yellow with blue-black head, black tail band, blue iris.
(b) Grey morph: Pale grey collar, breast and upper belly.

456. Black-crested Bulbul *Pycnonotus melanicterus* 19 cm (7.5 in)
Black head with long erectile crest, yellow plumage, yellow iris. **p. 294**

457. Scaly-breasted Bulbul *Pycnonotus squamatus* 16 cm (5.5 in)
Black scales on white breast and flanks, black head, white throat. **p. 294**

458. Grey-bellied Bulbul *Pycnonotus cyaniventris* 16 cm (5.5 in)
Bluish grey head and underparts, olive yellow upperparts, black lores. **p. 295**

459. Red-whiskered Bulbul *Pycnonotus jocosus* 20 cm (8 in)
Long black crest, brown above, white below, red undertail coverts. **p. 295**

460. Sooty-headed Bulbul *Pycnonotus aurigaster* 20 cm (8 in)
Black head, chin, and nape, yellow to orange-yellow undertail coverts. **p. 295**

461. Puff-backed Bulbul *Pycnonotus eutilotus* 23 cm (9 in)
Superficial crest, upperparts brown, tinged rufescent on wings and tail. **p. 296**

462. Yellow-vented Bulbul *Pycnonotus goaivier* 20 cm (8 in)
Black lores, brown coronal stripe, yellow undertail coverts. **p. 296**

463. Stripe-throated Bulbul *Pycnonotus finlaysoni* 19 cm (7.5 in)
Yellow streaks on forehead, chin and throat. **p. 297**

464. Olive-winged Bulbul *Pycnonotus plumosus* 20 cm (8 in)
Olive wings and tail, indistinct white streaks on ear coverts, reddish iris. **p. 297**

465. Streak-eared Bulbul *Pycnonotus blanfordi* 20 cm (8 in)
Pale brown upperparts, distinct white streaks on ear coverts, grey iris. **p. 297**

460

453
454
455(b)
455(a)
456
457
458
463
461
459
462
464
465

PLATE 55 TYPICAL BULBULS, BEARDED BULBULS, AND JUNGLE BULBULS

466. Cream-vented Bulbul *Pycnonotus simplex* 18 cm (7 in)
White iris, creamy centre of belly and undertail coverts. **p. 298**

467. Red-eyed Bulbul *Pycnonotus brunneus* 19 cm (7.5 in)
Red iris, dark brown plumage, paler below. **p. 298**

468. Spectacled Bulbul *Pycnonotus erythropthalmos* 16 cm (5.5 in)
Red iris, orange-yellow eyering, pale throat. **p. 298**

469. Sulphur-throated Bulbul *Alophoixus finschii* 17 cm (6.5 in)
Yellow throat, centre of belly and undertail coverts. **p. 299**

470. Ochraceous Bulbul *Alophoixus ochraceus* 23 cm (9 in)
Stocky, shaggy crest, white throat, dusky grey underparts. **p. 299**

471. Grey-cheeked Bulbul *Alophoixus bres* 22 cm (8.5 in)
Stocky, white throat, greyish sides of face, pale yellow centre of abdomen. **p. 299**

472. Yellow-bellied Bulbul *Alophoixus phaeocephalus* 20 cm (8 in)
Grey head, white throat, yellow underparts. **p. 300**

473. Hairy-backed Bulbul *Tricholestes criniger* 17 cm (6.5 in)
Brown upperparts, yellowish underparts. **p. 300**

474. Buff-vented Bulbul *Iole olivacea* 20 cm (8 in)
Greyish white iris, brownish olive upperparts. **p. 300**

475. Streaked Bulbul *Ixos malaccensis* 23 cm (9 in)
Dark olive upperparts, white streaks on throat and breast, white undertail coverts. **p. 300**

476. Mountain Bulbul *Hypsipetes mcclellandii* 24 cm (9.5 in)
Stocky, short crest, white streaks on crown, throat and breast, yellowish undertail coverts. **p. 301**

477. Ashy Bulbul *Hemixos flavala* 20 cm (8 in)
Short crest, black mask, grey upperparts, white throat. **p. 301**

466
467
468
469
470
471
472
473
474
476
475
477

PLATE 56 **GRASSHOPPER-WARBLERS, LEAF-WARBLERS, FLYCATCHER-WARBLERS, AND REED-WARBLERS**

478. Lanceolated Warbler *Locustella lanceolata* 13 cm (5 in)
Heavily streaked upperparts with finer streaks on breast, belly and undertail. **p. 302**

479. Rusty-rumped Warbler *Locustella certhiola* 15 cm (6 in)
Streaked crown and mantle, rufescent lower back and rump. **p. 302**

480. Dusky Warbler *Phylloscopus fuscatus* 12 cm (4.8 in)
Dark grey-brown above, pale supercilium, whitish below, grey wash on flanks and undertail. **p. 303**

481. Radde's Warbler *Phylloscopus schwarzi* 14 cm (5.3 in)
Thicker bill than 480, dull olive-brown upperparts, pale brownish-buff underparts tinged yellow on belly, whitish supercilium. **p. 303**

482. Yellow-browed Warbler *Phylloscopus inornatus* 11 cm (4.5 in)
Short bill, broad yellowish white wingbars, white edge of tertiaries. **p. 303**

483. Arctic Warbler *Phylloscopus borealis* 13 cm (5 in)
Yellowish white supercilium, pale horn legs, usually single wingbar. **p. 304**

484. Two-barred Warbler *Phylloscopus plumbeitarsus* 12 cm (4.8 in)
Entire upperparts olive green, two yellowish white wingbars. **p. 304**

485. Pale-legged Leaf-warbler *Phylloscopus tenellipes* 13 cm (5 in)
White supercilium, rufescent olive tinge on rump, pale flesh legs. **p. 304**

486. Eastern-crowned Warbler *Phylloscopus coronatus* 13 cm (5 in)
Green median coronal stripe, white below, yellow undertail coverts. **p. 305**

487. Mountain Leaf-warbler *Phylloscopus trivirgatus* 11 cm (4.5 in)
Yellowish median coronal stripe and supercilium. **p. 305**

488. Plain-tailed Warbler *Seicercus soror* 13 cm (5 in)
Grey crown, black supercilium, narrow yellow orbital skin. **p. 305**

489. Chestnut-crowned Warbler *Seicercus castaniceps* 10 cm (4 in)
Chestnut crown, grey head and breast, narrow white orbital skin. **p. 306**

490. Yellow-breasted Warbler *Seicercus montis* 10 cm (4 in)
Chestnut head, black supercilium, bright yellow rump and underparts. **p. 306**

491. Yellow-bellied Warbler *Abroscopus superciliaris* 11 cm (4.5 in)
Grey head, white supercilium, white throat, yellow lower breast and belly. **p. 306**

492. Black-browed Reed-warbler *Acrocephalus bistrigiceps* 14 cm (5.5 in)
Long pale buff supercilium bordering black lateral coronal stripe. **p. 307**

493. Oriental Reed-warbler *Acrocephalus orientalis* 20 cm (8 in)
Whitish supercilium, dark eyeline and tawny buff tinge on flanks. **p. 307**

494. Thick-billed Warbler *Acrocephalus aedon* 20 cm (8 in)
Short thick bill, whitish lores, pale mandible, graduated tail. **p. 307**

488
489
490
491
480
482
483
484
485
486
487
493
492
494
479
478

502. Rufous-fronted Babbler *Stachyris rufifrons* 13 cm (5 in)
Rufous forehead and crown and broad buff brown breast band. **p. 311**

503. Golden Babbler *Stachyris chrysaea* 13 cm (5 in)
Small yellow babbler with dark lores and fine black streaked crown. **p. 311**

504. Chestnut-winged Babbler *Stachyris erythroptera* 14 cm (5.5 in)
Greyish head and breast, blue orbital skin and rictal area, chestnut wings. **p. 312**

505. Grey-throated Babbler *Stachyris nigriceps* 14 cm (5.5 in)
Pale grey supercilium, grey throat with broad white malar stripe. **p. 312**

506. Grey-headed Babbler *Stachyris poliocephala* 15 cm (6 in)
Grey head with fine white streaks on crown and throat, yellow iris. **p. 312**

507. White-necked Babbler *Stachyris leucotis* 15 cm (6 in)
Black throat, buff lores and white spots behind grey ear coverts. **p. 313**

508. Black-throated Babbler *Stachyris nigricollis* 15 cm (6 in)
Broad blackish throat with white streaked lower border, white malar stripe. **p. 313**

509. Chestnut-rumped Babbler *Stachyris maculata* 18 cm (7 in)
Rufous chestnut lower back and rump, white breast with bold black spots. **p. 313**

525. Striped Wren-babbler *Kenopia striata* 14 cm (5.5 in)
White streaks on blackish crown and rufous brown upperparts. **p. 318**

526. Large Wren-babbler *Napothera macrodactyla* 19 cm (7.5 in)
Black ear coverts, upperparts streaked buff, marked with dark scales. **p. 319**

527. Marbled Wren-babbler *Napothera marmorata* 20 cm (8 in)
Rufous ear coverts and black underparts heavily scaled white. **p. 319**

528. Streaked Wren-babbler *Napothera brevicaudata* 13 cm (5 in)
Fine pale streaks on greyish brown sides of head, dark streaks on white throat. **p. 319**

529. Eye-browed Wren-babbler *Napothera epilepidota* 11 cm (4.5 in)
Blackish eyeline, long narrow buff supercilium. **p. 320**

530. Pygmy Wren-babbler *Pnoepyga pusilla* 9 cm (3.5 in)
Tiny size and tailless appearance. **p. 320**
(a) Whitish underparts with blackish scales.
(b) Buff underparts with blackish scales.

529

530(a)

530(b)

525
528
527
526
505
503
502
509
506
507
504
508

512. Puff-throated Babbler *Pellorneum ruficeps* 17 cm (6.5 in)
Rufous crown, pale buff supercilium, breast and flanks streaked dark brown. **p. 314**

513. Buff-breasted Babbler *Pellorneum tickelli* 15 cm (6 in)
Buff throat and breast, buff brown sides of head. **p. 315**

514. Black-capped Babbler *Pellorneum capistratum* 18 cm (7 in)
Black cap and malar stripe, white throat and rusty rufous underparts. **p. 315**

515. Moustached Babbler *Malacopteron magnirostre* 18 cm (7 in)
Greyish sides of head, black malar stripe and rufous brown tail. **p. 315**

516. Sooty-capped Babbler *Malacopteron affine* 17 cm (6.5 in)
Blackish crown, pale supercilium and greyish tinge on breast. **p. 315**

517. Scaly-crowned Babbler *Malacopteron cinereum* 17 cm (6.5 in)
Rufous crown with indistinct blackish scales, pinkish legs. **p. 316**

518. Rufous-crowned Babbler *Malacopteron magnum* 18 cm (7 in)
Rufous crown, greyish legs and greyish streaks on breast. **p. 316**

519. Grey-breasted Babbler *Ophrydornis albogulare* 15 cm (6 in)
Grey head with thin supercilium, white throat, grey breast and whitish underparts with tawny buff flanks. **p. 316**

520. White-chested Babbler *Trichastoma rostratum* 15 cm (6 in)
Dark brownish upperparts contrasting with white underparts. **p. 317**

521. Ferruginous Babbler *Trichastoma bicolor* 18 cm (7 in)
Relatively long-tailed with bright rufescent orange upperparts and white underparts. **p. 317**

522. Abbott's Babbler *Malacocincla abbotti* 17 cm (6.5 in)
Narrow grey supercilium with bright rufescent flanks and undertail coverts. **p. 317**

523. Horsfield's Babbler *Malacocincla sepiarium* 15 cm (6 in)
Short tail, robust bill, olive brown upperparts, greyish buff breast, flanks and underparts. **p. 318**

524. Short-tailed Babbler *Malacocincla malaccense* 15 cm (6 in)
Short tail, greyish sides of head with narrow dark malar stripe, white underparts and tawny flanks. **p. 318**

512
514
513
524
520
523
521
522
515
516
519
517
518

PLATE 59 RAIL-BABBLER, SCIMITAR-BABBLERS, TIT-BABBLERS, AND SONG BABBLERS

393. Rail-babbler *Eupetes macrocerus* 29 cm (11.5 in)
Long neck, legs and tail with long bill, black mask, broad white supercilium and narrow blue streak on sides of neck. **p. 265**

500. Large Scimitar-babbler *Pomatorhinus hypoleucos* 28 cm (11 in)
Large size, horn-coloured bill, olive brown ear coverts, whitish streaks on sides of neck and flanks. **p. 310**

501. Chestnut-backed Scimitar-babbler *Pomatorhinus montanus* 22 cm (8 in)
Black crown and broad eyeline, white supercilium, whitish bill, chestnut mantle, rump and flanks. **p. 311**

510. Striped Tit-babbler *Macronous gularis* 14 cm (5.5 in)
Pale yellow underparts with fine blackish streaks on breast. **p. 314**

511. Fluffy-backed Tit-babbler *Macronous ptilosus* 17 cm (6.5 in)
Black throat, blue orbital skin and rictal area with long plumes on lower back and flanks. **p. 314**

531. Collared Babbler *Gampsorhynchus torquatus* 24 cm (9.5 in)
White head with blackish band around neck, long tail with white tip. **p. 320**

533. White-crested Laughingthrush *Garrulax leucolophus* 30 cm (12 in)
White head with erect crest, broad black eyeline and bright rufous upperparts. **p. 321**

534. Black Laughingthrush *Garrulax lugubris* 27 cm (10.5 in)
Blackish slate plumage with orange bill and blue skin around eye. **p. 321**

535. Spectacled Laughingthrush *Garrulax mitratus* 23 cm (9 in)
Dark chestnut-brown cap, orange bill, white eyering and wingbar and chestnut undertail coverts. **p. 322**

536. Hwamei *Garrulax canorus* 25 cm (10 in)
Gingery brown plumage with white eyering and streak behind eye. **p. 322**

537. Chestnut-crowned Laughingthrush *Garrulax erythrocephalus* 27 cm (10.5 in)
Dark chestnut crown, greyish sides of head, bright olive wash on wings and tail,small black patch on wings. **p. 322**

539. Long-tailed Sibia *Heterophasia picaoides* 30 cm (12 in)
Slender with long tail, white wing patch and red iris. **p. 323**

510

511

501
500
533
534
535
536
537
539
531
393

PLATE 60 **SHRIKE-BABBLERS, ERPORNIS, AND SONG BABBLERS**

417. White-browed Shrike-babbler *Pteruthius flaviscapis* 17 cm (6.75 in)
Large headed with thickset bill. **p. 276**
(a) Male: Black head with broad white supercilium, grey mantle and reddish orange patch on tertiaries.
(b) Female: Grey head with less distinct whitish supercilium, bright greenish-olive mantle and wings.

418. Black-eared Shrike-babbler *Pteruthius melanotis* 11 cm (4.5 in)
Green crown and upperparts, bold black streak behind ear coverts and yellow underparts. **p. 277**
(a) Male: Rufous throat and breast, white wingbars.
(b) Female: Duller with pale buff wingbars.

419. White-bellied Erpornis *Erpornis zantholeuca* 13 cm (5 in)
Short erect crest, greenish yellow upperparts and whitish underparts with yellow undertail coverts. **p. 277**

497. Rufous-winged Fulvetta *Pseudominla castaneceps* 11 cm (4.5 in)
Narrow white supercilium, black line behind eye, narrow black malar stripe and rufous primaries. **p. 309**

498. Brown Fulvetta *Alcippe brunneicauda* 15 cm (6 in)
Greyish head and neck, rufescent brown upperparts. **p. 310**

499. Mountain Fulvetta *Alcippe peracensis* 16 cm (6.5 in)
Smokey grey head and mantle and broad curving black supercilium. **p. 310**

532. Cutia *Cutia nipalensis* 19 cm (7.5 in)
Chunky and short tailed with blue crown and wings with white underparts marked with bold black barrings on sides. **p. 321**
(a) Male: Black mask and bright rufous mantle.
(b) Female: Dull brown mask, olive brown mantle with dark spots.

538. Silver-eared Mesia *Leiothrix argentauris* 18 cm (7 in)
Black head, yellow bill and silvery white ear coverts. **p. 323**
Male: Upper and undertail coverts red.

540. Blue-winged Minla *Minla cyanouroptera* 16 cm (6.5 in)
Blue lining on wing and edge of tail and white iris. **p. 323**

541. Chestnut-tailed Minla *Minla strigula* 17 cm (6.75 in)
Black and white barred throat, greyish-yellow underparts and chestnut tail and wing patch. **p. 324**

532(a)
532(b)
417(a)
538
417(b)
540
418(a)
541
418(b)
497
498
499
419

555. Chestnut-cheeked Starling *Sturnus philippensis* 18 cm (7 in) **p. 330**
(a) Male: Purple mantle, chestnut cheek patch.
(b) Female: Greyish brown mantle, small whitish shoulder patch.

557. Rosy Starling *Sturnus roseus* 22 cm (8.5 in) **p. 331**
(a) Adult: Shaggy crest, pink bill and pink plumage, dark hood, tail, and wings.
(b) Juvenile: Light sandy-buff plumage with streaks on sides of breast, yellow bill with dark tip.

558. Black-winged Myna *Acridotheres melanopterus* 23 cm (9 in)
White head and underparts, pale grey mantle, yellow bill and orbital skin. **p. 331**

559. Common Myna *Acridotheres tristis* 25 cm (10 in) **p. 332**
(a) Dark brown plumage with blackish head, yellow orbital skin.
(b) Moult: Shows bare yellow skin of head and neck.

560. Jungle Myna *Acridotheres fuscus* 24 cm (9.5 in)
Short crest, dark grey-brown plumage, yellow bill with blue base. **p. 332**

561. Javan Myna *Acridotheres javanicus* 24 cm (9.5in)
Short crest, pale iris and dark grey plumage. **p. 333**

562. White-vented Myna *Acridotheres grandis* 26 cm (10.5 in)
Long wispy crest, uniform slaty black plumage, brown iris. **p. 334**

563. Crested Myna *Acridotheres cristatellus* 26 cm (10.5 in)
Floppy cresy, white bill, flesh pink legs. **p. 334**

564. Golden-crested Myna *Ampeliceps coronatus* 22 cm (8.5 in)
Glossy black with yellow orbital skin and pale yellow wing patches. **p. 335**
(a) Male: Crown, nape, orbital skin and throat patch bright yellow.
(b) Female: Yellow crown and throat patch much reduced.

565. Hill Myna *Gracula religiosa* 30 cm (12 in)
Large and thickset with yellow wattles, pinkish orange bill. **p. 335**

558
557(a)
555(b)
560
555(a)
557(b)
561
2
559(a)
559(b)
63
565

PLATE 63 ROCK-THRUSHES, WHISTLING-THRUSHES, AND TYPICAL THRUSHES

566. White-throated Rock-thrush *Monticola gularis* 19 cm (7.5 in) **p. 336**
(a) Male: Bright blue crown, orange rufous underparts, small white throat patch.
(b) Female: Narrow white throat patch, brown and blackish scaled upperparts.

567. Blue Rock-thrush *Monticola solitarius* 23 cm (9 in) **p. 337**
(a) Male: *M. s. madoci* and *M. s. pandoo*. Bright blue-grey plumage with whitish and blackish scalings.
(b) Female: Dark grey-brown plumage with grey and white scalings.

568. Malaysian Whistling-thrush *Myophonus robinsoni* 25 cm (10 in)
Plumage like 569 but smaller, squatter, and shorter necked and more rotund appearance with proportionately larger eye, and thinner bill. **p. 337**

569. Blue Whistling-thrush *Myophonus caeruleus* 33 cm (13 in)
M. c. dicrorhynchus :Large size with dark purplish blue plumage marked with lighter blue spangles. **p. 338**

570. Chestnut-capped Thrush *Zoothera interpres* 18 cm (7 in)
Small and short tailed with chestnut crown and nape. **p. 339**

571. Orange-headed Thrush *Zoothera citrina* 22 cm (8.5 in)
Orange rufous head and underparts distinctive. **p. 339**
(a) Male: Blue-grey upperparts.
(b) Female: Olive brown upperparts, dark vertical bars on side of head.

572. Siberian Thrush *Zoothera sibirica* 24 cm (9.5 in) **p. 339**
(a) Male: Slaty black plumage with white supercilium.
(b) Female: Black eyeline and long buff supercilium curving around ear coverts.
(c) Male and (d) Female in flight, showing typical *Zoothera* underwing pattern.

573. Eurasian Scaly Thrush *Zoothera dauma* 28 cm (11 in) **p. 340**
(a) Large and chunky with bold and golden buff scales.
(b) In flight, shows typical *Zoothera* underwing pattern.

574. Eyebrowed Thrush *Turdus obscurus* 24 cm (9.5 in)
White supercilium, pale orange rufous breast and flanks. **p. 340**
(a) Male: Grey head, throat and upper breast.
(b) Female: Olive brown head and dark streaks on whitish throat.

566(b)
567(a)
567(b)
566(a)
570
569
568
571(a)
571(b)
574(a)
572(b)
572(a)
574(b)

PLATE 64 **TITS, NUTHATCHES, AND THRUSHES**

444. Great Tit *Parus major* 13 cm (5 in)
White cheek patch black head with bold black line down centre of belly. **p. 288**

445. Sultan Tit *Melanochlora sultanea* 20 cm (8 in)
Shaggy sulphur yellow crown and crest, lower breast and belly. **p. 289**
(a) Male: Glossy black head, upper breast and upperparts.
(b) Female: Black parts tinged olive.

551. Velvet-fronted Nuthatch *Sitta frontalis* 12 cm (4.5 in)
Deep violet-blue above, velvety black forehead, red bill. **p. 328**
(a) Male: Narrow black supercilium.
(b) Female: Lacks narrow black supercilium.

552. Blue Nuthatch *Sitta azurea* 12 cm (4.5 in)
Whitish blue bill and orbital skin, black above, barred bluish grey. **p. 328**

575. Dark-throated Thrush *Turdus ruficollis* 25 cm (10 in) **p. 340**
(a) *T. r. ruficollis*—Male (breeding): Rufous supercilium, throat, upper breast.
(b) Female: Pale buff supercilium. rufous upper breast with whitish scales.

576. Lesser Shortwing *Brachypteryx leucophrys* 13 cm (5 in) **p. 341**
(a) Male: Dark slate grey with prominent white supercilium.
(b) Female: Rufescent olive brown above, shorter and narrower supercilium, buff brown below.

577. Common Stonechat *Saxicola torquatus* 14 cm (5.5 in) **p. 341**
(a) Male: Black head, white neck, wingbar, rump, orange rufous below.
(b) Female: Whitish throat, pale buff supercilium and pale orange rump.

578. Siberian Rubythroat *Luscinia calliope* 16 cm (6.5 in) **p. 342**
(a) Male: Red throat, white supercilium and malar stripe.
(b) Female: Whitish throat.

579. Rufous-headed Robin *Luscinia ruficeps* 15 cm (6 in)
Male: Orange-rufous crown and nape, white throat, slaty grey mantle. **p. 342**
Female: Dark brown sides of head and neck, scalings on throat and breast.

580. Siberian Blue Robin *Luscinia cyane* 15 cm (6 in) **p. 342**
(a) Male: Dull blue upperparts and white underparts.
(b) Female: Olive brown upperparts, bluish rump and uppertail coverts.

444
445(a)
445(b)
551(b)
551(a)
552
576(b)
576(a)
577(b)
577(a)
578(b)
578(a)
580(b)
579(b)
580(a)
579(a)

581. Oriental Magpie Robin *Copsychus saularis* 23 cm (9 in) **p. 343**
(a) Male: Glossy black head, breast and upperparts contrasting with white wingbar and underparts.
(b) Female: Black of head and breast replaced with dark grey.

582. White-rumped Shama *Copsychus malabaricus* 28 cm (11 in)
Long black wedge-shaped tail and white rump distinctive. **p. 343**
(a) Male: Glossy blue-black upperparts and head contrasting with deep orange rufous underparts.
(b) Female: Shorter tail with blue-black parts replaced with slaty grey, orange rufous parts paler.

583. Rufous-tailed Shama *Trichixos pyrropygus* 22 cm (8.8 in)
Bright orange rufous tail with dark terminal band diagnostic. **p. 344**
(a) Male: Blackish grey head, breast and upperparts, orange rufous rump, and short white supercilium in front of eye.
(b) Female: Upperparts browner, entire underparts orange rufous.

584. Chestnut-naped Forktail *Enicurus ruficapillus* 20 cm (8 in)
Chestnut crown and nape together with blackish scalings on breast diagnostic. **p. 344**
(a) Male: Chestnut crown, nape and upper back.
(b) Female: Chestnut crown, nape and entire back.

585. Slaty-backed Forktail *Enicurus schistaceus* 25 cm (10 in)
Slaty grey crown, nape and mantle, long tail. **p. 344**

586. White-crowned Forktail *Enicurus leschenaulti* 20 cm (8 in)
White forecrown, black back and breast. **p. 345**

587. White-tailed Robin *Myiomela leucura* 18 cm (7 in)
White crescentic lateral patches on tail diagnostic. **p. 345**
(a) Male: Blue-black plumage with glossy blue forehead and carpal joint.
(b) Female: Dark olive brown above, paler underneath.

581(b)
581(a)
582(b)
582(a)
583(b)
583(a)
587(b)
587(a)
584(b)
584(a)
586
585

PLATE 66 SMALLER FLYCATCHERS, JUNGLE-FLYCATCHERS, AND VERDITER FLYCATCHER

588. Taiga Flycatcher *Ficedula albicilla* 13 cm (5.25 in)
White edge of base of black tail distinctive. **p. 346**
(a) Breeding male: Brick red throat and breast.
(b) Female and non-breeding male.

589. Rufous-browed Flycatcher *Ficedula solitaris* 13 cm (5.25 in)
White throat, buff rufous forehead and lores. **p. 347**

590. Snowy-browed Flycatcher *Ficedula hyperythra* 12 cm (4.5 in) **p. 347**
(a) Male: Slaty blue upperparts with white supercilium.
(b) Female: Upperparts brown with pale buff supercilium.

591. Rufous-chested Flycatcher *Ficedula dumetoria* 12 cm (4.5 in) **p. 348**
(a) Male: Black above, white supercilium, white wing patch, orange-rufous breast.
(b) Female: Brown upperparts, thin buff wingbar and dull orange rufous breast.

592. Little Pied Flycatcher *Ficedula westermanni* 12 cm (4.5 in) **p. 348**
(a) Male: Black above, broad white supercilium, white wing patch and sides of tail.
(b) Female: Pale bluish grey above with rufescent tinge on rump and uppertail.

593. Yellow-rumped Flycatcher *Ficedula zanthopygia* 13 cm (5.25 in)
Yellow rump distinctive in both sexes. **p. 348**
(a) Male: Black above, white supercilium, white wing patch yellow below.
(b) Female: Olive brown above, thin pale wing bar, yellowish buff below.

594. Narcissus Flycatcher *Ficedula narcissina elisae* 13 cm (5.25 in) **p. 349**
(a) Male: Olive green above, yellow lores, and rump, white wing patch, black tail.
(b) Female: Duller and lacks yellow rump and white wing patch.

595. Mugimaki Flycatcher *Ficedula mugimaki* 13 cm (5.25 in) **p. 349**
(a) Male: Blackish above, short white supercilium and white wing patch, orange rufous throat, breast and upper belly.
(b) Female: Olive brown upperparts, dull orange rufous throat and breast.

607. Brown-chested Jungle-flycatcher *Rhinomyias brunneata* 15 cm (6 in)
Long bill with yellow mandible, brownish breast, whitish throat. **p. 354**

608. Grey-chested Jungle-flycatcher *Rhinomyias umbratilis* 15 cm (6 in)
All black bill, white throat sharply demarcated from greyish breast. **p. 355**

609. Verditer Flycatcher *Eumyias thalassinus* 17 cm (6.25 in) **p. 355**
(a) Male: Greenish-blue plumage and black lores.
(b) Female: Duller, dusky lores.

610. Dark-sided Flycatcher *Muscicapa sibirica* 13 cm (5.25 in)
Long wing tips, whitish throat and half collar, sooty grey brown flanks. **p. 355**

611. Ferruginous Flycatcher *Muscicapa ferruginea* 13 cm (5.25 in)
Rufous rump, uppertail coverts and buff rufous flanks. **p. 356**

612. Asian Brown Flycatcher *Muscicapa dauurica* 13 cm (5.25 in) **p. 356**
(a) *M. d dauurica*: Whitish below, pale grey-brown upper breast.
(b) *M. d williamsoni*: Rufescent tinge on uppertail, broad brown streaks on breast and flanks.

607
608
610
612(a)
612(b)
609(a)
611
595(b)
609(b)
594(b)
593(a)
594(a)
593(b)
595(a)
588(a)
589
590(a)
590(b)
588(b)
592(b)
592(a)
591(a)
591(b)

PLATE 67 NILTAVAS, BLUE-FLYCATCHERS, AND CANARY-FLYCATCHER

596. Blue-and-white Flycatcher *Cyanoptila cyanomelana* 18 cm (7 in) **p. 350**
(a) Male: Black face and throat, glossy blue upperparts, white belly.
(b) Female: Greyish brown above, pale brown breast, whitish throat and belly.

597. Large Niltava *Niltava grandis* 22 cm (8.5 in) **p. 350**
(a) Male: Dark blue plumage, iridescent blue crown and neck patch.
(b) Female: Rusty brown plumage, iridescent blue neck patch.

598. Rufous-vented Niltava *Niltava sumatrana* 15 cm (6 in) **p. 351**
(a) Male: Dark blue above, iridescent blue crown, neck and shoulder patches and orange rufous below.
(b) Female: Brown plumage, white crescentic patch on upper breast.

599. Pygmy Blue-flycatcher *Muscicapella hodgsoni* 10 cm (4 in) **p. 351**
(a) Male: Shining blue forecrown, orange rufous below, whitish centre of belly.
(b) Female: Rufescent rump and uppertail, Breast, throat and belly tinged rufous.

600. White-tailed Flycatcher *Cyornis concretus* 19 cm (7.5 in)
Robust with conspicuous white edges of spread tail. **p. 352**
(a) Male: Dark blue plumage with purplish tinge.
(b) Female: Rich buff brown with white gorget on throat and breast.

601. Pale Blue-flycatcher *Cyornis unicolor* 18 cm (7 in) **p. 352**
(a) Male: Dull pale blue upperparts and breast, pale greyish belly.
(b) Female: Brownish upperparts, greyish underparts.

602. Blue-throated Flycatcher *Cyornis rubeculoides* 15 cm (6 in) **p. 352**
(a) Male: Blue throat, narrow wedge of orange rufous extending upward from breast.
(b) Female: Whitish throat contrasting with pale orange rufous breast.

603. Hill Blue-flycatcher *Cyornis banyumas* 15 cm (6 in)
Gradual demarcation between orange rufous breast and white underparts **p. 353**
(a) Male: Upperparts dark blue
(b) Female: Tail and uppertail coverts rufescent brown.

604. Tickell's Blue-flycatcher *Cyornis tickelliae* 15 cm (6 in)
Sharp demarcation between orange rufous breast and white underparts. **p. 353**
(a) Male: Blue upperparts paler than male 603
(b) Female: Upperparts including tail tinged greyish blue.

605. Mangrove Blue-flycatcher *Cyornis rufigastra* 15 cm (6 in) **p. 353**
(a) Male: Blackish lores and chin.
(b) Female: Buff rufous lores, chin and throat.

606. Malaysian Blue-flycatcher *Cyornis turcosa* 13 cm (5.25 in) **p. 354**
(a) Male: Bright blue throat, blackish lores and shining blue rump.
(b) Female: Whitish buff throat.

613. Grey-headed Canary-flycatcher *Culicicapa ceylonensis* 13 cm (5 in)
Pale grey head and throat, olive yellow above, bright yellow below. **p. 357**

606(a) 606(b) 604(a) 604(b)

596(b)
596(a)
597(a)
597(b)
598(a)
598(b)
600(b)
601(a)
601(b)
600(a)
603(b)
602(a)
602(b)
613
603(a)
605(a)
605(b)
599(a)
599(b)

614. Asian Fairy-bluebird *Irena puella* 25 cm (10 in) **p. 357**
(a) Male: Velvety black head, throat and belly, shining blue crown, mantle, rump and undertail coverts.
(b) Female: Verditer blue plumage with blackish primaries.

615. Greater Green Leafbird *Chloropsis sonnerati* 21 cm (8 in) **p. 358**
(a) Male: Lacks yellow border along black throat.
(b) Female: Yellow throat and narrow eyering.

616. Lesser Green Leafbird *Chloropsis cyanopogon* 18 cm (7 in) **p. 358**
(a) Male: Narrow yellow border along black throat.
(b) Female: Green throat.

617. Blue-winged Leafbird *Chloropsis cochinchinensis* 19 cm (7.5 in)
Blue flight feathers and sides of tail. **p. 359**
(a) Male: Bright yellow crown and border along black throat.
(b) Female: Lacks yellow crown.

618. Orange-bellied Leafbird *Chloropsis hardwickii* 20 cm (8 in)
Orange-yellow belly and undertail coverts. **p. 359**
(a) Male: Blue shoulder patch, blackish flight feathers and upper tail coverts.
(b) Female: Green flight feathers and uppertail coverts.

616(b)
615(b)
616(a)
615(a)
617(b)
618(b)
617(a)
618(a)
614(a)
614(b)

619. Yellow-breasted Flowerpecker *Prionochilus maculatus* 10 cm (4 in)
Yellow underparts streaked olive green on sides of throat and upper flanks. **p. 360**

620. Crimson-breasted Flowerpecker *Prionochilus percussus* 10 cm (4 in) **p. 360**
(a) Male: Slaty blue upperparts, yellow underparts, crimson patch on breast.
(b) Female: Dull orange crown patch, yellowish streak down centre of breast and belly.

621. Scarlet-breasted Flowerpecker *Prionochilus thoracicus* 10 cm (4 in) **p. 360**
(a) Male: Black head, red breast, yellow underparts.
(b) Female: Grey head, white throat, yellow underparts with orange tinged breast.

622. Thick-billed Flowerpecker *Dicaeum agile* 10 cm (4 in)
Thick bill, reddish iris, whitish tip of tail feathers. **p. 361**

623. Brown-backed Flowerpecker *Dicaeum everetti* 10 cm (4 in)
Thick bill, pale yellow to orange iris, lacks whitish tip of tail feathers. **p. 361**

624. Yellow-vented Flowerpecker *Dicaeum chrysorrheum* 10 cm (4 in)
Whitish underparts with bold dark streaks, orange-yellow undertail coverts. **p. 361**

625. Orange-bellied Flowerpecker *Dicaeum trigonostigma* 9 cm (3.5 in) **p. 362**
(a) Male: Slaty blue upperparts, orange rump, lower breast and belly.
(b) Female: Dull orange rump.

626. Plain Flowerpecker *Dicaeum concolor* 9 cm (3.5 in)
Nondescript, slightly down-curved greyish bill. **p. 362**

627. Fire-breasted Flowerpecker *Dicaeum ignipectus* 9 cm (3.5 in) **p. 362**
(a) Male: Buff underparts, crimson breast patch with thin black line down centre of lower breast and belly.
(b) Female: Dull olive upperparts, yellowish buff underparts.

628. Scarlet-backed Flowerpecker *Dicaeum cruentatum* 9 cm (3.5 in) **p. 363**
(a) Male: Broad red band from crown to rump.
(b) Female: Red rump.

621(b)
620(a)
620(b)
621(a)
619
624
622
623
625(b)
626
625(a)
627(b)
628(b)
627(a)
628(a)

629. Ruby-cheeked Sunbird *Chalcoparia singalensis* 11 cm (4.5 in)
Bill shorter and straighter than *Anthreptes* sunbirds. **p. 364**
(a) Male: Pale rufous throat, iridescent coppery red ear coverts.
(b) Female: Pale rufous throat, olive green upperparts, yellow underparts.

630. Purple-naped Sunbird *Hypogramma hypogrammicum* 15 cm (6 in)
Large size, olive upperparts, yellowish white underparts with dark streaks. **p. 364**
(a) Male: Iridescent purple nape and rump.
(b) Female: Lacks purple nape and rump.

631. Plain Sunbird *Anthreptes simplex* 13 cm (5 in)
Bright olive green upperparts, dull greyish olive underparts. **p. 364**
(a) Male: Dark iridescent green forehead.
(b) Female: Lacks green forehead.

632. Brown-throated Sunbird *Anthreptes malacensis* 14 cm (5.5 in) **p. 365**
(a) Male: Pale brown throat and sides of head, dark iridescent green upperparts and yellow underparts.
(b) Female: Olive upperparts, bright yellowish underparts, lack of white in tail.

633. Red-throated Sunbird *Anthreptes rhodolaemus* 13 cm (5 in) **p. 365**
(a) Male: Light brick red throat, dark maroon-red sides of head.
(b) Female: Dull yellowish olive underparts with denser yellowish tinge down centre.

634. Purple-throated Sunbird *Leptocoma sperata* 10 cm (4 in)
Tiny size. **p. 366**
(a) Male: Iridescent purple throat, iridescent green cap, dull maroon breast and upper belly.
(b) Female: Yellowish white underparts, paler on throat than female 636.

635. Copper-throated Sunbird *Leptocoma calcostetha* 14 cm (5.5 in)
Graduated tail. **p. 366**
(a) Male: Iridescent coppery throat, iridescent violet-purple breast and belly.
(b) Female: Greyish head, white throat, yellow breast and belly.

636. Olive-backed Sunbird *Cinnyris jugularis* 11 cm (4.5 in) **p. 366**
(a) Male: Dark iridescent blue face, throat and breast, yellow breast and belly.
(b) Female: Lacks iridescent blue parts of male.

631(b)
632(b)
631(a)
632(a)
633(b)
629(b)
629(a)
633(a)
630(b)
634(b)
630(a)
634(a)
635(b)
636(b)
635(a)
636(a)

SUNBIRDS AND SPIDERHUNTERS

637. Black-throated Sunbird *Aethopyga saturata* 11 cm (4.5 in) **p. 367**
(a) Male: Maroon mantle, iridescent violet crown, rump and tail.
(b) Female: Olive grey upperparts, yellow rump.

638. Crimson Sunbird *Aethopyga siparaja* 11 cm (4.5 in) **p. 367**
(a) Male: Bright crimson head and mantle, iridescent green forehead, iridiscent violet tail.
(b) Female: Dark olive green upperparts, pale yellow underparts.

639. Scarlet Sunbird *Aethopyga temminckii* 11 cm (4.5 in) **p. 368**
(a) Male: Bright scarlet head, mantle and tail.
(b) Female: Scarlet tinge on wings and tail.

640. Little Spiderhunter *Arachnothera longirostra* 16 cm (6.25 in)
Whitish throat and supercilium, yellowish belly, white-tipped tail. **p. 368**

641. Thick-billed Spiderhunter *Arachnothera crassirostris* 17 cm (6.5 in)
Thick bill, greyish olive throat and breast, no white in tail. **p. 368**

642. Long-billed Spiderhunter *Arachnothera robusta* 22 cm (8.5 in)
Very long down-curved bill, yellowish green breast with faint streaks. **p. 369**

643. Spectacled Spiderhunter *Arachnothera flavigaster* 22 cm (8.5 in)
Large and robust with thick broad based bill, broad yellow eyering, yellow ear coverts. **p. 369**

644. Yellow-eared Spiderhunter *Arachnothera chrysogenys* 18 cm (7 in)
Proportionately longer and thinner bill than 643, prominent yellow ear coverts. **p. 369**

645. Grey-breasted Spiderhunter *Arachnothera affinis* 18 cm (7 in)
Pale greyish underparts with streaks on throat and breast. **p. 369**

646. Streaked Spiderhunter *Arachnothera magna* 19 cm (7.5 in)
Boldly streaked plumage, bright orange legs. **p. 370**

637(b)
639(b)
637(a)
639(a)
638(b)
640
638(a)
643
642
641
646
644
645

PLATE 72 **SPARROW, WEAVER, PARROTFINCHES, MUNIAS, AND JAVA SPARROW**

648. House Sparrow *Passer domesticus* 15 cm (6 in) **p. 371**
(a) Male breeding: Grey crown, black throat and upper breast.
(b) Female: Brownish above with pale supercilium

651. Streaked Weaver *Ploceus manyar* 13.5 cm (5.3 in)
Both sexes boldly streaked blackish. **p. 372**
(a) Male breeding: Yellow crown, black throat and sides of head.
(b) Female and non-breeding male: Brown ear coverts, whitish supercilium.

652. Tawny-breasted Parrotfinch *Erythrura hyperythra* 10 cm (4 in)
Dull grass green upperparts, tawny underparts, short and rounded tail. **p. 373**

653. Pin-tailed Parrotfinch *Erythrura prasina* 15 cm (6 in)
Bright grass green upperparts, pointed red tail. **p. 374**
(a) Male: Blue face, black lores, long pointed tail.
(b) Female: Pale greyish blue throat and upper breast, short pointed tail.

654. White-bellied Munia *Lonchura leucogastra* 11 cm (4.5 in)
Dark rump, flanks and undertail coverts, white lower breast and upper belly, pale yellowish uppertail. **p. 374**

655. White-rumped Munia *Lonchura striata* 11 cm (4.5 in)
Blackish brown upperparts, white rump, pointed and notched tail. **p. 374**

656. Javan Munia *Lonchura leucogastroides* 11 cm (4.5 in)
Black face, throat and upper breast, white lower breast and belly. **p. 375**

657. Scaly-breasted Munia *Lonchura punctulata* 11 cm (4.5 in) **p. 375**
(a) Adult: Brown scales on breast and flanks.
(b) Juvenile: Brown upperparts, paler underparts without scales.

658. Chestnut Munia *Lonchura malacca* 11 cm (4.5 in) **p. 375**
(a) Adult: Black head, bluish grey bill, rich chestnut plumage.
(b) Juvenile: Rich buff brown plumage, paler underparts, bluish grey bill.

659. White-headed Munia *Lonchura maja* 11 cm (4.5 in) **p. 376**
(a) Adult: White head, bluish grey bill, dark chestnut brown plumage.
(b) Juvenile: Whitish buff on ear coverts and underparts and bluish grey bill.

660. Java Sparrow *Padda oryzivora* 16 cm (6.25 in)
Black head, white cheek patch, red bill, grey upperparts. **p. 376**

653(b)
652
653(a)
655
660
657(b)
654
657(a)
656
658(b)
659(b)
658(a)
659(a)

PLATE 73 **WAGTAILS AND PIPITS**

661. Forest Wagtail *Dendronanthus indicus* 18 cm (7 in)
Black crescentic breast patch, white breast band, and wingbars. **p. 377**

662. White Wagtail *Motacilla alba* 19 cm (7.5 in) **p. 377**
(a) *M. a. leucopsis* (breeding male): Black throat, breast and mantle.
(b) *M. a. leucopsis* (non-breeding male): Black breast.
(c) *M. a. leucopsis* (female): Black breast and grey mantle.
(d) *M. a. ocularis* (non-breeding): Pale grey mantle, black breast, eyeline.

663. Citrine Wagtail *Motacilla citreola* 19 cm (7.5 in) **p. 378**
(a) Breeding male: Bright yellow head and underparts.
(b) Female: Yellow supercilium curving around margin of dark ear coverts.
(c) Juvenile: Buff forehead, grey crown and pale supercilium.

664. Yellow Wagtail *Motacilla flava* 18 cm (7 in) **p. 378**
(a) *M. f. simillima* (breeding male): Blue-grey head and white supercilium.
(b) *M. f. simillima* (female and non-breeding male): Pale underparts.
(c) *M. f. taivana* (non-breeding): Yellow supercilium and pale underparts.

665. Grey Wagtail *Motacilla cinerea* 19 cm (7.5 in) **p. 378**
(a) Breeding male: Black throat patch.
(b) Female and non-breeding male: Lacks black throat.

666. Paddyfield Pipit *Anthus rufulus* 16 cm (6.25 in)
Upright carriage, pale buff supercilium, indistinct, streaking on breast. **p. 379**

666.1. Richard's Pipit *Anthus richardi* 20 cm (8 in)
Larger and longer-tailed than 666 with heavier streaking on breast. Voice diagnostic (see text). **p. 000**

667. Red-throated Pipit *Anthus cervinus* 16 cm (6.25 in)
Heavily streaked upperparts and streaked breast and flanks. **p. 379**
(a) Breeding male: Pinkish supercilium, throat and breast.
(b) Female: Pale supercilium, pale throat, thin malar stripe.

668. Olive-backed-Pipit *Anthus hodgsoni* 16 cm (6.25 in)
Faintly streaked upperparts, white supercilium, and spot on ear coverts. **p. 380**

669. Blyth's Pipit *Anthus godlewski* 17 cm (6.5 in)
Like 666 but shorter legged with shorter, more pointed bill, heavier streaks above. **p. 380**

662(a)
662(b)
662(c)
662(d)
665(a)
665(b)
664(a)
664(b)
664(c)
661
663(b)
663(a)
668
663(c)
666.1
667(b)
667(a)

Introduction to this book

Peninsular Malaysia and Singapore are geographically part of the Malay Peninsula, which extends from the extreme southern part of Tenasserim (Myanmar) and the Isthmus of Kra in southern Thailand, south to Peninsular Malaysia and terminating in the island of Singapore. This includes all adjacent islands in the Straits of Melaka and the South China Sea (see Maps 1 and 2). The Malay Peninsula has a fairly rich avifauna with a total of 693 naturally occurring species, distributed in the various habitat types. At the turn of the twentieth century, almost the whole region was covered with equatorial rainforest while mangrove forest thrived along the west coast and sheltered estuaries on the east coast. Today, the region has lost approximately 60% of its forest cover to agriculture, the timber industry, and urbanization. What remains today is largely fragmented; the only exceptions are the 4343 km^2 Taman Negara (Peninsular Malaysia's first and largest National Park) and the montane forests of the Titiwangsa Range. In Singapore itself, less than 3% of the total land area is forest covered. This reduction in natural vegetation has greatly reduced the habitat of many forest dependent species, resulting in many resident forest species becoming locally extinct in Singapore since 1819 (see 'Conservation' chapter). At the same time, the creation of open country by forest clearance has portended well for open country species and disused mining pools and ricefields, too, have provided artificial habitats for many species of water birds. A total of 673 species has been recorded from both Peninsular Malaysia and Singapore as at 31 August 2010, an increase of 22 new species from 648 in the first edition. In addition, three species have been removed. The Red-billed Tropicbird and Saunders' Tern have been removed from the Peninsular Malaysia checklist, while the Grey-streaked Flycatcher, from the Singapore checklist, due to lack of evidence. Of the current 673, 440 are resident (including four Malay Peninsula endemics, 165 migrants, 50 vagrants, 14 feral species, and four species which have become locally extinct). This total is based on four published sources, namely *The Birds of the Thai-Malay Peninsula* Volumes One (Non-Passerines, published 1999, hereafter referred to as W1), and Two (Passerines, published 2007, W2) by David. R. Wells, *An Annotated Checklist of the Birds of Singapore* by Wang Luen Keng and Christopher Hails, published in 2007 (W and H), *A Checklist of the Birds of Peninsular Malaysia*, published in 2010 (MNS-Bird Conservation Council), and both published and unpublished reports of the Malaysian Nature Society Bird Conservation Council Records Committee, from 2009–2010. **Note:** the total number of resident species also includes 36 species which have migratory populations from higher latitudes; the latter will not be included under regular migrants, rare migrants, and vagrants which will comprise only true migrants which do not have resident populations.

A brief history of ornithology in both Peninsular Malaysia and Singapore

The current checklists of the birds of both Peninsular Malaysia and Singapore are the results of accumulated research based on collected specimens and observations by pioneer collectors, researchers, and birdwatchers in this region from the early nineteenth century until 2010. When Stamford Raffles established British influence in Singapore in 1819, he undertook the task of collecting of zoological specimens for the British Museum. Being a keen zoologist himself, Raffles had previously collected specimens in Java and Sumatra and was responsible for describing some of the birds which were then new to science; some notable examples being Raffles's Malkoha, Red-naped Trogon, Black-and-Yellow Broadbill, and Little Green-broadbill (Green Broadbill). Other notable collectors who worked in Singapore in the nineteenth century include Alfred Russel Wallace, A.O. Hume, and H.R. Kelham. H.N. Ridley, the father of Malaysia's rubber industry, also contributed greatly in the form of personal observations.

Although there was a fair trade in bird skins through Melaka in the early nineteenth century, collecting expeditions in Peninsular Malaysia (then Malaya) by Europeans did not begin until the late nineteenth century when the British established and consolidated their influence in Malaya in 1874 after the signing of the Pangkor Treaty. Collectors ventured into the interior to obtain specimens for museums. Among some of the collectors in the early twentieth century were J. Waterstradt, H.C. Robinson, C.B. Kloss, F.N. Chasen, and later C.A. Gibson-Hill. All specimens collected were prepared as skins and deposited both at the British Museum (Natural History) and the Raffles Museum in

Map 1 Peninsular Malaysia.

Singapore. Robinson and Chasen in particular were instrumental in making detailed descriptions of ornithological specimens collected and published their work in four volumes between 1927 and 1939. This excellent series, *The Birds of the Malay Peninsula*, though out of date, still remains as a major standard reference work on ornithology in this region

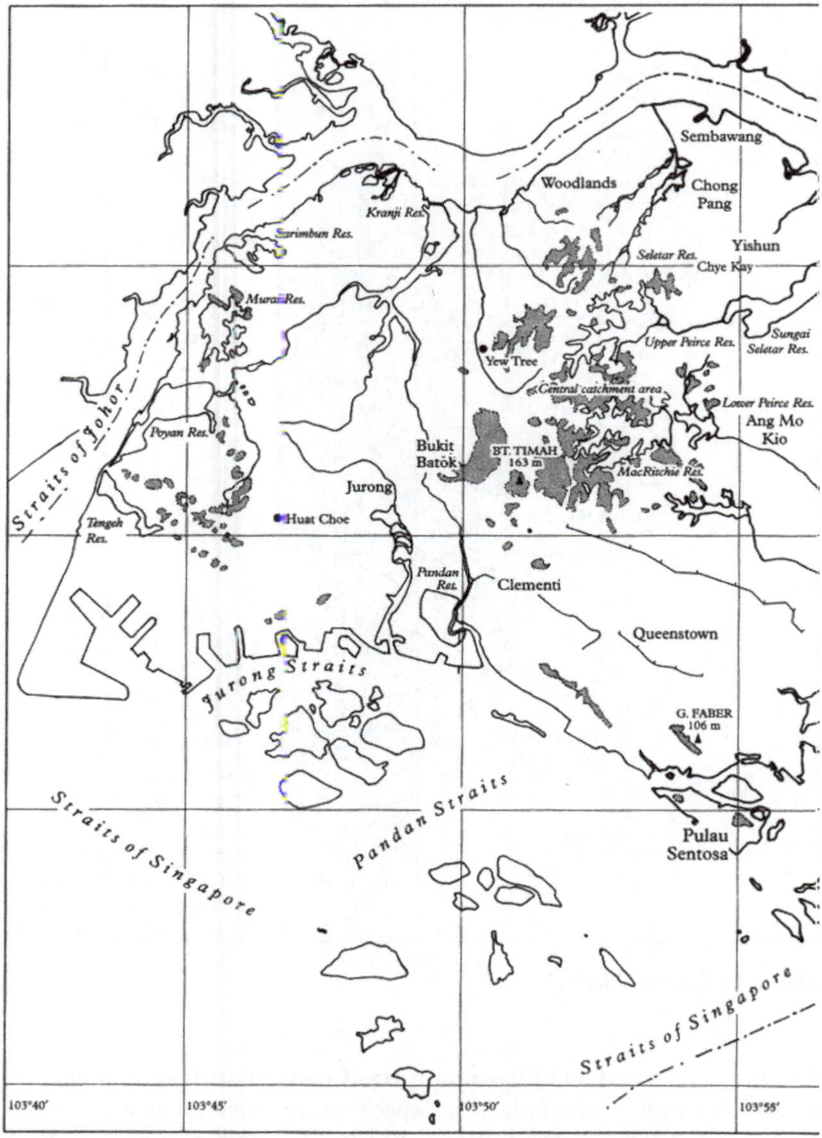

Map 2 Singapore.

Straits of Johor
Pulau Ubin
P. Tekong Res.
Pulau Tekong
Punggol
Port Changi
Kg. Pinang
Serangoon
Changi International Airport
Tampines
Singapore
Bedok Res.
Bedok
Geylang
1°25′
1°20′
1°15′
International border
Canal
Airport
Land over 40 metres
0 1 2 3 miles
0 2 4 km
104°
104°05′
104°10′

The formation of the Malayan Nature Society (MNS) (now the Malaysian Nature Society) in July 1940 was a great boost to conservation in Peninsular Malaysia and Singapore. Its publication, *The Malayan Nature Journal*, is a good source of information on birds especially on behavioural observations as well as birdwatching localities. Most of all it provided a forum for ornithologists and birdwatchers in this region to discuss and share their pooled knowledge. Among them were G.C. Madoc, A.T. Edgar James Cairns, and E.H. Bromley. F.G.H. Allen, B.D. Molesworth, and Loke Wan Tho also contributed greatly besides being the pioneers of bird photography in this region. The Singapore Branch of the MNS was formed in 1954.

The 1960s saw the study of birds in Malaya shift to a more scientific approach. This involved extensive ringing of migratory birds by the Migratory Animal Pathological Survey (MAPS) programme and the University of Malaya Bird Ringing Project (UMBRP) (see 'Migration' chapter for details). Scientific data were collated from studies by Lord Medway (now the Earl of Cranbrook), H. Elliot McClure, I.C.T. Nisbet, and David R. Wells. Since 1963, the *Malayan Nature Journal* has published an annual bird report, making it a valuable source of reference, especially on current updates. A successful expedition to Gunung Benom (Pahang) by the University of Malaya and the British Museum (Natural History) in 1967 yielded vital information on the altitudinal distribution of some of the species of birds there.

The increasing numbers of skilled and reliable observers from both Peninsular Malaysia and Singapore from the late 1960s until the present time have contributed significantly to ornithology and conservation in both countries. Notable among them are Dennis Yong Ghong Chong, Geoffrey W.H. Davison, Lim Kim Seng, Lim Kim Chuah, Mike H.N. Chong, Chacko Kurian, Ho Hua Chew, Christopher J. Hails, John Duckett, Glenda Noramly, Mano Tharmalingam, Frank Lambert, John Howes, Crawford Prentice, Taej Mundkur, Lim Kim Chye, Lim Swee Yian, Yang Chong, Choo Jin Wah, Theresa Ong, Wang Luan Keng. Kanda Kumar Doraisamy, Subaraj Rajathurai, David Bakewell, Ooi Beng Yean, Chiu Sein Chiong, Ooi Chin Hock, the late Laurence Poh Soon Pin, Michael S.N. Godfrey, Lian Godfrey, Clive Briffett, Lim Aun Tiah, Anthony C. Sebastian, Sutari Supari, Dr. Neoh Chin Boon, Neoh Hor Kee, Peter Kennerly, Morten Strange, Mah Teck Oon, Andrew J. Sebastian, Nazeri Abghani, Yeap Chin Aik, Melinda Tan, David Lai, Nina Cheung, Khoo Swee Seng, Caroline Ho, Mohamed Rafi Abdul Kudus, Angela Rafi Francombe, Anuar McAfee, Ravinder Kaur, Diya Rashidah Abu Rahman, Eileen Chiang, John and Jasmine Steed, Wilbur Goh, Alan Owyong, and Sunny Yeoh. Personnel of Peninsular Malaysia's Department of Wildlife and

National Parks who conducted significant research and shared new information include Louis Ratnam, Abdul Rahman Ismail, and Siti Hawa Yatim.

By and large, there have been many new sightings and published literature. New species have been added onto the list. The Mountain Hawk-eagle was only confirmed as a breeding resident on Pulau Langkawi in 2002 while the Short-tailed Shearwater and Dark-throated Thrush were added onto the Peninsular Malaysia checklist in 2010. Singapore being much smaller in area has been thoroughly covered by her pool of reliable observers and most birds there have been relatively well documented. Latest additions to the Singapore checklist include the Lesser Kestrel in 2001 and the Oriental Hobby in 2002. The formation of Interwader (later Asian Wetland Bureau and now Wetlands International) in 1983 has contributed significantly to the study of shorebirds as well as other aquatic birds in this region. Wetlands International conducts an annual waterfowl census in January every year, and observers from both Malaysia and Singapore regularly participate and this has yielded valuable information on especially migratory waterfowl and their habitats.

Both the Selangor and Singapore Branches of MNS had a bird study group each. These bird study groups organized numerous field trips as well as indoor talks and slide shows to boost enthusiasm. The Singapore Branch Bird Study Group was renamed the Singapore Bird Group in 1986 and publishes a regular ornithological bulletin, *Singapore Avifauna*, which publishes latest sightings, behavioural observations, and updates information in both Singapore and the Peninsular Malaysian state of Johor. It maintains a Records Committee to verify and record sightings in Singapore. The Selangor Branch Bird Group used to cover the whole of Peninsular Malaysia and used to publish a monthly ornithological bulletin, *Enggang* from January 1988 to June 1990. It had the same objectives of *Singapore Avifauna*. Due to unforeseen circumstances, *Enggang* ceased publication in 1990. The MNS Selangor Bird Group has been instrumental in helping to compile the bird lists during the Endau-Rompin expeditions of 1985 and 1989 and later the Belum Expedition in 1993. It also helps to update sightings of birds in the Kuala Selangor Nature Park which was established in 1987 and managed by MNS and the Selangor State Government.

The Singapore Branch broke away from the MNS in 1991 and became the Nature Society (Singapore) (NSS). Today, it is the authority on bird conservation in Singapore. In 1995 the MNS Bird Conservation Council (MNS-BCC) was formed primarily to help in the conservation of key habitats in the country besides identifying potentially threatened species. A Records Committee was formed to update and oversee new sightings for Malaysia (including Sarawak and Sabah). When MNS and BirdLife International (formerly International Council for Bird Preservation) became partners in 1995, the Bird Conservation Council was given the task of coordinating and updating information on endangered species in Malaysia in efforts to identify potentially threatened species to be included in the Red Data Book of BirdLife International. The NSS is also a partner of BirdLife International and performs the same task as MNS in this capacity. *Enggang* was revived in 1996 under the new title of *Suara Enggang* and continues to be a regular publication of the MNS-BCC. Since 2005, it has taken a new look with full-colour front and back covers and in 2011, will be in full colour on the same lines as established international birdwatching periodicals, thanks to generous funding from Schmidt Marketing (Malaysia) Sdn Bhd.

Being partners of BirdLife International, both the MNS and the NSS were actively involved in the Important Bird Area Programme (IBA) which was launched in 1996. This involved identifying key bird habitats within Malaysia and Singapore, documenting them, and working towards their conservation and sustainable management. This also involved working closely with the relevant governmental and non-governmental agencies in both countries to ensure that conservation goals are realized. A total of 18 IBAs was identified in Peninsular Malaysia and three in Singapore, their documentation is published in BirdLife's publication *Important Bird Areas in Asia: Key sites for conservation* (2004). Read this publication for details. The annual Raptor Watch, first started in 2000, is another important project of the MNS-BCC. The objectives are to educate the public on raptor migration and conservation as well as to highlight the importance of the Tanjung Tuan Forest Reserve as an important conservation area.

In Singapore, the Bird Ecology Study Group was formed in 2005 by Dr. Wee Yeow Chin, Subaraj Rajathurai, and Richard Hale. This group leads birdwatchers to study the behaviour of birds and document their interaction with the environment and other animals in a more systematic way. Sightings, breeding records, and behavioural observations are readily made available through its blog. Since March 2008, Malaysia has been part of Worldbirds global family of internet-based data collection and provision systems to capture bird records. Known as Bird I Witness Malaysia (BiW), it is managed by MNS and has an administrator. Its task will be to validate and collate collected data with the objective that such data can be used to effect conservation of birds and biodiversity. Local and visiting birdwatchers and scientists can register and participate in this project by recording all their personal observations which can then be made available to others. This project forms an invaluable database for all birds recorded in Peninsular Malaysia, Sabah, and Sarawak. Many new updates in this edition have been extracted from BiW.

The basis of this book

The aim of this book is to systematically identify all birds which occur in Peninsular Malaysia and Singapore as well as provide information on the status and distribution of every species in both countries. The unit of measurement used in this book is the metric system with the lengths of birds given in centimetres (cm). All elevations and heights are stated in metres (m) while distances are given in kilometres (km). Sibley and Monroe's Distribution and taxonomy of birds of the world, first published in 1990 and revised in 1993 reclassify bird families and genera based on DNA-hybridization studies. In the first edition, the classification of bird families in this book was based on Peter's Sequence which birdwatchers and field ornithologists were more familiar with then. In this second edition, we have opted to follow largely The Howard and Moore Complete Checklist of the Birds of the World 3rd edition (Dickinson, 2003). The taxonomic classification in this checklist follows the latest findings based largely on DNA-hybridization, and for the review area, begins with the family Phasianidae (Partridges, Quail, and Pheasants), and ends with the family Emberizidae (Buntings). Where appropriate we also follow The Birds of the Thai-Malay Peninsula Volumes One (non-passerines; W1) and Two (passerines; W2) as these are the two main authoritative sources for the review area. Splits followed by Robson (2008) will only be adopted where evidence justifying the splits is clear.

Family and species descriptions

The family and species description begins with the general introduction to each bird family and a short account of its general characteristics and habits. It also gives the total numbers of species in the world, followed by the total number in the review area. For this purpose, the world totals follow Howard and Moore 3.

A detailed treatment of each group and genera within most families is also given to show diversity, e.g. the family Ardeidae (Herons, Egrets, and Bitterns) is divided into six distinct groups and 12 genera. This is to familiarize the reader with the different groups to aid in quicker identification.

Species numbers

Each species is given a number in the main text, which is also indicated on the colour plates and caption pages for quick and easy reference. The species number is maintained throughout the main text when it is mentioned in relation to another species. This is to facilitate quick cross reference. When a bird family is mentioned, the whole species sequence within the family is given, e.g. storks (34–39). Likewise when a genus is mentioned, the whole species sequence within the genus is given, e.g. genus *Mycteria* (34, 35). When a subspecies of a species is still under review, pending elevation to species rank, it is indicated by a decimal number in the species account and on the plate and caption page, e.g. White-faced Plover, 146.1 because it is described in relation with the Kentish Plover which is number 146. Similarly when a hybrid occurs in the wild, it is also indicated by a decimal number. The Upland Buzzard x Long-legged Buzzard hybrid is given the number 108.1, because it is mentioned immediately after the Common Buzzard, which is number 108, as both belong to the same genus, *Buteo*. Neither the Upland nor Long-legged Buzzards has been recorded from the review area.

Common English names and scientific names

Each species has a common name in English, followed by the scientific name. In the first edition, the common and scientific names used followed King *et al.* (1975) and Medway and Wells (M and W, 1976). Current revisions in taxonomy and nomenclature have necessitated changes in this book. Scientific names follow *The Howard and Moore Checklist of the Birds of the World* (3rd Edition; Dickinson, 2003). These are all more or less standard globally. The English names used largely conform to *Birds of the World – Recommended English Names* by Frank Gill and Minturn Wright, published by Christopher Helm in 2006. The taxonomic sequence in this book follows Howard and Moore 3. Where we differ, we follow Wells' Volumes One and Two (W1 and W2). English names of birds are still in a state of flux with variations in different countries. While Gill and Wright have attempted to standardize all the English names of the birds of the world, there are still areas which could be refined. This book attempts to address this as far as Asian birds are concerned. In an attempt to standardize English names and move towards a more systematic form of coining names, a few changes have been made in this book (Jeyarajasingam, in prep., a and b). The use of patronyms for some common names has been changed to names which are more descriptive and reflective of the region, particularly if the species is a Malay Peninsula or Sunda endemic and if it has got a distinctive feature which readily identifies it. *Ciconia stormi* (Storm's Stork) is referred to as Sunda Stork because it is the only endemic stork in the Sunda subregion. *Alophoixus finschii* (Finsch's Bulbul) is called Sulphur-throated Bulbul because of its conspicuous bright yellow throat which is characteristic when puffed out. *Ixobrychus eurhythmus* (von Schrenk's Bittern) is called Chestnut Bittern because of its largely deep maroon-chestnut upperparts, a characteristic field mark. *Gorsachius melanolophus* (Malaysian Night-heron) is referred to as Rufous-necked Night-heron because of its bright rufous neck and sides of face. Its name of Malaysian Night-heron is a misnomer not only because it does not breed in Malaysia but occurs as a rare and localized winter visitor. *Psittacula longicauda* (Long-tailed Parakeet) is referred to as Reddish-cheeked Parakeet because all *Psittacula* parakeets have long tails and it is relatively difficult to determine visually, which species has the longest. On the other hand the reddish cheeks are a characteristic feature only of this species (varies from reddish pink to bright red throughout global range) and is therefore more appropriate. *Calyptomena viridis* is referred to as Little Green-broadbill. Its name of Asian Green Broadbill is also a misnomer because two other *Calyptomena* species in Borneo are also largely green and so is the Long-tailed Broadbill *Psarisomus dalhousiae*. The term green-broadbill is applied collectively to all three *Calyptomena species* (including two Bornean species which do not occur in the review area), which are all predominantly green and unique in having a tuft of feathers covering the maxilla. They should therefore ideally be called Black-throated Green-broadbill (Whitehead's Broadbill) *Calyptomena whiteheadi* and Blue-bellied Green-broadbill (Hose's Broadbill) *Calyptomena hosii*, because of the former's conspicuous black throat and the latter's characteristic pale blue belly. The Little Green-broadbill is named as such because it is the smallest of the three. As for the kingfishers, *Pelargopsis amauroptera* (Brown-winged Kingfisher) and *Pelargopsis capensis* (Stork-billed Kingfisher) are referred to as Brown-winged and Common Storkbills respectively. The Great-billed Kingfisher *Pelargopsis melanorhyncha* (not found here) has been named Sulawesi Storkbill, because it is endemic to Sulawesi. Calling all three *Pelargopsis* species storkbills, will clearly demonstrate their close relationship and common features (large size and relatively large stork-like bills) which readily distinguish them from other kingfisher groups. All changes made have taken into account practicality and long-term implications for future use and have been viewed from a global rather than a regional perspective (Jeyarajasingam in prep., a and b).

Vernacular names

All vernacular names are in *Bahasa Melayu*, the lingua-franca of Malaysia, Singapore, and Indonesia, where it is called Bahasa Indonesia. Vernacular names used in the first edition were based largely on the traditional vernacular names following Glenister (1951) and Madoc (1956a) as well as some of the names given in *Senarai Nama Burung-Burung di Semenanjung Malaysia dan Asia Tenggara* [A Checklist of the Birds of Peninsular Malaysia and South-East Asia] by Abdul Rahman Ismail, published in 1981. Names in the latter had been adapted and widely used in the first edition. This checklist attempted to systematically name all South-East Asian birds in Bahasa Melayu, but had limited success because it was hastily prepared and many of the names did not make any sense nor accurately described the birds.

The recent publication of *Khazanah Burung Bukit Fraser – Panduan Bergambar* by Mike H.N. Chong and Sutari Supari in 2010, attempts to, once again, try and systematically coin suitable *Bahasa Melayu* terms for birds. This book has succeeded in this objective, and many of the montane species have a specific *Bahasa Melayu* name. This was largely achieved by systematically translating and modifying established English names to convey meaning as accurately as possible. For example, the White-throated Fantail is called *Sambar-kipas Leher Putih*, literally a direct translation but meaningful enough to be applied in scientific texts in Bahasa Melayu. All fantails are generally referred to as *Murai Gila* (literally meaning 'mad thrush') because of the bird's habit of flitting around aimlessly. All *Bahasa Melayu* terms in this second edition will follow Chong and Sutari (2010) with suitable adaptations wherever possible. Most traditional names are onomatopoeic and vary from place to place. Other traditional vernacular names are based on colour, e.g. *Kunyit* (saffron) generally refers to most orioles and the Common Iora because of yellow in the plumage. Traditional names may be based on tail length or habits. The Asian Paradise Flycatcher is traditionally called *Murai Ekor Gading* (literally meaning 'ivory-tailed thrush') because of the long tail streamers of the male. Swifts and swallows because of similar morphological features and common aerial feeding habits are collectively referred to as *layang-layang* when they are completely unrelated. Chong and Sutari specify the swifts as *lelayang* and the swallows as *layang-layang* to show the difference. It must be stressed that traditional vernacular names although too generalized and difficult to be specifically applied in scientific literature in *Bahasa Melayu*, are still important. Therefore the traditional names will still be mentioned together with the specific *Bahasa Melayu* term so that it will be familiar with local readers. In groups where there are no traditional names such as in, petrels, storm-petrels, and jaegers, no vernacular names are given. The coinage of new terms for such groups has been resisted until a proper system is devised for such an operation. Thus the coinage of terms for these groups as stated in Abdul Rahman Ismail (1981) and used in Wells (1999) have not been included.

Description

Under description, the length of each species is given in centimetres (cm) together with the corresponding measurement in inches (in) stated in brackets. This is followed by the description of the general jizz of the species, plumage patterns, and colour as well as soft part colouration. Details on how to differentiate it from similar species are also given.

Voice

The description of voice for each species includes one or more of the usual utterances. Where species are generally silent during winter and vocalization is unknown, these are stated. As vocalization is a key factor in making positive identification where birds are difficult to observe, as in cuckoos, owls, frogmouths, and nightjars, this is given special emphasis. It must be stressed that many of the vocalizations are interpretations of both the author and the illustrator, based on field observations and sound recordings. Interpretations vary with individuals but descriptions are usually sufficient to make calls and songs recognizable in the field. As an example, the call of the Large-tailed Nightjar *Caprimulgus macrurus* is described as *tunk-tunk* by Robinson (1927), *tock-tock* by Madoc (1956a), *chonk* by King et al. (1975), and *chok* by Medway and Wells (1976).

Range, status, and distribution

The world range as well as the status and distribution of each species within Peninsular Malaysia and Singapore follow Medway and Wells (1976, W1, and W2). Where the names of towns or mountains are mentioned, the state or country (when referring to Singapore) is mentioned in brackets, e.g. the Larut Hills (Perak) or Tuas (Singapore). The incidence of occurrence is also given, and this follows two sources, namely *A Checklist of the Birds of Peninsular Malaysia* (MNS-Bird Conservation Council, 2010) and *An Annotated Checklist of the Birds of Singapore* (W and H). The following terms are used: abundant and widespread (a good chance of encountering the species in the right habitat), common and widespread (common in a variety of habitats), fairly common (fairly commonly encountered in the right habitat), locally common (a good chance of encountering the species in the right habitat), uncommon and localized (some sites where there is a moderate chance of recording it), rare and localized (one or two sites where there is a slim chance of recording it), and vagrant or very rare (less than one or two sightings over the years). Where a species is locally extinct in either Peninsular Malaysia or Singapore, this is stated in each species account. All information in these two areas have also been updated from published literature and unpublished field notes from various sources since 1998.

Global conservation status

A global listing of threatened birds is found in the International Union for the Conservation of Nature (IUCN) Red List 8 of Threatened Species (IUCN, 2009). All categories of threat were developed by IUCN using set criteria. These were adopted in 2000 and are used by BirdLife International. The categories used in this book are as follows. **NT** (Near Threatened), **VU** (Vulnerable), **EN** (Endangered), and **CR** (Critically Endangered). Where a bird falls under one of the above categories, the abbreviation is indicated in the species account under range, where the rationale of categorizing it as such is also explained. Species not threatened are categorized as **LC** (Least Concern) and are not stated in the range for such species.

References to published literature

Statements from some authoritative literature and other sources cited in the text, particularly relating to description, voice, and status and distribution, are stated in abbreviations (see 'Glossary' and 'Abbreviations') and can be looked up in the bibliography.

Habitat

In the description of habitat for each species, all the preferred habitats are given, beginning with the most preferred to the least. Forests refers to both primary and mature secondary forest where all or some of the different zones are in place (see 'Natural vegetation' under 'Introduction to Peninsular Malaysia and Singapore' chapter). The forest edge refers to clearings near roads and along rivers. Lowland and montane forests are denoted by the altitudinal range given under 'Status and distribution'. For the purpose of this book, lowland dipterocarp forest, peat swamp forest, and stands of *Melaleuca* will be collectively regarded as lowland forest. Upper montane forest will also be referred to as ericaceous forests. Freshwater swamps will include principally disused mining pools fringed by overgrown reedbeds and other thick aquatic vegetation, flooded ricefields, and ponds. Open country will cover both flooded and harvested ricefields, scrub, playing fields, and golf courses. Wooded areas will refer to very small patches of isolated secondary forest within urban environments (e.g. Bukit Tunku in Kuala Lumpur) as well as ornamental shade trees planted in towns and cities.

Habits

The habits of each species are based largely on behaviour and are given briefly but as much as possible to aid in identification.

Illustrations

Of the 673 species described, 669 are illustrated in 74 colour plates with each facing a caption page. Each caption page briefly describes the key identification points to check for each species in the corresponding colour plate. Measurements of birds are also included in the caption page to enable the reader to make a quick estimate of the size without having to refer to the main text. The five species not illustrated in colour are very rare species which are either vagrants or those which have not occurred for a long time. These include the Green Peafowl, White-rumped, Indian, Red-headed, and Cinereous Vultures. Black and white figures are also used in the family and species descriptions to show key identification features. These include a subspecies of the Whimbrel *Numenius phaeopus variegatus*, the different underwing pattern of the Wood and Green Sandpipers as well as the flight pattern of most hornbills.

Introduction to Peninsular Malaysia and Singapore

Geographical limits

This field guide covers Peninsular Malaysia and the Republic of Singapore, including all offshore islands within the territory of both countries. Both territories will be collectively referred to as the review area in this book.

Peninsular Malaysia is the southernmost extension of mainland Asia, bordering the Kingdom of Thailand in the north and the Republic of Singapore in the south. Singapore is separated from Peninsular Malaysia by the narrow 1.4-km Straits of Johor but they are linked together by two causeways. Both land masses were linked together during the ice ages at least 10,000 years ago before the sea level rose. Singapore was once part of Malaysia, the political union formed on 16 September 1963 between Peninsular Malaysia (then Malaya) and the British north Bornean territories of Sarawak and Sabah (often referred to as East Malaysia). Singapore broke away as an independent nation in 1965. Because of similarities in ethnic composition and a common vernacular language, Bahasa Melayu, both countries have maintained close cultural and economic ties and are both members of the Association of South-East Asian Nations (ASEAN). Both nations are also in close proximity to the Republic of Indonesia.

Physiography

Peninsular Malaysia

The territory of Peninsular Malaysia lies between 1° 15′ N–5° 40′ N: 99° 45 E–104°18 E and covers a total land area of 131,598 km^2. The territory is approximately 500 km from north to south and 300 km from east to west. The major islands are Pulau Pinang (296 km^2), Pulau Langkawi (478 km^2), and Pulau Tioman (136 km^2). Much of the land is hilly and mountainous with lowlands predominantly along the coasts as well as along the valleys of the river systems. The whole territory itself may be divided into four zones based on physical features. The general geology will also be discussed briefly here.

The central highlands

These consist of forest-covered ancient fold mountains which extend from Thailand. There are principally two sets of mountain ranges, all more or less running parallel; the northern ranges which lie staggered across the territory in NNE–SSW lines and the southern ranges which have much lower altitudes and lie in a NNW–SSE pattern. These ranges are composed largely of granitic rocks and today average between 1220 m and 2187 m, having been so extensively eroded by equatorial weathering throughout the ages that some of these rocks today are exposed and stand out as the greatest heights. Limestone outcrops are interspersed between these ranges particularly in Kedah, Perlis, Perak, Kelantan, Pahang, and Selangor. Igneous rocks are sparsely distributed especially from Kelantan, south to Perak, Pahang, and Johor. There are eight ranges:

(a) The Nakawan Range, a small 300-km^2 predominantly limestone range on the Perlis–Thailand border. Only 100 km^2 lies within Malaysia. It varies from sea level to 733 m with Gunung Perlis, a granite outcrop, the highest point on the Malaysian side. Gunung Pelarit, at 553 m is the highest point on limestone.
(b) The Western or Kedah-Singgora Range from Singgora in Thailand through central Kedah to Seberang Perai and also with minor outliers at Gunung Jerai (1213 m) and the island of Pulau Pinang.
(c) The Bintang Range from Thailand south to Bruas in Perak. The Larut Hills in Perak form part of this range with Gunung Inas (1801 m) the highest peak.
(d) The Kledang Range, a small low range between the Bintang and Titiwangsa Ranges, lying west of the Kinta District (Perak).
(e) The Titiwangsa Range, the longest and most continuous range of ancient fold mountains, stretching in an arc from Thailand, south to Melaka with Gunung Korbu (2183 m) the highest peak. Also called the Korbu or Main Range.
(f) The much eroded Benom-Ledang Range whose continuity is broken by east flowing streams from the Titiwangsa Range. Only Gunung Benom (2107 m) and Gunung Ledang or Mt Ophir (1276 m) stand out as isolated peaks.
(g) The Tahan Range consisting largely of sandstone and shales from Terengganu south to Johor. Much of it is eroded except for the isolated weathering-resistant sandstone quartzite blocks like Gunung Tahan (2187 m), Peninsular Malaysia's highest peak.
(h) The Eastern Range, a highly eroded range running from Kelantan to eastern Johor. Gunung Padang (1511 m) in Terengganu and Gunung Tapis in Pahang are part of this range.

The west coast plain A continuous stretch of uninterrupted flat alluvial land from the Kedah coast south to Johor. The sheltered coastline supports tidal mudflats, mangrove forests, and peat swamp forests further inland. The fertile soils have made it important for agriculture especially rice-growing in

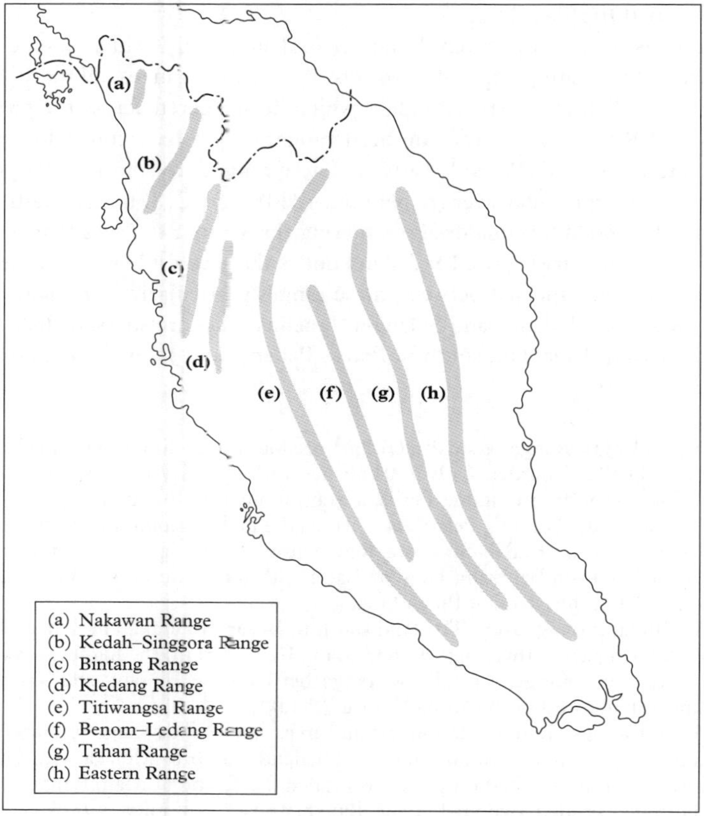

Map 3 Mountain ranges of Peninsular Malaysia.

the states of Kedah, Perak, and Selangor. Oil palm and rubber have also been cultivated extensively. Most of the open cast tin mines are in this area. This zone has the highest population density in Peninsular Malaysia and hence the highest utilization of land.

The east coast plain A narrow stretch of flat alluvial land, broken by many drainage systems. The unsheltered coast line and the action of strong waves and currents have resulted in sandy shores. Less fertile than the west coast plain with a lower human population density. Subject to heavy flooding during the north-east monsoon. Mangroves are confined to sheltered estuaries. Peat swamp forest is distributed in small patches. Casuarina woodlands are a more prominent feature of the beach.

The southern plain A largely eroded plain in Johor consisting of coastal alluvium, granite, quartzite, and shales. It is characterized by flat lowlands interspersed with low hills. Once covered by lowland forests which have been cleared and replaced by oil palm plantations as well as pineapple and pepper. Has a fairly high human population density.

Drainage Because of the high annual rainfall, most of the river systems in Peninsular Malaysia carry a great volume of water relative to their length or catchment areas. The rivers are generally short and fast flowing, emptying into the Straits of Melaka and the South China Sea. The Sungai Pahang is the longest river, flowing over 330 km from the Titiwangsa Range to the South China Sea. Other significant river systems include Sungai Kedah and Sungai Muda (Kedah), Sungai Perak and Sungai Kinta (Perak), Sungai Bernam, Sungai Selangor, Sungai Klang and Sungai Langat (Selangor), Sungai Muar, Sungai Johor and Sungai Endau (Johor), Sungai Rompin (Pahang), Sungai Terengganu (Terengganu), and Sungai Kelantan (Kelantan).

Singapore

The Republic of Singapore lies between 1 °09′ N–1 °29′ N: 103 °38′ E–104 °06′ E, covering a total land area of 699 km^2 (2010) which is increasing annually with continual coastal reclamation, and is expected to grow by another 100 km^2 by 2033. The main island is approximately 23 km from north to south and 42 km from east to west. There are altogether about 58 satellite islands, the largest ones are Pulau Tekong Besar (24 km^2). Pulau Ubin (10 km^2), and Sentosa Island (3.3 km^2). Much of the main island is low and undulating with Bukit Timah (163 m) the highest point.

Singapore has three principal rock formations. Igneous or granitic rocks make up much of the centre of the main island as well as the north and north-east. Sedimentary rocks, comprising largely sandstone and shale are distributed in the west, south, and southwest while alluvium covers much of the older rock structures in the east. Except for Pulau Ubin which is principally granitic, all of Singapore's satellite islands are composed of sedimentary rocks.

The rivers in Singapore are short and swift flowing and most originate from the high ground in the centre of the island. The longest river is the 15-km long Sungai Seletar. Other principal rivers include Sungai Kranji, Sungai Serangoon, Sungai Kallang, and Sungai Jurong. Both the Sungai Seletar and Sungai

Kranji have been dammed to form reservoirs. Most rivers are almost completely canalized with concrete banks.

Climate

Both Peninsular Malaysia and Singapore lie about 127–500 km north of the equator and because of this proximity, experience an equatorial type climate with high day temperatures and rainfall throughout the year. Day temperatures vary between 28–32 degrees Celsius. The flat lowlands are hottest while the highlands are much cooler because of the modifying effect of altitude. Night time temperatures tend to be lower especially towards the later part and average about 20–23 degrees Celsius at dawn in the lowlands and about 13 degrees Celsius or lower in the highlands.

The average annual precipitation is about 2000–3000 mm, varying from place to place. The wettest months are from November to January with the onset of the north-east monsoon which brings heavy rainfall to the east coast and the island of Singapore. The west coast states receive less rainfall because the Titiwangsa Range acts as a natural barrier. Monsoon winds crossing this range bring heavy rains on the eastern slopes and when they reach the western side, are devoid of much of their original moisture content. The south-west monsoon which occurs from June to July does not bring much rain because of a natural land barrier in the island of Sumatra which receives the bulk of the rainfall. Convectional torrential rainfall is the most characteristic rainfall of both countries, usually occurring on very hot days as afternoon thunderstorms.

Relative humidity is generally high at over 90 per cent on the average. It is usually higher at night and averages about 65–70 per cent during hot days. Relative humidity tends to be higher on the coasts than in the interior and in the highlands.

Natural vegetation

Lowland and hill dipterocarp forest

The former originally covered the valleys of the major river systems while the latter clothed hilly country and the lower slopes of mountain ranges. Primary dipterocarp forest has been greatly reduced due to logging and subsequent utilization of land for agriculture and urbanization. Save for Taman Negara, the remaining fragments are largely of the secondary type. Primary dipterocarp forest itself is a highly complex and diverse ecosystem, characterized by tall evergreen trees of which the Dipterocarpaceae dominates. This family is highly valued for timber and includes a variety of genera of which *Dipterocarpus* and *Shorea* predominate. Dipterocarp forest begins at sea level and extends up to 750 m. Five distinct vertical zones are recognized (see figure). The ground storey comprises the forest floor, which in its pristine state often has sparse undergrowth because very little sunlight reaches it. Above this is the lower storey which supports small woody plants, tree saplings, small palms, and ferns. The trunks of the larger trees and the crowns of smaller trees together form the middle storey. The canopy consists of closely joined crowns of the huge trees (mainly dipterocarps), thick-stemmed lianes, rattan, and fan palms. A number of emergents extend above the canopy; among them is the tualang *Koompassia excelsa* (Leguminoceae) which can attain a height of 50 m. The huge trees have buttressed roots to support their sheer weight and are often draped in lianes, woody epiphytes and strangling figs (*Ficus* sp.). Rich in flora, both lowland and hill dipterocarp forests support the richest and most varied diversity of birds, with each vertical zone supporting different species. These include partridges, pheasants, hornbills, babblers, and a variety of resident and migrant thrushes, warblers, and flycatchers. One Malay Peninsula endemic, the Malaysian Peacock-pheasant, occurs here. Bamboo clumps particularly in disturbed forest are frequented by the Pin-tailed Parrotfinch which is dependent on seeding bamboo as a principal food source and the Bamboo Woodpecker which nests in bamboo. In Singapore, primary lowland dipterocarp forest is confined to the nature reserves of Bukit Timah and the Central Catchment forest. The bulk of the two nature reserves, however, is made up of secondary forest. Primary patches are also found at MacRitchie Reservoir.

Lower montane rainforest

This begins from 750 m up to 1500 m and covers much of the Bintang and Titiwangsa Ranges as well as the other smaller mountain ranges and the isolated mountains. It differs from lowland and hill dipterocarp forests by having a lower and smaller canopy with fewer emergents. The trees have smaller buttressed roots. Woody and thick-stemmed lianes are scarce or absent while there is an abundance of vascular epiphytes. Dipterocarps become scarce and are absent above 1200 m. Oak (Fagaceae) and laurel (Lauraceae) predominate above this. Lower montane rainforest covers much of the mountain ranges as well as the isolated higher mountains. In terms of birdlife, it is less diverse than lowland and hill dipterocarp forests but supports species restricted to higher altitudes, including three Malay Peninsula endemics namely the Malaysian Partridge, Mountain Peacock-pheasant and the Malaysian Whistling-thrush.

Profile of the equatorial rainforest

The Crested Argus, though not a true endemic (it also occurs in Laos and Vietnam) occupies a very restricted range; on Gunung Tahan and other quartzite mountains near it. Much of the lower montane rainforest in Peninsular Malaysia is still pristine. The only settled areas in this zone are the various hill-stations.

Upper montane forest

This begins above 1500 m and covers the highest peaks of the Larut Hills and the Titiwangsa Range as well as Gunung Tahan and Gunung Benom. The vegetation in this zone consists largely of plants in the family Ericaeae which includes the rhododendrons. Other plants such as *Leptospermum flavescens* and *Baeckia frulescens* are characteristic features of this zone. Native conifers namely *Dacrydium* sp. and *Agathis* sp. also occupy this zone. The trees at this altitude become stunted and gnarled with increasing altitude, averaging about 10 m in height. Because of constant cloudiness, conditions are moist, allowing for the luxuriant growth of bryophytes which cover the boles and crowns of trees in great festoons. *Sphagnum* moss covers the moist ground. Also called ericaceous or elfin forest. In terms of birdlife, this zone is the least diverse, supporting only species restricted to very high altitudes; namely the Golden-throated Barbet, Chestnut-tailed Minla, Snowy-browed Flycatcher, Rufous-vented Niltava, Yellow-breasted Warbler, and Brown Bullfinch. The Hill Prinia is also not a true endemic but occupies a very restricted range and is confined to the *padang* (open grassy area) on the summit of Gunung Tahan.

Mangroves

Mangroves form a distinct ecological group of plants that have developed structural adaptations to root in mud and at the same time tolerate the daily flooding of their root system by saline tidal water. The trees are evergreen with shiny leathery leaves; these features helping to counter the high rate of transpiration. Mangroves develop well along muddy shores, and in Peninsular Malaysia, they thrive well along the west coast which is sheltered from the action of strong currents and waves by the island of Sumatra. Much of it is today is reduced and fragmented, having been exploited for fire wood, timber poles, and wood for charcoal as well as reclamation for aquaculture. In Singapore, mangrove development was greatest along the north and west coasts especially around Kranji, Mandai, Sungei Buloh, and Serangoon. These have now been greatly reduced due to land reclamation. Pulau Tekong has now the largest tract of mangrove left in Singapore. Mangrove forests in Peninsular Malaysia and Singapore contain some 30–40 species of plants, dominated by four principal genera. *Avicennia* and *Sonneratia* have extensive cable root systems with many tiny rootlets which stick out through the mud during low tide. *Rhizophora* has many roots supporting the trunk which begin above the mud level. *Bruguiera* has lateral roots, bending in loops called 'knees' which are exposed above the mud. The birdlife of the mangroves is not as rich and varied as the rainforests. Although shorebirds, herons, and egrets feed on the extensive mudflats fringing mangroves, roosting in them during high tide, they are fundamentally not mangrove birds as they are not entirely dependent on the niches within the mangrove forest itself. Within the mangrove bird community, a number of herons, kingfishers, woodpeckers, and warblers also occur in other habitat types. True mangrove specialists are the Brown-winged Storkbill, Mangrove Pitta, Great Tit, and Mangrove Blue-flycatcher.

Peat swamp forest

Peat swamp forests were formed because of rising sea levels and torrential rains which caused extensive alluvial lowlands to be frequently inundated. The soil is thick viscid mud covered with peat which in many areas averages 10 m in depth. Most trees, though lower than that of dipterocarp forests, have commercial value. The soggy ground has a thick growth of thorny plants, palms, and *Pandanus* screw pines which make it generally inaccessible. Stands of *Melaleuca* which can grow in up to 1 m of water, occur here. Peat swamp forests occur mainly near coasts and have a patchy distribution in Kedah, Perlis, Perak, Selangor, Terengganu, Pahang, and Johor. Much of this forest type has been exploited for timber. In terms of birdlife, peat swamp forest support a

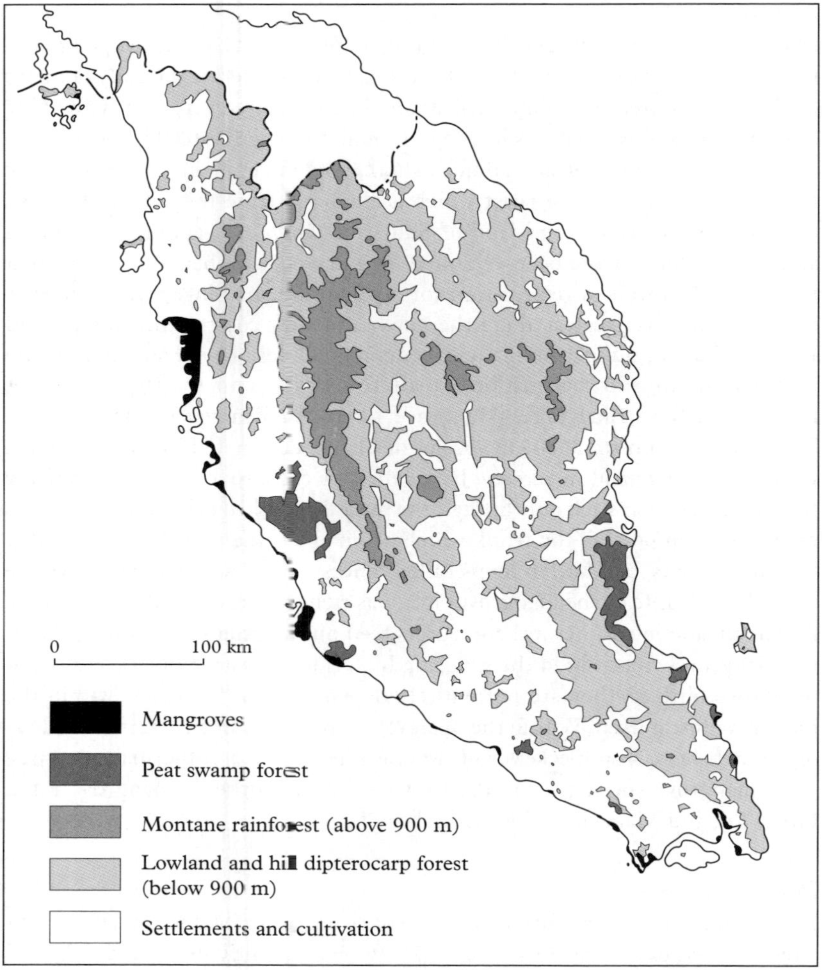

Map 4 Peninsular Malaysia (natural vegetation).

rich and diverse variety of the arboreal forest birds found in lowland dipterocarp forests.

Casuarina woodlands and beach scrub

This is a characteristic vegetation type on the east coast from Terengganu to Johor, with a patchy distribution in Singapore. *Casuarina* averages some 25 m

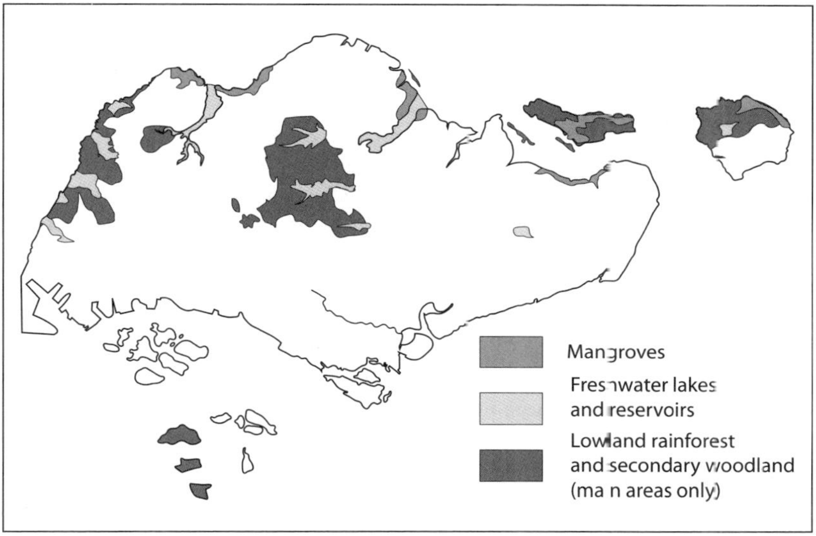

Map 5 Singapore (natural vegetation).

in height and has needle leaves. Other trees associated with this vegetation type include *Terminalia* and *Calophyllum*, both of which are shorter and gnarled. Behind these woodlands inland is usually scrub vegetation consisting largely of lallang grass *Imperata cylindrica* and the shrub *Melastoma malabathricum*. The birds of this habitat type are mostly open country birds as well as those which occur in gardens and parks such as Lesser Coucal, Spotted and Zebra Doves, Barred Buttonquail, Pied Triller, and Yellow-vented Bulbul. The Fiery Minivet and Oriental White-eye also seem to favour this habitat.

Limestone outcrops

Limestone outcrops occur sporadically throughout Peninsular Malaysia. Most limestone areas have very thin soil cover and thus support stunted trees except for the Nakawan range, which supports taller trees due to its thick moist soil cover. The vegetation type of this landscape is unique with many endemics. Clusters of wooly plants, *Paraboea* sp. grow only on limestone. Others include *Buxus malayana*, *Vitex siamica*, and the epiphyte *Schefflera musangensis*. In terms of birdlife, limestone outcrops support very few species such as cliff-nesting birds like the resident Peregrine Falcon and Dusky Crag-martin. Its associated caves also provide nesting sites to swiftlets, Striated Swallow, and Blue Whistling-thrush, especially at cave mouths. The Blue Rock-thrush also forages on limestone cliffs.

Other natural landforms

Inter-tidal mudflats Mudflats are a distinct feature of the west coast of Peninsular Malaysia and the north and west coasts of Singapore, which are all sheltered from strong wave action, allowing substrate to be deposited along the coastline. This substrate comprises largely silt, sand, clay, and decomposed organic matter which allows for the prolific growth of mangroves as well as small invertebrates and minute organisms to thrive. When exposed during low tide, mudflats offer good feeding places for a host of migratory shorebirds on 'refuelling' stops as well as large numbers of heron and egrets as well as gulls and terns which roost in large numbers on the exposed mud.

Sandy and rocky shores These shores are prominent features of Peninsular Malaysia's east coast as well as the south coast of Singapore. Sandy shores are much poorer in organic matter than mudflats and so do not support a great number of shorebird species. However some migrant shorebirds frequent sandy shores and these include Grey Plover, Kentish Plover, and Sanderling. The Malaysian Plover feeds and breeds only on sandy shores. Rocky shores support fewer species, amongst them the Pacific Reef-egret.

Offshore islands and rocky islets The offshore islands are largely forest covered as in Pulau Langkawi (Kedah), Pulau Pangkor (Perak), Pulau Redang (Terengganu), and Pulau Tioman (Pahang) or partially covered with mangroves and fringed with mudflats as in Pulau Ketam and Pulau Tengah (Selangor) and Pulau Ubin and Pulau Tekong (Singapore). The forested islands still retain some of their original cover but are poorer in terms of diversity of forest birds than the mainland. However, because of isolation, altitudinal zonation differs for some forest species on islands. The Hill Myna which is a lowland denizen on the mainland, is found up to above 900 m on Pulau Tioman. Though Pulau Langkawi and Pulau Pangkor have populations of Great Hornbill, the much reduced forest cover and the lack of contiguity with larger forested areas may compel hornbills to fly to the mainland to forage as observed on Pulau Pangkor. Island specialists such as the Pied Imperial-pigeon and the rarer Nicobar Pigeon occupy the canopy and ground storeys respectively. The mangrove islands still retain a less diverse population of birdlife but the adjoining mudflats are still important 'refuelling' stopovers for migrant shorebirds, especially Pulau Tengah.

Offshore rocky islets with sparse covering of vegetation are important breeding sites for a number of seabirds. Pulau Perak, situated some 142 km north-west of Pulau Pinang is the only breeding station for the Brown Booby and Brown Noddy; their populations have been decimated by fishermen who poach both eggs and young birds. Similarly other rocky islets in the Straits of Melaka, the South China Sea and off Singapore's north coast are used by the Black-naped Tern as nesting sites.

Disturbed vegetation

Secondary forest

When primary forest is logged, large gaps are created in the canopy, allowing sunlight to reach the ground storey, resulting in undergrowth to flourish luxuriantly. Bamboo clumps also thrive well in such a situation and is a characteristic feature of disturbed forest. In terms of composition, secondary forest does not exhibit the distinction vertical zonation of primary forest and is also floristically poorer. The birdlife is similar to that of primary forest but is poorer and less diverse especially when fragmented. Much of the lowland and hill dipterocarp forests in Peninsular Malaysia are largely of the secondary type, having been logged in the past.

Grassland and scrub

When large areas of forest are cleared and left idle for months at a time, tall grassland develops, comprising largely lallang *Imperata cylindrica* and other species of grasses. Barred Buttonquail, Yellow-bellied Prinia and munias are generally the first colonizers of such habitats. Within the passage of over a year, these grasslands develop into scrub when shrubs such as *Mimosa*, *Melastoma* and small trees such as *Albizia* take root. The Lesser Coucal and White-throated Kingfisher subsequently colonize such scrubland.

Man-made landforms

Artificial freshwater wetlands and reservoirs

Natural freshwater swamps are not a common feature of both Peninsular Malaysia and Singapore. Save for Tasik Bera and Tasik Chini in Pahang, there are no natural lakes. Even then these two lakes, because of the high acidity of the water, do not allow the growth and development of the diverse aquatic life essential for supporting a variety of waterbirds.

Most freshwater habitats are man-made. The former mining pools in Perak and Selangor, after being disused and allowed to remain untouched for years have developed large stands of *Phragmites* reedbeds surrounding them together with other aquatic vegetation such as *Typha*, *Cyperus*, and the sedge *Fimbristylis* as well as water hyacinth. The water supports aquatic life necessary

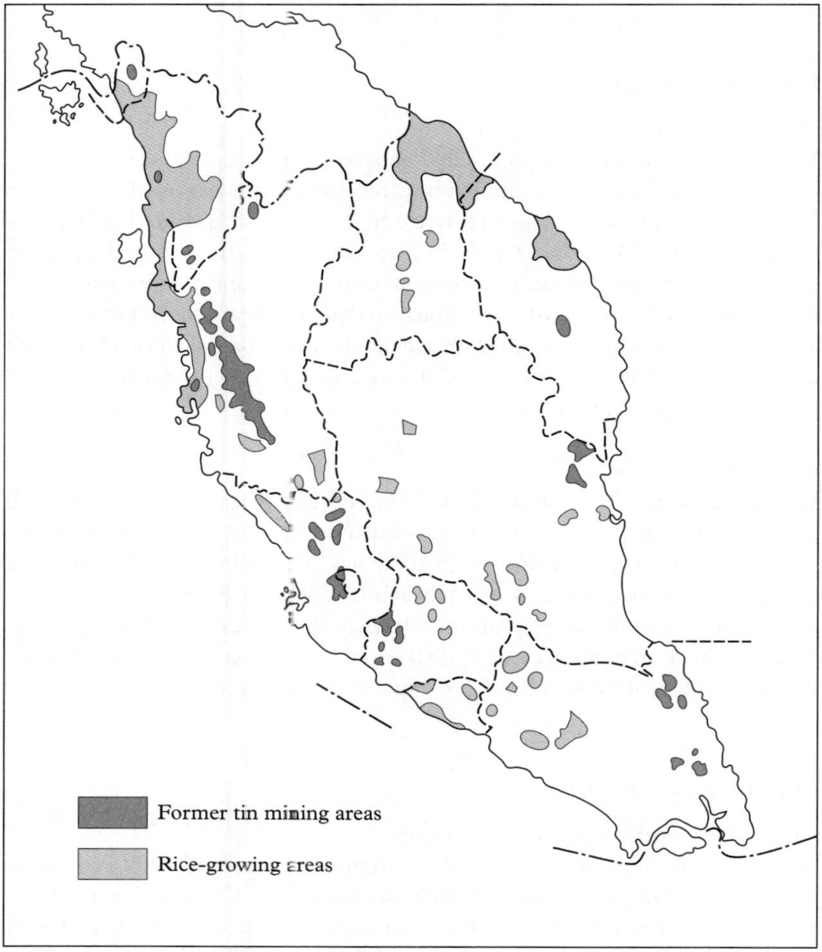

Map 6 Peninsular Malaysia (artificial wetlands).

for birds and as a result, these settled mining pools were colonized by grebes, herons, bitterns, jacanas, crakes, Lesser Whistling-ducks, and Cotton Pygmy-geese. The reedbeds also provided suitable habitat for reed-warblers as well as roosting sites for Cattle Egret, Barn Swallow, and Yellow Wagtail. Likewise the freshwater ponds and marshes in Singapore, created largely by damming of rivers also support a variety of aquatic birds. Unfortunately many of these pools and marshes in Peninsular Malaysia and Singapore have been filled up for land use. Freshwater reservoirs created by damming rivers and estuaries which are usually shallow and have vegetation cover, attract a host of waterbirds from herons and bitterns to ducks and crakes. The concrete reservoirs in towns and cities are deep and usually kept clear of vegetation and so do not attract birds except for swifts and swallows which skim the water surface to drink.

The extensive ricefields in Kedah, Perak, Selangor, Melaka, and Kelantan are important artificial freshwater wetland habitats. Flooded ricefields with paddy seedlings attract a variety of aquatic birds such as herons, egrets, bitterns, crakes, terns, kingfishers and a host of migratory shorebirds, especially Pacific Golden Plover, Wood Sandpiper, Greenshank, Marsh Sandpiper, and Black-winged Stilt. The Black-winged Kite feeds on the numerous rats in the ricefields. When the paddy ripens, munias and weavers come in flocks to feed but do not cause significant losses to rice farmers. When the paddy is harvested and the fields left fallow, migratory raptors like Greater Spotted, Steppe, and Eastern Imperial Eagles together with Eastern Marsh-harriers and Pied Harriers arrive. The Oriental Pratincole also prefers such habitat. This habitat type is fairly stable but the use of pesticides and rodent poison may not portend well for the birdlife.

Coastal power-station ash-ponds, sewage oxidation ponds, and palm oil factory sludge ponds

This refers specifically to the Sultan Salahuddin Abdul Aziz Power Station, situated on the coast at Kapar in the Klang District (Selangor). This power-station is managed by *Tenaga Nasional Berhad* (TNB) (the National Energy Limited), and the daily operation to produce electricity results in ash as a by-product and this together with water is channelled into ponds within the premises which form important high tide roosts for wintering shorebirds and terns as well as those on passage. These ash-ponds have offered a secure roosting place for tide dependent waterbirds since 1988, and have the capacity to support up to 20,000 birds. These ash-ponds are continually monitored by members of the MNS Bird Conservation Council as well as the Selangor Branch Bird Group with the cooperation of TNB, and form an important shorebird conservation area in the region. Sewage oxidation ponds in towns and cities as well as palm oil factory sludge ponds in oil palm plantations are rich in organic matter, making them good breeding grounds for a number of insects, among them dragonflies and damselflies. Aquatic birds such as terns, herons, and egrets frequently congregate here to feed on insect larvae and a variety of other aquatic invertebrates and vertebrates.

Plantations

Throughout the twentieth century, lowland forests throughout Peninsular Malaysia have been cleared to make way for plantations, namely rubber, oil-palm, and, to a lesser extent, coconuts. In Singapore, a few rubber and coconut plantations still exist. Plantations are poor in birdlife and contain species such as Laced Woodpecker, Common Flameback, Greater Coucal, White-throated Kingfisher, Spotted and Zebra Doves, the Magpie Robin, tailorbirds, and prinias. The Lineated Barbet has adapted well to coconut plantations and the Red Junglefowl to tea plantations in Cameron Highlands and oil-palm plantations elsewhere, where fallen palm fruit forms an important part of its diet. The Blue-rumped Parrot and Reddish-cheeked Parakeet have also adapted to a diet of oil palm fruit, feeding on the mesocarp, also occur on oil-palm plantations. The Barn Owl and Spotted Wood-owl appear to be thriving well in oil palm plantations due to the abundance of the Malayan Wood Rat *Rattus tiomanicus* and nest box schemes. Where plantations are overgrown with vegetation and adjoin forests or mangroves, they may contain birds from these two habitat types which may not necessarily be able to exploit the limited niches in plantations.

Gardens and parks

These are largely the feature of many towns and cities throughout Peninsular Malaysia and Singapore. Being largely landscaped with the planting of many exotic tropical shade trees like the Rain Tree, Jacaranda, and Flame of the Forest. Such habitats provide very few suitable niches for native birds to exploit and are therefore very poor in terms of birdlife. Flowering bushes also make up part of the landscape. Weeds and other vegetation fringing ponds are periodically cleared together with aquatic vegetation, making them unattractive to water birds. The usual array of birds here include the opportunistic types which are fairly numerous like Yellow-vented Bulbul, White-vented and Common Mynas, Asian Glossy Starling, Black-naped Oriole (especially in fruiting *Ptycosperma* palms), and Zebra Dove. The resident Paddyfield Pipit and Yellow Wagtail can be found on playing fields and golf courses. The Banded Bay Cuckoo may sometimes forage in the tree tops.

Urban environments

Cities and towns with their huge concrete structures and buildings have the least diversity in birds. The available green spaces are scattered lawns and roadside trees. Birds found here are usually commensals of man like the Rock Pigeon, House Crow, and Eurasian Tree Sparrow. The House Crow usually frequents roadside stalls, wet markets and rubbish tips, feeding on scraps of food. Common and Javan Mynas also proliferate and form noisy communal roosts in the available trees. Natural cliff nesters like the House Swift and Pacific Swallow have adapted well in cities and towns, fully utilizing concrete buildings and bridges as nesting sites. The Asian Glossy Starling too utilizes the eaves of roofs to nest. Where there are electricity or telephone wires, these are utilized by thousands of Barn Swallows as communal roosting sites. The Common Iora, Black-naped Oriole, and Golden-bellied Gerygone have also successfully utilized the available niches in the trees within the towns and cities.

The hill-stations

The four hill-stations in Peninsular Malaysia—Bukit Larut in the Larut Hills (Perak), Cameron Highlands, Fraser's Hill and Genting Highlands (Pahang) in the Titiwangsa Range—are the only human settlements in the mountains. While Cameron Highlands was developed extensively for vegetable farming and tea planting, the other hill-stations served as holiday resorts with bungalows and neat, well-kept gardens.

Since these hill-stations occupy only a small area in the mountain ranges, they are surrounded by forests. Many exotic trees and shrubs have been planted in these settled areas and they include a variety of conifers such as *Pinus* and *Juniperus* as well as *Eucalyptus*. In terms of birdlife, the hill-stations are fairly rich with a number of montane species like the Silver-eared Mesia, Long-tailed Sibia, and Spectacled and Chestnut-crowned Laughingthrushes regularly visiting gardens to feed on food scraps. The Streaked Spiderhunter and Black-throated Sunbird regularly visit flower beds while the Large Niltava and Little Pied Flycatcher hawk for insects along roadside clearings. It must be noted that the hill-station birds are fundamentally forest birds that have taken advantage of the feeding opportunities in the hill-stations. The exotic conifers have also provided nesting sites for the Cutia and White-browed Shrike-babbler. A few birds associated with human settlements in the lowlands have followed man to the hill-stations. Among these are the Magpie Robin, Eurasian Tree Sparrow, Yellow-vented Bulbul and Common Myna (only at Cameron Highlands).

Conservation

The urgent need for conservation

Of the 440 indigenous resident species of birds recorded from both Peninsular Malaysia and Singapore, 14 are seriously threatened globally (ranked **VU**, **EN**, and **CR**) due to primarily habitat loss and to a lesser extent, hunting and trapping. In Peninsular Malaysia, the once fairly common Green Peafowl has been extirpated. Singapore has lost 62 species of her approximately 115 resident forest birds (53.9%) due to large-scale clearing of forests since its establishment as a British trading post in 1819.

Most forest-dependent birds have presumably been reduced by an amount proportionate to the reduction of forest cover. Forest now covers about 40% of Peninsular Malaysia. The loss of lowland and hill dipterocarp as well as peat swamp forests is therefore the key factor in the reduction of many resident forest bird species. Lowland and hill dipterocarp forests are highly diverse ecosystems on their own, the various vertical zonations offer specific feeding and breeding niches to different species. Much of South-East Asia's dipterocarp forests contain valuable timber species which have been greatly exploited through logging When forests are logged, the various ecological niches are also reduced. Frugivores, particularly pigeons, hornbills, and barbets have been hard hit since these families are highly dependent on fruiting figs *Ficus* sp. as primary food sources. Different species of *Ficus* have different intervals between fruiting cycles, and so removal of forests would mean the depletion of *Ficus* sp. since these grow as epiphytes on large trees. The Large Green-Pigeon because of its dependence on large figs is now less numerous due to the reduced lowland forest. Hornbills require large tracts of forest to not only feed but to breed as well since many forest trees have natural crevices which are suitable nesting sites. Sunda Stork and Wrinkled Hornbill which are typical lowland species (inhabiting both lowland dipterocarp and peat swamp forests) are endangered species which will likely to disappear from Peninsular Malaysia (except Taman Negara) if more dipterocarp and peat swamp forests are continually lost.

Ground-dwelling pheasants and partridges too are sensitive to ecological changes. Since most species require a relatively clear forest floor to forage on fallen fruit and to scratch the soil for termites and other invertebrates, these opportunities are greatly reduced in secondary forest which has dense undergrowth covering the forest floor.

When forests are logged and cleared, often fragmented blocks remain. Smaller blocks tend to lose more species than larger blocks because of their limited carrying capacity and the small bird populations within them may not be able to sustain viable breeding populations. This explains the disappearance of the Rhinoceros and Helmeted Hornbills, pheasants, partridges, and a number of babblers from Singapore. The smaller forest passerines like a number of babblers, thrushes, and flycatchers also require primary forest to fully exploit the various ecological niches. While small passerines are able to adapt better to habitat changes than larger birds, further fragmentation will result in smaller non-viable populations that will eventually die out.

The montane forests covering the Titiwangsa Range, the Bintang Range, and the other smaller ranges on the east coast as well as the isolated mountains are largely pristine, though in many parts, hill dipterocarp forests on their lower slopes have been logged. Montane forests contain a total of 68 species that specialize in exploiting ecological niches above 900 m. Among these are the three Peninsular Malaysian endemics, a variety of babblers, some thrushes, warblers, and flycatchers. Because montane forests have proportionately less trees of commercial value than lowland and hill dipterocarp forests, they have not been exploited fully.

Fortunately Peninsular Malaysia's population of 22.5 million (2010) is concentrated along the coastal plains and the lowlands. In neighbouring Indonesia, Thailand, and Myanmar, because of high human population densities, lower montane forests have been cleared for agriculture and settlements, leaving only upper montane forests intact. Native tribes in Thailand and Myanmar who practise shifting cultivation, employing slash and burn methods have caused much forest degradation both in the lowlands and the hills. In Peninsular Malaysia, montane forests are relatively unpopulated besides the hill-stations. The small populations of aboriginal tribes inhabiting the mountain ranges engage largely in gathering forest produce like rattan and hunting for food, causing hardly any adverse effects on the forest ecosystem. The mountain ranges are also the watersheds of some of the major river systems in Peninsular Malaysia. The Pahang and Selangor rivers have their sources in the Titiwangsa Range. Clearing of montane forests will not only cause heavy siltation of these river sources but also erosion on the slopes and also the eventual loss of habitat for a number of montane bird species which will die out as the forests become fragmented. Montane forests, because of their restricted range and the rather specialized species they hold, should be a top priority for conservationists and must be left intact and pristine.

Mangrove forest today occupies only a fraction of its once extensive range in Peninsular Malaysia and Singapore with much of it today fragmented. Exploited largely for mangrove poles and timber, large trees have been removed, reducing the nesting sites for two declining species that are dependent on mangroves to a great extent, the Milky Stork and the Lesser Adjutant. Reclamation of coastal mudflats has virtually wiped out opportunities for mangroves to develop. Though mangroves hold a less diverse avifauna than dipterocarp or montane forests, some species are greatly dependent on the mangrove ecosystem and these include the Brown-winged Storkbill, Mangrove Pitta, Great Tit, and Mangrove Blue-flycatcher. Besides the relatively large Matang Mangrove Forest Reserve in Perak which is the largest expanse of managed mangrove forest in Peninsular Malaysia, other mangroves exist in small patches and should be conserved and allowed to regenerate to be able to support viable breeding populations of animal life. In Singapore, the Sungei Buloh Wetland Reserve together with the Sungai Mandai mangroves in the north hold the largest mangrove forest on the mainland besides Pulau Ubin and Pulau Tekong which both support Singapore's last remnant population of Mangrove Pitta. These should be wisely conserved to be part of Singapore's natural heritage.

Migratory shorebirds, though large in numbers face two major threats: coastal land reclamation as well as excessive hunting and trapping. Throughout South-East Asia, mudflats and adjoining mangroves have been greatly reduced due to growing human population and the corresponding demand for land. This has reduced the number of 'refuelling' or staging points along migratory routes between breeding and wintering grounds.

The illegal shooting and trapping of shorebirds for food in Thailand and Indonesia, and to a lesser extent Malaysia, take a heavy toll every year on migratory shorebirds both during spring and autumn passage. The actual numbers that return to their North-East Asian breeding grounds each spring must represent only much reduced numbers of the original populations that set out in autumn. Particularly vulnerable are endangered species like the Asian Dowitcher, Spotted Greenshank, and Spoon-billed Sandpiper which have restricted breeding ranges in North-East Asia.

Feral species

Deliberately introduced species of birds in their non-indigenous ranges have had adverse effects. In Singapore, a number of resident species of birds are feral, having established themselves due to frequent cage escapees from lively bird trade in Singapore as well as from the Jurong Bird Park and Mandai Zoo. Some of these feral species have now established breeding populations throughout the island and 17 are now on the official Singapore checklist. When a bird is deliberately introduced into a non-indigenous range, two things can happen: either it quickly dies out or proliferates and spreads.

Feral species compete with native species for food sources and nesting sites. They also bring diseases which may affect native species. The House Crow which was introduced to Peninsular Malaysia and Singapore from India and Sri Lanka, is a very opportunistic species and has expanded its range, particularly in Kuala Lumpur and Selangor. Not only has it displaced the indigenous Large-billed Crow in these areas but also preys on the eggs and young of nesting native birds. The Javan Myna too was introduced to Singapore from Java in the 1900s and has now spread northwards into most of the southern half of Peninsular Malaysia and has steadily displaced both the Common and Jungle Mynas where they overlapped. Feral species should be promptly identified and exterminated if potential threats to the native bird population are to be checked. The feral colonies of Painted Stork in Kuala Lumpur and Selangor too must be carefully monitored by the National Zoo. If colonies spread to the Selangor coast, they could jeopardize potential reintroduction programmes for the Milky Stork there through the threat of hybridization. The Mandai Zoo in Singapore too should monitor its free-flying Painted Storks. Free-flying Milky Stork should be captured and secured in enclosed aviaries while positively identified hybrids should be destroyed.

Forest management policies

Marginally, Malaysia still retains proportionately a higher percentage of forest cover than most South-East Asian countries. This is largely due to sound forest management policies in the early 1900s which set aside large tracts of forest and game reserves as well as the establishment of Taman Negara in 1939. The communist insurgency in Malaya in 1948 followed by a state of emergency from that year until 1960 made forests out of bounds to non-military and police personnel, hence curbing logging activities. It was only from the 1960s through the 1980s that logging activities were intensified for the purpose of extracting tropical timber for export. Besides, large areas of forest were also cleared during this period for land settlement schemes under the Federal Land Development Authority (FELDA). Illegal logging has also taken its toll. Today protected forest areas make up approximately 36% of the total land area in Peninsular Malaysia.

With the formulation of the National Forest Policy of Malaysia (NFPM) in 1977, the way was paved for greater uniformity in implementing strategies for achieving forest conservation, management and development of the country. The aim of this policy was to fulfil environmental and conservation needs besides meeting economic production goals. A National Policy on Biological Diversity was developed after the Earth Summit in 1992 and was endorsed by the government in 1997. The policy addresses biological diversity issues across sectors and contains conservation strategies and action plans.

Singapore passed a National Parks Act in 1990, which established the National Parks Board in order to develop and manage Singapore's national parks and other protected areas. This board comes under the Ministry of National Development. Currently protected areas in Singapore total 29.79 km2, which is approximately 4.5% of the country's total land area.

Legislation

To enable Malaysia to effectively implement sustainable forestry practices, a National Forestry Act was passed by Parliament in October 1984. This has been revised recently with increased fines and jail terms for illegal logging and forest encroachment. This Act also empowers the Police and Armed Forces to assist in its enforcement and this has helped to curb illegal logging.

The Protection of wildlife in Peninsular Malaysia is governed by the Wildlife Conservation Act 2010, which repealed the Protection of Wildlife Act 1972 (Act 6/72). This Act also covers the Federal Territory of Labuan. Under this Act, most birds are totally protected Under the previous Act, many species of game birds which could be hunted under licence are now totally protected. The limited number of species which can be hunted include the Red Junglefowl and White-breasted Waterhen. They, however, may not be trapped or reared in captivity. This Act provides for increased penalties for poaching, including heavy fines and jail terms. It also gives enforcement agencies more clout in dealing with wildlife crimes. It also covers all globally threatened species which are listed by the Convention on International Trade in Endangered Species (CITES). This Act is subject to review from time to time to include species which are threatened. The government agency responsible for the enforcement of this Act is the Department of Wildlife and National Parks (Peninsular Malaysia).

In Singapore, four wildlife laws exist. Under the Wild Animals and Birds Act 1965, all wild vertebrate animals with the exception of the Rock Pigeon, House Crow, Asian Glossy Starling, Purple-backed Starling, Common and Javan Mynas are totally protected throughout the year from hunting or capture. This law is enforced by the Agri-Food and Veterinary Authority (AVA). The Public Utilities Board Act which is enforced by the Public Utilities Board

(PUB) gives protection status to all catchment areas around all reservoirs, including the Central Catchment Nature Reserve which contains four reservoirs (MacRitchie, Lower Peirce, Upper Peirce, and Upper Seletar). The Parks and Trees Act 2006 covers all National Parks (Singapore Botanic Gardens and Fort Canning), Nature Reserves (Bukit Timah, Central Catchment, Labrador, and Sungei Buloh) plus over 300 parks in Singapore (as well as planted roadside trees). It is enforced by the National Parks Board, which is established under the National Parks Board Act 2005. Together, the National Parks Board Act and the Parks and Trees Act provide for strict penalties, making it even illegal to possess a trapping device within a park or reserve without a permit. Applications for research permits can be made on-line, on the National Parks Board website (see Appendix 5).

Trapping and conducting research on restrained birds—legal implications

It should be emphasized that trapping, mist-netting, colour-marking, and wing-tagging wild birds in Peninsular Malaysia can only be carried out under licence from the Department of Wildlife and National Parks, Peninsular Malaysia. Under the Wildlife Conservation Act 2010 (Peninsular Malaysia), handling restrained wild birds without a licence constitutes an offence, punishable by heavy fines or prison terms or both. This also applies to collecting and exporting specimens from the country. Local researchers and those from overseas are therefore cautioned that the licences can only be issued by the department concerned; local universities and other non-governmental organizations are not empowered to do so. In Singapore, permits to trap and ring in areas under the jurisdiction of the National Parks Board must be obtained from the board. In all other areas, such permits must be obtained from the Agri-Food and Veterinary Authority as well as from the owner or administrator of the land concerned.

The avifauna

The Malay Peninsula, which is zoogeographically part of the Sunda subregion, lies in the Oriental region and is situated between mainland South-East Asia and the Greater Sunda islands, which comprises Sumatra, Borneo, Java, and Bali together with all intervening islands west of Wallace's line. The Sunda subregion itself is an archipelago of tiny islets to very large islands, straddling the equator in an arc covering some 5000 km between mainland Asia and Australia. The whole region itself was connected to mainland Asia during the Great Ice Age of the Pleistocene epoch some 8000 to 3 million years ago before the climate became warmer, melting the ice and causing the sea levels to rise and inundate low lying areas, leaving much of the high ground and mountains standing out as islands. Prior to the last connection about 10,000 years ago, the present Greater Sundas were colonized by birds from mainland Asia which stayed on after the separation. This explains the presence of the Wreathed Hornbill and the White-browed Shrike-babbler in Sumatra, Java, and Borneo, which are separated by several hundred kilometres of water from mainland Asia where these species range from India to the Malay Peninsula. The Sunda subregion itself has a very rich avifauna of over 830 recorded species, of which about 560 are resident land birds. Because of isolation from the mainland for a very long time, the Sunda subregion has a relatively high level of endemism of about 150 species or 27% of the total number of resident species of land birds.

Although situated in mainland Asia, the avifauna of the Malay Peninsula contains a proportionately high percentage of Sundaic species which are absent from other continental South-East Asian countries. As examples, the Rhinoceros and Helmeted Hornbills, Gold-whiskered and Yellow-crowned Barbets, Black-and-Crimson Oriole, Blue Nuthatch, and Spectacled Laughingthrush are widely distributed in the Greater Sundas while on mainland Asia, their distribution is restricted only to the Malay Peninsula. Because of the lack of long periods of isolation from other parts of mainland Asia, the level of endemism is very low in the Malay Peninsula with only four species and these comprise three montane species, namely the Malaysian Partridge (see Map 7), Mountain Peacock-pheasant (Map 8), and Malaysian Whistling-thrush (Map 10), and one lowland forest species, the Malaysian Peacock-pheasant (Map 9). Two

Map 7 Distribution of Malaysian Partridge, *Arborophila campbelli*.

Map 8 Distribution of Mountain Peacock-pheasant, *Polyplectron inopinatum*.

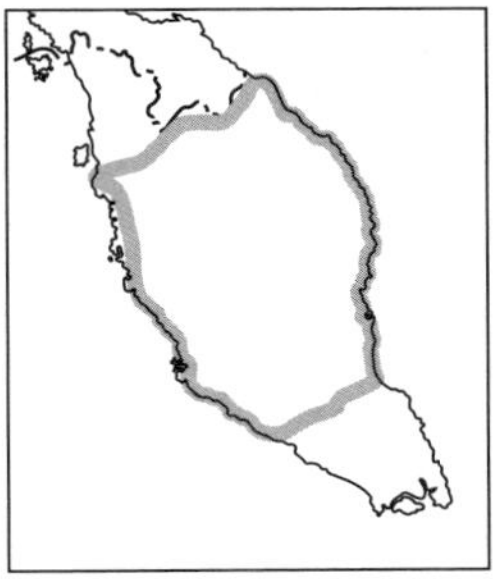

Map 9 Distribution of Malaysian Peacock-pheasant, *Polyplectron malacense*.

Map 10 Distribution of Malaysian Whistling-thrush, *Myophonus robinsoni*.

Map 11 Distribution of Crested Argus, *Rheinardia ocellata*.

Map 12 Distribution of Hill Prinia, *Prinia atrogularis*.

species—the Crested Argus (Map 11) and the Hill Prinia (Map 12)—occupy very restricted ranges; the former is confined to the sandstone quartzite mountains of Gunung Tahan, Gunung Perlis, Gunung Gagau, Gunung Mandi Angin, and Gunung Rabong, and also Gunung Gelemat on the Eastern Range, while the latter is found only on the ericaceous summit of Gunung Tahan. The Brown Bullfinch is the only equatorial representative of a genus of largely temperate species. Having arrived here during the Ice Ages, this species stayed on in the highest mountains when the climate became warmer at the end of the Pleistocene epoch.

While most resident seabirds, coastal mudflat and mangrove birds, and open country birds are distributed in the various habitat types at low elevations, forest birds are found from low elevations in lowland dipterocarp forest to higher elevations in hill dipterocarp forest with many species extending into lower montane forest. An example is the Blue-eared Barbet which ranges from low elevations up to 1000 m. Of the resident forest birds only 62 occur in montane forest, usually above 900 m but some have lower altitudinal limits. Sixteen genera, i.e. *Psilopogon, Picumnus, Cissa, Pseudominla, Gampsorhynchus, Garrulax, Cutia, Leiothrix, Heterophasia, Pteruthius, Minla Phyllegartes, Brachypteryx, Myiomela, Niltava*, and *Pyrrhula* specialize in exploiting niches above 900 m and are absent below this altitude; their distribution being restricted to the mountain ranges or large isolated mountains. There exists a definite altitudinal zonation among some bird families to ensure minimal competition for food resources and maximum utilization of the various ecological niches. For example, within the *Megalaima* barbets, the Red-crowned Barbet occurs at very low elevations while the Gold-whiskered Barbet occurs in hilly country from low elevations up to 1100 m. The Black-browed Barbet occurs between 760–1500 m in the mountains while the Golden-throated Barbet has the highest altitudinal limit, occurring above 1500 m. All four species are fruit eaters and each utilizes the resources within a specific altitude although there may sometimes be overlaps as in the case of the Gold-whiskered and Black-browed Barbets or the Black-browed and Golden-throated Barbets. Another example is the family Rhipiduridae (Fantails) which is insectivorous. The White-throated Fantail exploits niches above 700 m. The Spotted Fantail occurs from low elevations up to about 760 m, while the Pied Fantail does not occur above 450 m.

Most of the migratory birds occurring in Peninsular Malaysia and Singapore breed throughout much of the Palaearctic Region; the populations wintering here breed largely in the eastern Palaearctic. There are, however, some exceptions. Some birds which have breeding populations in the western Palaearctic do not migrate to Africa which is the nearest tropical wintering quarters as one would suppose them to do but instead winter in South-East Asia. A notable example is the Arctic Warbler which has breeding populations as far west as Europe; the entire population is known to winter only throughout southern China and South-East Asia. The large numbers of shorebirds utilizing both coastal mudflats and freshwater habitats are found throughout the Palaearctic; some species have rather restricted breeding ranges in the eastern Palaearctic. Notable examples are Spotted Greenshank, Asian Dowitcher, and Spoon-billed Sandpiper which have rather narrow breeding ranges in North-East Russia. The large numbers of migratory birds of prey seen here also breed throughout much of the Palaearctic.

Among migratory passerines, the thrushes, flycatchers, warblers, shrikes, starlings, and buntings are fundamentally eastern Palaearctic species. All these occupy niches in a variety of habitats from forests, mangroves, and plantations to wooded areas, open country, and gardens.

Migration

Migration has always been a subject of great interest and fascination to those interested in the study of birds. Birds with their power of flight have a great advantage over other animals in being able to travel great distances over natural geographical barriers and international boundaries.

The process of migration

Migration is clearly instinctive, involving largely internal hormonal changes in the birds. This is triggered by decreasing lengths of daylight during autumn in northern temperate regions. The onset of the northern hemisphere winter with freezing conditions would mean limited food resources. Under such conditions, migratory birds will feed to accumulate large reserves of body fat (which can amount to half their body weight), essential to act as fuel to provide much needed energy for the long and often hazardous journey. When fat is converted to energy, water is produced, so birds do not need to drink during migration. Birds breeding in northern temperate regions fly southwards in autumn to new feeding grounds in subtropical and tropical regions, returning north in spring to breed.

Migration takes place during both day and night with some species migrating only during the day. The presence of shorebirds and migrant terns in very large numbers along coastal mudflats and the presence of large numbers of raptors thermalling overhead are all visible signs of migration. Most migrants utilize a series of internal cues to navigate, recognizing landmarks and determining direction by using solar navigation by day and stellar navigation at night. It is also thought that the presence of iron molecules in the brains of some birds help to detect the earth's magnetic field to aid navigation. Since vision plays an important role in navigation, bad weather may cause disorientation in migrating birds, causing them to drop onto the nearest land. Bright lights also cause disorientation among night-flying migrants, attracting them and causing casualties when they crash into lighted buildings, especially light houses at sea and buildings at the hill-stations. Gas flares on offshore oil rigs also attract birds on migration. Exceptionally strong storms can cause migrating birds to veer off course from their normal flight path and land as vagrants in areas normally outside their passage route or wintering range. The Little Curlew and Chestnut-cheeked Starling—which both migrate far to the east through the Philippines—have occurred in Peninsular Malaysia and Singapore, probably due to heavy storms which might have blown them further west.

Raptors, because of their large wing surface areas, cannot flap continuously for long periods as this requires much energy expenditure. They rely largely on thermal uplifts during the day to soar and migrate only during the day. Rising thermals during the day allow raptors to gain altitude without having to flap laboriously. Once altitude is gained, they break into a long glide along their intended route towards the next thermal. This method is not only efficient but energy saving. The best soaring conditions are during hot days. Thermals generally build up after 10.00 hours when the rising sun's rays have heated the surface of the earth. On the other hand, during overcast or rainy days there is little or no thermal build up, so raptors do not migrate under such conditions.

Smaller birds such as shorebirds and passerines which have much smaller wing surface areas, are energetically more efficient than the large raptors and because they do not depend on thermals, are capable of maintaining long non-stop flights day and night before stopping to rest and feed. Because of the need to stay on land in case of bad weather, traditional migration routes avoid large ocean crossings, passing close to coasts and through island chains; crossings are usually made at the narrowest stretch of water between lands. The Malay Peninsula, being the southernmost extension of the continent of Asia, forms a natural flight path for migratory birds which converge here before spreading out further south to the Indonesian Archipelago and Australasia.

In Europe and northern temperate Asia, birds breeding west of the ninetieth meridian as a rule, migrate to spend winter in Africa while birds breeding east of this line migrate to winter throughout much of South and South-East Asia and Australasia. Insectivores such as the Barn Swallow, some flycatchers, and some warblers which breed in temperate Asia migrate south to tropical regions when insects become scarce in temperate regions because of winter. Shorebirds which breed in the far eastern Arctic tundra in spring and summer have to migrate south during autumn before their feeding grounds become frozen in winter, drastically limiting their food supply. In tropical regions, the coastal mudflats, estuaries, and the variety of freshwater wetlands provide shorebirds with an abundance of new food resources. Migrants usually start to arrive in both Peninsular Malaysia and Singapore as early as July and August, but in very large numbers from September through November, the bulk of which passes through to more southerly wintering grounds. From March to late May, the migrant populations suddenly swell with returning northward-bound spring migrants. A number of migratory species, particularly shorebirds, overwinter in tropical regions and are thus seen throughout the year, though in very small numbers. A few southern hemisphere migrants also reach the Malay Peninsula during the southern winter. A notable example is Horsfield's Bronze-cuckoo, an Australasian species which has occurred sporadically in both Peninsular Malaysia and Singapore. Ospreys seen outside the northern winter months are presumed to be wintering birds from the southern hemisphere.

Besides the influx of migrants from northern temperate zones, a number of species within tropical and subtropical zones also undertake southward migration. Resident species in Peninsular Malaysia and Singapore are represented in higher latitudes in mainland Asia by populations which are migratory, notable examples being the Little Heron, Asian Koel, Ruddy Kingfisher, Dollarbird, and Black-naped Oriole. During winter, these migratory populations move south and augment the resident population. The Blue-throated Bee-eater breeding populations in both Peninsular Malaysia and Singapore regularly disperse to Sumatra outside breeding. Some seabirds which migrate to Peninsular Malaysian waters are merely non-breeding visitors which have dispersed over a wide area from their breeding grounds. A notable example is the Great Crested Tern, a tropical species. Populations seen in the Straits of Melaka breed mainly in the Mergui Archipelago in southern Myanmar. The Christmas Island Frigatebird, which breeds only on Christmas Island in the eastern Indian Ocean, disperses north to Malaysian waters outside the breeding season.

Migration routes

Most birds travel along well-defined flight paths or flyways during migration. These tend to follow wind directions, sea currents, land contours, and coastlines. Migration is a very tedious and energy consuming process, so migratory birds will have to stop over at points along migration routes to rest and feed to maintain their reserves of fat. A large influx of shorebirds is seen on the mudflats and wetlands of both Peninsular Malaysia and Singapore from March to April when migrants move north. During these peak periods, most of the shorebirds seen are passage migrants which winter as far south as southern Australia and New Zealand; they merely utilize the mudflats to feed and replenish their much needed fat reserves. Birds can sometimes fly non-stop for a number of days in between staging sites along migration routes.

An important route for migratory shorebirds is the East Asia–Australasia Flyway. From Siberia in the north, it stretches eastwards to Alaska, and in the south it encompasses Australia and New Zealand. In between these two areas, it covers much of eastern Asia from China, Japan, and Korea south through

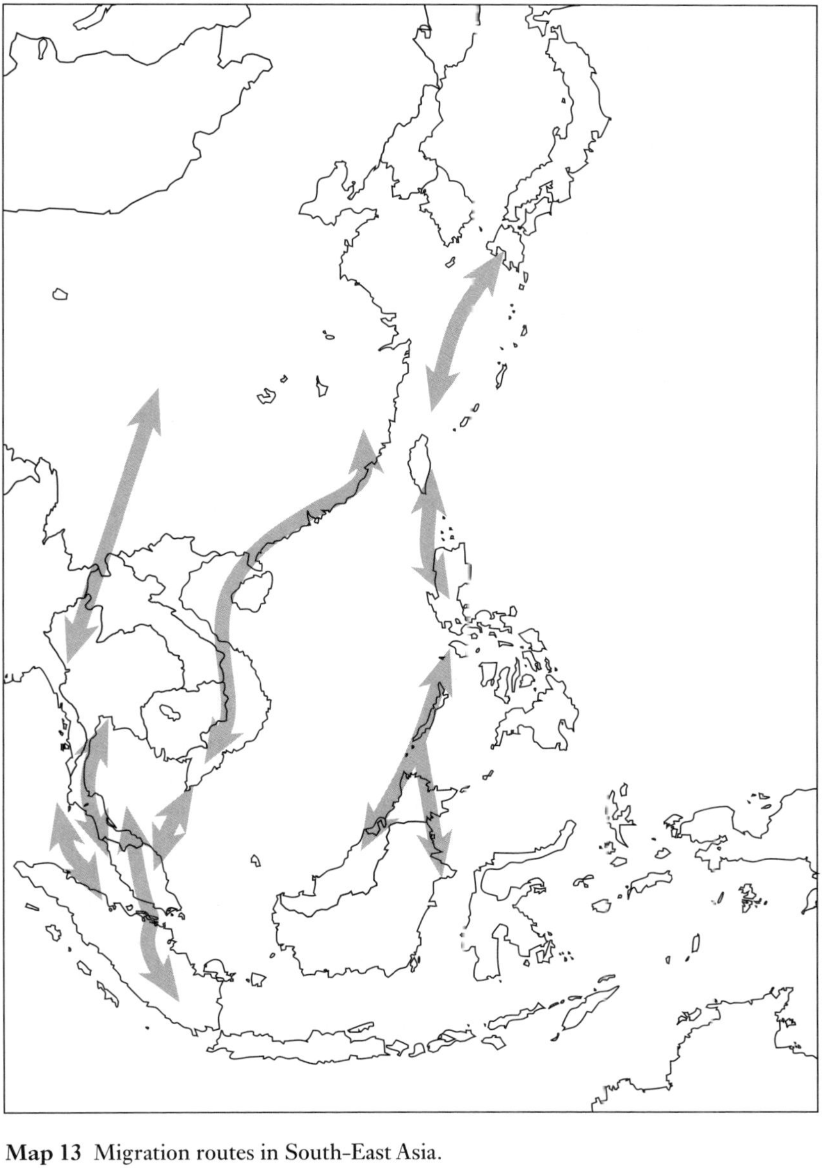

Map 13 Migration routes in South-East Asia.

continental and insular South-East Asia and the western Pacific (Map 13). Shorebirds use staging points along this route to feed and replenish their fat reserves to continue their gruelling journey southwards or northwards. This flyway passes through approximately 22 countries and so it is vital that the governments of these countries jointly cooperate in efforts to preserve breeding and wintering for shorebirds as well as their staging points in between.

Within Peninsular Malaysia, migrants moving south follow two major routes. An important route follows the western coastal plain. This route is traditionally used by migrating raptors which converge onto Tanjung Tuan (Melaka), the narrowest point on the Straits of Melaka between Peninsular Malaysia and Sumatra before crossing it into Sumatra. This route is also taken by the Blue-throated Bee-eater on annual non-breeding dispersion to Sumatra.

A major route taken by nocturnal migrants follows the Titiwangsa Range and other parallel ranges along most of the central part of Peninsular Malaysia. Another route used by migrants that have crossed the South China Sea from Indo-China follows the east coast plain. Singapore, lying at the southern tip of Peninsular Malaysia is the point of convergence from these two streams on their way to wintering grounds further south.

Studying migratory birds

Ringing

Ringing or banding is an important and efficient way of getting information about not only the breeding and wintering grounds of migratory birds but their migratory routes as well. Birds are caught by the use of mist nets or traps and vital information such as sex, subspecies, wing length, gape width, weight, moult, date and site of trapping are all recorded before the birds are ringed and released.

Amateur ringing of birds began on a small scale in both Peninsular Malaysia and Singapore in the 1950s. Not until 1963 did the ringing of migratory birds begin on an extensive scale by the US Army Medical Research Unit (USAMRU). Under the supervision of H. Elliot Mc Clure, ringing was carried out under the Migratory Animal Pathological Survey (MAPS) programme. The University of Malaya (Zoology Department) Bird Ringing Project (UMBRP) which was formed in 1968 carried on ringing activities when MAPS ceased operations. The Department of Wildlife and National Parks, Peninsular Malaysia (DWNP) also carries out regular ringing activities. In Singapore, the only ringing of any continued duration in recent times have been on a personal level by Peter Kennerley and N. Sodhi (Behavioural Studies of the Zoology Department, National University of Singapore). The Sungei Buloh Wetlands Reserve also conducts regular ringing.

Ringing stations were set up along points where migratory birds converged. These included selected feeding and roosting sites of shorebirds on the west coast. Other sites included the now non-existent Sungai Way reedbeds (Selangor) for a variety of wetland birds and the Pasoh Forest Reserve (Negeri Sembilan) for birds wintering in forests. The regular ringing of nocturnal migrants during both autumn and spring passage was carried out at the former telecommunications station along Pine Tree Hill Road, Fraser's Hill (Pahang) by both UMBPR and DWNP. Mist nets were set up along the perimeter of the station overlooking the valley and spot lights were set up to attract the migrants. A variety of migrants (non-passerines and passerines) were mist-netted and ringed at this station between 1964 and 1988. The Telecommunications Department closed this site in 1988 to construct a holiday bungalow for its employees, bringing an end to more than two decades of ringing.

Mist-netting and ringing results have yielded valuable scientific information. From netting results, we have a wealth of information on the migrants that pass through both Peninsular Malaysia and Singapore. Rarities like the Besra, Rufous-headed Robin, Siberian Rubythroat, and Plain-tailed Warbler are all known through netting. These would otherwise have been difficult to observe in the field. The netting of Tufted Duck and Masked Finfoot at the Fraser's Hill ringing station show these birds undertake nocturnal migration at high altitudes. Netting has also shown that Blue-winged Pittas form a higher percentage of migrant pittas caught in October while in November, Hooded Pittas form the bulk. This indicates that both species probably migrate at different times though they may share a common breeding range, and demonstrates the importance of even the simplest information such as the date of occurrence.

An extensive programme of ringing over 70,000 Barn Swallows on their communal roosts in Raub and Bentong (Pahang) between 1964 and 1969 has yielded valuable information when some of these were actually recovered on their East Asian breeding grounds. Out of 105 Brown Shrikes ringed at the Sungai Way reedbeds in the 1960s, 14 were retrapped at the same site in subsequent years.

It should be emphasized that mist-netting and ringing are very delicate operations requiring skills and should not be attempted by individuals unfamiliar with them. Mist-netting and ringing should therefore be supervised by trained professionals to minimize casualties due to stress and mishandling. A licence is required without which the trapping of most birds is an offence (see 'Conservation' chapter).

Colour marking

Colour marking is a technique employed to aid easy recognition of birds, particularly herons, egrets, terns, and migratory shorebirds because of their large numbers and also because they are easy to locate on open coastal mudflats and the periphery of freshwater swamps. In 1983 and again in 1990 and 1991, hundreds of shorebirds were netted in Singapore during autumn migration and colour dyed blue. The purpose was to monitor their individual duration of stay as well as local movements between feeding and roosting sites. Most important, it was also to try and plot the extent of their southernmost wintering grounds as well as an estimated duration of time of their arrival at a point, south or north from the location they were colour marked.

Wing-tagging

Wing-tagging has never been carried out in both Peninsular Malaysia and Singapore except in 1988 at a Blue-throated Bee-eater breeding colony at Sungai Buloh (Selangor) by Paul Trevor Green, from the Department of Zoology, University of Malaya. Individuals were netted at their burrows, ringed, and fixed with coloured wing tags with numbers. The purpose was to enable individuals to be visually identified either on their non-breeding range or in other places within Peninsular Malaysia. No wing-tagged individual from the Sungai Buloh colony was ever recovered but an individual ringed there in April 1987 was recovered the following year in southern Sumatra (Green, 1990).

Breeding

The commencement of breeding

The onset of breeding for most resident birds in Peninsular Malaysia and Singapore usually varies from year to year but generally begins around late January or early February, reaching a peak in April or May, before declining gradually in June and July. Breeding is not governed by distinct seasonal changes as in temperate regions. Because of the equatorial climate, rainfall is received throughout the year and so no distinct dry season is experienced as in northern Thailand, Myanmar, and Indo-China where breeding commences with the start of the dry season because of the abundance of insects.

For most species of birds, the commencement of breeding is variable with each individual species but to a large extent coincides with the breeding of insects, food for most resident birds to feed their fledglings. In some species such as the Zebra Dove, breeding takes place throughout the year. Seabirds tend to start breeding in late May or June, and stretching to about September or October. This is probably due to the necessity of avoiding the full force of the monsoons, especially the north-east monsoon (November to March), which bring heavy rain and strong winds, both of which will undoubtedly take a heavy toll on the eggs and young of seabirds which are usually placed in open and exposed areas. A number of herons and raptors are also known to breed mainly during the second half of the year.

The establishment of territories and courtship

The start of breeding is marked by the establishment of territories. In most cases these territories are well defined and variable in size with different species. Territoriality is very marked and pronounced and in most cases accompanied by periods of intense vocalization. Cuckoos call persistently from the canopy, the Yellow-vented Bulbul sings its bubbling cadence from atop bushes or trees while the Large-tailed Nightjar utters its monotonous chonk notes for long periods, usually from the top of fence posts or tree stumps. Singing not only functions to advertise an individual's territory and warn off other males but also serves to attract potential mates.

The Magpie Robin strongly exhibits territoriality. Once a territory has been established, the male sings loudly and melodiously from open perches and will always chase away other males which intrude. The Yellow-vented Bulbul and Paddyfield Pipit are also highly territorial and will chase away intruders. Among non-passerines, raptors are also known to display a strong sense of territorialism. The male White-bellied Sea-eagle is known to harry other intruding males. The male Great Argus Pheasant demonstrates territorialism by establishing a 'dancing ground' on the forest floor and calling frequently. Birds which nest communally like herons, terns, swiftlets, and weavers have relatively small territories within the breeding colony and in the case of herons and terns, often within pecking distance. Pigeons, though gregarious outside breeding, usually nest solitarily (except the Rock Pigeon) and also exhibit territorialism.

Males attract females to an established territory by performing a number of elaborate actions called courtship display. The male Great Argus Pheasant frequently calls on its dancing ground and struts around it. When a female responds, the male performs a variety of elaborate actions by spreading out its wings with tail raised and pointed upwards. The male Paddyfield Pipit engages in a lilting flight, uttering a high-pitched *chik-a-chik-a-chik* on the downward glide. The male Magpie Robin utters a repertoire of loud and melodious musical phrases, often with tail cocked. When a female responds, he will pursue her, often in a dippy flight, low over the ground. When she descends, he will hop towards her, constantly cocking its tail.

Courtship display is enhanced further in some species by a number of attractive physical changes. The resident Great Egret develops long and fine plumes on its back while its bill turns from yellow to black and the facial skin from yellow to blue. The male Common Tailorbird develops a long tail while the Rufescent Prinia develops a dark grey head and dark rufescent brown upperparts (both head and upperparts are dark rufescent brown during non-breeding). The Black-naped Tern develops a rosy flush to its white plumage while the Little Tern develops a black crown with a white forehead, the bill and legs turning yellow from black.

Nesting

Nest building begins once a pair bond has been established. The process itself is very tedious, involving the efforts of both the male and female. A variety of material is used. Raptors, herons, and pigeons use sticks while swifts and swiftlets utilize their own saliva together with aerially gathered material and bits of their feathers to construct their nests. Most passerines use leaves, blades of grass, cobwebs, bark, and lichen. Swallows utilize mud which they gather at puddles. Nests are usually constructed in a variety of sites, usually in inconspicuous places or in inaccessible areas to counter the threat from potential predators. These include the forks and branches of trees, the tips of branches, palm fronds and bamboo, ferns and undergrowth close to the ground, sides of boulders, tree stumps, cliffs, steep banks, tree holes or cavities, burrows along paths or riverbanks, and the eaves of buildings within human dwellings.

Nest shape and size vary with different families and species. Non-passerines tend to construct rudimentary or primitive structures. Pheasants and partridges do not build nests but lay their eggs in mere scrapes on the ground. Nightjars deposit their eggs on the bare ground. Most species of resident terns which nest in colonies lay their eggs on bare rock or sandy ground. A number of non-passerine families like hornbills, rollers, barbets, and woodpeckers are hole nesters, using natural cavities in trees while kingfishers and bee-eaters excavate burrows to lay their eggs. Most raptors build huge nests, consisting of piles of sticks placed on the branches of tall trees or on cliffs. These may be used in subsequent years with more sticks added on to the pile. Falcons do not build nests but lay their eggs on the bare surfaces of cliffs or ledges or in tree holes as in falconets. Pigeons and doves construct flimsy platforms of small sticks, usually placed on branches or in low bushes. Parasitic cuckoos do not construct nests but lay their eggs in the nests of passerines, ejecting the host's eggs in the process. Non-parasitic cuckoos like malkohas and coucals build nests; the former piles up sticks while the latter makes globular structures of vegetation with a side entrance.

Most passerine nests are elaborately woven structures with the exception of the Asian Glossy Starling, mynas, and the Eurasian Tree Sparrow which pile up loose material in untidy heaps in a variety of places from the eaves of buildings and the top of lamp posts to holes in trees. Bulbuls and babblers construct neat cup-shaped structures of dead vegetation, some of which have an external covering of bryophytes; these are usually placed in low vegetation or branches of trees. The nests of ioras are also cup-shaped but are usually covered externally with cobwebs, giving them a whitish appearance. Flycatchers also construct cup-shaped structures which in most cases are externally felted with moss and lichen. Broadbills build huge gourd-shaped structures of dead vegetation, suspended from the tips of branches or palm fronds, in many cases above water. Sunbirds also build similar suspended structures which are much smaller. Pittas build dome-shaped structures of dead vegetation on or near the ground. Orioles and drongos build rather loose cradle-shaped structures which are suspended between the forks of branches. Most tailorbirds stitch large green leaves into tubular structures, building their tiny cup-shaped nests within them. Spiderhunters also stitch their cup-shaped nests on the underside of broad leaves, mainly banana leaves. Prinias and munias construct spherical structures of grass and other vegetation with side openings; prinias place them in tall grass while munias place them in trees. The Baya Weaver is probably the most artistic nest builder, constructing a beautiful and elaborate flask-shaped structure which is usually suspended from branches or the tips of coconut palm fronds.

Egg shape, size, and colour

Avian eggs are usually oval-shaped and vary in size with different birds. Large birds like the larger herons and eagles lay large eggs while white-eyes and flowerpeckers lay tiny eggs. While this may be the case, as a rule, the larger the bird the smaller the egg in proportion to its body size. Hence a Scarlet-backed Flowerpecker lays a proportionately larger egg than a White-bellied Sea-eagle. Clutch sizes are highly variable; raptors and pigeons lay a single egg, sometimes two while for most passerines, the average clutch is three or four.

The colour of eggs also varies from plain pale colours to lightly speckled or heavily blotched with darker colours. Eggs with blotched patterns offer excellent camouflage against potential predators. Hole nesters like kingfishers, bee-eaters, and woodpeckers lay whitish eggs and it is thought that the light colour helps the nesting bird to locate the eggs in the abrupt darkness of the hole or burrow in contrast with the brightness outside. Ground-nesting birds such as nightjars also lay pale coloured eggs. These stand out very conspicuously during the day but in most cases are usually concealed by brooding adults or placed in very concealed positions to avoid detection by predators.

Incubation and fledging

There is considerable variation in the egg incubation of different species of birds; the larger the bird, the longer the incubation period. For example, the White-bellied Sea-eagle may take up to five weeks to incubate its eggs. In the Red Junglefowl, it is up to three weeks while the Yellow-Vented Bulbul only takes two weeks to complete incubation, starting on completion of the clutch. In raptors, incubation usually starts when the first egg is laid. If there are two eggs in a clutch, they will hatch at different times and the older fledgling, because of its superior size and strength will successfully compete for all the food the parents bring. This sibling rivalry usually results in the death of the younger fledgling due to starvation and malnutrition.

Egg incubation is performed mainly by the female because of the presence of a brood patch. In many passerines, however, males have also brood patches and assist in egg incubation. Exceptions are polyandrous species such as the Small and Barred Buttonquails and the Greater Paintedsnipe, where both egg incubation and the raising of the young is carried out entirely by the male as the female in these three species are promiscuous and mate with a number of different males during one breeding season before laying the eggs.

In nidifugous species like pheasants, ducks, and rails, the young have a covering of down and are able to move around and follow the parents to feed. On the other hand, nidicolous birds like birds of prey, woodpeckers, hornbills, and all passerines are naked and helpless on hatching and require constant care by the parents.

Fledglings are constantly fed on an assortment of food by both parents. When the parents approach with food, fledglings respond by quivering their bodies and fully stretching out their necks with gapes widely open. The bright orange-red colour of the gape interior serves to direct the parents to deposit the food into the gapes of the young. Food items vary from fish and meat in birds of prey to small vertebrates in kingfishers. Most passerines feed their young almost exclusively on insects. Pigeons feed their young on a diet of 'pigeon milk', a nutritious fluid with a high fat content from the crop. To avoid predator detection by odour, both parents regularly remove the faecal sacs of the young from the nest. The brooding of the young at night in the nest is usually carried out by the female; the male usually roosting nearby.

The fledgling period that is the time from which the young hatch until they leave the nest varies with different species. In the large raptors, it may stretch up to six weeks or more. Passerines normally fledge within two weeks. In most cases, newly fledged birds have different plumage pattern from adults which they gradually acquire as they mature. Newly fledged birds will still be dependent upon their parents for food for a period of time before they finally learn to forage on their own. Young raptors especially need to be conditioned before they are able to fend for themselves. Prior to actually leaving the nest, young raptors spend time exercising their wings by constant flapping as they watch their parents fly. Hunting skills are also picked up by watching their parents and these are gradually sharpened and perfected as they grow older.

Compiling and collating data on breeding

Data on the breeding biology of individual resident species in both Peninsular Malaysia and Singapore can be better obtained and collated if a systematic and efficient nest record card scheme is regularly maintained. The existing information on nesting we have today are largely the works of G.C. Madoc, A.T. Edgar, E.H. Bromley, B.D. Molesworth, C.A. Gibson-Hill, and in Singapore,

R.J. Spittle. There was also a wealth of information in the now defunct University of Malaya Zoology Department (now School of Biological Sciences) Nest Record Card Scheme.

The nests, eggs, and young of many resident species are often encountered but these are often seldom reported and hence very little information is available on the breeding biology of many species. The nests and eggs of some species such as the Brown-winged Storkbill, Chestnut-tailed Minla, and Black-eared Shrike-babbler from Peninsular Malaysia have not been described although they are known from other parts of their world range. Since egg colour of different races of the same species may vary throughout its world range, it would be very important to discover and describe the nest and eggs of resident species hitherto undiscovered. Even with common species like the Yellow-vented Bulbul, Asian Glossy Starling, Common Myna, and Eurasian Tree Sparrow, much information still needs to be gathered to compile comprehensive data on their breeding biology. While observing and recording information on nesting birds, eggs, and fledglings requires the observer to be present at the nest site, care and skill must be exercised in handling the eggs and young and keeping the nest concealed. Careless handling can induce stress in the parents and young and may expose the nest to predators

The MNS Bird Conservation Council set up a Nest Record Card Scheme in 1999 to gather long-term information on the breeding biology of resident birds. Today it is managed by the Science and Conservation Division of the society. It is hoped that local birdwatchers and those from overseas will contribute to this scheme by filling details of nesting activities on the nest record cards online which may be accessed from the society's website. Details of nesting and breeding may also be submitted to Bird I-Witness (BiW) which helps to compile and collate all data. All records from Singapore should be sent to the Singapore Bird Group of The Nature Society (Singapore). With more data on nesting and breeding, average egg clutch sizes as well as incubation and fledging periods of individual species may be worked out. Behavioural observations as well as information on food items fed to fledglings may also be compiled.

The ringing of resident birds

Like migratory birds, resident birds are also ringed to establish whether populations are sedentary or disperse over wide areas. It also helps to gather information on longevity. Both adults and nestlings are ringed for the purposes of gathering such information. A Black-crowned Night-heron nestling ringed at Kuala Gula in the Matang Mangrove Forest Reserve (Perak) in November 1968 was recovered as an adult 160 months later in Bukit Mertajam (Seberang

Prai) about 50 km north. A Barn Owl ringed as a nestling at Batang Berjuntai (Selangor) on 27 November 1987 was found dead 96 months later at Ulu Bernam (Perak), some 40 km north-west, on 2 December 1995. An adult Yellow-vented Bulbul ringed at Rantau Panjang (Selangor) was retrapped at the same site 127 months later, showing that this species is sedentary.

Birdwatching in Peninsular Malaysia and Singapore

Birdwatching is an interesting and absorbing pastime, requiring both patience and skill to derive the maximum benefit. Birdwatching knows no bounds. A six-year-old novice who makes his or her humble attempts in the garden will soon be amazed at how much can be achieved in terms of gaining knowledge and improving one's skill in observing and listening. A 70-year-old can also begin birdwatching with remarkable achievement too. With true dedication and effort, one will soon expand his or her horizons beyond the backyard. A birdwatcher with a sound knowledge of the avifauna of his or her country can make significant contributions to ornithology by keeping good field notes and records and making these available to scientific and ornithological organizations.

A pair of binoculars—the essential tool

A pair of binoculars is undoubtedly the single most important tool in a birdwatcher's arsenal. It makes birdwatching not only enjoyable but also rewarding. Staring wide-eyed at birds flitting about in the canopy or skulking in the undergrowth will only bring you exhaustion, frustration, and a strained neck.

Your binoculars not only magnify objects but also the essential details needed to ascertain positive identification. A Crested Serpent-eagle perched high in a tree may just appear like a plain brownish blob with the naked eye but on examination through your binoculars, the whitish spots on the breast, the yellow cere, and iris become apparent. A Spectacled Laughingthrush moving through a tangle of ferns at the hill-stations may only offer fleeting glimpses of something darkish but through your binoculars, the orange-yellow bill, the white eyering and wingbar may be picked up, thus proving its identity. Below are some useful pointers on how to choose and use a pair of binoculars with maximum results.

- Invest in the most affordable pair. Avoid cheap pairs with poor-quality optics as these are only meant for general pleasure and not birdwatching. A pair of binoculars with high-quality optics is best as this will allow for details and accurate judgement of colour, all of which are very important in positive identification. A good quality pair can be expensive but it is a lifetime investment if you take good care of it. Make it a point to try out a few pairs from a range before making a purchase. Always consult an experienced friend.

- Always attach a strap to your binoculars and have it around your neck. Avoid very narrow straps as the weight of the binoculars may cause such straps to become tight around the back of your neck, causing skin abrasion and discomfort. Wide straps spread the weight of the binoculars and are more comfortable. Ideally the binoculars should be kept at chest level for quick and efficient use.
- The best binoculars for birdwatching are those which have a combination of moderate magnification and good brightness. A 8 × 40 is therefore a better choice than a 6 × 25. The first number is the factor of magnification while the second number refers to the width of the diameter of the front glass in mm. Ideally the best specifications for use in the rainforest are 8 × 40 and 10 × 40 as these not only allow for a reasonably high magnification but also allow for good light and brightness, as well as a good field of view. Magnification should not exceed 10×, as above this the angle of view becomes much narrower and movements will be apparent because of high magnification, hindering rather than aiding you. A 12 × 50 is therefore unsuitable for birdwatching.
- Choose a pair that is of moderate weight and stable. A very heavy pair will put pressure around your neck, causing discomfort.
- Make sure that your focusing rings are all aligned to ensure that nothing is missed whilst focusing.
- When viewing birds, scan for movements first with your naked eye because your eyes have a better field of view. When you have located something, put your binoculars to your eyes. Whatever you had noticed will suddenly come in view, much enlarged.
- Humidity and heat are enemies of optics in the tropics. Store your binoculars in a dessicator full of silica gel when not in use. Do not forget to regenerate the silica gel once it is saturated. In the field, protect your pair of binoculars from exposure to excessive moisture. Put it in a waterproof bag when it rains or when you are crossing rivers on foot. Although most modern high-quality binoculars have waterproof properties, do not take any chances. You will be amazed at how long your pair will last if you take good care of it.
- Always keep the lenses of your binoculars clean and dust free. Use a blower brush to gently clean the surface of the lens. Do not frequently clean the lens with lens cleaning fluid and tissue as this as this will remove the lens coating, causing the lens to lose its quality.

A spotter scope—an optional tool

A spotter scope is a useful accessory but it is not a necessity. It ranks second in priority to your binoculars. A spotter scope magnifies objects from 20× or more depending on the eyepiece. It is relatively heavy and has to be mounted on a sturdy tripod for good support. It is excellent for use on the mudflats, open country, and the forest edge where lighting conditions are good but in the poorly lit forest, especially on hill slopes, it can be relatively cumbersome.

A spotter scope complements your binoculars; it can never replace it. Some obvious advantages of a spotter scope are:

- It greatly magnifies birds, allowing details to be observed closely.
- It is especially useful for observing and identifying shorebirds and terns on the mudflats.
- It is useful when making prolonged observation of nesting behaviour, particularly when the bird species is shy. With a spotter scope, one can safely observe from a distance without causing any disturbance.

Making your birdwatching successful

To derive the best from birdwatching, the following tips are recommended.

- Dress appropriately in sober colours such as browns, greens, or greys which blend with the environment. Avoid white or bright colours.
- Do not make noise or talk unnecessarily as this may frighten off birds, particularly the shy ones.
- Walk slowly and avoid unnecessary or abrupt movements as these may frighten off birds.
- When birdwatching in the rainforest, keep in small groups of three or four to minimize movements which may scare off birds. Larger groups may be formed when birdwatching at the forest edge or in open country.
- When birdwatching in the forest, it is always best to start early in the morning to make the best of the day. Birds are very active from dawn with peak feeding activities up to about late morning. Activity usually declines by midday when the sun reaches its zenith. Birds become active again in the late afternoon to dusk when they roost for the night.
- When watching shorebirds, make it a point to check the tide times because the feeding and roosting of shorebirds are governed by tide movements.
- In the forest, be alert for birds. Keep your eyes on the trails and the trees above for movement. Where there are breaks in the canopy, look up at the sky to see if there are any birds flying overhead. Be especially alert for the 'bird wave', a phenomenon where insectivorous birds band together and move through the forest, stirring up insects which they quickly feed on. This phenomenon is a characteristic feature of the equatorial rainforest and occurs at all storeys.
- Birds tend to be more easily observed where there are breaks in the vegetation rather than in the forest where the poor light and vegetation make observation difficult. Where roads or paths cut through forest, stick to these as you have better chances of seeing most forest species from here unless if you are specifically seeking partridges, pheasants and pittas which are best observed along forest trails.
- Look for fruiting figs at the forest edge or in the forest as these will be exploited by a variety of frugivorous birds, allowing good opportunities to observe them. Fallen fruit often attract partridges and pheasants which may be observed if one is concealed and waits patiently.

- When you have located a bird, show it to all members of your party. Point gently towards the general location and give specific directions in a low voice. Avoid shouting or screaming in excitement.

Some birdwatching etiquette

When going birdwatching, it is essential that birdwatchers stick to a set code of ethics to ensure the welfare of the birds as well as maximum benefit for everyone. Below are a few pointers to consider seriously.

- Do not be selfish. Be prepared to share your knowledge and experience with others.
- Respect the privacy and property of others. Do not trespass on private property. If you need to, please ask the owner's permission.
- When observing birds nest building, keep at a safe distance as disturbances may cause some birds to desert.
- When observing birds incubating eggs or feeding young at the nest, keep at a safe distance. Do not tamper with the vegetation around the nest as this may expose the nest to predators.
- Though a portable tape recorder is encouraged to record bird calls, playback of calls in the field, particularly territorial songs, often induces stress in birds and should therefore be avoided. Try 'pishing' by making squeaky sounds to entice birds to come closer.

Some considerations on safety when going birdwatching

- When going on overnight trips in the forest, bring along a compass and a basic first aid kit with relevant medication.
- Inform your friends or relatives of your intended date of return. In remote areas, it will be useful to inform the nearest police station of your whereabouts, leaving a list of the members of your party.
- Stick closely to marked trails. If you need to wander away from trails, do not go too far.
- In the forest, wear long-sleeved shirts and long trousers and carry adequate insect repellents. It is a good idea to put on leech socks to prevent leech bites.
- When camping overnight in the forest or in very remote places, sleep under a mosquito net.
- Avoid probing with bare hands under fallen logs or into thick undergrowth to prevent being bitten by venomous snakes. Two species, the Equatorial Spitting Cobra *Naja sumatrana* and the King Cobra *Ophiophagus hannah* are highly venomous and their bites especially the latter's, could be fatal if not treated promptly.

The importance of keeping good field notes

It must be stressed that the forest avifauna of Peninsular Malaysia is still relatively little known. Although scientific expeditions have been organized to Gunung Benom, Endau-Rompin, and Belum-Temengor, there is still much more information to be gathered on the status and distribution, habits, behaviour, and the breeding biology of many resident birds. The behaviour of many wintering species is also poorly known because of the lack of observations. Always keep good field notes on what you observe as these are important sources of information which will be essential in compiling data on every species. Make it a point to publish your observations in journals, magazines, or ornithological bulletins as these will become important sources of reference. Also compare notes with other birdwatchers who have observed the same species as this will help to bridge some of the gaps in your knowledge. Below are some important hints one should seriously consider in keeping and maintaining good field notes and records.

- Always have a notebook handy. Keep it in a waterproof bag.
- Carry a pencil or ballpoint pen. Avoid felt tip or fountain pens as the ink smudges when wet.
- Make a list of all the birds seen or heard at a particular locality. Do not ignore the common ones.
- Make a note of the date, time, and place where your sightings were recorded. It would be a wasted exercise if these are neglected.
- Make detailed notes of unusual or rare sightings. If the bird is calling, try and describe the call accurately. Is it nasal, sonorous, or plaintive? Is it high or low pitched? Is it a single note or a cadence? If you have a portable tape recorder, try and record the call.
- Record all migrants seen as this will be important in compiling data to work out extreme dates of arrival and departure.
- Note the condition of the plumage. Is it fresh or worn out? Is the bird in moult? Is it a male, female, or juvenile? Is it in breeding or non-breeding plumage? Record all other relevant details.
- Record all details of breeding birds, making a note on nesting behaviour, feeding of young and fledging. If nestlings have been taken by a predator, note the predator species.
- Make a note of all kinds of unusual behaviour.

It is imperative that all relevant notes be taken on the spot when details are fresh in one's mind. Do not wait until you get home as you may forget a good deal of information. Your notes are very important in the compilation and collation of data about individual species. Detailed field notes also help you to subsequently identify birds you were unable to immediately recognize in the field.

How to identify birds

Birds may be identified in several ways using several distinctive features such as the jizz or general shape, relative size, plumage colour, voice, and general behaviour. The habitat type as well as the altitude of the area are also essential in ascertaining identification. Diagnostic features of a bird also aid identification. Look for distinguishing features or field marks such as bill shape and colour, supercilium, wingbar, subterminal tail band, tail length and shape, tail streamers, crest, and colour of orbital skin. Very often a conspicuous feature may be remembered and other features overlooked. When checking details in the field guide, an observer becomes aware that the conspicuous feature recorded may be shared by two or even more species within the same family. For example, when viewing green-pigeons (genus *Treron*) at a fruiting tree, an observer may note the orange breast of a male from below and quickly conclude that it is Pink-necked Green-pigeon, based on its relative abundance. However, the male Little and Orange-breasted Green-pigeons also share orange breasts, so the observer has to note other features such as relative size, colour of head and upperparts, tail patterns, and the associated females before coming to a conclusion. The issue may be complicated if two or more species with similar features are feeding in the same tree, so great care has to be taken when making observations.

When an unfamiliar bird or a new species is seen, it is imperative that a relevant sketch together with supporting notes be made in your notebook. Your sketch need not necessarily be elaborate but sufficient to record details such as jizz, length and shape of bill, as well as description of the plumage colour and important field marks. Other details like call, behaviour, habitat, and locality are useful in ascertaining identification. An unknown bird may resemble a known species and in such instances the distinguishing features should be noted to tell them apart. For example, the highly endangered Silvery Woodpigeon, a Sundaic species which used to occur as close as the Riau Archipelago has not been recorded from both Peninsular Malaysia and Singapore. It closely resembles the Pied Imperial-pigeon in plumage colour and pattern and also shares a similar habitat. This species could turn up here and therefore its key features such as the greyish tinge in the white plumage, red orbital skin and red base of bill should be noted (the Pied Imperial-pigeon lacks these). When such relevant features are noted, they can be checked in the field guide for identification.

The leaf-warblers (genus *Phylloscopus*) are sometimes very difficult to identify without close examination. However an observer will gradually become familiar with them once the distinguishing physical features, voice, and habitat are gradually learnt. This is an ongoing process which takes time, patience, and experience. If you are not certain of the identity of a bird, do not hazard a guess.

Voice is a very important feature in the positive identification of birds. While many species may share similar plumage patterns, voice is usually distinctive. Each species has a series of calls from contact notes and alarm calls to territorial songs and these are usually diagnostic. It is essential that the birdwatcher be familiar with calls as this will help in identification when the bird is not seen. Once mastered, calls and songs can almost instantly identify a bird. The Black-capped Babbler which is a skulker is difficult to identify visually in the thick underbrush but when it calls, its identity is clear. There are numerous tape recordings and compact discs of bird calls and songs from Peninsular Malaysia and Singapore. A birdwatcher should acquire these and learn the calls and songs, comparing them to what is heard in the field.

Nightbirds because of their cryptic plumage are often difficult to locate at their daytime roosts and are usually flushed as with nightjars or discovered by accident as in owls and frogmouths. At dusk when they become active, the poor light makes it impossible to see the colour and distinguishing features. Watching and identifying nightbirds is a skill. When a calling nightbird is located, use a powerful torchlight or spotlight to illuminate the bird. Take note of the colour and other features. Please remember that tungsten lighting may not show accurate colour but it is still sufficient to show other details. The colour of the iris may not be a factor here because the light often causes a red reflection in the eyes of nightbirds. For watching and identifying nightbirds, it is best to work in pairs with each person taking turns to operate the spotlight and watching the bird.

Photographing birds

Taking good digital photographs of birds today does not necessarily require elaborate or expensive equipment. Although digital single-lens reflex cameras (DSLRs) produce very high-quality images, good digital images can also be produced using simple digital cameras attached to the eyepieces of spotter scopes. This process, called digiscopy was discovered by the late Laurence Poh Soon Pin, who perfected it to such an art, that reasonably good quality images could be produced. Unlike in the past when film was used, digital images can be viewed instantly. These could be sent to fellow birdwatchers for identification. These images could also be uploaded onto websites for feedback and sharing information.

Since 1998, digital images have provided a wealth of information on the birds in the review area. In Peninsular Malaysia, the first confirmed records of the Little Buttonquail, Pied Cuckoo, Ashy Woodswallow, Red-throated Thrush, and Radde's Warbler, were accompanied by digital images, which provided visual and supporting evidence to the field notes. Besides, digital cameras can also be used in the video mode to record bird behaviour. While photographing birds is greatly encouraged, as a means of gathering more information, it should not be misused to the extent of jeopardizing the welfare of birds. Operate at a safe distance, away from the comfort space of birds.

Submitting records

It is important that detailed notes and relevant sketches accompany descriptions of new species or sightings. All records for Peninsular Malaysia should be submitted to the Records Committee of the Malaysian Nature Society Bird Conservation Council while records for Singapore should be passed on to the Bird Records Committee of the Nature Society (Singapore) (see Appendix 5). Both bodies will deliberate on whether a record should be accepted or rejected depending on the evidence and details submitted. Your description should include the following.

1. Observer's name and witnesses.
2. Name of species seen.
3. Date of observation.
4. Time of day.
5. Location (include a relevant sketch-map if possible).
6. Include details of optics used (binoculars or spotter scope).
7. The habitat, including the estimated elevation.
8. Light and weather conditions.
9. Estimated distance between observer and bird.
10. Detailed physical description of the bird including size, jizz, details of plumage and soft parts.
11. Behavioural notes such as posture, gait, feeding methods, flight, and voice.
12. Description and notes on associated species if any.
13. Additional supporting information including digital images, recordings of calls, or details on weather conditions like storms or unusual environmental conditions like haze from regional forest fires which may hamper visibility and disrupt the regular flight paths of migratory birds.

Key birdwatching sites

Peninsular Malaysia (Map 14)

1. Perlis State Park, Perlis (Lowland dipterocarp and deciduous forest and limestone (5075 ha)

Situated in the Nakawan Range, this park consisting of the Mata Air and Wang Mu forest reserves, borders the Thaleban National Park in Thailand. It has a visitor centre and a good network of trails. Some 200 species of birds have been recorded here and these include Great and Rhinoceros Hornbills,

Streak-breasted Woodpecker, Puff-throated Babbler, Banded, Blue-winged and Hooded Pittas. The watch tower in the park is a good vantage point for watching raptors on passage.

2. Timah Tasoh Reservoir, Perlis (Freshwater lake and surrounding scrub)
The shallow man-made lake which is about 17 km^2 supports a variety of aquatic birds from the Little Grebe and Common Moorhen to Lesser Whistling-duck and Cotton Pygmy goose. Cinnamon and Yellow Bitterns as well as rails and crakes frequent the vegetation at the water's edge. A variety of egrets and the garganey are found here during the winter months. The scrub and overgrown rubber plantations surrounding the reservoir are also worthwhile exploring; the interesting species here include the Lineated Barbet, Streak-breasted Woodpecker, and Puff-throated Babbler.

3. Pulau Langkawi, Kedah (Mangroves, rainforest, and open country) (478 km^2)
The main island together with about 98 other smaller islands form an island group. Much of the islands have been developed for tourism but still have good forest reserves like the Machinchang Forest Reserve and Gunung Raya. Over 220 species of birds have been recorded here. The Mountain Hawk-eagle occurs on Gunung Raya and its foothills and is the only site in the review area for this species. The two other specialities here are the Brown-winged Storkbill in the mangroves and the rarer Black-hooded Oriole, both of which do not occur in mainland Peninsular Malaysia. The Oriental Pied Hornbill is fairly common in the open scrubby areas while the Great Hornbill may be seen in the forested areas. The ricefields are good to see the Red-Wattled Lapwing. Pulau Langkawi is linked by air and sea to Peninsular Malaysia and has a fine network of roads.

4. Ulu Muda Forest Reserve, Kedah (Lowland and hill dipterocarp forest) (115 km^2)
Travel within this reserve is principally by boat along the Sungai Muda. Around 195 species of birds have been recorded in this reserve. Ten species of hornbill, including the Plain-pouched Hornbill have been seen here. The Masked Finfoot has been seen on the lake of the Ulu Muda Dam, while the Dusky Eagle-owl has been seen on the upper reaches of the Sungai Muda.

5. Pulau Perak, Kedah (Bare rock with sparse vegetation)
Situated approximately 148 km north-west of Pulau Pinang, it holds the only known breeding populations of the Brown Booby and Brown Noddy in

Peninsular Malaysian waters. The populations of both species have been drastically reduced by marauding fishermen who poach both the eggs and young. To get there fishing vessels may be hired from Pulau Pinang.

6. Matang Mangrove Forest Reserve, Perak (Coastal mudflats and mangroves) (30 km^2)

This is the largest single expanse of managed mangrove forest in Peninsular Malaysia. The coastal mudflats fringing the mangroves are frequented by thousands of shorebirds and egrets during the winter months while a number of migrant terns also occur along the coast, among them White-winged, Common, and Great Crested Terns. The Brown-headed Gull also occurs in fair numbers during the winter months. Holds the only population of the Milky Stork in the country. Fair numbers of the Lesser Adjutant are also found here. There is also a large colony of Black-crowned Night-herons, just north of Kuala Kurau. The mangrove forests also hold populations of kingfishers, woodpeckers and a number of passerines which include the Great Tit and Mangrove Whistler. Access to the reserve is by road from Kuala Kurau or Simpang Empat Semanggol. Travel within the reserve is by boat which may be hired at Kuala Kurau or Kuala Gula.

7. Royal Belum State Park/Temengor Forest Reserve, Perak (Lowland and hill dipterocarp forest) (1342 km^2)

A large area of largely logged forest near the border with Thailand. It has been the site of a major expedition by the Malaysian Nature Society in 1993. Has a good selection of forest species from pheasants and partridges to barbets, woodpeckers, and a variety of forest-dwelling passerines. Has over ten species of hornbills and is especially good to see the Plain-pouched Hornbill. Has a limited network of trails. Access is by boat from Banding.

8. Bukit Larut (Maxwell's Hill), Perak (Lower montane rainforest)

The oldest hill-station in Peninsular Malaysia with the least development on the environment. An excellent place to watch montane rainforest species which includes a variety of pigeons, babblers, and flycatchers. The Malaysian Partridge may sometimes be seen in the clearings along the main road. There are no extensive systems of trails, so birdwatching will have to be done along the main road. Access to Bukit Larut is by four-wheel drive vehicle from the foot of the hill at Taiping.

9. Tanjung Karang–Sekinchan, Selangor (Ricefields)

An extensive area of ricefields north of Kuala Selangor. Very good for bitterns, shorebirds, terns, and kingfishers. The Black-winged Kite is the common resident raptor. After harvest when the fields are fallow, large numbers of Oriental Pranticole may be seen together with large raptors such as the Greater Spotted and Steppe Eagles. Also good for the Eastern Marsh-harrier and Pied Harrier. A good place to see the White-breasted Woodswallow.

10. Kuala Selangor Nature Park, Selangor (Coastal mudflats, mangroves, and scrub) (2.6 km^2)
Established in 1987, this park is jointly managed by the Malaysian Nature Society and the Selangor State Government. There is a good system of trails within the park. A boardwalk runs through the mangroves to the mudflats where a variety of shorebirds, egrets and sometimes, the Lesser Adjutant may be seen. A good place to see the Mangrove Pitta, Great Tit, and Mangrove Blue-flycatcher. Two observation hides overlook a man-made lake which attracts the Grey Heron (which breeds in the park) Great, Little, and Plumed Egrets as well as the Purple Heron and Chinese Pond-heron. A variety of sunbirds, tailorbirds, and other scrubland passerines may be seen in the scrub. There is also a small resident population of Red Junglefowl in the scrub.

11. Pasoh Forest Reserve, Negeri Sembilan (Lowland dipterocarp forest) (20 km^2)
Comprising largely selectively logged forest with an inner core of primary forest this reserve is an International Biological Programme lowland forest study area and is managed by the Forest Research Institute of Malaysia. This reserve is practically surrounded by oil palm plantations. There is a good network of trails. A great variety of lowland forest species including the Malaysian Peacock-pheasant and the Garnet Pitta occur here. A number of migratory passerines are also seen here during the winter months. It is especially good for owls and frogmouths. Access is by road from the town of Simpang Pertang.

12. Tanjung Tuan, Melaka
A small forested cape with a lighthouse approximately 10 km south of Port Dickson. A good place to watch migrating raptors on both autumn and spring passage. These include the Oriental Honey-buzzard, Black Baza, Grey-faced Buzzard, and Japanese Sparrowhawk. The best times are in the months of September and March on very hot days.

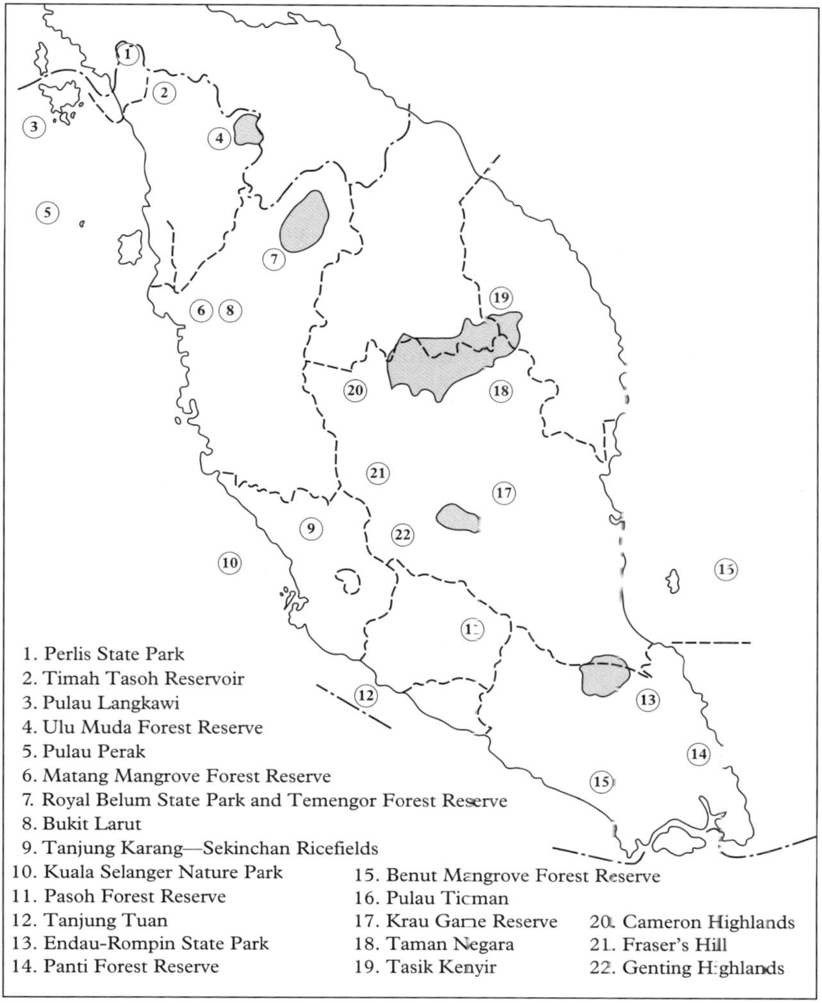

Map 14 Key birdwatching sites in Peninsular Malaysia.

13. Endau-Rompin State Park, Johor/Pahang (Lowland dipterocarp forest) (800 km^2)
Comprises largely logged forest. There are a few trail systems as well as old logging tracks that are good for watching a variety of lowland forest birds. This area has been the site of a major expedition by the Malaysian Nature Society in 1985. Access to the park is by road from Rompin (Pahang) or Kahang (Johor).

14. Panti Forest Reserve, Johor (Lowland and hill dipterocarp forest) (275 km^2)
This reserve comprises largely logged forest and is a good site to watch a variety of lowland rainforest birds. Over 250 species have been recorded here and these include partridges, pheasants, trogons, hornbills, and a variety of resident and migratory passerines. There is a system of trails. Access to the area is via the town of Kota Tinggi.

15. Benut Mangrove Forest Reserve, Johor (Coastal mudflats and mangroves)
A small coastal strip approximately 7 km^2. The extensive mudflats are good for a variety of migratory shorebirds and egrets. Resident birds here include the Grey and Purple Herons as well as a fair number of Lesser Adjutants. The mangrove forests here too have a rich avifauna. Access is from the towns of Benut or Pontian.

16. Pulau Tioman, Pahang (Lowland forest, open sea, and rocky stacks)
With limited development and good forest, some good birds may be seen here. The forests shelter populations of Pied Imperial-pigeons. The White-bellied Sea-eagle and the Brahminy Kite are the dominant raptors. The rocky slets off Tioman are important nesting sites for both the Black-naped and Bridled Terns; their breeding colonies are constantly raided for eggs by fishermen. A good place to see both Christmas Island and Lesser Frigatebirds, both of which roost on Pulau Rengis, a small islet on the west coast.

17. Krau Game Reserve, Pahang (Lowland and hill dipterocarp forest, lower and upper montane forest) (624 km^2)
Comprises largely primary lowland and hill dipterocarp forests with montane forests covering Gunung Benom. Has a good trail system around the reserve headquarters at Kuala Lompat. A good place to see a variety of lowland forest birds. Also good for owls and frogmouths, especially around Kuala Lompat. Access is by road from Kuala Krau, a town along the Mentakab-Jerantut road.

18. Taman Negara, Pahang/Kelantan/Terengganu (Lowland and hill dipterocarp forest, lower and upper montane forest) (4343 km^2)
The largest single expanse of primary lowland dipterocarp forest in Peninsular Malaysia, located in the headwaters of the Tembeling, Relai-Aring-Lebir, and Trenggan river systems. Over 250 species of resident and migratory birds have been recorded here. There is a good network of trails. Very good for pheasants, partridges, barbets, pigeons, babblers, warblers, and flycatchers. The Masked Finfoot may often be seen on the Tahan river during the winter months. Montane residents may be seen on Gunung Tahan which also houses two of Peninsular Malaysia's birds with restricted distribution, the Hill Prinia which inhabits its summit as well as the Crested Argus which is also found on Gunung Perlis, Gunung Gagau and Gunung Mandi Angin. Access to Taman Negara is by boat from Kuala Tembeling in the east and from Merapoh in the west. Both are in Pahang. Taman Negara may also be reached from Tasik Kenyir (Terengganu) and Kuala Koh (Kelantan). All travel within Taman Negara, especially in the Tembeling valley is by boat or on foot.

19. Tasik Kenyir (Terengganu) (Lowland and hill dipterocarp forest)
South-East Asia's largest man-made lake, formed after the creation of a hydroelectric dam on the Sungai Terengganu. Much of the forest surrounding the lake is a mixture of primary and selectively logged forest. The southern portions of the lake fall within Taman Negara. There is no network of trails so birdwatching is best done along the paved roads that pass through the forest and skirt around the lake edge near Gawi Jetty. Over 230 species have been recorded here. It is especially good for a variety of lowland rainforest birds especially raptors which include Wallace's Hawk-eagle, Rufous-bellied Eagle, and Bat Hawk. Both the Lesser and Grey-headed Fish-eagles can be found around the lake as well. The area is also good for hornbills, of which it is possible to see nine species in a single day.

20. Cameron Highlands, Pahang (Settled areas, tea plantations, lower and upper montane forest)
Most of the forested areas contain a variety of montane bird species which sometimes enter gardens in the settled areas. The Silver-eared Mesia, Spectacled Laughingthrush, and Long-tailed Sibia are relatively common. Along rocky streams, the Slaty-backed Forktail may be seen. A good place to see the White-tailed Robin. A variety of upper montane species such as the Golden-throated Barbet, Chestnut-tailed Minla, Yellow-breasted Warbler and

Rufous-vented Niltava and Brown Bullfinch may be seen along the road from Brinchang town to Gunung Batu Brinchang.

21. Fraser's Hill, Pahang (Settled area, lower montane forest)
A good and popular birdwatching destination with numerous opportunities to see a great variety of montane resident species. There is a good network of trails but birds are best observed along the roads. The Mountain Peacock-pheasant and Malaysian Partridge may sometimes be seen along trails. The 8-km road from Fraser's Hill to The Gap (Genting Semangkok) below, offers numerous opportunities to look down into the forest canopy. The forest along this road and at The Gap itself allows opportunities to observe both montane and lowland forest species.

22. Genting Highlands, Pahang (Settled area, lower and upper montane forest)
Much of this area is developed with many high rise buildings but the forest patches around the town centre and below around the Awana Resort are good for birdwatching, with opportunities to see a good variety of montane residents though not as varied or in great numbers as Fraser's Hill. The road leading to the pumphouse is however a good place to see hornbills and a mixture of lowland and montane species. Both the Chestnut-naped and Slaty-backed Forktails may be seen along this road. Further up on Gunung Ulu Kali, which is the highest point here, upper montane species such as the Golden-throated Barbet, Snowy-browed Flycatcher, Chestnut-tailed Minla, and Brown Bullfinch may be observed.

Singapore (Map 15)
1. Central Catchment Nature Reserve (Lowland rainforest) (20 km^2)
More than half of this area consists of primary and mature secondary forest and supports some of Singapore's remnant forest specialists which include the Red-crowned Barbet, Short-tailed Babbler, Chestnut-bellied Malkoha, Blue-eared Kingfisher, and Asian Fairy-bluebird. Also found are the Banded Yellownape, Collared Scops-owl, and Brown Boobook, Crimson and Purple-throated Sunbirds, Grey-headed Fish-eagle, Violet Cuckoo, and Chestnut-winged Babbler. Winter migrants which include a number of cuckoos, flycatchers and warblers are also encountered here.

2. Bukit Timah Nature Reserve (Coastal hill dipterocarp forest) (1.6 km^2)
This reserve consists mostly of primary dipterocarp forest with *Shorea curtisii* dominating. It has a good network of trails. A variety of birds, both resident

and migrant occur here. Resident forest specialities here include species that are nearly extinct locally such as the Scarlet Minivet, Blue-rumped Parrot, Yellow-vented Flowerpecker, and Buff-vented Bulbul. Some of the migrants here include Hooded Pitta, Orange-headed Thrush, Mugimaki Flycatcher, as well as Siberian Blue Robin.

3. Sungei Buloh Wetland Reserve (Mixed coastal habitat) (0.9 km^2)
This site has a good mixture of mangrove, brackish and freshwater ponds, and abandoned agriculture that attracts a good variety of birds. During winter, up to 6000 shorebirds utilize the reserve as a high tide roost. This includes the endangered Asian Dowitcher as well as Black and Bar-tailed Godwits, Broad-billed Sandpiper, Greater Sand Plover besides the various egrets and bitterns. Resident species here include White-browed Crake, Common Storkbill, Sunda Woodpecker, and Copper-throated Sunbird. The rarer Mangrove Whistler and White-chested Babbler also occur in the park. Has a few heronries with nesting Grey, Purple, and Little Herons as well as Black-crowned Night-herons.

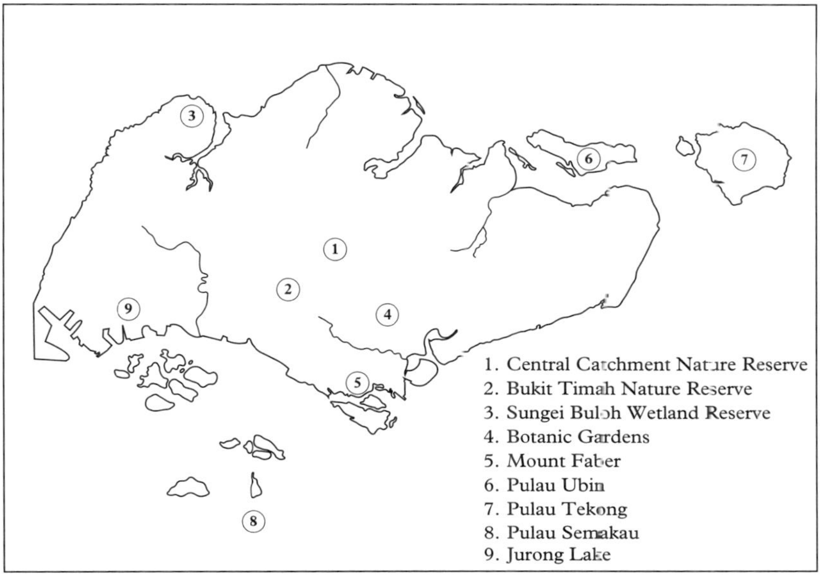

Map 15 Key birdwatching sites in Singapore.

4. Singapore Botanic Gardens (Parkland) (0.3 km^2)
A good place to see a variety of local species such as the Asian Palm-swift, Coppersmith Barbet, Large-tailed Nightjar, Golden-bellied Gerygone, Olive-backed Sunbird, and Scarlet-backed Flowerpecker. Due to the presence of an adjacent woodland patch, some woodland species also occur here such as the Banded Yellownape and Greater Racquet-tailed Drongo. Other species to look out for include Blue-crowned Hanging-parrot, Red-legged Crake, and Spotted Wood-owl.

5. Mount Faber
A high ridge in southern Singapore, covered with secondary forest and overlooks the sea. A very good spot to watch migratory raptors such Oriental Honey-buzzard, Black Baza, and Japanese Sparrowhawk during autumn and spring passage. The best times are hot days with rising thermals.

6. Pulau Ubin (Rural countryside and mangrove) (10 km^2)
This area consists largely of villages, fish ponds, orchards, and abandoned rubber estates. Good mangrove is also found especially around the Sungai Besar. There are dirt tracks leading to all parts of the island. Good residents include Red Junglefowl, Straw-headed Bulbul, Laced Woodpecker, Common Storkbill, Copper-throated Sunbird, Pacific Reef-Egret, Oriental Pied Hornbill, Reddish-cheeked Parakeet, and for the more fortunate Cinnamon-headed Green-pigeon, Mangrove Whistler, and Mangrove Pitta. On the boat crossings, the Great Crested, Lesser Crested, Black-naped, Little and White-winged Terns may be seen.

7. Pulau Tekong (Mangroves) (24 km^2)
Access to the island is limited because of military installations. Holds the largest single expanse of mangrove forest in Singapore and is a good site for typical mangrove species such as the Ruddy Kingfisher, Mangrove Pitta, Mangrove Whistler, and Mangrove Blue-flycatcher. Also a good site for the Black-naped Monarch.

8. Pulau Semakau (Mangroves and shallow reef flats)
A good place in Singapore to watch the Great-billed Heron and the Pacific Reef-egret.

9. Jurong Lake (Public parks and gardens)
A site increasingly known to bird photographers, especially for migrant passerines, with the adjacent lake attracting herons and egrets.

Family and species descriptions

PARTRIDGES, QUAILS, AND PHEASANTS

ORDER *Galliformes* **FAMILY** *Phasianidae*

World: *179 species* **Peninsular Malaysia and Singapore:** *15 species*

A diverse group of small to large terrestrial birds characterized by rather plump bodies, unfeathered tarsi and short strong bills. Sexually dimorphic, the males of most species are brightly coloured and elaborately decorative while the females are dressed in mottled brown and buff camouflage. All are ground nesters while many species roost in trees. The males of many species have elaborate nuptial displays during breeding. Forage on the ground for insects, seeds, and fallen fruit. Being very shy and secretive, the habits of many species are poorly known. Flight strong and direct on short rounded wings over short distances. Prefer to run from danger. Many species in the review area have declined greatly due to loss and fragmentation of lowland rainforest as well as hunting and in global terms, rank between **NT** and **VU** to **EN**; these are stated appropriately in the species account below. Represented by the following groups and genera: **Partridges,** genera *Rhizothera* (1) *Melanoperdix* (2) *Arborophila* (4, 5), *Caloperdix* (6) and *Rollulus* (7). Medium size with plump bodies and short tails. Move in small coveys on the forest floor. Utter distinct whistling calls. Sexes alike or almost so in most species. **Quails,** genus *Coturnix* (3). Small size with proportionately shorter tail and legs than partridges. Inhabit open country. Usually solitary or in pairs. **Gallopheasants:** Large size with bright facial skin and wattles. Have long heavy tails. Males have sharp spurs above hind toe. Genus *Lophura* (9, 10). The males of both species in review area show a flame orange patch on rump and lower back. Generally unobtrusive and quiet but has rather noisy wing-whirring displays. Inhabit forests. Genus *Gallus* (8). Smaller and more slender than *Lophura*. Inhabits forest edge and scrub. **Peacock-pheasants,** genus *Polyplectron* (11, 12). Large size with long wedge-shaped tails. Lack facial skin and wattles. Males have one or more sharp spurs above hind toe and are elaborately marked with iridescent ocelli on the wings and tail. Have loud calls. **Argus Pheasants:** Very large with long tails. Lack both facial wattles and spurs on legs. Males have special 'dancing' grounds to carry out elaborate courtship displays. Have loud and distinct calls. Consists of two monotypic genera. Genus *Rheinardia* (13). Has prominent crest. Male has very broad and long tail, marked with white spots. Genus *Argusianus* (14). Male has large and elongated secondaries which are marked with iridescent ocelli. **Peafowl,** genus *Pavo* (15). Very large with long erect crest. Male has long train, elaborately marked with large iridescent ocelli, which can be spread in a large fan during courtship. Has spurs above hind toe.

1. Long-billed Partridge, *Rhizothera longirostris* Plate 1
Burung Selanting

Description: 36 cm (14 in). Large and robust brown partridge with characteristic large, long and slightly down-curved bill. From *Arborophila* partridges (4–5) by all rusty brown head and throat and the lack of black and white markings in plumage. From Ferruginous Partridge (6) by size and unmarked rufous buff flanks and belly. **Male:** Greyish breast and collar around neck diagnostic. **Female:** All rufous breast and collar. Has brown ear coverts. Crown and nape dark brown in both sexes **Voice:** A series of rising bell-like notes which overlap when uttered antiphonally by pairs, the whole sequence repeated. Contact call is a single soft whistle (Madoc, 1956 and Robinson, 1928). **Range:** Tenasserim (Myanmar) through the Malay Peninsula to Sumatra and Borneo **NT. Status and distribution:** Rare and localized resident from low elevations up to 1500 m on the slopes of Gunung Benom (Pahang) (W1), south to Johor. Frequently heard at the hill-stations, especially Fraser's Hill (Pahang). **Habitat:** Forests. **Habits:** Usually solitary or in pairs, keeping to deep forest gullies and ravines. Frequently duets at dawn, late afternoon, after rain, and dusk, sometimes at night. Roosts at mid storey level. Very shy and secretive.

2. Black Partridge, *Melanoperdix niger* Plate 1
Burung Bertam

Description: 25 cm (9.5 in). Relatively small dark partridge, with deep and stout bill. **Male:** Glossy black plumage with blackish brown wings diagnostic. Bill black. **Female:** Overall dark chestnut with throat and belly whitish buff. Scapulars marked with broad black bars. Bill horn. Legs greyish in both sexes. **Voice:** A disyllabic whistle, similar to Crested Partridge (7). **Range:** The Malay Peninsula, Sumatra, and Borneo. **NT. Status and distribution:** Very rare resident at low elevations, recorded from the Pondok Tanjung Forest Reserve, Larut District (Perak), south to Johor. Locally extinct in Singapore since 1895 (L and G). **Habitat:** Closed canopied primary or mature secondary forest. **Habits:** Usually solitary or in pairs. Shy and secretive. Other habits very poorly known.

3. Blue-breasted Quail, *Coturnix chinensis* Plate 1
Pikau

Description: 15 cm (6 in). **Male:** Black throat, white sides of head, white upper breast with narrow black border, slaty blue head, breast and flanks together with bright chestnut belly and undertail coverts diagnostic. **Female:** From male Barred Button-quail (134) by presence of hind toe, whitish throat and the lack of bright rufous buff flanks, belly and undertail coverts. Plumage buff brown with fine blackish barrings across breast and flanks. Legs yellow in both sexes. **Voice:** A pleasing trisyllabic whistle, *ti-ti-eu, ti-ti-eu*. When flushed, gives a soft *tirru, tirru, tirru* in flight (W1). **Range:** The Indian subcontinent and southern China through South-East Asia, the Greater and Lesser Sundas to New Guinea and Australia. **Status and distribution:** Uncommon and localized resident at low elevations, south to Singapore, including Pulau Ubin and Pulau Tekong (W and H). Individuals presumably undertaking local nocturnal movement were also mist-netted at 1400 m at Fraser's Hill (Pahang). Less common than Barred Buttonquail. **Habitat:** Open country (including cultivation, dry grassland, and marshy areas). **Habits:** Usually solitary or in pairs. Skulks in thick cover and seldom comes out unless flushed; takes off with fast whirring wingbeats but quickly drops back to cover. Partly nocturnal.

4. Malaysian Partridge, *Arborophila campbelli* Plate 1

Sang Seruk Gunung

Description: 25 cm (10 in). Brown upperparts mottled greyish, grey underparts with whitish centre of belly, black and rufous bars on flanks, small white patch on forecrown, larger white patch on cheek and throat together with broad red orbital skin diagnostic. Legs reddish. **Voice:** A rising disyllabic whistle, *wut-wut, wut-wut*... persistently uttered, often antiphonally by pairs. Also low peeping sounds, *wut-wit, wut-wit*... usually uttered by foraging parties. **Range** Endemic to the Malay Peninsula. **Status and distribution:** Uncommon and localized resident above 1000 m in the Larut Hills (Perak) and the Titiwangsa Range, south to Ulu Langat (Selangor). At the hill-stations, may sometimes be observed along forest paths and in gullies. **Habitat:** Forests. **Habits:** Moves on the forest floor in small coveys, usually foraging near thick cover. Sometimes appears along grassy verges of roads or at rubbish dumps at the hill-stations. Shy and will quickly scurry to cover when disturbed.

5. Chestnut-necklaced Partridge, *Arborophila charltonii* Plate 1

Sang Seruk Rimba

Description: 28 cm (11 in). From Malaysian Partridge (4) by diagnostic dark chestnut breast band (bordered above by narrow black band which merges into black side of neck), greenish yellow bill with red base and yellow legs (altitudinal ranges of the two do not overlap). Crown and nape dark brown; throat and narrow supercilium white with fine blackish spots; upperparts brown with fine blackish and buff markings; lower breast and flanks pale tawny buff marked with bold blackish bars. Belly and undertail coverts whitish. **Voice:** A series of loud melodious whistling notes at a steady pitch and volume. Usually uttered in couplets or in triplets, often antiphonally in pairs. **Range:** Southern China and Indo-China south to the Malay Peninsula, Sumatra, and Borneo. **NT. Status and distribution:** Very rare and localized resident at low elevations, recorded south to central Perak and northern Pahang. No recent sightings. **Habitat:** Forests. **Habits:** Forages on the forest in small coveys. Other habits little known.

6. Ferruginous Partridge, *Caloperdix oculeus* Plate 1

Sang Seruk Perang

Description: 26 cm (10.5 in). Dark chestnut crown, rufous buff head with black eye-line and unmarked bright rusty rufescent underparts contrasting with fine black and white scales on upper back and flanks diagnostic. Wings brown with blackish spots. Lower back blackish with small V-shaped rufous markings. Bill black. Legs greenish with short spur above hind toe in the male. **Voice:** A distinct trill, beginning slowly and accelerating (not unlike that of the *Eurylaimus* broadbills (382, 383)) before terminating abruptly in a series of harsher notes; *pi-pi-pi-pi-pipipipipi, di-dwit, di-dwit*... **Range:** Tenasserim (Myanmar) through the Malay Peninsula to Sumatra and Borneo **NT. Status and distribution:** Rare and localized resident from low elevations up to 1200 m, south to Ulu Langat (Selangor). **Habitat:** Forests. **Habits:** Usually solitary or in pairs, preferring thick cover or gullies and ravines. Shy. Very vocal. Other habits little known.

7. Crested Partridge, *Rollulus rouloul* Plate 1

Burung Siul

Description: 26 cm (10.5 in). **Male:** Dark glossy blue plumage with dark brown wings, bright red orbital skin together with broad maroon crest diagnostic. **Female:** Green plumage contrasting with chestnut wings and grey head and throat diagnostic. Both sexes have red legs and sport a long, wispy black crest on forecrown. **Voice:** A series of high pitched upturned whistles, *su-il*, persistently uttered. **Range:** Tenasserim (Myanmar), the Malay Peninsula, Sumatra, and Borneo **NT. Status and distribution:** Locally common resident from low elevations up to 1200 m, south to Johor. **Habitat:** Forests. **Habits:** Moves about in large coveys of up to 15, feeding on grubs amongst litter on the forest floor. Also gathers under fruiting trees to feed on fallen fruit. Calls persistently especially at dawn and during breeding.

8. Red Junglefowl, *Gallus gallus* Plate 2

Ayam Hutan/Ayam Beroga

Description: 42–75 cm (17–30 in) of which 27.5 (11 in) is tail length of male. The evolutionary ancestor of all domestic breeds of chicken. Peterson and Brisbin (1998) on the basis of examined recent and historical museum specimens suggest that the genetically pure breed may be critically endangered or extinct in the natural state. Determining genetic purity would require a thorough knowledge of the morphological, ecological and genetic characteristics of both present-day and historical junglefowl populations (Brisbin *et al.*, 2002). On the basis of morphology, Delacour (1977) recognizes two principal characteristics of genetic purity; a complete moult into an overall dark/black eclipse plumage after breeding and the complete absence of a comb in the adult female. Genetically purer breeds show long, slender, and streamlined body in both sexes. Tail is carried horizontally in both sexes; female sometimes cocks tail vertically in a fan shape. Both sexes also display dark greyish legs and white ear patches (usually absent in very young birds); these features are however not necessarily diagnostic of the pure breed. **Male (Breeding):** Red bare facial skin, comb and lappets below throat distinctive. Bright orange-yellow hackles cover neck, breast and upper back. Lower back maroon contrasting with reddish orange rump. Secondaries largely chestnut; primaries blackish. Underparts dull brownish black. Long black tail has iridescent greenish tinge with white patch at base (conspicuous in flight). **Male (Eclipse) (not illustrated):** Crown and neck blackish, lacking the hackles, tail shorter, comb and lappets, shrivelled and duller red. Individuals may still show remnants of neck hackles in early stages of eclipse moult. **Female:** Crown, head and neck reddish chestnut with dark brown and buff streaks on neck. Upperparts dark olive brown; breast chestnut brown merging into olive-brown belly. Bare facial skin dull red. Comb absent or very rudimentary in purer breeds; more prominent in genetically swamped individuals. Often intra-breeds with domestic chicken especially in rural settlements near forests; the resultant genetically swamped offspring, though retaining much of the plumage characteristics of their wild progenator, are usually larger and plumpish. Genetically swamped offspring may also display the grey legs, white ear patch and white feathers at base of tail and males have larger and more prominent combs and lappets. The tail is carried more vertically. **Voice:** Male crows like domestic rooster, *kruk-kukuu-kuk*, but is sharper and higher pitched with the last note abbreviated. This is usually preceded by flapping of wings. Also utters clucking alarm calls which are higher pitched in females. Young chicks have plaintive peeping notes like domestic chicks. **Range:** North India and Nepal east through southern China throughout South-East Asia, south to Sumatra, Java and Bali. Widely introduced in the Phillippines and the Pacific. Domestic strains are found worldwide. Being highly adaptable to changes in habitat and having withstood hunting pressures through a highly successful breeding rate, wild populations are still not globally threatened. However genetic swamping by domestic varieties remains the single most threat to genetically pure populations. **Status and distribution:** Fairly common resident principally at low elevations up to 1500 m, south to Singapore, but also up to 2031 m, on Gunung Batu Brinchang, Cameron Highlands (Pahang). Pure breeds may still occur in primary and mature secondary forest far from human habitation and domestic chicken. An individual displaying the characteristics of an eclipse male was seen near the summit of Gunung Batu Brinchang, Cameron Highlands, on 15 June 2006 (Yang Chong and Wilbur Goh, pers. comm). Small populations of genetically purer breeds still hold out in isolated wooded areas within the city of Kuala Lumpur, where in the absence of domestic chicken, face no risk of genetic swamping today. Elsewhere throughout Peninsular Malaysia, wild populations in rural areas have constantly come in contact with domestic chicken and have been genetically swamped. Has adapted very well to oil-palm plantations, feeding on fallen palm fruit, and today are more likely to be observed in oil-palm plantations than on the margins of tall forest. In Singapore, populations probably originating from escaped or released captives are confined mainly to Pulau Ubin where it is common (W and H). On mainland Singapore, populations are increasing due to reintroductions and colonization from Johor, utilizing secondary habitats, especially in the Changi area (Yong, 2010). Also occurs at the Sungei Buloh Wetland Reserve. **Habitat:** The forest edge, the mangrove edge, rubber, oil palm and tea plantations, scrub, open grassland. **Habits:** Gregarious, moving about in small parties of up to 10, consisting largely of females and young. Scratches the soil for ants, termites and other grubs. Active especially in the early morning and late afternoon when they forage at forest clearings, retiring to thick cover during the heat of the day. The male is pugnacious, having mating rights over a number of females. Wild males tend to mate with domestic females at the forest edge, the resulting half wild offspring scattering into the forest and further contaminating the pure gene pool by further intra-breeding. Very shy, quickly scurrying for cover at the slightest disturbance. Flight strong and powerful for a pheasant and is usually over a short distance. When pursued, will even fly very high up and perch on tall trees. Roosts high up in trees but nests on the ground. **Note:** The term 'intra-breeding' is preferred to 'inter-breeding or 'hybridizing' because both the wild and domestic varieties belong to a single species.

9. Crestless Fireback, *Lophura erythrophthalma* Plate 2
Ayam Pegar Mata Merah/Merah Mata

Description: 49 cm (20 in). Bare red facial skin and greyish legs distinctive in both sexes. **Male:** Dark glossy blue plumage (tinged greyish on mantle and secondaries) with broad rufous buff tail with blackish base together with flame-orange patch on rump and lower back diagnostic. **Female:** Similar to male but entire body and tail dark glossy blue and no orange on rump. Head brownish and throat buff brown. **Voice:** A repeated low-pitched *tak-takrau* (M and W). Alarm call is a loud *kak*. **Range:** The Malay Peninsula, Sumatra and Borneo **VU. Status and distribution:** Rare and localized resident at low elevations, south to Johor. Locally extinct in Singapore since 1895 (L and G). **Habitat:** Forests. **Habits:** Gregarious, feeding on the forest floor in small parties. During courtship, male gives a loud standing wing-whirring display. Shy. Other habits similar to Crested Fireback (10).

10. Crested Fireback, *Lophura ignita* Plate 2
Ayam Pegar Mata Biru/Ayam Pegar

Description: 67 cm (27 in). From respective sexes of Crestless Fireback (9) by bare blue facial skin, upright crest and red legs. **Male:** Overall dark bluish violet with white streakings on sides of breast and white central tail feathers diagnostic. Patch on lower back flame orange, merging into maroon on rump. **Female:** Overall chestnut brown with fine white streaks on breast. Flanks blackish with white scales. **Voice:** A low vibrant *chukun-chukun* (W1), and a loud *kak* when alarmed. **Range:** Tenasserim (Myanmar) through the Malay Peninsula to Sumatra and Borneo **NT. Status and distribution:** Locally common resident at low elevations, south to Melaka and north-east Johor. Especially common and tame around trails at the Taman Negara Headquarters at Kuala Tahan (Pahang). **Habitat:** Forests. **Habits:** Gregarious, moving around in groups of 5–6. Feeds mainly by scratching the soil for ants and termites. Also feeds on fallen fruit. Generally silent but males often whirr their wings especially during courtship. Highly territorial during breeding.

11. Mountain Peacock-pheasant, *Polyplectron inopinatum* Plate 2
Merak Pongsu Gunung/Kuang Cermin

Description: 45.7–66 cm (18–26 in). Rather nondescript pheasant with chestnut-brown upperparts marked with iridescent green ocelli on wings and tail. Underparts blackish with fine white spots on throat and sides of head. **Male:** Has very superficial crest and long tail. **Female:** Lacks crest, has shorter tail and smaller ocelli. **Voice:** Male utters a series of loud bursts when advertising (W1). Also utters low chicken like whinnying sounds when foraging or loud chicken-like clucks when alarmed (W1), **Range:** Endemic to the Malay Peninsula. Classified as **VU** because of small, fragmented range, and population which is believed to be slowly declining. **Status and distribution:** Rare and localized resident from 820–1600 m in the Titiwangsa Range from Cameron Highlands, south to Genting Highlands (Pahang). Also on Gunung Benom and Gunung Tahan (Pahang). Fairly common at Fraser's Hill (Pahang) especially along forest trails and gullies. **Habitat:** Forests. **Habits:** Forages on the forest floor in small parties, usually keeping to gullies and ravines. Generally shy. When disturbed, will fly up and perch on high branches.

12. Malaysian Peacock-pheasant, *Polyplectron malacense* Plate 2

Merak Pongsu Rimba/Merak Pongsu

Description: 40–53 cm (16–21 in) of which 25 cm (10 in) is tail length of male. From Mountain Peacock-pheasant (11) by prominent glossy blue-green crest and bare orange facial skin (altitudinal ranges of both species do not overlap). **Male:** Upperparts warm dark buff brown marked with blue-green ocelli (narrowly bordered buff) on wings, back and tail. **Female:** Like male but duller with shorter tail. Ocelli smaller and black. **Voice:** Male utters a disyllabic whistle, the first note brief and the second rising. A rasping *tchorrr*, repeated every few seconds for about three minutes (W1). Also an explosive series of cackles which tapers off in a long series of clucking notes. **Range:** Endemic to the Malay Peninsula **VU. Status and distribution:** Uncommon and localized resident at low elevations up to 300 m, south to Negeri Sembilan. Labis (north Johor) was the southernmost limit for Peninsular Malaysia. Locally extinct in Singapore since 1895 (L and G). **Habitat:** Forests. **Habits:** Usually solitary or in pairs. During courtship displays, crest of male often droops over bill, almost covering it. Very vocal especially during breeding. Shy.

13. Crested Argus, *Rheinardia ocellata* Plate 2

Kuang Raya Gunung

Description: Male: 238 cm (94 in) of which 175 cm (69 in) is tail length of male; female 76 cm (30 in). Long bushy crest, pale bill, greyish legs and buff supercilium diagnostic. **Male:** Long crest when erect is predominantly white, resembling a powder-puff; when held flat against head, blackish feathers on forehead conceal most of white parts. Has very large, broad and long tail. Entire crown blackish; throat pale buff. Plumage largely dark blackish brown marked with white spots on upperparts and tail. **Female:** Has much shorter tail and crest; the latter is largely pale buff with superficial strands of black in front. Dark brown plumage, tinged rufescent with black bars (with fine buff vermiculations) on mantle, wings and tail. Upper throat pale buff. **Voice:** A loud and resonant *oo-ki-iau-ki-iau* uttered up to 12 times. Also an 8-note cadence *oo-kia-wau* (W1). Both calls are less powerful than Great Argus (14). Also a low *kok-kok-kok*...(SHY). Alarm call is a loud *owk-owk-owk* (SHY). **Range:** Northern Vietnam and Laos, discontinuously south to the Malay Peninsula. Suspected to be undergoing a rapid population decline due to unsustainable exploitation and reduction of its habitat. **NT. Status and distribution:** Rare and localized resident above 900 m, on the sandstone mountains of Gunung Tahan, Gunung Perlis (Pahang), Gunung Rabong (Kelantan), Gunung Mandi Angin and Gunung Gagau (Terengganu). Also between 850–900 m on the flanks of Gunung Gelemat, on the Eastern Range (Terengganu) (W2, Ismail and Mohd, 1998). **Habitat:** Forests. **Habits:** Usually solitary or in pairs. Male constructs 'dancing' ground like Great Argus to perform courtship displays. While calling, will ruffle head feathers and spread white crest. Vocal both day and night. Other habits in the wild little known.

14. Great Argus, *Argusianus argus* Plate 2
Kuang Raya/Burung Kuang

Description: Male: 200 cm (80 in) of which 140 cm (57 in) is tail length of male; female: 75 cm (30 in). From Crested Argus (13) by much shorter crest, bare blue skin on head and neck, red legs and the lack of pale supercilium. **Male:** Long central tail feathers and lengthened secondaries with buff brown ocelli distinctive. Plumage dark rusty brown (more rufous on underparts) with fine black and buff markings. **Female:** Like male but tail much shorter. Lacks the lengthened secondaries. **Voice:** Territorial call of male is a loud disyllabic *kuau-waau*, the second note longer and on an upward inflection. Both sexes also utter a long cadence of up to 30 evenly spaced *wau* notes, the last few on an ascending scale. **Range:** Tenasserim (Myanmar) through the Malay Peninsula south to Sumatra and Borneo. Besides habitat loss, also affected by hunting **NT. Status and distribution:** Locally common resident from low elevations up to 900 m, south to Johor. Also on Pulau Pangkor (Perak). Locally extinct in Singapore since 1895 (L and G). **Habitat:** Forests. **Habits:** Usually solitary or in pairs. Highly territorial, breeding males construct circular 'dancing' grounds (kept bare of debris) on the forest floor to perform nuptial displays which consists of raising and fanning its large secondaries. Calls both during night and day.

15. Green Peafowl, *Pavo muticus*
Merak Hijau

Description: Male: 245 cm (96 in); Female: 102 cm (40 in). Brilliant green plumage, long and narrow erect crest, blue and yellow facial skin together with pale buff rufous primaries diagnostic. In flight, buff rufous primaries contrast with darker body plumage. **Male:** During breeding, the uppertail coverts are extended into a long, broad train, elaborately marked with large iridescent blue ocelli, surrounded by a broad buff border. This is frequently abraded and lost completely outside breeding. **Female:** Like male but duller and lacks train. **Voice:** The male's territorial call is a loud and far carrying disyllabic trumpet like *kay-yaw, kay-yaw,* uttered persistently. **Range:** North-East India and southern China through continental South-East Asia and the Malay Peninsula discontinuously to Java. Virtually extirpated throughout most of its range because of fragmented population and high hunting levels together with rapid conversion of its habitat for human use **EN. Status and distribution:** At the beginning of the twentieth century, fairly common and widespread resident in the lowlands south to Johor. Excessive hunting and habitat destruction have caused its decline, marked especially during the 1940s. Small populations were known from Kemaman (Terengganu) and Rompin (Pahang) at least in the early 1960s with more recent reports from the Besut District (Terengganu). There have been no sightings since the 1970s and is currently presumed locally extinct. **Habitat:** Forests (especially along riverine areas), coastal scrub, mature plantations. **Habits:** Usually solitary or in pairs, foraging in open grassy or wooded areas. The male has elaborate courtship displays, fanning out its huge train. Very territorial and will chase away other intruding males. Very vocal especially at dawn and dusk. Roosts in tall trees. **Note:** Not illustrated in colour (see figure).

Green Peafowl, *Pavo muticus* (15)

DUCKS

ORDER *Anseriformes* **FAMILY** *Anatidae*

World: *161 species* **Peninsular Malaysia and Singapore:** *11 species*

(This cosmopolitan family also includes swans and geese which do not occur in the review area. For this purpose only the ducks are described below.)

Small to medium sized water birds with plump bodies, broad flattened bills, narrow pointed wings, short tails and short legs with fully webbed feet, making them strong swimmers. Flight strong and direct with moderate to rapid continuous wing beats and fully extended necks. Strictly aquatic, obtaining food in shallow water by surface feeding or diving; some species graze on banks of wetlands. Sexually dimorphic in many species, the males have bright breeding plumage; outside breeding they have an eclipse plumage which is similar to the sober coloured females. The males of many temperate breeding species are mostly in eclipse plumage in tropical wintering quarters. Gregarious, gathering in large flocks where common. Represented by the following groups and genera: **Whistling-ducks,** genus *Dendrocygna* (16, 17). Whistling-ducks differ from typical ducks in having longer necks and squarish heads, longer legs, broader and rounded wings, and a distinctive upright goose-like posture when alert. Tails held above water. Take off vertically from water surface. In flight, fully extended necks droop below main axis of body with back hunched and legs trailing beyond tail. Flight slower than typical ducks with moderately swift flapping. Calls consist of loud plaintive whistling notes, frequently uttered in flight. Whistling-ducks are principally herbivorous, feeding on vegetable matter sieved from shallow water but will also take small aquatic vertebrates and invertebrates. Highly gregarious, usually congregating in large flocks. Usually nest on floating vegetation and tree holes. Sexes alike **Perching ducks:** Surface feeders in shallow water. Perch on trees. Tail held above water. Take off vertically from water surface.

Genus *Cairina* (18). Large size and heavy bodied inhabitants of forested water bodies. Genus *Nettapus* (19). Small size with largely green and white plumage. **Dabbling ducks**, genus *Anas* (20–25). Elongated bodies. Surface feeders, dabbling in shallow water, tipping tail-up to reach aquatic plants and other food material. Many species have coloured speculum at trailing edge of wing, conspicuous in flight. Take off vertically from water surface. **Diving ducks:**, genus *Aythya* (26). Shorter bodies than *Anas*. Tail held close to water surface. Feed in deep water by diving. Patter along water surface before taking off.

16. Wandering Whistling-duck, *Dendrocygna arcuata* Plate 3

Belibis Merantau

Description: 45 cm (18 in). From Lesser Whistling-duck (17) by larger size and more robust build, richer chestnut flanks with prominent creamy white flank plumes marked with dark streaks, blackish spots on breast and the lack of the narrow yellowish eyering. Blackish-brown crown extends down to eye level, contrasting with rufous-buff face and neck. Has broad blackish band running down from crown through nape and entire hindneck. In flight, shows white or fawn U-shaped patch across rump; in the Lesser, the white or fawn markings are limited to either side of the tail. Bill and legs black. **Juvenile (not illustrated):** Plumage duller with paler belly, white flank markings less distinct. **Voice:** A series of shrill twittering whistling notes and high pitched whistles, distinct from those of the Lesser Whistling-duck and uttered mostly in flight. **Range:** The Philippines, Borneo, Java and the Lesser Sundas, and Sulawesi through New Guinea, south to northern Australia and the western Pacific. **Status and distribution:** In Singapore, scarce feral resident, originating from free-flying stock from the Jurong Bird Park with the first record from Marina South on 1 October 1989 (W and H) where breeding was recorded. With the subsequent clearing of Marina South, populations declined with the last few birds confined to Serangoon. Also at Marina City Park and the Singapore Botanic Gardens (W and H). In Peninsular Malaysia, sightings of 30 at Pinji Estate, Ipoh (Perak) on 4 May 1999 and two at Bidor (Perak) on 9 February 2010 (BiW) need to be assessed further as their origins are unknown. **Habitat:** In review area, freshwater marshes, ponds, and disused tin mining pools. **Habits:** Similar to Lesser Whistling-duck with which it associates. **Note:** W and H have included this species on the Singapore list.

17. Lesser Whistling-duck, *Dendrocygna javanica* Plate 3

Belbis Kecil

Description: 41 cm (16 in). Crown and nape dusky grey; face, neck and breast greyish buff grading into tawny rufous on belly and flanks; upper flanks creamy; undertail coverts white; wing coverts and tail dark chestnut contrasting sharply with dark grey mantle with narrow rufous bars especially in upperwing flight pattern. Dark underwing contrasts sharply with rufous underparts in flight. Bill and legs dark grey. Has narrow yellowish eyering, discernible at close range. **Juvenile (not illustrated):** Duller and paler. **Voice:** A high pitched wheezy plaintive *whi-whee*, uttered persistently especially in flight. **Range:** India (including the Andaman and

Nicobar Islands) and southern China through South-East Asia, the Greater and Lesser Sundas east to Sumbawa and north-east to the Ryukyus. **Status and distribution:** The most common wild duck in review area. In Peninsular Malaysia, scarce to locally common resident at low elevations. In Singapore, rare resident, often supplemented by free-flying stock from the Jurong Bird Park. Distribution much reduced due to the filling in of freshwater ponds and former tin mining pools together with shrinking small-scale rice economies. Illegal hunting also takes a substantial toll. **Habitat:** Disused mining pools, flooded ricefields, freshwater ponds. Usually absent from saline or brackish wetlands. **Habits:** Where common, gathers in groups of up to 40 or more; up to 82 at Kampung Teluk Rimba, Muar (Johor) on 5 November 2008 (BiW). Otherwise small flocks of up to 10 would be common at present times. To an extent, nocturnal, feeding largely on small aquatic plants. Will also graze on land near the water edge. Perches on trees. Bobs head up and down when alarmed. Readily takes to flight when disturbed, circling and calling before alighting further away. Shy where persecuted.

18. White-winged Duck, *Cairina scutulata* — Plate 3

Serati Hutan

Description: 76 cm (30 in). Large size. Sexes virtually alike; female smaller and duller. At rest, white head (heavily spotted with black on crown and hind neck), orange-yellow bill and legs, white scapular patch and pale grey speculum diagnostic. Upperparts and breast black with greenish sheen; flanks dull brown; iris red. In upperwing flight pattern, broad pale grey speculum contrasts with black flight feathers and white wing coverts. **Voice:** No information from review area. **Range:** North-East India through South-East Asia discontinuously to Sumatra and Java. Declining greatly due to its small fragmented population and loss and disturbance to riverine habitats **EN.** **Status and distribution:** In Peninsular Malaysia, vagrant, known from a single record (date unknown but prior to 1900), said to be collected near Ipoh (Perak). **Habitat:** In the Malay Peninsula, seasonal freshwater swamp and flooded rice fields at the margin of forests (W1). **Habits:** Unknown in review area.

19. Cotton Pygmy-goose, *Nettapus coromandelianus* — Plate 3

Itik Kapas /Itik Air /Belabas

Description: 33 cm (13 in). The smallest duck in review area. **Male breeding:** Glossy dark green upperparts and band around breast and neck diagnostic. Crown black; head, neck and underparts white with fine black vermiculations on flanks; vent black; iris red. In flight, broad white trailing edge of wing contrasts with black underwing and green upperwing. **Female:** Dark narrow eyeline distinctive. Crown brownish; upperparts dull brown; band around neck, breast and neck scaled blackish; faint, narrow brownish bars on foreneck and flanks; iris brown. In flight, underwing dark with very thin white trailing edge of wing; upperwing brown with green bar on wing coverts. **Eclipse Male** (not illustrated): Like female but underwing pattern same as breeding male. Bill and legs blackish in male, brownish grey in female. **Juvenile (not illustrated):** Like female but duller with broader eyeline. **Voice:** Male utters a soft grating crackle, female soft nasal quacks. **Range:** India and China through South-East

Asia to New Guinea and Australia. **Status and distribution:** In Peninsular Malaysia, uncommon and localized resident throughout at low elevations. Occasionally common in certain localities; up to 500 at Timah Tasoh reservoir (Perlis) in January 1996. In Singapore, rare non-breeding dispersant, usually between November and July; the largest count being 35 at the Poyan reservoir on 16 May 1992. Distribution reduced due to the filling in of ponds and former mining pools. **Habitat:** Disused mining pools, freshwater ponds, flooded ricefields, large forested river courses. **Habits:** Feeds in the shallows mainly on seeds and buds of aquatic plants. Sometimes perches on trees. Shy, taking to flight when disturbed. During courtship, a few males may indulge in noisy display flights and pursue a single female.

20. Gadwall, *Anas strepera* — Plate 3

Itik Kelabu

Description: 50 cm (20 in). **Male breeding:** Dark greyish plumage with all black bill distinctive. Neck greyish brown; breast boldly marked with blackish crescents; flanks finely marked with black vermiculations; belly white; tail grey; uppertail and undertail coverts and patch near vent black. In upperwing flight pattern, white speculum contrasts with black and chestnut bars. **Female:** Brown with dark brown markings on back and flanks. Bill black with orange-yellow sides. **Eclipse Male (not illustrated):** Like female (including bill colour) but darker. Legs orange-yellow in both sexes. **Voice:** No information from review area. **Range:** Breeds in northern temperate North America, Europe and Asia. Winters in southern temperate zones as well as tropical South America, Africa, South and South-East Asia. **Status and distribution:** Vagrant in Singapore, a female at Punggol on 19 January 1989. **Habitat:** Freshwater ponds. **Habits:** Keeps close to waterside vegetation when feeding.

21. Eurasian Wigeon, *Anas penelope* — Plate 3

Itik Bersiul

Description: 48 cm (19 in). At rest short neck, rounded head and small bill distinctive. **Male breeding:** Chestnut head and neck, yellow forehead and crown and vinous breast diagnostic. Upperparts and flanks grey with fine black vermiculations; folded wing coverts and patch bordering vent white; vent black. In upperwing flight pattern shows green speculum and broad white rectangular wing coverts. **Female:** Deep brown with greyish brown wing coverts. In upperwing flight pattern, greyish brown rectangular wing coverts conspicuous. **Eclipse Male (not illustrated):** Like female but wing coverts white. Legs dark grey, bill light bluish grey with dark tip in both sexes. **Voice:** Male utters a loud melodious whistle, *wheeoh,* female a growling *warr, warr, warr*... **Range:** Breeds throughout northern temperate Europe and Asia. Winters in southern temperate zones as well as tropical Africa, South and South-East Asia to Borneo, the Philippines and Sulawesi. **Status and distribution:** Vagrant in Peninsular Malaysia, recorded on the Matang coast and Batu Gajah (Perak) and Mersing (Johor). Vagrant in Singapore, a female at Sungei Buloh from December 1986 to February 1987. **Habitat:** Disused mining pools, freshwater ponds. **Habits:** Keeps to shallow waterside vegetation, feeding on small aquatic plants. Will also graze on land at the water edge.

22. Northern Shoveler, *Anas clypeata*

Plate 3

Itik Paruh Bajak

Description: 50 cm (20 in). Heavy build. From other ducks by large spatulate bill. **Male breeding:** Black bill, dark glossy green head, chestnut belly and flanks diagnostic. Upperparts white with black markings; breast, sides of vent scapulars and tail white; vent black; iris yellow. In upperwing flight pattern, shows green speculum with narrow bar above bordering broad bluish grey wing coverts. **Female:** Brown plumage with darker brown markings on mantle, breast and flanks. Narrow eyeline black; bill blackish with orange sides; iris dark. **Eclipse male: (not illustrated):** Like female but iris yellow with breast and flanks tinged rusty buff. Legs orange in both sexes. **Voice:** Usually silent in wintering quarters. **Range:** Breeds in northern temperate North America, Europe and Asia. Winters in southern temperate zones as well as tropical South America, Africa, South and South-East Asia to Borneo and the Philippines. **Status and distribution:** Rare winter visitor at low elevations, south to Singapore, with records from Serangoon and Lorong Halus. **Habitat:** In review area, disused mining pools, freshwater ponds, brackish prawn ponds, coastal mudflats. **Habits:** Feeds in very shallow water, using its bill to sift through the mud for minute aquatic animals and seeds. Will also feed on small aquatic plants.

23. Northern Pintail, *Anas acuta*

Plate 3

Itik Ekor Runcing

Description: 53 cm (21 in). Slender body with relatively long neck and pointed tail distinctive. From eclipse male and female Gadwall (20) by darker bill, longer neck and longer pointed tail. **Male breeding:** Long black tail, chocolate brown head and hind neck with narrow white stripe from breast to side of head diagnostic. Back and flanks grey; foreneck, breast and underparts white; undertail coverts black; sides of vent creamy yellow; bill black with bluish sides. In upperwing flight pattern (in both breeding and eclipse plumages) shows green speculum with narrow buff bar in front and white bar behind. **Female:** Brown with dark brown markings on back. In flight shows brown speculum with white bar behind. **Eclipse male (not illustrated):** Like female but darker. Sides of bill bluish as in breeding. Legs grey in both sexes. **Voice:** Male utters a weak rolling *proop-proop*, female a low *quack* (Extralimital data). **Range:** Breeds throughout northern temperate North America, Europe and Asia. Winters in southern temperate zones as well as tropical South America, Africa, South and South-East Asia to Java, Borneo and the Philippines. **Status and distribution:** Rare winter visitor at low elevations, south to Singapore. **Habitat:** Disused mining pools, freshwater ponds. **Habits:** Keeps to shallow water, feeding largely on minute aquatic animals and plants.

24. Garganey, *Anas querquedula*

Plate 3

Itik Kening Putih

Description: 41 cm (16 in). Small size. From female and eclipse male Eurasian Teal (25) by narrow pale supercilium and narrow white patch at base of bill. **Male breeding:**

Broad white supercilium curving over eye to nape diagnostic. Entire cap (including hindcrown) dark brown, head purple-brown with fine white streaks; belly white with fine grey vermiculations on flanks; wing coverts blue-grey. In upperwing flight pattern shows green speculum with conspicuous white bars on either side contrasting with blue-grey wing coverts in front. **Female:** Black eyeline. Upperparts mottled brown; wing coverts greyish brown. **Eclipse male: (not illustrated):** Like female but wing coverts blue-grey. **Voice:** Male utters a crackling *rrrp,* female low quacks. **Range:** Breeds in northern temperate Europe and Asia. Winters in southern temperate zones as well as tropical Africa, South and South-East Asia to New Guinea and Australia. **Status and distribution:** Uncommon and localized winter visitor at low elevations, south to Singapore. The most numerous migratory duck. **Habitat:** Disused mining pools, freshwater ponds, sea coasts. **Habits:** Feeds in small groups in shallow water, often together with Lesser Whistling-duck (17). Shy. Flies fast with rapid wing beats. Often rests on sea during migration.

25. Eurasian Teal, *Anas crecca* Plate 3

Itik Kecil

Description: 37 cm (15 in). Small size and short bill distinctive. **Male breeding:** Pale greyish body with fine black vermiculations on flanks. Rusty red head with broad curving iridescent dark green patch and white horizontal scapular line diagnostic. Breast pale buff with fine dark speckles; patch near vent cream yellow with narrow black border; bill and legs black. **Female:** Mottled brown with dark eyeline, white belly. **Eclipse male (not Illustrated):** Like female but eyeline less distinct. Both sexes show a green speculum in flight. Bill and legs dark grey. **Voice:** Male utters a short ringing *kreek*, female a high pitched nasal *quack* (Extralimital data). **Range:** Breeds throughout northern temperate North America, Europe and Asia. Winters in southern temperate zones as well as tropical South America, Africa, South and South East Asia to the Philippines. **Status and distribution:** Vagrant with two confirmed records; a female shot near Kuala Lumpur in April 1912 and one with Garganeys (24) on the Jurong River (Singapore) on 29 October, 1978 (W1, W and H). **Habitat:** Disused mining pools, freshwater ponds, rivers. **Habits:** Flies very fast with ease and dexterity, twisting and turning like shorebirds. Feeds close to the water edge.

26. Tufted Duck, *Aythya fuligula* Plate 3

Itik Jambul

Description: 40 cm (16 in). Drooping nuchal crest usually indistinct in field. **Male breeding:** Glossy black head, black breast, upperparts and wing coverts contrast sharply with white flanks, belly and undertail coverts. **Female:** Black parts replaced by dull dark brown, white parts by pale brown, white base of bill. **Eclipse male (not illustrated):** Flanks mottled greyish white. In both sexes, iris yellow, bill pale grey with dark tip, legs dark grey. In flight both sexes show broad white bar on trailing edge of wing. **Voice:** Silent in wintering quarters. **Range:** Breeds throughout northern temperate Europe and Asia. Winters in southern temperate zones as well as tropical Africa, South and South-East Asia to Borneo and the Philippines. **Status and distribution:** Vagrant, with four confirmed records; a female at Batu Berendam

(Melaka) from 3–7 February 1972, a female mist-netted at 1400 m at Fraser's Hill (Pahang) on 10 December 1975, one at Pinji Estate, Ipoh (Perak) on 16 January, 1999, and one at Tanah Merah (Singapore) on 25 December, 1999. **Habitat:** In review area, freshwater ponds, disused mining pools, sandy mudflats. **Habits:** Feeds in deeper water than other ducks. Obtains buds and seeds of submerged aquatic plants by diving. Seldom flies off when disturbed, preferring to drift out to deeper parts in centre of water body.

PETRELS AND SHEARWATERS

ORDER *Procellariiformes* **FAMILY** *Procellariidae*

World: *79 species* **Peninsular Malaysia and Singapore:** *4 species*

A distinct group of seabirds characterized by sleek cigar-shaped bodies, long narrow wings, short necks and short rounded or wedge-shaped tails. Bill slender and hooked at tip with tubular external nostrils above; the latter an adaptation to excrete excess salt ingested by drinking saline water. Feet webbed. Flight pattern distinctive with sporadic rapid stiff-winged flaps interspersed with long periods of banking, wheeling, and gliding very low over the ocean surface. Most species are clad in dark shades of brown or grey above and white below. Some are wholly dark. Feed largely on small squid, fish, crustaceans, cephalopods and plankton, obtained on the water surface or by diving. Often rest on water surface where their shorter necks and shorter and less protruding tail and wingtips distinguish them from dark morph jaegers (212–214). Take off by pattering along the water surface. Strictly pelagic, they spend most of the time at sea outside breeding, descending on oceanic islets to breed in burrows or on rocky ledges. Sexes alike. Gregarious. Individuals sometimes get blown off course into coastal waters or land by storms. Field identification of this group may pose difficulties and great caution has to be taken to correctly identify a species. For more information on critical identification, see Enticott and Tipling (1996), Harrison (1997), and Onley and Scofield (2007).

27. Streaked Shearwater, *Calonectris leucomelas* Plate 4

Description: 48 cm (19 in). Larger than other shearwaters in review area. Distinguished from them by pale head contrasting with dark brownish streaks (perceivable only at close range) on hindcrown and nape. Upperparts greyish brown; underparts white contrasting with darker primaries, wingtips and carpal area. Bill horn-coloured with darker tip; legs flesh-coloured. **Voice:** No information from review area. **Range:** Breeding on islands in temperate North-West Pacific; wintering south to the tropical Pacific and adjacent seas. **Status and distribution:** Vagrant, with sightings of flocks in the Straits of Melaka on 8 December 1973 and 25 January 1980, far off the coasts of Selangor and Perak respectively. **Habitat:** Oceans, open sea. **Habits:** Flies over the ocean surface with leisurely flapping, sometimes soars with stiff wings in windy conditions. Does not follow fishing vessels.

28. Wedge-tailed Shearwater, *Puffinus pacificus* Plate 4

Description: 43 cm (17 in). Long wedge-shaped tail and flesh-coloured legs distinctive. Toes do not extend beyond tail in flight. **Pale morph (not illustrated):** From Streaked Shearwater (27) by smaller size and proportionately longer tail. Upperparts (including head and sides of neck) dark brown. Underparts all white except dark tip of tail. Underwing all white with dark margins. Bill pale with dark tip. **Dark morph:** Overall uniformly sooty brown, including underwing. Bill greyish. See Short-tailed Shearwater (29) **Voice:** No information from review area. **Range:** Breeding on islands in tropical Indian and Pacific Oceans and adjacent seas. **Status and distribution:** Rare non-breeding visitor; about 100 pale morph individuals in the Straits of Melaka (exact location unknown) in January 1980. In Singapore, vagrant, a dark morph individual inland at Woodlands on 22 June 1998 (W and H). **Habitat:** Oceans, open sea. **Habits:** Like other shearwaters, skims low over the ocean surface in long glides, making swoops and banking from side to side. Follows fishing vessels. Congregates with other seabirds while foraging.

29. Short-tailed Shearwater, *Puffinus tenuirostris* Plate 4

Description: 41 cm (16 in). More slender than other shearwaters in review area. From dark morph Wedge-tailed Shearwater (28) by smaller size, more narrow-winged with much shorter tail which is square or rounded with toes extending beyond tail tip. Underwing has pale markings. Upperparts dark brown with whitish chin with paler breast and belly. Shows conspicuous white markings along centre of underwing which shows contrast with dark base and border. **Voice:** No information from review area. **Range:** Breeds off south-east Australia (including Tasmania), dispersing northwards to the Pacific ocean and adjacent seas during the southern hemisphere winter. **Status and distribution:** Very rare non-breeding visitor; 10 off the coast of Tanjung Dawai (Kedah) on 10 and 16 May, 2010. **Habitat:** Oceans, open sea. **Habits:** Usually gathers in small to large groups, sometimes resting on the water surface. Observed using wings, to propel itself underwater (Bakewell, 2010; RCS). Follows fishing vessels and visits trawler nets. Other habits like Wedge-tailed Shearwater.

30. Bulwer's Petrel, *Bulweria bulwerii* Plate 4

Description: 27 cm (10.5 in). Overall sooty brown with whitish band on secondaries, conspicuous in upperwing flight pattern. Bill all black; legs flesh-coloured. From Swinhoe's Storm-petrel (32) by larger size, long wedge-shaped tail and typical petrel flight pattern. Brown Noddy (211) has whitish crown, pointed bill and lacks pale markings on upperwing. **Voice:** No information from review area. **Range:** Breeding on islands in the Pacific and Atlantic Oceans; Pacific populations regularly disperse southwards, reaching the Indian Ocean and adjacent seas. **Status and distribution:** Very rare non-breeding visitor; 15 south-west of Pulau Tioman (Pahang) on 5 May 1986. **Habitat:** Oceans, open sea. **Habits:** Usually in small groups. Skims low over the ocean surface in an erratic flight, zig-zagging, gliding and making swoops.

STORM-PETRELS

ORDER *Procellariiformes* **FAMILY** *Hydrobatidae*

World: *21 species* **Peninsular Malaysia and Singapore:** *2 species*

A distinct group of largely blackish seabirds, similar to shearwaters and petrels 27–30) but much smaller and shorter-winged with a more fluttery flight pattern. Bill small and slender with terminal hook and conspicuous tubular nostrils on the basal part of the culmen, which are fused together and have a single opening. Readily distinguished from other seabirds by erratic fluttery flight low over the ocean surface with webbed feet pattering the water. Tails square-cut or forked. Wings long with pale stripe on wing coverts. Some species have white rumps. Feeding is largely carried out on the wing and prey items include small squid and crustaceans which are picked up from the water surface. Outside breeding, strictly pelagic. Breed in rock crevices on oceanic islands. Often follow ships. Many species are extremely difficult to identify and therefore close observation of tail and wing patterns as well as colour of rump would be extremely useful (see Enticott and Tipling (1996) and Harrison (1997) for more information on identification). Gregarious. Sexes alike.

31. Wilson's Storm-petrel, *Oceanites oceanicus* Plate 4

Description: 17 cm (7 in). Very small. Plumage sooty black with conspicuous white rump, black square-cut tail and yellow webbed feet (disernible only at very close range and in the hand) diagnostic. Proportionately long legs extend slightly beyond tail in flight, giving the impression of a short streamer. **Voice** Usually silent at sea. Sometimes soft peeping notes by feeding birds. **Range:** Breeding in the Antarctic; dispersing northwards to temperate and tropical oceans worldwide. **Status and distribution:** A rare and localized non-breeding visitor to Peninsular Malaysian waters, recorded only in the Straits of Melaka off Pulau Pinang on 20 and 22 July, near the Arua islands, off Selangor on 19 and 30 September, and between mainland Kedah and Pulau Langkawi on 12 December (W1). **Habitat:** Oceans, open sea. **Habits:** Usually in small groups, making short glides, fluttering and moving side to side low over the water surface. Hovers when feeding, pattering along the water surface. Frequently follows ships.

32. Swinhoe's Storm-petrel, *Oceanodroma monorhis* Plate 4

Description: 20 cm (8 in). Very small. Plumage overall sooty brown with obscure pale band on secondaries (See Bulwer's Petrel (30)).Webbed feet dark. From Wilson's Storm-petrel (31) by moderately forked tail and all dark rump. **Voice:** Usually silent in wintering quarters. **Range:** Breeding on islands of the temperate North-East Pacific, off northern China, Korea and Japan; wintering south to the South China Sea and tropical Indian Ocean. **Status and distribution:** A rare and localized but occasionally numerous passage migrant through the waters of Peninsular Malaysia and Singapore. Large flocks, estimated at the rate of about 1000 per hour have been observed passing through the Straits of Singapore in early September (W1; W and H). **Habitat:** Oceans, open sea. **Habits:** Usually in small groups. Flight pattern and feeding behaviour distinguish it from Wilson's Storm-petrel; flight pattern consists of swoops and bounds in the manner of terns. Does not patter along water surface. Follows ships less frequently.

GREBES

ORDER *Podicipediformes* **FAMILY** *Podicipedidae*

World: *21 species* **Peninsular Malaysia and Singapore:** *1 species*

Small and duck-like with pointed bills, short rounded wings and very short tails which give them a 'tailless' appearance. Strictly aquatic, they have broadly lobed feet which enable them to swim with ease above and below the water surface. Feed on fish and small aquatic vertebrates and invertebrates. Efficient divers, often submerging to escape danger. Flight is weak and direct with rapid wingbeats, undertaken after pattering along the water surface. Both neck and legs are outstretched in flight. Sexes alike. Often in small groups.

33. Little Grebe, *Tachybaptus ruficollis* Plate 3

Description: 25 cm (10 in). Smallest duck-like bird with rather stocky body and characteristic pointed bill. Yellow iris and spot at base of bill diagnostic. **Breeding:** Sides of head and neck bright rufous-chestnut. **Non-breeding:** Crown and upperparts greyish brown contrasting with sandy-buff sides of head and flanks. **Juvenile:** Like non-breeding but throat paler with bold blackish stripes on head and neck. **Upperwing flight pattern:** All greyish-brown upperparts with narrow white trailing edge to secondaries. Birds in the review area are almost always in breeding plumage throughout the year. **Voice:** Territorial call is a high pitched whinnying trill, *krik-ki-ki-ki-ki*, reminiscent of those of the smaller *Porzana* crakes (123–126). Uttered persistently especially during breeding. **Range:** Europe and Africa through much of temperate Asia, India, China, South-East Asia to New Guinea and the Solomons. **Status and distribution:** In Peninsular Malaysia, locally common resident in suitable freshwater bodies west of the Titiwangsa Range, including on Pulau Langkawi (Kedah), but very scarce east of it; the only known sightings are of individuals at Seri Jaya and Lanchang (Pahang) respectively on 12 January and 30 May 2009 (BiW). A probable winter visitor in northern Peninsular Malaysia. (W1). Absent from Tasik Bera and Tasik Chini (Pahang) and also from wetlands subject to constant draining and flooding such as ricefields. In Singapore, very rare resident, first recorded at Punggol in December 1992. Small numbers were established in Lorong Halus by May 1994, declining from the mid 1990s to 2 in May 2004 (W and H). Current distribution in review area much reduced due to the filling in of overgrown mining pools and freshwater ponds. **Habitat:** Still waters from large disused mining pools, to smaller fish ponds, reservoirs, and sewage oxidation ponds. Avoids saline and brackish water. **Habits:** Found singly or in small groups, persistently uttering its shrill cry as it moves on the water. Active during breeding, pairs often indulge in pursuit with noisy pattering on the water. Seldom flies when alarmed, preferring to dive underwater before resurfacing a few metres away.

STORKS

ORDER *Ciconiiformes* **FAMILY** *Ciconiidae*

World: *19 species* **Peninsular Malaysia:** *6 species*

Large wading birds with long legs, long necks, broad wings and long pointed bills. In flight, necks outstretched and slightly drooped (except Lesser Adjutant (39)) with legs trailing beyond short tails. Frequently soar on thermals. Forage solitarily or in pairs, sometimes in small loose groups, feeding on a variety of invertebrates and small vertebrates. During breeding gregarious, nesting in loose colonies. Sexes alike. Represented by the following groups and genera. **Wood Storks,** genus *Mycteria* (34, 35). Large size with bare facial skin and predominantly white plumage. Bill long, slightly down-curved, tapering and rounded at tip. **Openbills,** genus *Anastomus* (36). Small size with thickset pincer-shaped bill with conspicuous open space between maxilla and mandible. **Typical Storks,** genus *Ciconia* (37–38). Smaller than Wood Storks. Face feathered and bills proportionately shorter and much straighter. **Adjutants,** genus *Leptoptilos* (39). Very large with unfeathered head and neck. Bill massive and wedge shaped.

34. Milky Stork, *Mycteria cinerea* — Plate 5

Upih Bakau

Description: 97 cm (38 in). Predominantly white plumage with black flight feathers and tail. In underwing flight pattern, black flight feathers and tail contrast with white wing coverts and all white underparts. **Breeding:** Facial skin dark red; legs dark red; bill bright yellow. **Non-breeding (not illustrated):** Facial skin pinkish red; legs greyish; bill dull yellowish. **Juvenile (not illustrated):** Dark greyish brown plumage with white rump. **Voice:** Generally silent except for bill clattering. **Range:** Southern Vietnam and Cambodia discontinuously to the Malay Peninsula, Sumatra, Java, and Sulawesi. Declining due to loss and degradation of mangroves, human disturbances, and hunting **VU. Status and distribution:** Rare and localized resident; in the wild, the only sizeable population was confined to the Matang Mangrove Forest Reserve in the Kerian District (Perak) where there were approximately 115 in August 1983, declining to 9 in October 2004, and to 4 in January 2006 (Li *et al.*, 2006). No successful breeding in recent years although unsuccessful attempted nesting was reported there in November 1989 (W1). Former range was more extensive along the mangrove fringed west coast from Kedah and Perak through Selangor, south to the Benut Forest Reserve (Johor). A breeding colony existed on Pulau Ketam (Selangor) at least up to 1935 (Madoc, 1956) but has was wiped out by poaching (W1). A flock of 10 captive bred birds at the Kuala Selangor Nature Park (Selangor), obtained from the National Zoo in 1998 as part of a reintroduction programme escaped in May 2003. At least 6–7 still utilized the lake systems within the park as well as the habitats fringing the mangroves here up to December 2004. Breeding was attempted but not successful. **Habitat:** In review area, restricted to mature mangroves and associated inter-tidal mudflats. **Habits:** Usually solitary or in small loose groups, often together with egrets and the Lesser Adjutant (39). Feeds principally on mud skippers *Periophthalmus spp,* obtained by probing into mud skipper burrows during low tide but will also take molluscs, crustacea, and a variety of small aquatic animals. Like Painted Stork (35), probes for prey by submerging bill in water and moving it from side to side. Rarely feeds in inland freshwater swamps. Frequently soars over the mangroves. Very shy and ecologically sensitive to human activities, quickly taking off on approach. Successful breeding would therefore require large areas of protected tall mature mangroves with low levels of human disturbance. **Note:** Individuals at the National Zoo in Hulu Klang (Selangor) are kept in enclosures but a few free flying individuals sometimes join the free flying colony of Painted Storks to disperse and feed in wetlands around Kuala Lumpur (see Painted Stork (35) and 'Conservation' chapter). The Mandai Zoo has free-flying stock, which were introduced in 1987, so sightings from Singapore should be treated with caution. Look for coloured rings on legs of some birds. Hybrids of Milky x Painted can occur (see Painted Stork (35)) and any further release of captive bred stocks needs extreme caution.

35. Painted Stork, *Mycteria leucocephala* — Plate 5

Upih Sawah

Description: 102 cm (40 in). Shape (including bill) like Milky Stork (34) but differs by larger size and closely barred broad blackish breast and wing coverts. In underwing flight pattern, blackish breast band distinctive and blackish grey wing coverts do not contrast with black flight feathers. Scapulars, secondaries and wing coverts have pink flush. Tail black. Facial skin more extensive than Milky Stork, and is orange-red during breeding. **Juvenile (not illustrated):** Head and neck dull brown; back pale brown; rump and uppertail coverts white. **Voice:** Generally silent except for bill clattering. Juveniles in a feral colony uttered a hoarse and nasal *eaaah*, with bill open and tilted upwards, and a sharp and abrupt *kuk* on the return downwards; both wings outstretched during vocalization. **Range:** The Indian subcontinent through much of South-East Asia. Declining due to drainage and pollution of habitat as well as hunting **NT. Status and distribution:** At the beginning of the twentieth century, migrant in small numbers to Perlis and Kedah (including Pulau Langkawi) Further south, vagrant, to Kuala Lumpur. Presently feral and locally common resident with free-flying populations successfully breeding at the National Zoo in Hulu Klang (Selangor). This population is free ranging and regularly disperses to feed in wetlands outside the zoo and are frequent at the Pantai Dalam sewage oxidation ponds (Kuala Lumpur) and the Putrajaya wetlands where a small satellite breeding colony has formed. Individuals have also been sighted north at Rawang and south at Bangi, both in Selangor. Birds seen in Singapore are most likely to be free flying individuals from the Mandai Zoo which were introduced in 1987 (W and H) and may occasionally stray across the Straits of Johor to feed in wetlands around Johor Bahru (Johor). **Habitat:** In review area, man made freshwater ponds and sewage oxidation ponds. **Habits:** Feeds in small groups in shallow freshwater or sewage oxidation ponds by probing and moving bill from side to side. Frequently soars on rising thermals. Perches on roadside lamp posts and also roosts on tall buildings within Kuala Lumpur city. **Note:** In the National Zoo, has hybridized with Milky Stork, confirmed by DNA tests. One hybrid closely resembled Milky Stork but had small black spots on the white wing and a difficult to see pink tinge to the tail feathers. Another hybrid showed a much paler wing colour than Painted Stork (Li *et al.*, 2006). See 'Conservation' chapter.

36. Asian Openbill, *Anastomus oscitans* Plate 5

Upih Paruh Sepit

Description: 81 cm (32 in). From Milky Stork (34) by smaller size, feathered face and shorter dull greenish-horn bill, the pincer shape is diagnostic when viewed laterally and the tip is specially adapted for cutting open the operculum of snails. Plumage white (often appears dirty white); scapulars, primaries, secondaries, and tail black. Facial skin blackish, legs and feet flesh-pink. When breeding, appears whiter with legs becoming redder. In underwing flight pattern, black flight feathers and tail contrast sharply with white wing coverts and underparts. **Juvenile (not illustrated):** Similar to non-breeding adult but plumage mainly brownish grey on head, neck, upperparts, wings, and breast. Bill brownish. **Voice:** No information from review area. **Range:** The Indian subcontinent (including Sri Lanka), Central Thailand, Cambodia, and Cochinchina (Vietnam). Vagrant in SW Myanmar and the Malay Peninsula. But recent records in review area suggest a southward range extension may be happening. **Status and distribution:** Vagrant. An adult in harvested sugar cane field at Chuping (Perlis) on 1 and 2 March 2008 (Lim, 2009) and 8, including 4 juveniles at Kampung Permatang Nibong, Seberang Frai (Pulau Pinang) between 1 January–27 February 2010 (BiW). **Habitat:** In review area, freshwater canal in a sugar cane plantation (Chuping) and ricefields (Kampung Permatang Nibong). **Habits:** A specialized mollusc feeder, probes for principally gastropod snails in suitable wetland habitats. The Chuping individual was observed foraging on the ground in a sugar cane field but no food item was observed and recorded though it is also known to dig molluscs out of dried mud (W1). Frequently soars on rising thermals.

37. Woolly-necked Stork, *Ciconia episcopus* Plate 5

Upih Leher Putih

Description: 91 cm (36 in). Black body plumage (with purple or greenish blue sheen) contrasting with white nape, throat, neck and upper breast distinctive. Crown black; bill blackish; lower belly and undertail coverts white; legs red. **Juvenile (not illustrated):** Like adult but glossy black parts, dark brown. In flight, all dark underwing and belly contrast sharply with white neck, breast and undertail coverts. **Voice:** No information from review area. **Range:** Africa, South and South-East Asia discontinuously to Java, Sulawesi, the Flores and the Philippines. **Status and distribution:** Vagrant, known only from Pulau Langkawi (Kedah). No recent records. **Habitat:** Freshwater swamps, ricefields, open country. **Habits:** Unknown from review area.

38. Sunda Stork, *Ciconia stormi* Plate 5

Upih Rimba

Description: 89 cm (34 in). From Woolly-necked Stork (37) by smaller size, orange-red bill with small knob at base of culmen (when breeding only) and yellow orbital skin. White on head confined to throat, nape and hind neck. Legs orange. In flight, shows all dark underwing and breast. **Voice:** No information from review area. A cow-like *krauu* on take off (Myers, 2009). Clatters bill. **Range:** The Malay Peninsula, eastern Sumatra and Borneo. World population is largely fragmented and declining due to loss and fragmentation of lowland rainforest, including peat swamp. **EN. Status and distribution:** Rare and local resident at low elevations, recorded around Sitiawan (Perak), Krau Game Reserve, Ulu Tekam Forest Reserve, the upper Pahang headwaters (within Taman Negara) (all in Pahang (W1)), Kuala Juram and Sungai Relau, Merapoh, Kuala Lipis District, and Tasik Bera (Pahang) (BiW), south to the lower reaches of the Sungai Rompin and Sungai Endau (Pahang/Johor) and the Panti Forest Reserve (Johor) (W1). **Habitat:** Forested streams and rivers, freshwater and peat swamp forests. **Habits:** Usually solitary or in pairs, occasionally in small loose groups, feeding in margins of swampy forest. Frequently soars on rising thermals. Sometimes appears at salt licks to hunt in the boggy gaps caused by the hooves of large mammals. Roosts high up in forest trees.

39. Lesser Adjutant, *Leptoptilos javanicus* Plate 5

Upih Botak /Burung Botak

Description: 114 cm (45 in). Hunched appearance distinctive. Has massive yellowish wedge-shaped bill with light corrugations on base of mandible. Unfeathered head and neck yellowish orange with sparse growth of hair like feathers. Upperparts black with slight gloss; underparts white. During breeding **(not illustrated)**, upperparts show dark greenish sheen in good light with coppery magenta spots on wing coverts. Flies with neck retracted in an 'S' shape like herons (42–61). In flight, upperwing all black except for small white triangular patch at base of wings. **Juvenile (not illustrated):** Head and neck more feathered; upperparts less glossy. **Voice:** Silent, except for clattering of bill at nest. **Range:** The Indian subcontinent and southern China through South-East Asia to the Greater Sundas. Declining due to habitat loss and hunting pressure **VU. Status and distribution:** Locally common to scarce resident at low elevations, south to the Kukup Mangrove Forest Reserve, Johor. Distribution concentrated mainly on the mangrove fringed west coast where it is fairly common in the Matang Mangrove Forest Reserve, Kerian District (Perak), the Selangor coast, and Parit Jawa, Muar District (Johor) where breeding was recorded in May 2002 (BiW). On the east coast, occurs sporadically from the Setiu estuary (Terengganu) south to the Pekan-Nenasi peat swamp forests (Pahang) where it is fairly common with up to 120 during a single sighting at Kampung Medan, Pekan (BiW). Rarer inland, with sporadic sightings along the Pahang river, the Ulu Dedap ricefields (Perak) and Panti Forest Reserve (Johor). Locally extinct in Singapore since 1882 (L and G), the presence of a juvenile and an adult respectively on 27 June and 9 August 2000 at the Sungei Buloh Wetland Reserve could be strays from Johor (W and H). Other sightings in Singapore are probably free-flying stock from the Mandai Zoo. **Habitat:** Mangroves, mudflats, peat swamp forests, estuaries, large rivers. **Habits:** Usually solitary or in small loose groups but rarely in large congregations. Strides along the mudflats, or wades in flooded ricefields, feeding on molluscs, fish and crustacea. During breeding, nesting pairs will often clatter bills while tilting necks back. Young are fed on regurgitated prey items. Often perches on tall dead trees. Frequently soars on thermals.

IBISES

ORDER *Ciconiiformes* **FAMILY** *Threskiornithidae*

World: *33 species* **Peninsular Malaysia and Singapore:** *2 species*

(This family also includes the spoonbills which do not occur in the review area although the Black-faced Spoonbill *Platalea minor* has been recorded as a vagrant to the west coast of the Malay Peninsula (Krabi Province, Thailand).)

A distinct group of medium-sized to rather large wading birds closely allied to storks (34–39) but generally smaller with long, thin down-curved bills which are specially adapted for probing into mud or water. Feed largely on fish, crustacea and other small aquatic animals. In flight, necks are fully outstretched and slightly drooped with legs trailing beyond short tails. Flight consists of moderately rapid flapping, alternated with short glides. Sexes alike. Gregarious.

40. Black-headed Ibis, *Threskiornis melanocephalus* Plate 5

Burung Sekendi Kepala Hitam

Description: 76 cm (30 in). White plumage with long down-curved black bill, featherless head and grey tertials diagnostic. Legs black In flight, shows bright narrow patch of red skin along inner leading edge of underwing. **Juvenile (not illustrated):** Shows some brownish to greyish white on head and neck with black edges and tips to outer primaries. **Voice:** No information from review area. **Range:** The Indian subcontinent and China east to Japan and south through South-East Asia to the Greater Sundas. Currently undergoing population reduction due to hunting as well as drainage and conversion of foraging habitats to agriculture. **NT. Status and distribution:** Previously a non-breeding visitor in small numbers along the mangrove fringed west coast, south to Pulau Ketam (Selangor) at least up to 1934. Currently vagrant with three sightings, two from Kuala Gula (Perak); a juvenile on 2 October 2005, and an adult on 2 May 2006 (BiW), and a juvenile at Kampung Permatang Nibong, Seberang Prai (Pulau Pinang) on 28 November 2010. (Chan Kai Soon, pers.comm). **Habitat:** Mangroves, mudflats, estuaries. **Habits:** Where common, feeds in small parties, probing soft mud or stalking aquatic prey. Roosts in trees together with herons and egrets. Partly nocturnal. **Note:** The Jurong Bird Park (Singapore) has free-flying individuals together with the similar looking Afro tropical Sacred Ibis *T. aethiopicus*. The latter is distinguished by having long black plumes on the lower back and black tips forming a narrow black trailing edge along secondaries. Both species sometimes occur at the Sungei Buloh Wetland Reserve. Sightings in Singapore should therefore be treated with caution.

41. Glossy Ibis, *Plegadis falcinellus* Plate 5
Burung Sekendi Kilat

Description: 64 cm (25 in). Small dark ibis with feathered head. Plumage rich dark chestnut especially on head, neck, mantle and rump, heavily glossed with green and purple. Underparts deep purple-brown. Facial skin dark with distinct thin white border above and below. In poor light, appears all black. In non-breeding plumage, shows distinct whitish streaks on head and neck. **Juvenile (not illustrated):** Paler, the rich dark chestnut parts replaced with dull brown. Throat whitish with dense whitish mottling on head and neck. Lacks white facial border of adult. **Voice:** No information from review area. **Range:** Tropical and subtropical latitudes nearly worldwide. Northern populations move south during non-breeding season. **Status and distribution:** Very rare non-breeding dispersant with two records from Singapore, a wild flock of 11, consisting of both adults and juveniles at the Serangoon estuary from 12–15 June 1984 (Wells, 1990), and 2 at Sungei Buloh on 21 May 1989 (W and H). **Habitat:** In review area, sludge-drying beds of a sewage farm (Serangoon) and tidal mudflats (Sungei Buloh). **Habits:** Like Black-headed Ibis but strictly diurnal. **Note:** The Jurong Bird Park has free-flying individuals. Sightings in Singapore should therefore be treated with caution and verified.

HERONS, EGRETS, AND BITTERNS

ORDER *Ciconiiformes* **FAMILY** *Ardeidae*

World: *65 species* **Peninsular Malaysia and Singapore:** *20 species*

A distinct group of small to large wading birds with long legs and short to very long necks with long pointed dagger-like bills. Flight is strong with regular wing-beats on long, broad and rounded wings. In flight, necks are tucked back in an 'S' shape with legs trailing beyond short tails. Feed largely on small fish as well as small vertebrates and invertebrates. When hunting, stride slowly and cautiously in shallow water or reed beds, catching victims with quick and powerful stabs of their bills. Most species nest in small to large colonies in trees. Nests are shallow platforms of sticks. Calls harsh and loud croaks or squawks. Gregarious or solitary. Sexes alike or nearly so. Represented by the the following groups and genera: **Bitterns,** genera *Botaurus* (42) *Ixobrychus* (43–45) and *Dupetor* (46). Comparatively short necked and short legged with brownish or blackish plumage. Solitary skulkers in reed beds and thick, aquatic vegetation. When alarmed, 'freeze' and align their bills and neck upwards with the reeds to avoid detection. Some species crepuscular. **Night-herons,** genera *Gorsachius* (47). and *Nycticorax* (48). Stocky with comparatively short neck and legs. Nocturnal in habits. The former is solitary and secretive, the latter highly gregarious. **Little Heron,** genus *Butorides* (49). Small size with comparatively short neck and legs. Crown dark, plumage greyish. **Pond-herons,** genus *Ardeola* (50–52). Comparatively short necked and short legged. In non-breeding plumage in all species, streaked head, breast and brownish mantle contrast with all white wings and tail in flight. During breeding, all species in the review area assume brightly coloured necks, breasts and mantles which are distinct in each species **Large Herons,** genus *Ardea* (54–56). Large size with very long slender neck distinctive. Develop long nuchal plumes when breeding. **Egrets,** genera *Bubulcus* (53), *Casmerodius* (57), *Mesophoyx* (58), and *Egretta* (59–61). Wholly white plumage (except dark morph Pacific Reef-egret (60)). Neck lengths vary from short to very long and kinked. Highly gregarious, often associating in mixed species flocks. Develop fine plumes on breast and back during breeding. **Note:** This is an arbitrary grouping based on wholly white plumage and not on any taxonomic classification.

42. Great Bittern, *Botaurus stellaris* Plate 8
Pucung Paya

Description: 76 cm (30 in). The largest bittern in the review area. From juvenile Black-crowned Night-heron (48) by much larger size, more thickset yellowish bill, and lighter sandy buff plumage with fine black and brown mottles on upperparts. Crown and malar stripe blackish; throat whitish, sides of head rufous-buff, legs greenish. In upperwing flight pattern, broad-winged profile distinctive with sandy buff wing coverts contrasting with blackish barred flight feathers. **Voice:** No information from review area. **Range:** Breeding across Europe, North Africa and temperate Asia, also southern Africa; wintering south to tropical Africa, South and South-East Asia through the Malay Peninsula to the Philippines. **Status and distribution:** A very rare non-breeding visitor at low elevations, south to Singapore. **Habitat:** Reedbeds, inland freshwater swamps. **Habits:** Solitary. Usually keeps hidden in tall reeds, emerging at the edge to feed on fish and other aquatic prey. Very shy and secretive.

43. Yellow Bittern, *Ixobrychus sinensis* Plates 6 and 7
Pucung Kuning

Description: 38 cm (15 in). **Male:** Crown and uppertail coverts black; mantle pale brown; underparts and wing coverts pale yellowish buff. **Female (not illustrated):** Like male but crown reddish brown with underparts lightly streaked brown. **Juvenile:** Bold brownish streaks on upperparts and paler underparts. In flight, blackish flight feathers contrast sharply with pale buff wing coverts. **Voice:** Usually silent. Sometimes a soft and sharp staccato *kakak-kakak* in flight when flushed. **Range:** India, China, eastern Russia and Japan discontinuously through South-East Asia, south-east to New Guinea and Micronesia. Northern populations are migratory. **Status and distribution:** Locally common resident, passage migrant, and winter visitor at low elevations, south to Singapore where numbers have declined since the 1990s (W and H). Occurs above 900 m only during spring and autumn passage. **Habitat:** Reedbeds, flooded ricefields, settled mining pools, inland freshwater swamps. **Habits:** Usually solitary or in pairs. Moves through reeds in search of prey. May sometimes feed in the open amongst floating aquatic vegetation. Unobtrusive most of the time and takes to flight only when approached too close. Flies with moderate wingbeats, usually low over water or reedbeds.

44. Chestnut Bittern, *Ixobrychus eurhythmus*

Plates 6 and 7

Pucung Gelam

Description: 38 cm (15 in). From Yellow Bittern (43) by darker plumage and less contrast between dark flight feathers and upperwing coverts. **Male:** Crown and uppertail coverts black; head, neck and mantle deep mahogany chestnut. Underparts and wing coverts pale yellow buff with dark chestnut stripe from throat to breast. In upperwing flight pattern, greyish black flight feathers show slight contrast with mantle and wing coverts. **Female:** Like male but duller with upperparts marked with small white spots. Wing coverts almost concolorous with mantle. Underparts marked with bold dark streaks. In upperwing flight pattern, pale grey flight feathers show very little contrast with mantle and wing coverts. **Juvenile (not illustrated):** Like female but more heavily streaked. **Voice:** No information from review area. **Range:** Breeding in eastern Russia, Japan, Korea, and China; wintering south to southern China through South-East Asia to the Greater Sundas, Sulawesi and the Philippines. **Status and distribution:** Uncommon and localized passage migrant and winter visitor at low elevations, south to Singapore. Occurs above 900 m only during spring and autumn passage. **Habitat:** Reedbeds, disused mining pools, the mangrove edge, inland freshwater swamps, reservoir edges. **Habits:** Similar to Yellow and Cinnamon Bitterns (45).

45. Cinnamon Bittern, *Ixobrychus cinnamomeus*

Plates 6 and 7

Pucung Bendang

Description: 38 cm (15 in). From all bitterns in all plumages by bright uniform cinnamon rufous wings and upperparts both at rest and in flight. **Male:** Dark bright cinnamon rufous above; underparts paler rufous. Conspicuous narrow white streak on side of neck. **Female (not illustrated):** Like male but underparts streaked dark brown. **Juvenile:** Like female but duller with underparts more boldly streaked and upperparts boldly mottled dark brown. **Voice:** Territorial call is a soft repeated *kok*, *kok-kok-kok-kok*.... **Range:** Breeding in the Indian subcontinent and China through South-East Asia to the Greater Sundas, Sulawesi, and the Philippines. Northern populations are migratory, wintering to the southern limits of its range. **Status and distribution:** Locally common resident, passage migrant and winter visitor, south to Singapore where numbers have decreased since the 1990s (W and H). Occurs above 900 m only during spring and autumn passage. **Habitat:** Reedbeds, settled mining pools, flooded ricefields, inland freshwater swamps, overgrown drains and ditches. **Habits:** Usually solitary or in pairs, keeping to the dense waterside vegetation. Sometimes feeds in the open. Usually flies low over reedbeds and water bodies but during courtship, pairs may fly higher in the sky.

46. Black Bittern, *Dupetor flavicollis*

Plates 6 and 7

Pucung Hitam

Description: 54 cm (21 in). **Male:** Black crown, hindneck and upperparts contrasting with conspicuous buff yellow stripe running down side of neck diagnostic. Throat and breast whitish buff with dark streaks on the latter. **Female:** Like male but black parts replaced with blackish brown. **Juvenile (not illustrated):** Like female but upperparts dark brown, fringed with buff, giving a scaly appearance. **Voice:** No information from review area. **Range:** Breeding in the Indian subcontinent and southern China discontinuously through South-East Asia to Australia. Northern populations are migratory. **Status and distribution:** Uncommon and localized passage migrant and winter visitor at low elevations, south to Singapore. Occurs above 900 m only during spring and autumn passage. **Habitat:** Reedbeds, dense vegetation fringing settled mining pools and inland freshwater swamps. **Habits:** Usually solitary, keeping to thick cover by day and emerging to feed at dusk. When alarmed, extends neck in typical bittern fashion. Shy and secretive.

47. Rufous-necked Night-heron, *Gorsachius melanolophus* Plates 6 and 7
Pucung Harimau

Description: 48 cm (19 in). From *Ixobrychus* bitterns (43–45) by stocky appearance, short legs and short stubby bill. Neck and breast bright rufous; chin whitish with blackish streaks running down centre of foreneck to breast; upperparts dull chestnut with fine black vermiculations. Flanks and undertail coverts mottled black and white. In upperwing flight pattern, blackish flight feathers with white tips to primaries contrast with chestnut wing coverts and narrow chestnut trailing edge of wing. **Juvenile:** Crown black with fine white barrings; upperparts dull brown, finely mottled black; breast finely barred black. Bill shape and body colouration separate it from juvenile Black-crowned Night-heron (48). **Voice:** No information from review area. Territorial call in the breeding area is a monotonous cadence of deep *oo* notes on a level note, interspersed by brief intervals (extralimital data). A captive individual uttered harsh croaking notes when handled. **Range:** India and southern China, discontinuously through South-East Asia to the Greater Sundas and the Philippines. **Status and distribution:** Rare and localized passage migrant and winter visitor at low elevations, south to Singapore. Occurs above 900 m only during spring and autumn passage. **Habitat:** Forests, secondary growth, reedbeds. **Habits:** Usually solitary. Roosts in trees by day, emerging at dusk to feed on small vertebrates and invertebrates in more open areas. Usually found near forested streams but not strictly attached to water. When approached, a captive individual raised crown feathers in a crest and partly outstretched wings. Very shy and secretive but sometimes tame and allows fairly close approach.

48. Black-crowned Night-heron, *Nycticorax nycticorax* Plates 6 and 7
Pucung Kuak

Description: 61 cm (24 in). Stocky and short-necked. Crown and mantle black with greenish tinge; bill black; wings and tail pale grey; underparts, including underwing, white; iris red; legs pinkish. Two or three long white nuchal plumes. In upperwing flight pattern, black crown and mantle contrast with pale grey wings and tail. **Juvenile:** Dark brown with whitish buff streaks on underparts and whitish buff spots on upperparts. Iris yellow; bill pale pinkish. **Voice:** A throaty *kuak*. **Range:** Breeding at temperate and tropical latitudes worldwide except Australia. Northern populations are migratory. **Status and distribution:** Fairly common to uncommon resident along the mangrove fringed west coast, south to Singapore where it is uncommon. The only large breeding colony in Peninsular Malaysia is at the Matang Mangrove Forest Reserve (Perak) where large flocks nest in the *Avicennia* zone. Elsewhere breeding colonies existed at Yong Peng and Johor Bahru city (Johor). Further inland there were breeding colonies near former tin mining wetlands near Batu Gajah and Malim Nawar (Perak) as well as in screw-pine swamp forest near Bandar Tasik Puteri, Rawang (Selangor), where there were over 600, between 2001–2005. Rarer along the east coast. Also seen along the Sungai Tembeling (Pahang). The National Zoo at Hulu Klang (Selangor) has free-flying stock which may range outside the zoo premises. In Singapore, breeeding colonies existed at the Yishun mangroves. Today the largest colony is in the grounds of the Jurong Bird Park, consisting of both free flyers and wild birds (W and H). **Habitat:** Mangroves, ricefields, inland freshwater swamps, reservoirs, prawn ponds. **Habits:** Largely nocturnal but also active by day. Roosts and breeds colonially in mangroves and other wetlands but feeds at night mostly in freshwater habitats and also on mudflats at low tide. Breeding colonies are not stable and shift due to habitat destruction and disturbances by human activities. At dusk, circles roost in noisy flocks and flies in V-formations to feeding grounds, returning at dawn. During breeding season, very active at day, collecting nest material and indulging in a variety of aerial displays.

49. Little Heron, *Butorides striata* — Plates 6 and 7
Pucung Keladi

Description: 46 cm (18 in). Small and stocky grey-blue heron with typical hunched appearance at rest. Bill, crown, nuchal plumes and narrow malar stripe black. Wing coverts dark brown with greenish gloss marked with yellowish white streaks. Throat and ear coverts whitish. Lores and legs yellow. **Juvenile:** Dark brown with buff streaks on underparts and buff spots on wing coverts. **Voice:** A penetrating *keeyow*, when flushed. **Range:** Temperate and tropical latitudes worldwide, except Europe. Northern populations are migratory. **Status and distribution:** Fairly common resident and winter visitor from low elevations up to 240 m, south to Singapore. Within Kuala Lumpur city, seen at most man-made freshwater ponds. Occurs above 900 m only during spring and autumn passage. **Habitat:** Mangroves, mudflats, riverbanks, large drains, sewage oxidation ponds. **Habits:** Usually solitary, stalking prey in shallow water or at the water's edge. Larger groups may feed together, especially with the arrival of migratory populations during winter. At low tide, strides across the mudflats to pick up small aquatic animals. In fishing villages, perches on jetties or sterns of fishing boats, feeding on discarded fish thrown into the water. When alarmed, will raise crown feathers.

50. Indian Pond-heron, *Ardeola grayii* — Plate 7
Pucung Danau India

Description: 46 cm (18 in). At rest, non-breeding individuals of all three species in review area appear all brown with white wings concealed. In upperwing flight pattern, all white wings, tail and rump conspicuous, contrasting with darker back and head. Belly and undertail coverts white in all seasons. **Breeding:** From Chinese (51) and Javan Pond-herons (52) by diagnostic brownish-buff head, neck, and breast together with dark brownish-maroon mantle and scapulars. Chin, throat, and wings white. Displays long white nuchal plumes. Bill yellow, tipped black. **Non-breeding:** Like Chinese (51) and Javan Pond-herons (52). **Voice:** No information from review area. **Range:** The Persian Gulf region and the Indian subcontinent (including Sri Lanka) east to Myanmar and the Malay Peninsula. **Status and distribution:** Uncommon and localized non-breeding visitor, recorded from Pulau Langkawi (Kedah) and Pulau Pinang, south to Perak. **Habitat:** In review area, flooded ricefields and inland freshwater swamps. **Habits:** Like Chinese and Javan Pond-herons with which it associates. **Note:** In Singapore, an individual in breeding plumage at Senoko on 24 March 1994 (W and H) was probably a dispersant from the free-flying stock from Mandai Zoo.

51. Chinese Pond-heron, *Ardeola bacchus* Plate 6 and 7
Pucung Danau Cina

Description: 46 cm (18 in). **Breeding:** Head and neck dark chestnut; breast rich brownish maroon; back black. Lores and base of bill slate blue. Usually acquired before spring passage. Bill yellow with dark tip **Non-breeding:** Head, neck and breast whitish buff with bold brown streaks; back brownish. **Voice:** A low deep *gwa* when flushed. **Range:** Breeding from Japan and North-East China, south to northern Indo-China, Myanmar and North-East India; wintering south through South-East Asia and the Malay Peninsula to Borneo. **Status and distribution:** Common and widespread winter visitor at low elevations, south to Singapore where it is uncommon. **Habitat:** Mangroves, inland freshwater swamps, flooded ricefields, oil palm factory sludge ponds, sewage oxidation ponds. **Habits:** Usually in small loose flocks while feeding but gathers in large numbers when commuting between feeding and roosting sites. Generally keeps to thick cover, stalking prey at the water's edge. Feeds largely on small vertebrates and insects. Roosts communally in small trees.

52. Javan Pond-heron, *Ardeola speciosa* Plates 6 and 7
Pucung Danau Jawa

Description: 46 cm (18 in). **Breeding:** Like Chinese Pond-heron (51) but head and nape pale beige, merging into rich buff breast. Head plumes distinctively pale. **Non-breeding:** Like Indian (50) and Chinese Pond-herons (51). **Voice:** A low deep *gwa* when flushed. **Range:** Breeding in Tenasserim (Myanmar), central Thailand, and Indo-China discontinuously to Java, South-East Borneo, and the Lesser Sundas. Outside breeding, disperses to the Malay Peninsula, southern Sumatra and most of Borneo. **Status and distribution:** Locally common non-breeding visitor recorded in breeding plumage from Kedah, Seberang Prai (Pulau Pinang), and Perak, south to Selangor. Far less numerous and less widespread than Chinese Pond-heron. **Habitat:** Mangroves, inland freshwater swamps, flooded ricefields. **Habits:** Similar to Indian and Chinese Pond-herons. **Note:** In Singapore, individuals in breeding plumage at the Sungei Buloh Wetland Reserve in March 2003 and April 2004 are strongly believed to be free-flying individuals from the Mandai Zoo (W and H).

53. Cattle Egret, *Bubulcus ibis* Plate 8
Bangau Kerbau

Description: 51 cm (20 in). From other white egrets by stocky appearance, rounder head, comparatively shorter neck and short yellow bill. Throat more puffed out, giving a 'jowled' appearance. Lores yellowish; legs dark. **Breeding:** Develops orange-yellow plumes on head, breast and back. For part of the breeding season, bill, lores and legs become red. **Voice:** Silent in wintering quarters. **Range:** Temperate and tropical latitudes worldwide; range still expanding. **Status and distribution:** Uncommon resident to fairly common winter visitor at low elevations, south to Singapore. Has bred at the Kinta Nature Park in Batu Gajah (Perak). Between September and April, fairly common on pasture land and ricefields. Free-flying stock from the Jurong Bird Park in Singapore are often seen in western Singapore where they are difficult to differentiate between migratory populations (W and H) which are known to regularly winter at the Serangoon Sewage Works. **Habitat:** Ricefields, pasture land, reedbeds, sewage oxidation ponds. **Habits:** Not particularly attached to water. Avoids saline water and mudflats, occasionally appearing there during autumn and spring passage. Gathers in small to large flocks, associating largely with grazing cattle, feeding on insects disturbed by the animals' trampling hoofs. Flies in formation between feeding and roosting grounds. Roosts communally on trees and in reedbeds fringing freshwater swamps and disused mining pools.

54. Grey Heron, *Ardea cinerea* Plate 8
Pucung Seriap

Description: 102 cm (40 in). Grey upperparts, white head and neck with black streaks on foreneck diagnostic. Black line above eye joins black nuchal plumes. Underparts paler grey, almost whitish. In upperwing flight pattern, grey wing coverts and upperparts contrast with black primaries and secondaries together with white patch at base of primaries. Black shoulder patch conspicuous at rest. During breeding, bill and legs dark orange to brilliant pink; otherwise pale yellow to flesh. **Juvenile:** Like adult but duller and more grey on neck. **Voice:** A harsh *kraank*, usually uttered in flight. **Range:** Breeding in Africa, Europe, temperate Asia, east to eastern Russia, Japan, and China and south to South and South-East Asia to the Greater and Lesser Sundas and the Philippines. Northern populations are migratory. **Status and distribution:** Locally common to rare resident, south to Singapore. Breeding colonies concentrated mainly in the Matang Mangrove Forest Reserve (Perak), Kuala Selangor and Kapar (Selangor) and the Benut Mangrove Forest Reserve (Johor). Has also bred inland at the Kinta Nature Park, Batu Gajah (Perak) and Shah Alam Lake (Selangor). Elsewehere sparsely distributed with individuals at Teluk Mahkota and Mersing (Johor) on the east coast. In Singapore, breeding colonies exist in both mangrove and freshwater habitats, especially at Tanah Merah and the Sungei Buloh Wetland Reserve. **Habitat:** Mangroves, mudflats, freshwater swamps, coastal power station ashponds. **Habits:** Usually solitary, sometimes in small loose flocks. Largely stalks prey in shallow water

55. Great-billed Heron, *Ardea sumatrana* Plate 8
Pucung Lembu

Description: 114 cm (45 in). Large dark heron with dusky grey plumage and long stout dagger shaped bill diagnostic. In flight from Lesser Adjutant (59) by all grey wings and the lack of soaring. During breeding, develops silvery white nuchal plumes with whitish plumes on breast and scapulars, giving it a streaky appearance. Bill two toned; blackish maxilla and whitish to chrome yellow mandible. Legs blackish. **Juvenile (not illustrated):** Pale rusty brown without the whitish plumes of adult. From juvenile Purple Heron (56) by much heavier build, and proportionately shorter neck, and heavier bill. **Voice:** No information from review area. In Australia, deep, prolonged, resonant guttural roar (P and K) and harsh croaks in alarm (P and K). **Range:** The Andamans through the coasts of South-East Asia, the Greater and Lesser Sundas east to Sulawesi and the Philippines and south to New Guinea and northern Australia. **Status and distribution:** Rare and localized coastal resident, south to Singapore. Also in the Langkawi archipelago (Kedah), Pulau Pangkor (Perak), Pulau Redang (Terengganu), Pulau Tioman (Pahang) including the satellite islets of Seri Buat and Tinggi and the Merambong shoals at the estuary of the Sungai Pulai, Pontian District (Johor). In Singapore, occurs on most of the offshore south islands of Pulau Hantu, Pulau Salu, Pulau Semakau, and Pulau Sudong (W and H) with breeding recorded only on Pulau Anak Bukom (W1). Also twice on the mainland at Changi and Tuas (W2). **Habitat:** Mangroves, mudflats, rocky islets. **Habits:** Usually solitary, stalking prey at the edge of mudflats or estuaries. Flight leisurely, with slow wingbeats.

56. Purple Heron, *Ardea purpurea* Plate 8
Pucung Serandau

Description: 97 cm (38 in). From other *Ardea* herons, by rufous neck and underparts. In dense waterside vegetation with only neck exposed, may resemble Oriental Darter (72) but lacks the long white streak on side of neck. Neck proportionately longer and more slender than Grey Heron (55). In flight, neck bulge more pronounced than Grey Heron. Throat whitish; sides of neck, throat and breast streaked black; crown and nuchal plumes black. Wings grey. Bill and legs yellowish. In flight, grey wing coverts contrast with black flight feathers and rufous wing lining. **Juvenile:** Like adult but wing coverts brownish with head and neck sandy rufous. **Voice:** Harsh croaking notes, typical of the family. **Range:** Africa and Europe discontinuously through temperate Asia east to eastern Russia and China and south to South and South-East Asia through the Greater and Lesser Sundas to Sulawesi and the Philippines. Northern populations are migratory. **Status and distribution:** Fairly common resident and winter visitor at low elevations, south to Singapore. The most common large heron occurring in inland.freshwater habitats. In Singapore, breeds in the Kranji marshes and the Sungei Buloh Wetland Reserve. **Habitat:** Mangroves, freshwater swamps, settled mining pools, sewage oxidation ponds. **Habits:** Usually solitary. Feeds mostly in freshwater, slowly and silently stalking prey in shallow water with neck cocked before rapidly spearing its victim. Very rarely feeds at mudflats or intertidal estuaries.

57. Great Egret, *Casmerodius albus* Plate 8
Bangau Besar

Description: 90 cm (35 in). Large size and long kinked neck diagnostic. In flight, neck bulge pronounced. Gape extends beyond eye. **Breeding:** Bill black; facial skin blue; tarsi reddish; long fine plumes on lower back. **Non-breeding:** Bill and facial skin yellow; legs black. **Voice:** A dry throaty rattle, *kraaaak,* when flushed. **Range:** Temperate and tropical latitudes worldwide. Northern populations are migratory. **Status and distribution:** In the first half of the twentieth century, fairly common resident along the mangrove fringed west coast. Currently rare and localized resident along the Perak and Selangor coasts. Elsewhere along west coast, fairly common passage migrant and winter visitor, south to Singapore. Some migrants overwinter in the review area. Rarer inland and along east coast. **Habitat:** Mangroves, mudflats, flooded ricefields, inland freshwater swamps. **Habits:** Usually solitary or in small loose flocks, often with the other smaller egrets. Flight graceful with slow wingbeats. When feeding in shallows, strides slowly with neck sometimes fully outstretched, surveying the water below. Makes quick strikes on prey below.

58. Intermediate Egret, *Mesophoyx intermedia* Plate 8
Bangau Kendi

Description: 71 cm (28 in). From Great Egret (57) in winter by much shorter bill with gape extending to eye, shorter legs and proportionately shorter neck without the kink. Legs and feet black. **Breeding:** Bill black with varying amounts of yellow at base; plumes on breast and lower back. **Non-breeding:** Bill yellow with dark tip. **Voice:** Usually silent but utters low croaks when disturbed. **Range:** Breeding in Africa and South Asia east to South-East Asia north-east to China and Japan and discontinuously south to New Guinea and Australia. Populations at high latitudes are migratory, wintering south to South-East Asia and the Greater and Lesser Sundas. **Status and distribution:** Locally common passage migrant and winter visitor, south to Singapore where it is uncommon. **Habitat:** Mudflats, inland freshwater swamps, flooded ricefields, disused mining pools, sewage oxidation ponds, coastal power station ashponds. **Habits:** Usually in small loose flocks when feeding. Sometimes solitary. Associates with other egrets. Flies in compact flocks between feeding and roosting grounds.

59. Little Egret, *Egretta garzetta* Plate 8
Bangau Kecil

Description: 61 cm (24 in). From other egrets by small size and thin black bill. Legs black. Feet yellow in *E. g. garzetta* from mainland Asia and black in *E. g. nigripes* **(not illustrated)** from Java and the Lesser Sundas. **Breeding:** Two long nuchal plumes as well as plumes on breast and back, with toes becoming orange and bare facial skin from bill-base to eye turning bright violet magenta. **Voice:** A harsh throaty *gwaa* in flight. **Range:** Breeding in Africa, southern Europe, southern temperate Asia through South and South-East Asia discontinuously to New Guinea and Australia. Northern populations are migratory, wintering south to tropical latitudes. **Status and distribution:** Uncommon resident to common and widespread passage migrant and winter visitor, south to Singapore. *E. g. nigripes* is very rare. Breeding recorded near Malim Nawar (Perak) in November 2001 and at screw pine swamp forest at Bandar Tasik Puteri, Rawang (Selangor), in March 2005. **Habitat:** Mangroves, mudflats, flooded ricefields, disused mining pools, sewage oxidation ponds, oil palm factory sludge ponds, coastal power station ashponds. **Habits:** Usually in small loose flocks, often with other egrets. Sometimes solitary. Typically dashes about in the shallows with wings partly outstretched, to stir up prey. Will also pick up insects from the surface of sewage oxidation ponds.

60. Pacific Reef-egret, *Egretta sacra* Plate 8

Bangau Batu

Description: 58 cm (23 in). Grey facial skin and greenish legs distinctive. During breeding, sports short nuchal crest and plumes on breast and lower back. In flight, only toes protrude beyond tail. **Dark morph:** Dark slaty grey plumage with inconspicuous narrow white mark on throat diagnostic. From Great-billed Heron (55) by much smaller size, shorter neck and proportionately shorter legs. Bill blackish with varying amounts of yellow on mandible. **Pale morph:** All white with yellow mandible and dark maxilla and tip. Some have black markings on breast and back. **Voice:** No information from review area. **Range:** Coasts and islands from China, Japan and Korea through South-East Asia to Australia and the western Pacific. **Status and distribution:** Locally common coastal and offshore resident, south to Singapore, where it is uncommon with breeding recorded on Pulau Bukom (W and H). More common on the smaller offshore islands than on the mainland. The dark morph appears to be more common. **Habitat:** Rocky and sandy shores, coral reefs, mudflats. **Habits:** Usually solitary or in pairs. Exclusively coastal in habitat preference but on very rare occasions, may occur solitarily in flooded ricefields inland. Pursues fish by pacing over exposed ground in shallows on the shoreline. Often rests on one leg on rocks.

61. Chinese Egret, *Egretta eulophotes* Plate 8

Bangau Cina

Description: 69 cm (27 in). From Little Egret (59) by larger size and proportionately shorter and thicker neck, and in winter by yellow base to mandible and greenish brown legs. From pale morph Pacific Reef-egret (60) in winter by more exposed tibia and progressively narrow bill from base to tip. In the Reef-egret, tibia is feathered up to just above knee; bill thick from base, narrowing just near tip. **Breeding:** Bare skin on lores blue; bill orange-yellow; shaggy white nuchal crest and plumes on breast and back; legs black with yellow feet. **Non-breeding:** Lacks nuchal crest and plumes. Skin and lores greenish grey; bill dark on maxilla and pinkish on mandible. **Voice:** Usually silent in wintering quarters. Low croaks when alarmed. **Range:** Breeding in Korea and parts of China; wintering south to South-East Asia, Sumatra, Borneo, Sulawesi and the Philippines. Declining due to reclamation of tidal mudflats and estuarine habitats for industry, infrastructure development and aquaculture. **VU. Status and distribution:** Uncommon and localized winter visitor, south to Singapore, where it is fairly regular at the Serangoon and Seletar estuaries. **Habitat:** Mudflats, estuaries.

Great Egret, *Casmerodius albus* (57)

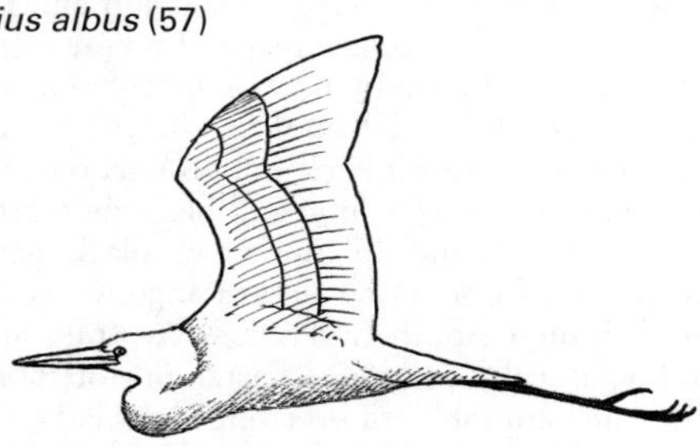

Intermediate Egret, *Mesophoyx intermedia* (58)

Little Egret, *Egretta garzetta* (59)

Cattle Egret, *Bubulcus ibis* (53)

Habits: Solitary or in small loose flocks, Usually pursues small fish in shallows on the shoreline. Feeding methods similar to Little Egret.

FRIGATEBIRDS

ORDER *Pelecaniformes* **FAMILY** *Fregatidae*

World: *5 species* **Peninsular Malaysia and Singapore:** *3 species*

Large tropical seabirds with long, narrow, pointed angular shaped wings, long forked tails (often closed and pointed), long and slender bills, hooked at the tip. Feet partially webbed. Magnificent soarers, staying aloft for hours on end without flapping. Feed mostly on squid and flying fish obtained by skimming low over the water and plucking them off the surface. Kleptoparasitises other feeding seabirds by vigorously pursuing and harassing them until their victims disgorge their food which the frigatebirds will skilfully catch before it hits the water. Sexually dimorphic, males are largely black with inflatable red sacs at the base of bill (inflated only during breeding); females are larger with white around the neck and on the breast and belly. Complex series of juvenile plumage patterns sometimes pose identification problems. Long distance wanderers, frigatebirds have been sighted far inland over the Titiwangsa Range (W1) in apparent attempts to cross the peninsula. Gregarious, breeding in colonies and forming large communal roosts outside breeding.

62. Christmas Island Frigatebird, *Fregata andrewsi* — Plate 9

Simbang Pulau Christmas

Description: 95 cm (37 in). **Male:** From other male frigatebirds by white belly on all black plumage. Deflated red sac at base of bill inconspicuous in the field in all 3 species. **Female:** Black head with white breast and belly prominent blackish spurs on upper breast and white spurs on base of underwing diagnostic. **Juvenile:** Head tawny buff with complete blackish breast band. **Voice:** Usually silent in non-breeding grounds. **Range:** East Indian Ocean to the South China Sea. Breeds only within a tiny area on Christmas Island; this restricted breeding range together with its small declining population of about 1171 breeding pairs in 2003 make it a highly endangered species. **CR. Status and distribution:** Rare and localized non-breeding visitor to Peninsular Malaysia; more frequent on the east coast. Pulau Rengis, off the west coast of Tioman (Pahang) in the South China Sea is the known roosting site for this species which it shares with the more numerous Lesser Frigatebird (64). In Singapore, a male over the main island on 30 May 1986 (W1). **Habitat:** Oceans, oceanic islands, sea coasts. **Habits:** Typically seen soaring aimlessly in small groups, often following ships. Often perches on *kelong* poles. Kleptoparasitism on a White-bellied Sea-eagle (88) has been observed.

63. Great Frigatebird, *Fregata minor* — Plate 9

Simbang Besar

Description: 95 cm (37 in). **Male:** From males of both Christmas Island (62) and Lesser Frigatebirds (64) by all black plumage with pale brown band on upperwings. **Female:**

Restriction of white to the breast separates it from female Christmas Island Frigatebird (16). Whitish throat and lack of white spurs on underwing diagnostic. **Juvenile:** Head tawny or white; blackish breast band complete or partial. **Voice:** Usually silent in non-breeding grounds. **Range:** Pantropical oceans. **Status and distribution:** Very rare non-breeding visitor to Peninsular Malaysia; an individual off Pulau Babi Besar (Johor) on 5 May 1986. Nearest known breeding stations are Christmas Island and the Cocos Keeling Islands (eastern Indian Ocean) and the Paracels (South China Sea). **Habitat:** Oceans, oceanic islands, sea coasts. **Habits:** Soars aimlessly on rising thermals. Like other frigatebirds, skims low over the sea to feed. Follows ships.

64. Lesser Frigatebird, *Fregata ariel* — Plate 9
Simbang Kecil

Description: 78 cm (31 in). Smallest frigatebird (size apparent only when seen with larger congeners). **Male:** All black plumage with white patch on base of underwing diagnostic. **Female:** Black throat, concave hind border of white breast diagnostic. Shows white spurs on underwing. White underparts less extensive than female Christmas Island Frigatebird (62). **Juvenile:** Head rusty buff, breast band partial or complete; whitish spurs on underwing and concave hind border of white breast. **Voice:** Usually silent on non-breeding grounds. **Range:** Pantropical oceans. Nearest breeding station is the Cocos Keeling Islands (eastern Indian Ocean). **Status and distribution:** Locally common non-breeding visitor off both coasts of Peninsular Malaysia. The most numerous frigatebird. Roosts communally on Pulau Rengis with Christmas Island Frigatebird. **Habitat:** Oceans, oceanic islands, sea coasts. **Habits:** Soars in small groups on rising thermals, often following ships. Frequently perches on *kelong* poles. Often seen picking off flying fish from the sea surface. Also takes discarded fish from fishing vessels.

PELICANS

ORDER *Pelecaniformes* **FAMILY** *Pelecanidae*

World: *7 species* **Peninsular Malaysia and Singapore:** *2 species*

Large chunky short-tailed waterbirds with huge bills and large distensible gular pouches. Legs short with fully webbed feet. Swim well but walk clumsily on land. Flight laboured and powerful on long and broad wings with necks tucked back deeply into shoulders. Often soar on thermals. Feed principally on fish. Gregarious. Sexes alike. At present very scarce non-breeding visitors to the review area; the last sizeable flock seen was 15 (species unidentified) in December 1960, off Punggol Point (Singapore).

65. Great White Pelican, *Pelecanus onocrotalus* — Plate 4
Undan Putih

Description: 157cm (60 in). At rest, from Spot-billed Pelican (66) by white plumage with pinkish skin around eye bright yellow to orange gular pouch and pinkish legs and in flight by uniform black primaries and secondaries contrasting sharply with white wing coverts in both upperwing and underwing patterns. **Breeding:** Sports short shaggy nuchal crest. White plumage becomes tinged pink with yellow patch on breast. **Non-breeding (not illustrated):** Gular pouch dull pale yellow. **Voice:** No information from review area. **Range:** Breeding in sub-Saharan Africa, South east Europe through temperate and subtropical northern Asia; northern populations migrate into China, northern India, and Myanmar with stragglers reaching Java and Bali. **Status and distribution:** Vagrant, with sporadic records south to Melaka. Confirmed records include an individual each at Puchong (Selangor) in November 1956 and on coastal mudflats at Penaga, Seberang Perai (Pulau Pinang) on 19 August 1988. **Habitat:** Sea coasts, estuaries, disused mining pools. **Habits:** Engages in cooperative fishing, swimming in small groups and scooping up fish. **Note:** The bird parks in Kuala Lumpur and Jurong (Singapore) and the National Zoo in Hulu Klang (Selangor) have pinnioned individuals although free-flying individuals are possible. Sightings should therefore be treated with caution. Look for coloured rings on legs.

66. Spot-billed Pelican, *Pelecanus philippensis* Plate 4

Undan Paruh Titik

Description: 140 cm (55 in). From Great White Pelican (65) at rest by smaller size, greyish tinged white plumage especially on upperparts and flesh coloured gular pouch (tinged pale bluish or purple) with dark spots on maxilla (perceivable only at close range). Neck lightly tinged brown; legs dark grey to dark brown. In flight, blackish primaries and secondaries show much less contrast with greyish white wing coverts especially in underwing pattern. **Breeding:** Whitish skin around eye; short brownish nuchal crest with pinkish lower back, rump and undertail coverts. **Voice:** No information from review area. **Range:** The Indian subcontinent and southern China through South-East Asia discontinuously to Sumatra and Java. Threatened throughout range due to drainage and pollution of habitat. **NT. Status and distribution:** Vagrant, with sporadic records from Perak and Selangor south to Melaka. An individual on Pulau Kelumpang in the Matang Mangrove Forest Reserve (Perak) in January 2005 is the most recent record (BiW). **Habitat:** Sea-coasts, estuaries, inland freshwater pools. **Habits:** Like Great White Pelican.

BOOBIES

ORDER *Pelecaniformes* **FAMILY** *Sulidae*

World: *9 species* **Peninsular Malaysia and Singapore:** *3 species*

Large stocky seabirds with cigar-shaped bodies, long pointed wings, wedge-shaped tails, strong pointed bills and webbed feet. Flight strong and graceful. Feed largely on fish and squid obtained by plunge diving headfirst from the air into the water. Sexes virtually alike (soft part colouration and voice separate the sexes). Gregarious, forming loose colonies during breeding.

67. Masked Booby, *Sula dactylatra* Plate 9

Dendang-Laut Besar

Description: 86 cm (34 in). Has a more upright posture than other boobies. Distinctive white plumage with black flight feathers, greater wing coverts and tail. Bare skin on face and throat black; bill yellowish; legs vary from yellow and orange to greenish or dark slate blue (never red). From pale morph Red-footed Booby (68) by much larger size, proportionately shorter tail and lack of red feet. **Juvenile:** From juvenile Brown Booby (69) by paler brown upperparts and white collar around neck. **Voice:** Usually silent at sea. During breeding females utter loud honks, males a thin descending whistle (extralimital data). **Range:** Pantropical oceans. **Status and distribution:** Very rare, recorded on Pulau Perak (Kedah) in the Straits of Melaka where occasionally a pair or two may occur together with the more numerous Brown Booby. Nesting and egg laying has been reported there once. Breeding status now uncertain. No other sightings from the west coast. On the east coast, vagrant with two records; an individual off Mersing (Johor) in 1974 and another off Terengganu in October 1982. **Habitat:** Oceans, oceanic islands, sea coasts. **Habits:** Usually solitary or in pairs. Flies high with bill pointed downwards when hunting. Dives straighter and deeper than other boobies. Nests on the ground.

68. Red-footed Booby, *Sula sula* Plate 9

Dendang-Laut Kaki Merah

Description: 70 cm (28 in). Red legs diagnostic. **Pale morph:** Like Masked Booby (67) but much smaller with tail and greater wing coverts all white. **Dark morph:** All brown except for white rump, lower abdomen, tail coverts and tail. Bare facial skin blue; bill light blue with dark tip and red base. **Juvenile (not illustrated):** All brown, the head and neck being paler or of the same colour as rest of underparts. **Voice:** Usually silent at sea. **Range:** Pantropical oceans **Status and distribution:** Vagrant, known from two sight records in the South China Sea; an individual (no date) off Pulau Redang (Terengganu) and a dark morph specimen off Pulau Aur (Johor) in May 1996 (A.C. Sebastian, pers. comm.). **Habitat:** Oceans, oceanic islands, sea coasts. **Habits:** Like other boobies, hunts by plunge-diving. A tree nester, spending a lot of time perched in trees when not aloft. Seldom seen on rocky stacks or ledges.

69. Brown Booby, *Sula leucogaster* Plate 9

Dendang-Laut Coklat/Angsa Laut

Description: 74 cm (29 in). Distinctive dark chocolate-brown head and upperparts (including tail), contrasting sharply with white belly, wing lining and undertail coverts. Bill whitish. Bare facial skin and base of bill blue in male and whitish in female. **Juvenile:** Like adult but white parts pale brown, often suffused with white (very clearly demarcated from permanent brown parts). **Voice:** Usually silent at sea. Males utter wheezy whistles, females throaty honks. A variety of repertoires from crow like caws and duck like quacks to goose like hisses have also been recorded. **Range:** Pantropical oceans. **Status and distribution:** In Peninsular Malaysia, rare and localized resident only on Pulau Perak (Kedah) where the colony has been greatly reduced by the poaching of young and eggs by fishermen. No other sightings from the west coast. On the east coast, vagrant;an individual off Terengganu in October 1982. The other nearest known breeding station is Terumbu Layang-Layang in the South China Sea. Vagrant in Singaporean waters. **Habitat:** Oceans, oceanic islands, sea coasts. **Habits:** Hunts in small groups, plunging intermittently into the water to catch fish. Frequently rests in small numbers on rocky stacks in between hunting. Tame and approachable but shy where persecuted.

CORMORANTS

ORDER *Pelecaniformes* **FAMILY** *Phalacrocoracidae*

World: *30 species* **Peninsular Malaysia:** *2 species*

A group of predominantly black fish-eating birds characterized by long necks, long stiff tails and slender bills with a strong terminal hook. Swim well with webbed feet, usually with bill slightly tilted upwards. Dive well to pursue prey underwater. Highly gregarious during breeding, nesting in dense colonies on cliffs or in waterside trees. Less gregarious outside breeding. Flight strong and direct with long necks outstretched. Fly low over water surface and high above land, usually in V-formations. Lack preen oil to insulate feathers which therefore become waterlogged after prolonged exposure to water. Often perch in the sun with outstretched wings and spread tail in order to dry themselves. Sexes alike.

70. Little Cormorant, *Phalacrocorax niger* — Plate 4
Pependang Kecil

Description: 52 cm (20 in). From Great Cormorant (71) by much smaller size, slender build and proportionately smaller and more slender bill. Plumage black with bronze sheen on mantle. **Breeding:** All black head with whitish sheen on head and neck. **Non-breeding:** Develops white throat and lacks whitish sheen on head and neck. **Juvenile** (not illustrated): Like adult but paler and browner with whitish throat, neck and upper breast. **Voice:** Usually silent in wintering quarters. **Range:** Breeding throughout the Indian subcontinent through South-East Asia discontinuously to Borneo and Java. Population along west coast of Peninsular Thailand appears to be expanding, increasing possibility of future records along the west coast of Peninsular Malaysia. **Status and distribution:** A very rare non-breeding visitor in small numbers at low elevations, south to Melaka, recent sightings in 2007 and 2009 from Sungai Cenang and Pantai Cenang, Pulau Langkawi (Kedah). Also an individual at Bidor (Perak) on 25 January 2007. **Habitat:** Open freshwater swamps, settled mining pools, fish ponds. **Habits:** Like other cormorants, gathers in small groups where common. Frequently dives to catch fish. Other general habits similar to Great Cormorant.

71. Great Cormorant, *Phalacrocorax carbo* — Plate 4
Pependang Air

Description: 82 cm (32 in). Large size and all white throat distinctive. Plumage black with dark bronze sheen on mantle. Heavy bill is blackish with yellow skin at base of mandible. Lores yellow. **Breeding:** Develops dense growth of whitish feathers on thighs and around sides of head and neck. **Non-breeding:** Lacks whitish thigh and sides of head and neck. **Voice:** Usually silent in wintering quarters. **Range:** Breeding in eastern North America east to Africa, Europe and temperate Asia through South and South-East Asia discontinuously to Australia and New Zealand. Northern populations are partially migratory, dispersing south to the Malay Peninsula and the Greater and Lesser Sundas. **Status and distribution:** A very rare non-breeding visitor at low elevations, south to Johor. Recent sightings include individuals from Bagan Belat (Pulau Pinang) on 14 September 2004, Kuala Terengganu (Terengganu) on 28 December 2008 and 10 January 2009, and Johor Bahru (Johor), where 4 were seen on 9 October 2008, with a pair actively engaged in courtship rituals (BiW). **Habitat:** Sea-coasts, inland freshwater swamps, large rivers. **Habits:** Where common, gathers in small groups. Constantly dives to catch fish. Often swims with body partly submerged. Typically seen on open waterside perches with fully outstretched wings, drying waterlogged feathers.

DARTERS

ORDER *Pelecaniformes* **FAMILY** *Anhingidae*

World: *4 species* **Peninsular Malaysia:** *1 species*

Large, long-necked and long-tailed aquatic birds. Differ from cormorants (70, 71) by long, slender neck with long dagger-like bills and long fan-shaped tails. Swim low in the water with bodies fully submerged, exposing only fully outstretched necks. Dive well and pursue prey under water. Feed principally on fish which they spear with bill before swallowing them whole. Like cormorants, darters do not have preen oil and so feathers become waterlogged after prolonged exposure to water. Spend long periods in the sun with outstretched wings, drying themselves. Flight strong and direct with powerful flapping and gliding with neck fully outstretched. Will also soar. Sexes alike.

72. Oriental Darter, *Anhinga melanogaster* Plate 4
Burung Kosa

Description: 91 cm (36 in). Long, slender and distinctively kinked snake-like neck with long and pointed yellow bill distinctive (see Purple Heron (56)). **Non-breeding:** Head and neck brown with long white streak running from chin down side of neck. Upperparts black, marked with white streaks on wings and long white plumes on wing coverts. Tail blackish brown. Legs, greyish-black. Flight sillhouette is distinctive with long outstretched neck and long tail. **Voice:** No information from review area. **Range:** The Indian subcontinent east through South-East Asia to the Greater and Lesser Sundas, Sulawesi and the Philippines. Declining due to hunting, as well as pollution, and drainage of habitat. **NT. Status and distribution:** Presently very rare; the latest records, individuals at disused tin mining pools at Puchong (Selangor) on 14 December 1991 (W1), and at the Kinta Nature Park, Batu Gajah (Perak) on 13 December

2009 (Chiu, 2009), presumably long distance dispersants on transit. **Habitat:** Inland freshwater swamps, disused tin mining pools. **Habits:** Where common, occurs in small flocks, roosting communally in waterside trees and on the ground. Typically sits on exposed branches with fully outstretched wings. Often soars.

FALCONS

ORDER *Falconiformes* **FAMILY** *Falconidae*

World: *64 species* **Peninsular Malaysia and Singapore:** *7 species*

Small to medium sized birds of prey with distinctive long, slender and pointed wings and long narrow tails. Have hooked bill and tooth-like projection at edge of maxilla. Females generally larger than males. Flight strong and swift with rapid but shallow wingbeats. When soaring, wings of some species may sometimes appear rounded, resembling *Accipiter* hawks (101–106). Prey comprises largely birds and insects, most often seized in mid-air. The falconets, genus *Microhierax* (73) are tiny and shorter winged than typical falcons. Nest in tree hollows and spend long periods of time perched. Typical falcons, genus *Falco* (74–79) are much larger and longer winged and nest on cliff ledges or utilize old nests of large birds. Usually solitary. Cere, eyering, and legs yellow in most species in review area. Sexes alike in some species

73. Black-thighed Falconet, *Microhierax fringillarius* Plate 10

Falko Kecil Peha Hitam/Helang Rajawali /Helang Belalang

Description: 16 cm (6 in). Tiny size with small hooked dark greyish bill, blackish upperparts, broad blackish patch on ear coverts separated from mantle by narrow white stripe which is joined to narrow down-curving supercilium diagnostic. Underparts white with rufous undertail coverts and inner thighs as well as buff suffusion on lower breast and flanks. Outer thighs blackish. **Juvenile (not illustrated):** Like adult but forehead, supercilium and cheek patch rufous. **Voice:** A shrill, high-pitched *kli-kli-kli-kli* in flight. **Range:** Tenasserim (Myanmar) and the Malay Peninsula to the Greater Sundas. **Status and distribution:** Locally common resident from low elevations up to 2031 m on Gunung Batu Brinchang, Cameron Highlands (Pahang), south to Singapore, where it is currently a rare non-breeding dispersant (W and H). **Habitat:** Forests, the forest edge, plantations, sparsely wooded open country, wooded gardens. **Habits:** Usually in pairs. Sometimes in small groups of up to five, huddled together on bare branches of dead trees. Frequently darts out rapidly from a perch to capture flying insects. Will also attack small birds.

74. Lesser Kestrel, *Falco naumanni* Plate 11

Falko Kestral Kecil

Description: 29–32 cm (11.5–13 in). Both male and female resemble respective sexes of Common Kestrel (75) but are smaller and noticeably slimmer, with proportionately shorter tail (broadly fanned when hovering) and diagnostic whitish claws. **Male:**

From Common Kestrel by unmarked bluish-grey crown, nape, and sides of head, and unmarked rufous-chestnut mantle with bluish-grey tertiaries, greater coverts and median covert tips, buff underparts with lesser black spots, and lacking the dark malar stripe, characteristic of the former. **Female:** Has more diffuse malar stripe than female Common with finer streaks on crown and nape and narrower, more V-shaped dark markings on upperparts. Both sexes also have grey bills with dark tip and yellow cere, eyering and legs. **Voice:** Unknown in review area. **Range:** Breeding in southern Europe and North Africa throughout much of the Palaearctic, wintering principally in sub Saharan Africa, rarely in South and South-East Asia. **Status and distribution:** Vagrant with sightings of 2–4 individuals at Changi Coast Road, Changi International Airport, East Hagen, and Simei (Singapore) from 4 January to 16 March 2001 (W and H) and 1 male at Changi Cove (Singapore) from 6–18 March 2010 (SINAV 24: 3). **Habitat:** Open country. **Habits:** Usually in small groups. Flight more buoyant than Common Kestrel. Hovers while hunting. Other habits little known within the review area.

75. Common Kestrel, *Falco tinnunculus* Plate 10

Falko Kestral Biasa

Description: 31–35 cm (12–14 in). Small largelybrown falcon with proportionately long graduated tail, which is broadly fanned when hovering. **Male:** Greyish head and nape together with bluish grey tail contrasting with rufous-chestnut upperparts, marked with black spots diagnostic. Underparts buff with blackish streaks or spots particularly on the breast, belly, and flanks. Tail has black subterminal band and narrow white tip. **Female:** Upperparts all brown with dark streaks on crown and nape and distinct triangular or bar-shaped markings on mantle and uppertail. Both sexes have grey bills with dark tip and yellow cere, eyering and legs with black claws. Also show diffuse blackish malar stripe. **Juvenile (not illustrated):** Both sexes resemble female but have more heavily streaked underparts. **Voice:** A high-pitched *ki-ki-ki-ki-ki*... **Range:** Breeding in Africa, Europe, the Middle-East, and temperate Asia as well as parts of India and northern Indo-China; northern populations are migratory, wintering south to tropical Africa, South and South-East Asia to Sumatra, Borneo, and the Philippines. **Status and distribution:** Rare and localized winter visitor in small numbers principally at low elevations, south to Singapore. **Habitat:** Open country, scrub. **Habits:** Usually solitary, perching on dead trees and fence posts. Flies with shallow wingbeats and also soars and circles in the air, stopping to hover momentarily before plunging down to catch prey (principally rodents) which is devoured on the ground or on a perch.

76. Amur Falcon, *Falco amurensis* Plate 11

Falko Amur

Description: 28–31 cm (11–12.2 in). Adult of both sexes differ from all *Falco* falcons in review area by red eyering, cere and legs. **Male:** Unmistakeable with slate-grey head and upperparts with pale grey underparts together with rufous-chestnut thighs and undertail coverts. In underwing flight pattern, white wing coverts contrasting with black flight feathers diagnostic. **Female:** Resembles adult Eurasian Hobby (77) but soft parts red, dark grey upperparts (including tail) marked with dark bars, thighs and undertail coverts white, and malar stripe narrower. In underwing flight pattern, shows whiter base colour of coverts and more obvious dark pale bars on undertail. **Juvenile:** Eyering and cere pale yellow. From juvenile Eurasian Hobby by reddish legs and more prominent and broader buff fringes on upperparts and wing coverts, with pale parts of head and underparts (including underwing) noticeably whiter. **Voice:** Unknown from review area. **Range:** Breeding in the eastern Palaearctic from eastern Siberia and Ussuriland, north and north-east China, northern Korea, wintering south-west principally throughout sub-Saharan Africa, less often in South or South-East Asia. **Status and distribution:** Vagrant, with three confirmed sightings in Peninsular Malaysia; one each at Sungai Balang, Muar (Johor) on 6 January 2002, at Tasik Temengor (Perak) on 24 December 2005, and Kuala Ibai, Kuala Terengganu (Terengganu) on 13 November 2007 (BiW). In Singapore, a female at Changi Coast on 21 November 2007 (SINAV 22: 5). **Habitat:** In review area, open scrubland fringing ricefield, and forest, and beach side casuarinas. **Habits:** Often hunts and catches insects on the wing. Sometimes hovers while hunting. Other habits little known.

77. Eurasian Hobby, *Falco subbuteo* Plate 10

Falko Peha Merah

Description: 31–35 cm (12–14 in). Small, long and pointed-winged falcon. From migratory subspecies of Peregrine Falcon (79) by much smaller size, less stocky appearance, proportionately broader white side of neck, streaked (not barred) underparts and unbarred uppertail. Black head and broad malar stripe, blackish upperparts, white throat and sides of neck together with chestnut thighs and undertail coverts diagnostic. Bill grey with yellow cere. Legs yellow. **Juvenile:** Upperparts browner with paler fringes; thighs and undertail coverts buff. **Voice:** A high-pitched *kiu-kiu-kiu-kiu* **Range:** Breeding throughout Europe and temperate Asia through the Himalayas and China to Indo-China; wintering south to Africa, the Indian subcontinent and South-East Asia to Java. **Status and distribution:** Vagrant, known from two confirmed sightings; an adult at sea level at Melaka town (Melaka) on 8 November 1982 (Wells, 1990) and a juvenile along the East-West Highway (Perak) at 1050 m on 26 January 2002 (BiW). **Habitat:** In review area, open country fringing intertidal mudflats and roadside scrub at the edge of forest. **Habits:** Flies fast over open country with wings swept far back, giving the impression of a large swift. Largely hawks for flying insects, sometimes in the late evening. Will also attack other smaller birds. Does not hover.

78. Oriental Hobby, *Falco severus* Plate 11

Falko Perut Merah

Description: 28–33 cm (11–13 in). Small falcon like Eurasian Hobby (77) in size and shape but smaller and stockier with shorter wings. At rest, all black hood without contrasting white cheek markings together with unmarked reddish-rufous breast, belly, and undertail coverts distinctive. Throat and fore collar buff white. Upperparts, including tail, slate-grey marked with paler fringes, with darker flight feathers. Bill greyish, tipped black. In underwing flight pattern, shows distinctive unmarked reddish-rufous wing coverts. **Juvenile:** Like adult but breast and belly marked with blackish streaks. Upperparts and wing coverts darker and browner. In underwing flight pattern, shows black-streaked reddish-rufous wing coverts. **Voice:** Unknown in review area. **Range:** North and north-east India, Nepal, and Bhutan east through south-west and southern China to the Greater Sundas and the Philippines, east to Sulawesi, New Guinea, and northern Melanesia. **Status and distribution:** Vagrant, a juvenile at Pulau Ubin (Singapore) on 14 July 2002 and seen hunting there on 19 July 2002 (W and H). Another juvenile over Kranji and the Sungei Buloh Wetland Reserve on 4 January 2010 (SINAV 24: 1) Reports from Singapore on 9 December 1997 and 28 October 2005 could not be confirmed (W and H) **Habitat:** Open country. **Habits:** Flight swift like Eurasian Hobby. Hunts for insects in the air. Other habits little known in review area.

79. Peregrine Falcon *Falco peregrinus* Plate 10

Falko Perigrin

Description: 38–48 cm (15–19 in). Large, compact, broad and pointed winged falcon. Upperparts slate-grey; bill blackish with yellow cere, and yellow legs. Shows narrow yellow orbital skin. *F.p. ernesti* is a dark bird with black on head showing as a hood with white sides of neck very much reduced. Underparts lavender grey marked with bold dark bars. In poor light from Bat Hawk (84) by less pointed wings. **Juvenile:** Upperparts dark brown with buff fringes, underparts whitish with black bars, cere blue-grey. *F.p. japonensis* **(not illustrated)** is paler with pale buff underparts marked with narrow dark bars. Has faint dark streaks on upper breast. Head black with broad black malar stripe contrasting with white sides of neck. **Voice:** A chattering *kek-kek-kek-kek-kek*. **Range:** Breeding worldwide on all continents. Populations from northern latitudes are migratory, wintering south to South America, Africa, South and South-East Asia to New Guinea. **Status and distribution:** Occurs from low elevations up to 2000 m. *F. p. ernesti* is a locally common resident, confined to limestone outcrops south to Batu Caves (Selangor). Individuals may disperse over a wider area outside breeding. *F. p. japonensis* is a fairly common passage migrant and winter visitor, south to Singapore. **Habitat:** Limestone cliffs, forests, the forest edge, open country, mangroves, sea-coasts. **Habits:** Usually solitary or in pairs. Flies very fast, making steep dives to knock out birds in the air before swooping down to retrieve them. Often displays great dexterity and acrobatic skills in flight. Appears often at high tide roosts of shorebirds in mangroves and mudflats. **Note:** *F. p. peregrinator* from the Indian subcontinent has rufous underparts and could turn up in the review area. May be confused with Oriental Hobby *Falco severus* (78) but is much larger and more robust and shows a broader white patch on sides of neck, in addition to streaks on underparts.

OSPREY, HAWKS, KITES, EAGLES, AND VULTURES

ORDER *Falconiformes* **FAMILY** *Accipitridae*

World: *242 species* **Peninsular Malaysia and Singapore:** *37 species*

A large and varied family of small to very large predatory birds, characterized by strong, hooked bills and powerful talons which are specialized for catching and dismembering prey. Plumage highly variable (colour morph, adult, subadult and juvenile) in many species but distinct size, body shape, length, breadth and position of wings in flight as well as tail length and shape are important field characteristics for positive identification. Since many are seen mostly in flight, a good knowledge of underwing and tail patterns (distinct in each species) is also essential. Wing and tail shapes however may vary according to what the bird is doing; tails may be spread or held close together when soaring or gliding on level wings. Wings may be slightly angled backwards when gliding slowly or steeply angled backwards in a fast glide. Most of the larger species have long, broad wings which are specially adapted for soaring on rising thermals. Most hunt live prey while some scavenge on refuse and carrion. Many species are migratory. Nests are usually untidy piles of sticks, constructed in tall trees. Sexes alike in most species but females are considerably larger than males. Solitary or in pairs but migrate in large flocks. May be divided into the following groups and genera according to distinct physical and behavioural characteristics: **Osprey,** genus *Pandion* (80). Medium sized. Differs from other raptors by having long, narrow angled wings which allow it to efficiently plunge-dive into water to catch fish on which it feeds exclusively. Being piscivorous, its feet have rough scales and sharp claws which are specially adapted for catching and gripping fish. Lives near large water bodies and along water courses. Solitary. Sexes alike. **Bazas,** genus *Aviceda* (81, 82). Small to medium size with long erectile crest and broad wings. Like falcons, have tooth-like projection at edge of maxilla. Hunting behaviour like hawks. **Oriental Honey-buzzard,** genus *Pernis* (83). Medium sized with distinctive longish neck and tail and highly variable plumage patterns. Drooping crest varies in length. Northern latitude populations are migratory while southern latitude birds are sedentary. Unaggressive in nature it mimics the plumage patterns of the more powerful and aggressive hawk-eagles in their south and south-east Asian wintering grounds as a means of protection against other more powerful raptors. **Bat Hawk,** genus *Macheiramphus* (84). Black with long, pointed wings and long wispy crest. Crepuscular. **Kites,** genera *Elanus* (85), *Milvus* (86), and *Haliastur* (87). Small to medium size with long, angular wings distinctive. Flight buoyant and graceful. Inhabit primarily open country and lightly wooded areas. **Eagles:** Medium size to large and powerful raptors. Smaller and shorter necked than vultures, with feathered head and lacking neck ruff. Larger than buzzards with proportionately larger heads and stronger and more robust bills. **Sea-eagle,** genus *Haliaeetus* (88) and **Fish-eagles,** genus *Ichthyophaga* (89, 90). Medium to large size, long broad wings, wedge-shaped tails and unfeathered distal half of tarsi distinctive. Feed largely on fish and live near water. **Vultures,** genera *Gyps* (91–93), *Sarcogyps* (94), and *Aegypius* (95). Very large masterful soarers on long, broad wings, usually held level in flight. Tails short and wedge-shaped. Heads and rather longish necks featherless or sparsely down covered. Have ruff at base of neck. Feed largely on carrion. All Asian vultures (except Himalayan Griffon) have experienced drastic population declines in the Indian subcontinent as a result of mortality from feeding on livestock carcasses treated with the veterinary drug diclofenac which causes renal failure as a result of visceral gout (BirdLife International, 2010). In South-East Asia, the decline is mainly due to the demise of wild ungulates due to much reduced suitable habitats as well as proper and more modern disposal of dead livestock (BirdLife International, 2010). All affected species globally rank **CR** and are appropriately indicated in the species account below. **Snake-eagles,** genera *Circaetus* (96) and *Spilornis* (97). The former is large, heavy and broad-winged with unfeathered tarsi while the latter is smaller and darker with long, broad wings, marked with distinct white band. Tail marked with black and white bands. **Harriers,** genus *Circus* (98–100). Medium size with long slim body (including tail), long wings and owl-like facial disc distinctive. Inhabit open country. Hunt by quartering low over ground, holding wings upwards in a shallow 'V' with head protruding. Tarsi unfeathered. Sexes differ. **Sparrow-hawks,** genus *Accipiter* (101–106). Small to medium size. with long banded tail and distinctive short, rounded wings. Prey largely on smaller birds, captured through relentless aerial pursuit. Also hunt largely from perches. Flight consists of rapid wingbeats interspersed with short glides. Sometimes soar. **Buzzards,** genera *Butastur* (107) and *Buteo* (108). Medium size with long wings and proportionately small heads and bills. The latter is polymorphic with pale and dark morphs. Tail shapes square-cut to broad and rounded. Soar high in great arcs. Tarsi unfeathered. **Large dark eagles:** Genus *Ictinaetus* (109), large and dark with long, broad wings and characteristic long, square-cut narrow tail. Inhabits forests. Genus *Aquila* (110–112), large size with black or dark brown plumage with long, broad wings and broad wedge-shaped tail. Tarsi feathered. Inhabit largely open country. **Smaller eagles,** genera *Hieraaetus* (113) and *Lophptrorchis* (114). Medium size with long wings with striking dark and light underwing pattern, the latter sports a short crest. Tarsi feathered. **Hawk-eagles,** genus *Nisaetus* (115–118). Medium to large size with very broad wings and long rounded tails. Sport short to long erectile crests. Tarsi feathered.

80. Osprey, *Pandion haliaetus* — Plate 12
Helang Tiram

Description: 55–61 cm (20–23 in). White head and neck with broad black eyeline, blackish-brown nape and upperparts together with white underparts with blackish brown streaked breast band (sometimes not prominent) diagnostic. Shows superficial nuchal crest. Uppertail marked with alternate blackish brown and paler greyish-brown bands with broad blackish brown terminal band. Iris yellow, legs greyish. In flight, long narrow wings are bent back at the carpal joint. Underwing all white with prominent black patch on carpal joint, blackish flight feathers and band along margin of coverts. Undertail whitish with dark bands. **Juvenile (not illustrated):** Upperparts paler brown with white fringes. Lacks broad blackish terminal band of uppertail. **Voice:** A series of loud whistles, *kiu-kiu-kiu-kiu*... **Range:** Breeding throughout temperate and sub-tropical latitudes worldwide; northern populations are migratory, wintering south to tropical and equatorial regions. **Status and distribution:** Locally common non-breeding visitor at low elevations, south to Singapore. Fairly common during and between autumn and spring migration from October to March. Some individuals, particularly juveniles stay throughout the year. Birds seen from May to August are presumably migrants from Australasia during the southern hemisphere winter. **Habitat:** Sea-coasts, mangrove creeks, brackish ponds, fish ponds, disused mining pools, forested waterways. **Habits:** Typically seen in flight over water courses, flapping with slow wingbeats, pausing to hover before folding back long wings and plunge-diving heavily into water to catch fish. Sometimes soars with slightly arched wings. Often utilizes exposed waterside perches as vantage points as well as places to feed.

81. Jerdon's Baza, *Aviceda jerdoni* Plate 12

Helang Baza Perang

Description: 46 cm (18 in). Long erectile white-tipped crest, dark brown to tawny-golden upperparts, paler on head and upper breast together with white lower breast, belly and thighs marked with broad rufous bars diagnostic. Has dark mesial stripe. Tail marked with alternate blackish brown and pale rufous bands. In underwing flight pattern, shows whitish wing coverts, pale rufous and whitish wing lining, black and white wingtips and dark subterminal tail band. From Wallace's Hawk-eagle (118) (which it mimics) at rest by folded wingtips reaching more than halfway down the tail and unfeathered lower tarsi. **Juvenile:** Underparts paler, marked with brownish streaks. **Voice:** A shrill and distinct disyllabic whistle, *pee-weeow*. **Range:** The Indian subcontinent east to Myanmar and Thailand, discontinuously to Sumatra, Borneo, Sulawesi, and the Philippines. **Status and distribution:** Rare and localized resident at low elevations, south Singapore, where it is a probable dispersant. **Habitat:** Forests, the forest edge. **Habits:** Keeps to tree-tops, hunting from perches. Other habits little known.

82. Black Baza, *Aviceda leuphotes* Plate 12

Helang Baza Hitam

Description: 33 cm (13 in). Small black and white raptor with long erectile crest diagnostic. When perched shows maroon patch on secondaries; perceivable at close range. Wings broad and rounded. In underwing flight pattern, shows black wing coverts, wingtips and undertail coverts which contrast with grey flight feathers and tail together with white breast band and whitish underparts, marked with narrow rufous bars. In upperwing flight pattern **(not illustrated)**, all black with large white band on secondaries. **Voice:** A disyllabic *chu-weep*. Also a soft squeal or whistle. **Range:** Breeding in India, Myanmar and northern Thailand; wintering south to Sri Lanka, South-East Asia and Sumatra. **Status and distribution:** Locally common passage migrant and winter visitor at low elevations, south to Singapore. Common in large numbers at Tanjung Tuan (Melaka) during autumn and spring passage. **Habitat:** Forests, the forest edge, plantations, heavily wooded areas. **Habits:** Usually in small loose groups, frequenting the canopy and crowns of tall trees. Sometimes sits on open perches. Flight pattern distinctive, consisting of short crow-like flaps interspersed with glides; the wings are usually held flat. Migrates in large compact flocks, soaring high on rising thermals.

83. Oriental Honey-buzzard, *Pernis ptilorhyncus* Plate 13

Helang Lebah Asia

Description: 51–61 cm (20.4–24.4 in). A polymorphic species with highly variable plumage types and two subspecies in review area. Longish tail and neck together with small pigeon-like head distinctive in all subspecies and plumage types. Wings long and broad, usually held horizontal when soaring or slightly down-curved with upturned tips when gliding. When perched, shows drooping crest which varies in length. Legs and feet yellow. Migrant *P. p. orientalis:* Crest very short, appearing as a tuft on hindcrown, mimicking Changeable Hawk-eagle (115). Upperparts vary from dark to rusty brown or creamy white, usually marked with dark bars or streaks. **Male:** Head greyish, showing dark malar stripe and broad pale throat (sometimes bordered black) with dark mesial stripe. Underparts vary from dark brown to white (dark and pale morphs) but marked with dark or rufous bars on breast, belly, greater wing coverts with dark trailing edge of wings and wingtips. Iris dark brown or orange. Bill and cere dark grey. Tail has two complete, well spaced bold dark bands with a broad white central band like Crested Serpent-eagle (97). **Female:** Head brownish, iris pale yellow and tail shows narrower alternate dark and pale bands. **Juvenile: (not illustrated):** Plumage varies from dark brown to creamy buff or whitish underparts and underwing coverts. Barrings usually finer. Some individuals show large pale patch at base of primaries. Cere yellow. Resident *P. p. torquatus*: Crest taller and wispy. Also has pale or white throat and dark mesial stripe. **Typical morph**: Largely brown, showing little variation in plumage. **Adult male:** Sides of face grey, upper breast all brown with fine barrings on lower breast, belly, and flanks. Tail marked with broad black terminal band with a similar width white band in the centre and another broad black band at the base. **Female: (not illustrated):** Sides of face brownish. Alternate white and brown barrings from chin to rest of underparts. Tail marked with narrow dark terminal band, with a broad white band in the centre and smaller black band at the base. Iris orange yellow in both sexes. **Juvenile (not illustrated):** At 12 months, upperparts like adult, underparts paler with dark streaks on breast, and rufous barrings on belly and thighs, cere yellow, and iris brown, which gradually turns yellow by 16 months (Chiu, 2005). **Dark morph (not illustrated):** Like typical morph but overall dark brown with whitish barrings on underparts. **Tweeddale morph:** Black head and upperparts, white underparts with blackish streaks on upper breast and blackish bars on lower breast, belly, flanks, thighs, and undertail coverts, recalling Blyth's Hawk-eagle (117). **Voice:** *P. p. torquatus*: A nesting adult uttered a loud plaintive cadence *kwik-kwik-kwik-kwik-kwik-kwik*... with all notes level. A young still in nest uttered a wailing and repeated sequence, *peee-eeu*, *peee-eeu*, with the second note descending. There is no information on the call of *P. p. orientalis* from the review area. **Range:** Breeding throughout northern temperate Asia, India, and South-East Asia discontinuously to the Greater Sundas and the Philippines. Temperate populations are migratory, wintering south to tropical latitudes. **Status and distribution:** *P. p. orientalis* which breeds from Siberia to Japan winters largely in the Indian subcontinent, continental South-east Asia and the Indonesian archipelago and occurs as a fairly common passage migrant at both low and high elevations south to Singapore. Very common in very large numbers at Tanjung Tuan (Melaka) during autumn and spring passage. *P. p. orientalis* is a rare and localized resident from low elevations up to 1400 m, south to Johor, occurring as an uncommon dispersant to Singapore. **Habitat:** Forests, heavily wooded areas, open country (during migration). **Habits:** The migratory *P. p. orientalis* tends to thermal at great heights in large flocks during passage. Will perch on trees in between passage. Migrating flocks during spring passage at Tanjung Tuan have been observed with open bills, gasping for air while riding the thermals on arrival at the promontory, after tedious flapping over the sea. Will also twist tail in flight to change direction. The resident *P. p. torquatus* is usually solitary, gliding high over forests. Performs the 'sky dance', an elaborate aerial courtship display involving constant upward swoops with flapping at the apex and followed by downward stalls. Feeds principally on the larvae of honey bees, wasps, and hornets which it obtains by breaking up their nests with its bill. Also takes honeycombs from bee-hives. Will also take small vertebrate prey.

84. Bat Hawk, *Macheiramphus alcinus* Plate 10
Helang Kelawar

Description: 46 cm (18 in). From Peregrine Falcon (79) by bulkier build, longer and broader wings which are pointed at tips and black plumage. Has long, wispy crest, held horizontally over nape while perched. Throat whitish with broad dark mesial stripe, iris yellow; short supercilium and streak below eye white. In underwing flight pattern, pale throat contrasts with dark mesial stripe and all dark underparts. **Voice:** A cadence of short high-pitched whistles, *kwik-kwik-kwik-kwik*. **Range:** Tropical Africa (including Madagascar) east to South and South-East Asia, from Tenasserim (Myanmar) to the Malay Peninsula, Sumatra, Borneo, and New Guinea. **Status and distribution:** Uncommon and localized resident from low elevations up to 1200 m, south to Singapore, where it is a rare dispersant.. **Habitat:** Forests, the forest edge, usually near limestone outcrops. **Habits:** Frequents the canopy and middle storey, solitarily or in pairs. Wingbeats slow and deep, but displays great dexterity when pursuing bats which form its main food item. Principally crepuscular but will sometimes fly about during the day.

85. Black-winged Kite, *Elanus caeruleus* Plate 12
Helang Tikus

Description: 28–33 cm (11–13 in). Medium sized pale and long-winged raptor. Pale grey crown, mantle and wings with black shoulder patch together with white underparts and blackish wingtips and flight feathers diagnostic. Iris red; legs yellow. Wings raised when gliding. **Juvenile:** Upperparts tinged brown and marked with whitish scales; breast faintly streaked brownish; iris yellow. **Voice:** A soft whistle. **Range:** North and South America, Africa, the Iberian Peninsula, the Indian subcontinent and southern China through South-East Asia and the Indonesian Archipelago to eastern New Guinea and Australia. **Status and distribution:** Fairly common resident at low elevations, south to Singapore. Has been observed flying west across the Straits of Melaka at Tanjung Tuan (Melaka) and may undertake local dispersion to Sumatra. **Habitat:** Open country, ricefields, the mangrove edge, inland freshwater swamps. **Habits:** Usually solitary or in pairs. Flight buoyant with moderate wingbeats. Often hovers while hunting. Feeds largely on rodents. Utilizes electricity posts and cables as vantage points to spot for prey. Frequently nests in coconut palms.

86. Black Kite, *Milvus migrans* Plate 12
Helang Ekor Cabang

Description: 61–66 cm (24.4–26 in). Dark brown plumage with long angular wings and forked tail distinctive. When broadly fanned in flight, tail often appears almost square-cut. In underwing flight, shows conspicuous whitish patch on base of primaries. In flight, wings slightly droop at carpal joint or held horizontal. Subspecies *M. m. govinda* **(not illustrated)** is more reddish brown and has yellow cere and legs.

Subspecies *M. m. lineatus* is larger, has bluish grey cere and legs and larger whitish wing patch on underwing. **Voice:** Silent in wintering quarters. **Range:** Breeding in Africa, Europe, temperate Asia, South and South-East Asia discontinuously to New Guinea, and Australia. Northern temperate populations are migratory, wintering south to Africa, South and South-East Asia. **Status and distribution:** Locally common winter visitor in small numbers at low elevations, south to Singapore, where it is uncommon. *M. m. lineatus* is more common. **Habitat:** Open country, ricefields, beach scrub. **Habits:** Usually solitary or in small groups, soaring over open country. Flight slow with floppy wingbeats. Feeds principally on carrion, occasionally taking live prey. Kleptoparasitises other raptors. Where common, roosts communally in tall trees.

87. Brahminy Kite, *Haliastur indus* — Plate 12
Helang Merah

Description: 43–51 cm (17.2–20 in). White head, neck and breast, streaked blackish brown contrasting with bright chestnut upperparts diagnostic. Wings held forward and angled at carpals in flight. In underwing flight pattern, black wingtips and chestnut wing coverts contrast with rufous flight feathers. **Juvenile:** From Black Kite (86) by smaller size and more rounded tail. Overall dull brown with buff spots and streaks; older birds show pale buff head. In underwing flight pattern, shows large pale patch at base of primaries. **Voice:** A loud and long querulous *mew*, falling, rising and then tapering off on a falling tone. **Range:** The Indian subcontinent, southern China and South-East Asia to the Indonesian Archipelago, Australia, and the Solomon Islands. **Status and distribution:** Common and widespread resident at low elevations in coastal districts and offshore islands south to Singapore. Very common at harbours and fishing villages. Also occurs inland in the suburbs of Kuala Lumpur. **Habitat:** Mangroves, beach scrub, ricefields, freshwater swamps and reservoirs in coastal districts. **Habits:** Usually in small groups, soaring over mangroves and fishing villages, where it scavenges largely on offal and discarded fish from fishing boats. Will also take live prey. Flies with rather slow and floppy wingbeats. Often perches on *kelong* poles out at sea. Roosts in tall trees.

88. White-bellied Sea-eagle, *Haliaeetus leucogaster* — Plate 14
Helang Laut/Helang Siput

Description: 60–69 cm (24–28 in). Large pale eagle with long, broad wings (narrower at tips) and short, wedge-shaped tail. Assumes horizontal posture while perched. When soaring, wings are held forward and raised in a shallow 'V'. White head and underparts contrasting with grey upperparts diagnostic. In underwing flight pattern, white wing coverts contrasts with black flight feathers. Also shows a narrow black band at base of white tail. **Juvenile:** Pale brownish plumage with creamy underwing coverts, often marked with patches of brown. In underwing flight pattern, shows pale patch at base of primaries (reduces with age) as well as buff tail with dark terminal band. **Voice:** A series of loud sonorous nasal goose-like honks, *kank-*

kank-kank-kank..., uttered both in flight and while perched. **Range:** The Indian subcontinent, southern China and South-East Asia to the Indonesian Archipelago, Australia and the western Pacific. **Status and distribution:** Fairly common resident at low elevations in coastal districts, including most offshore islands, south to Singapore. Sometimes occurs inland. Also seen at 1500 m at Tasik Habu, Cameron Highlands (Pahang). Less common than Brahminy Kite (87). **Habitat:** Sea-coasts, large rivers, inland freshwater swamps, settled mining pools, reservoirs, open country. **Habits:** Usually in pairs, usually soaring on thermals. Feeds largely on fish, plucked from just below water surface. Sometimes scavenges on refuse near fishing villages. Has been observed harrying Osprey (80) and robbing it of catch. (see Christmas Island Frigatebird (62)). Nests in tall trees or on rocky ledges and sometimes on tall telecommunication masts. Very vocal.

89. Lesser Fish-eagle, *Ichthyophaga humilis* Plate 14
Helang Kanguk Kecil

Description: 51–64 cm (20.4–26 in). Like Grey-headed Fish-eagle (90) in shape and colour but much smaller with all dark tail, marked with little white mottling at base. **Juvenile:** Like juvenile Grey-headed Fish-eagle but has less prominent whitish streaks on breast. Shows darker brownish mottles on underwing coverts. **Voice:** Loud cries, *hak-hak*..., distinct from Grey-headed Fish-eagle. **Range:** India through South-East Asia to Sumatra, Borneo and Sulawesi. Being dependent on forested waterways with good stocks of fish, it has undergone a moderately rapid decline due to deforestation, overfishing, and pollution. **NT. Status and distribution:** Locally common resident from low elevations up to 300 m, south to Johor. More common than Grey-headed Fish-eagle. **Habitat:** Forested waterways. **Habits:** Usually solitary, often perching on waterside trees. Flies with slow steady wingbeats low over water surface. Catches fish from the water surface. Seldom soars.

90. Grey-headed Fish-eagle, *Ichthyophaga ichthyaetus* Plate 14
Helang Kanguk Besar

Description: 61–74 cm (24.4–30 in). Large eagle with short broad wings and short rounded tail. Mantle and wings dark brownish grey, contrasting slightly with pale grey head and sharply with white lower belly and undertail coverts. Breast and upper belly brownish. Tail white with black terminal band, conspicuous while perched and in flight. In underwing flight pattern, shows all dark wings and breast. **Juvenile:** Pale brown head and breast marked with whitish streaks on the latter. Shows pale buff supercilium. Basal third of tail brownish, showing less contrast with dark terminal band. **Voice:** A loud *u-wak, u-wak, u-wak*...Also a wailing *oaaaar-oaaaar*, with the second syllable on a falling tone. **Range:** India through South-East Asia to the Greater Sundas, Sulawesi and the Philippines. Like Lesser Fish-eagle (89) is also dependent on forested waterways with good stocks of fish, and is undergoing a moderately rapid decline due to deforestation, overfishing, and pollution. **NT. Status and distribution:** Uncommon and localized resident at low elevations, principally in freshwater habitats, south to Singapore, where it is rare. **Habitat:** Forested waterways, coastal ricefields, reservoirs, sea-coasts. **Habits:** General habits similar to Lesser Fish-eagle. Very vocal especially at dawn and dusk and sometimes at night.

91. White-rumped Vulture, *Gyps bengalensis*
Hereng Pinngul Putih

Description: 75–85 cm (30–34 in). Large size. From all *Gyps* vultures by blackish plumage, white lower back and rump and white underwing coverts contrasting with blackish flight feathers (conspicuous in flight). Secondaries slate-grey; head and neck dark greyish brown; ruff white; maxilla pale bluish grey with blackish tip; mandible and cere blackish; legs dark grey. Soars with wings held in a shallow 'V'. **Juvenile:** Dark grey-brown plumage with all brown lower back and rump, brownish ruff and whitish streaked underparts. Bill and cere blue-black. In underwing flight pattern, grey-brown wing coverts contrast with narrow whitish patagial bar and blackish brown flight feathers and tail. **Voice:** Usually silent except for harsh grunts and hisses at carcasses. **Range:** South-east Iran and southern Afghanistan to the Indian subcontinent, southern China and South-East Asia. **CR. Status and distribution:** Locally extinct. Formerly common resident at low elevations, south to Kelantan and northern Perak with wandering dispersants reaching Johor and Singapore. Bred in Kedah at least up to 1950. **Habitat:** Open country, lightly wooded areas. **Habits:** Usually solitary or in small groups, soaring high on rising thermals. When carcass is sighted, quickly descends to feed. Noisy feeder. Roosts in tall trees. **Note:** Not illustrated in colour (see figure).

92. Indian Vulture, *Gyps indicus*
Hereng Paruh Panjang

Description: 80–95 cm (32–38 in). From White-rumped Vulture (91) by creamy brown plumage, dark bluish grey neck, yellowish bill and cere. Ruff white. When soaring, wings are held more level than White-rumped Vulture. In both upperwing and underwing flight patterns, blackish brown flight feathers and tail contrast with creamy brown upperparts, underparts and greater wing coverts. Shows pale creamy buff rump in flight. **Juvenile:** Browner and darker, lacking ruff. Paler than juvenile White-rumped Vulture. **Voice:** Usually silent except for harsh hisses and grunts at carcasses. **Range:** The Indian subcontinent and South-East Asia.**CR. Status and distribution:** Vagrant, known from two specimens collected in Perak and Ulu Terengganu (both pre-1950s but exact dates and locations unknown). **Habitat:** Open country, lightly wooded areas. **Habits:** Like White-rumped Vulture. **Note:** Not illustrated in colour (see figure).

93. Himalayan Griffon, *Gyps himalayensis* Plate 14
Hereng Himalaya

Description: 115–125 cm (46–50 in). Very large pale vulture with long, broad wings (narrower at tip) and short wedge-shaped tail distinctive. Soars with wings held in shallow 'V'. Upperparts pale buff brown (including wing coverts); flight feathers blackish; ruff pale buff with white streaks; underparts pale buff. Head and neck sparsely covered with whitish feathers. Bill and cere pale yellowish, often tinged grey in the latter. Legs pale pinkish. In underwing flight pattern, pale buff underparts and greater wing coverts contrast with blackish flight feathers and tail. **Juvenile:** Darker plumage with long, narrow whitish streaks on underparts and entire upperparts. Ruff brown with long, lanceolate feathers. Bill and cere blackish. **Voice:** Usually silent except for grunts and hisses at carcasses. **Range:** Breeding at high altitudes in central Asia, Afghanistan, North Pakistan, the Himalayas, southern Tibet, and central China; juveniles are known to regularly winter at low altitudes in North India and Nepal. The occurrence of individuals in Thailand, Peninsular Malaysia and Singapore is a point that needs further investigation. **Status and distribution:** Vagrant, first known from 5 juveniles at Bukit Timah (Singapore) in January 1992. A juvenile each in a ricefield at Muar (Johor) in January 1995 and another in flight over MacRitchie reservoir (Singapore) on 15 January 2010 (SINAV 24: 1) besides other sporadic records. **Habitat:** Open country. **Habits:** Soars high on rising thermals. Other habits little known.

94. Red-headed Vulture, *Sarcogyps calvus*
Hereng Kepala Merah

Description: 81 cm (32 in). Large blackish vulture with broad wings (narrower at tip), short wedge-shaped tail, red head, white ruff, red legs and oval white thigh patches diagnostic. When soaring, wings are level and only slightly raised. In underwing flight pattern, shows whitish patagial bar and greyish white band on base of secondaries. **Juvenile:** Plumage browner with brownish ruff, pinkish head and white lower belly and undertail coverts. **Voice:** Hisses and grunting notes at carcasses. Otherwise usually silent. **Range:** The Indian subcontinent and South-East Asia. **CR. Status and distribution:** Formerly resident at low elevations south to Taiping (Perak) and Kemaman (Terengganu), with dispersants reaching Singapore. Has greatly declined since the 1950s, the last known record was of a probable dispersant in Ampang (Selangor) in February 1964. **Habitat:** Open country, lightly wooded areas. **Habits:** Like other vultures. **Note:** Not illustrated in colour (see figure).

95. Cinereous vulture, *Aegypius monachus*
Hereng Hitam

Description: 100–110 cm (40–44 in). Very large black vulture with blackish bill and bluish grey cere; blackish ruff; pale blue skin on neck and pale greyish blue legs diagnostic. Always soars on straight wings, showing all dark upperparts and underparts. **Juvenile:** Plumage browner; head blackish. **Voice:** Usually silent except for croaks, hisses and mews at carcasses. **Range:** Breeding from southern Europe through the Middle-East and southern temperate Asia; wintering south to Africa and South Asia with stragglers reaching South-East Asia. Its small population is undergoing a decline in Asia despite the fact that the European population is increasing. **NT. Status and distribution:** Vagrant, a juvenile at a golf course in Kuantan (Pahang) on 14 August 1978. **Habitat:** Open country. **Habits:** Like other vultures, soars high on thermals. Other habits little known. **Note:** Not illustrated in colour (see figure).

Red-headed Vulture, *Sarcogyps calvus* (94)

Cinereous Vulture, *Aegypius monachus* (95)

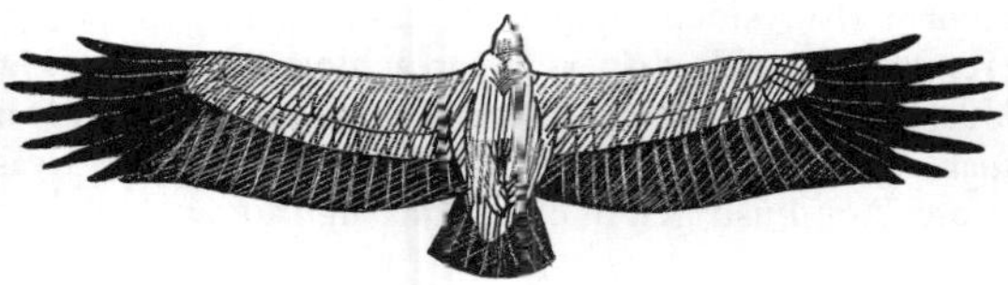

Indian Vulture, *Gyps indicus* (92)

White-rumped Vulture, *Gyps bengalensis* (91)

96. Short-toed Snake-eagle, *Circaetus gallicus* Plate 15
Helang Ular

Description: 60–81 cm (24–32 in). Large, pale and robust eagle with rounded head. When soaring, very long wings are held straight and lifted slightly. Upperparts greyish brown with pale buff fringes; head brownish with dark brown lores and ear coverts; underparts white with brownish throat and breast which sometimes have dark streaks. Has faint dark bars on belly. Tail marked with alternate blackish and brownish bars with broad blackish terminal band. In upperwing flight pattern, greyish brown upperparts and wing coverts contrast with blackish flight feathers and wingtips. In underwing flight pattern, shows white underparts with faint dark streaks and indistinct dark trailing edge of wing. **Voice:** No information in review area. **Range:** Breeding in Africa, Europe, the Middle-East through the Indian subcontinent discontinuously to the Lesser Sundas. Northern populations are migratory. **Status and distribution:** Rare and localized passage migrant at low elevations, south to Singapore. **Habitat:** Wooded grassland. **Habits:** Usually solitary, soaring and circling at great heights. Flies with soft flapping. Often hovers. Feeds principally on snakes.

97. Crested Serpent-eagle, *Spilornis cheela* Plate 15
Helang Ular Berjambul /Helang Kuik

Description: 53 cm (21 in). Stocky, short-tailed, and short-necked dark eagle with blackish nuchal crest, marked with white spots. When soaring, wings are held in a shallow 'V'. Plumage dark brown, slightly paler below with white spots on lower breast and belly (not visible when feathers are ruffled on wet birds). Tail marked with alternate black and white bands with broad white central tail band and very narrow white tip. Bill blackish with yellow cere. Legs yellow. In underwing flight pattern, shows dark brown greater wing coverts spotted white and white band bordering narrow black trailing edge of wing. **Juvenile:** Upperparts duller with distinct whitish fringes; underparts whitish with dark streaks. **Voice:** A disyllabic plaintive *kwee-kwi*, the last note on a slightly falling tone. Also a loud and fast paced *hak-hak-hak-hak . . . kwi-kwi-kwi-kwi*. **Range:** The Indian subcontinent and southern China through South-East Asia to the Greater Sundas, Bali and the Philippines **Status and distribution:** Common and widespread resident from low elevations up to 1900 m, south to Singapore where it is very rare. Also on Pulau Pinang, Pulau Langkawi (Kedah), and Pulau Tioman (Pahang). Probably the most common large eagle. **Habitat:** Mangroves, scrub, forests, the forest edge. **Habits:** Usually solitary, frequenting the canopy and crowns of tall trees. Often utilizes tall dead trees as vantage points. Frequently calls in flight. Raises crest when alarmed. Feeds principally on snakes and other vertebrates but will also take poultry in rural settlements.

98. Eastern Marsh-harrier, *Circus spilonotus* Plate 16
Helang Sawah Besar

Description: 48–56 cm (19–22 in). The largest harrier with broader wings and more protruding head. **Male:** Upperparts blackish with whitish fringes; head, neck and breast white with blackish streaks; belly white; tail pale grey. Shows broad blackish mask. In upperwing flight pattern, blackish mantle and wing coverts together with blackish wingtips contrast with narrow white strip on forewing and pale grey flight feathers and uppertail which contrast slightly with narrow greyish-white rump patch. In underwing flight pattern, shows rufous bars on secondaries and black barred primaries on white underparts. **Subadult male:** Like adult male but mantle and upperwing coverts marked brown with blackish bands on pale tail. **Female:** Upperparts dark rufous brown marked whitish; crown pale buff; rump white; underparts bright rufous; tail brown, sometimes marked with indistinct bars. In upperwing flight pattern, pale strip on forewing contrasts with dark upperwing. In underwing flight pattern, shows broad whitish patch on base of primaries. In both sexes bill black with yellow cere, legs yellow. **Juvenile (not illustrated):** Like female but darker with all dark cap. **Voice:** Usually silent in wintering quarters. **Range:** Breeding throughout temperate Asia east to Japan; wintering south to South and South-East Asia to Sumatra. **Status and distribution:** Locally common winter visitor at low elevations south to Singapore, where it is uncommon. The most common harrier in open country, particularly rice-growing schemes. **Habitat:** Ricefields, inland freshwater swamps, disused mining lands, wet grasslands. **Habits:** Flies low over open country with short flaps interspersed by long glides. Sometimes hovers. Flight generally heavier and less buoyant than Pied Harrier (100). Roosts communally on the ground especially in harvested ricefields. Females sometimes displace males for roosting space.

99. Hen Harrier, *Circus cyaneus* — Plate 16

Helang Sawah Kelabu

Description: 43–51 cm (17–20 in). **Male:** Head, breast and upperparts pale grey apart from white rump; belly and undertail coverts white. In underwing flight pattern, pale greyish head and black flight feathers contrast with rest of white wings, underparts and tail. **Female:** From female Eastern Marsh-harrier (98) by smaller size, heavily barred underwing, barred tail and the lack of rufous underparts. Mantle and wings dark brown, contrasting with white rump. Shows pale buff hind border to facial disc. In both sexes, bill black with yellow cere; legs yellow. **Voice:** Usually silent in wintering quarters. **Range:** Breeding throughout temperate North America, Europe and northern temperate Asia with a separate breeding population in southern South America; wintering south to North Africa, the Middle-East, South and South-East Asia to the Malay Peninsula. **Status and distribution:** Very rare winter visitor at low elevations, south to Singapore. **Habitat:** Ricefields, inland freshwater swamps. **Habits:** Like other harriers.

100. Pied Harrier, *Circus melanoleucos* — Plate 16

Helang Sawah Hitam-Putih

Description: 43–46 cm (17–18 in). **Male:** Unmistakable. Black head, breast, upperparts and primaries with grey wings and tail together with white rump and underparts diagnostic. In upperwing flight pattern, shows conspicuous white bar on wing coverts which contrast with narrower black bar on wing. In underwing flight pattern, shows all white except for black head and primaries. **Female:** Similar to female Hen

Harrier (99) but head browner and lacks pale hind border to facial disc; upperparts less streaked; underparts whiter. In upperwing flight pattern, shows dark brown mantle and wing coverts and pale bluish grey flight feathers marked with blackish bands and narrow white 'V'-shaped rump patch. In both sexes, bill black with yellow cere. **Juvenile: (not illustrated):** Upperparts dark brown; underparts and underwing coverts dark cinnamon buff. **Voice:** Usually silent in wintering quarters. **Range:** Breeding in the eastern Himalayas and much of temperate Asia; wintering south to India, southern China and South-East Asia to the Malay Peninsula, Borneo and the Philippines. **Status and distribution:** Uncommon and localized winter visitor at low elevations, south to Singapore, where it is rare. Occurs in rice-growing areas but less common than Eastern Marsh-harrier (98). **Habitat:** Ricefields, inland freshwater swamps, disused mining lands. **Habits:** Usually quarters low over open country. Flight buoyant and graceful. Roosts communally in harvested ricefields, often with Eastern Marsh-harrier.

101. Crested Goshawk, *Accipiter trivirgatus* — Plate 17

Helang Sewah Besar

Description: 40–46 cm (16–18 in). Large and powerful accipiter with short crest and long tail distinctive. Upperparts dark brown with slaty grey crown and sides of head. Underparts whitish with bold dark mesial stripe, rufous brown streaks on breast and bold dark bars from lower breast below, becoming narrower and more compact towards thighs. Has four dark bands on tail. In flight, shows rounded wings and well protruding head. In underwing flight pattern, shows all white wing with sparse dark spots and dark wingtips; tail usually well spread, showing alternate dark and pale bands. Iris yellow. **Juvenile:** Like adult but head brown and crest marked with whitish buff streaks; belly marked with bold dark spots and not bars. **Voice:** A high-pitched strident *keeeee*. **Range:** The Indian subcontinent and southern China through South-East Asia to the Greater Sundas and the Philippines. **Status and distribution:** Fairly common resident principally at low elevations up to 250 m, south to Singapore where it is rare. Has also been recorded above 1200 m at Fraser's Hill and Cameron Highlands (Pahang) (W1). **Habitat:** Forests, the forest edge, plantations. **Habits:** Frequents the canopy and middle storey, usually keeping to thick foliage, from which it hunts. Very vocal especially during courtship.

102. Shikra, *Accipiter badius* — Plate 17

Helang Sewah Shikra

Description: 30–36 cm (12–14 in). **Male:** Unmistakable with bluish grey upperparts, greyish cheeks, whitish throat and underparts marked with pale rufous bars. Tail has indistinct subterminal band on central tail feathers. In underwing flight pattern, dark wingtips and faint dark barrings on undertail coverts contrast with paler underparts and wings. **Female:** Like male but upperparts tinged brownish; barrings on underparts broader and subterminal tail band on central tail feathers more distinct. From female Besra (105) by the lack of mesial stripe, dark streaks on upper breast and alternate dark and pale bands on tail. In underwing flight pattern, shows more distinct dark barrings on wing and more rufous barrings on underparts. Iris yellow to brown in both sexes. **Juvenile** Blackish brown upperparts marked with rufous and white streaks on nape. Has white supercilium and distinct dark mesial stripe. Whitish underparts marked with broad streaks on breast and short broad bars on sides of head, flanks and thighs. **Voice:** A high-pitched *keeeew*, repeated up to five times. **Range:** Breeding in tropical Africa, South Asia, southern China through South-East Asia to the Malay Peninsula to the Malay Peninsula. **Status and distribution:** Rare and localized non-breeding visitor at low elevations, south to Johor. **Habitat:** Open country, scrub. **Habits:** General habits like other accipiters. Known to take insects from the ground.

103. Chinese Sparrowhawk, *Accipiter soloensis* Plate 17

Helang Sewah Cina

Description: 30–36 cm (12–14 in). Uniform dark slaty grey upperparts together with whitish underparts with orange suffusion on breast distinctive. Lacks dark mesial stripe or barrings on underparts which are characteristic in many accipiters. Central tail feathers usually uniform dark slaty-grey but some females may show dark bands. In underwing flight pattern, shows pale unmarked wing and underparts which contrast with blackish wingtips. Also shows faint dark bands on central tail feathers. Iris red. **Juvenile:** Generally like most juvenile accipiters but crown and nape darker slaty grey (usually indiscernible in field) and underparts whiter. Lacks dark mesial stripe. **Voice:** Harsh shrill notes. **Range:** Breeding in south Ussuriland, Korea and China; wintering through South-East Asia and the Indonesian Archipelago to New Guinea. **Status and distribution:** Locally common passage migrant and winter visitor in small numbers principally at low elevations, south to Singapore, where it is uncommon. **Habitat:** Forests, the forest edge, wooded areas. **Habits:** Solitary. Like other accipiters, circles overhead and scrutinizes ground for prey. Usually hunts from perch.

104. Japanese Sparrowhawk, *Accipiter gularis* Plate 17

Helang Sewah Jepun

Description: 25–30 cm (10–12 in). The smallest accipiter. From Besra (105) by indistinct mesial stripe and the lack of dark streaks on upper breast (apparent when viewed closely). Upperparts dark grey-brown, throat and underparts white with faint mesial stripe. Iris yellow to red. **Male:** Breast and belly pale rufous with indistinct whitish bars. **Female:** Dark grey-brown upperparts (darker on crown) with underparts marked with narrow dark brown bars. In underwing flight pattern, both sexes show distinct narrow dark bands on underparts and wings. **Juvenile:** Belly whiter than juvenile Besra and Shikra (102). Has indistinct mesial stripe. Underparts marked with bold dark streaks on breast and narrower dark bars on remaining underparts. **Voice:** High-pitched shrill notes. **Range:** Breeding in far eastern Russia, China and Japan; wintering south through Japan, China, South-East Asia to the Greater and Lesser Sundas, and Sulawesi. **Status and distribution:** Fairly common passage migrant and winter visitor from low to high elevations, south to Singapore. **Habitat:** Open country, scrub, mangroves, forests, the forest edge. **Habits:** Usually solitary, keeping to the crowns of trees. Often circles overhead, scrutinizing the ground for prey. Migrates in large flocks.

105. Besra, *Accipiter virgatus* Plate 17

Helang Sewah Besra

Description: 30–36 cm (12–14 in). Dark upperparts with distinct dark mesial stripe distinctive. Both sexes show fine dark streaks on upper breast. Dark bands on uppertail coverts show little contrast. Whitish undertail contrasts with dark bands. Iris yellow. **Male:** Blackish head and dark slaty upperparts contrasting with deep rufous underparts, with narrow whitish bars on belly and flanks. **Female:** Upperparts dark brown; underparts marked with brownish bars. In underwing flight pattern, shows narrow dark barrings on breast and wings. **Juvenile:** Dark brown upperparts with prominent mesial stripe together with bold dark streaks on breast distinctive. **Voice:** Shrill notes typical of the genus. **Range:** Breeding in the mountains of the Indian subcontinent, the Himalayas and southern China discontinuously through South-East Asia to the Greater Sundas, Bali, Flores and the Philippines. Northern populations are migratory, wintering south to the plains of India and South-East Asia. **Status and distribution:** Very rare winter visitor, a female mist-netted at Tuas (Singapore) on 16 November 1996 is the first official record from the review area. Another four records from Singapore (both sexes) date from 20 September 1998–3 November 2002 (W and H). **Habitat:** Forests, the forest edge. **Habits:** Solitary. Usually hunts from perches. General habits like other accipiters. **Note:** May be a regular migrant, probably overlooked as Japanese Sparrowhawk (104) due to general similarities in plumage patterns.

106. Eurasian Sparrowhawk, *Accipiter nisus* Plate 11

Helang Sewah Eurasia

Description: 28–38 cm (11–15 in). At rest appears rather long-tailed and in flight, appears both long-winged and long tailed. **Male:** From male Japanese Sparrowhawk (104) and Besra (105) by larger size and slaty grey upperparts, including uppertail, which is marked with faint darker bands. Sides of head washed orange-rufous, with narrow orange-rufous streaks on throat. Lacks the obvious mesial stripe which is characteristic of its smaller congeners, but shows a faint, narrow, pale supercilium. Breast, belly and flanks marked with faint orange-rufous bars. **Female:** Larger with upperparts tinged brown with more pronounced pale supercilium as well as darker and more obvious uppertail bands. Markings on underparts darker and more pronounced. **Juvenile:** Relatively larger size, heavy rufous chestnut to blackish markings below and prominent uppertail bands distinguish it from the juveniles of its smaller congeners. **Voice:** Unknown from review area. **Range:** Breeding throughout much of the Palaearctic. **Status and distribution:** Vagrant, a female at 1400 m at Fraser's Hill (Pahang) on 23 March 2007 (Dion Hobcroft, RCS). **Habitat:** In review area, lower montane forest. **Habits:** Perched on a dead tree, this individual was seen making prospecting flights to search for prey and was seen harrying a Little Cuckoo-dove (219) in flight. Other habits unknown.

107. Grey-faced Buzzard, *Butastur indicus* Plate 16

Helang Kepala Kelabu

Description: 41–46 cm (16–18 in). Long-winged buzzard with wingtips almost reaching tail tip when perched. In flight, wings usually held flat with tail broadly fanned.

Grey ear coverts, sharply demarcated from white throat with prominent dark mesial stripe distinctive. Upperparts brown with blackish streaks and bars. Breast and belly rufous-brown or grey-brown with whitish barrings. Bill grey with yellow cere. Iris and legs yellow. **Juvenile:** Like adult but head browner with whitish streaks. Shows white supercilium. Underparts white with narrow brown streaks. **Voice:** A disyllabic *chi-kwee*, with an upward inflection on the second note. **Range:** Breeding in far eastern Russia, northern China, Korea and Japan; wintering south to South-East Asia and the Indonesian Archipelago to New Guinea. **Status and distribution:** Locally common passage migrant and winter visitor at low elevations, south to Singapore, where it is uncommon. **Habitat:** Open country, lightly wooded areas. **Habits:** Usually solitary, hunting from tree perches. Migrates in very large flocks.

108. Common Buzzard, *Buteo buteo* Plate 16
Helang Gempal

Description: 50–56 cm (20–22 in). Stocky build with broad, rounded wings and tail distinctive. Tail well fanned in flight and when hovering. Upperparts dark brown; underparts highly variable but usually pale buff or whitish with breast streaked rufous, belly and flanks dark brown, lower belly whitish. In upperwing flight pattern, shows all brown wings with whitish patch at base of primaries. In underwing flight pattern, shows pale wings with black patches at carpal joints as well as black trailing edge to primaries. **Voice:** A drawn out plaintive whistle, *peeee-ay*. **Range:** Breeding throughout Europe and temperate Asia; wintering south to Africa, South and South-East Asia to Java and the Philippines. **Status and distribution:** Uncommon and localized passage migrant and winter visitor at low elevations, but also up to 1500 m at Genting Highlands (Pahang), (W1) south to Singapore. **Habits:** The forest edge, open country, lightly wooded areas. **Habits:** Usually solitary. Flies high, occasionally flapping labouriously. Often hovers.

108.1. Long-legged Buzzard *(Buteo rufinus)* x Upland Buzzard, *(Buteo hemilasius)* Plate 16

Description: No measurements available. A medium sized pale raptor, a juvenile moulting into first winter plumage. Larger and plumper than Brahminy Kite (87) with rather largish white head (bigger than 108) streaked brown on the crown and diffused blackish malar stripe and line behind eye. Back and rump brownish with whitish tail marked with dark barrings; underparts white marked with dark streaks and patches of rufous, belly rufous with dark streaks, 'trousers' and flanks dark rufous. In upperwing flight pattern, wings brownish with pale patch on blackish primaries, with secondaries marked with barrings. In underwing flight pattern, showed white underwing with black and brownish carpal patches, slightly rufous median and greater wing coverts, with whitish secondaries marked with faint barrings. The following soft part colouration were noted; blackish bill with yellow cere, pale yellow gape, yellow iris, and yellow unfeathered tarsi. This individual appeared to be a Long-legged Buzzard but because of too much brown instead of rufous plumage tones (an Upland Buzzard trait), it was deemed a possible hybrid. The very striking white on upperwing, at the 'hand' area were more typical of the Upland Buzzard but the tarsi were bare and unfeathered, a characteristic of the Long-legged Buzzard (feathered in Upland Buzzard). Long gape line, big head, strong bill as well as strong feet, long tarsi and long tail, carpal patch and dark thigh feathers contrasting with rest of underparts all point to Long-legged Buzzard. Although Upland and Long-legged Buzzards have been known to hybridize in Asia, where ranges overlap (Pfander and Schmigalew, 2001), with the unavailability of evidence based on mitochondrial and DNA sequences, a confirmed hybrid between the two species cannot be conclusive of this individual though morphology and plumage pattern seem to suggest this. **Voice:** Unknown from review area. **Range:** Long-legged Buzzard: Northern Africa, south-eastern Europe, west and central Asia, east to China and the Indian subcontinent. Upland Buzzard: Central Asia from Kazakhstan, Tajikistan and Uzbekistan east to Russia, Iran, the Indian subcontinent, China, Korea, Mongolia, and Japan. Neither species has occurred in the review area. **Status and distribution:** An individual at Chuping (Perlis) from 17–24 February 2008. (Hum, 2008). **Habitat:** In review area, cleared sugar cane field. **Habits:** The flight of this individual was described as sluggish and low over the ground. Assumed a rather upright stance when perched on ground, with long wingtips touching tail.

109. Black Eagle, *Ictinaetus malayensis* Plate 18
Helang Hitam

Description: 69–78 cm (28–31 in). Large, long and broad-winged dark eagle. In flight, broad wings distinctive with tail usually held close, appearing narrow and square-cut. Wings usually held level or slightly bowed when gliding or soaring. Plumage blackish, showing very little contrast with flight feathers and underwing coverts. In underwing flight pattern at close range, shows indistinct pale barrings on tail and yellow legs contrasting sharply with dark plumage. Bill black with yellow cere. **Juvenile:** Upperparts browner streaked buff. Tail marked with alternate dark and pale bands. **Voice:** A distinct plaintive *kleeee-kee*. **Range:** The Indian subcontinent and South-East Asia to the Greater Sundas, Bali, and Sulawesi. **Status and distribution:** Locally common resident principally in hilly country from low elevations up to at least 1800 m, south to Johor. **Habitat:** Forests. **Habits:** Frequents the canopy, usually in pairs. Hunts by quartering low over the canopy.

110. Greater Spotted Eagle, *Aquila clanga* Plate 18
Helang Bintik

Description: 66–74 cm (26–30 in). Large, long and broad-winged dark eagle with short, slightly rounded tail. From Steppe (111) and Eastern Imperial Eagles (112) by shorter wings and tail as well as smaller and less robust bill with head less protruding in flight. When soaring or gliding, wings are held flat and horizontal with tips slightly drooped; short tail when fully spread appears distinctly wedge-shaped. Plumage blackish brown with pale crescentic patch on rump. In underwing flight pattern, wing coverts darker than primaries. Greyish bill does not extend beyond eye. Cere yellowish green. **Juvenile (not illustrated):** Scapulars and upperwing coverts marked with large whitish spots with pale crescentic patch on rump broader. In upperwing flight pattern, shows whitish spots on entire wing coverts with two whitish bands. **Voice:**

A barking *kyak, kyak, kyak*, seldom heard in wintering quarters. **Range:** Breeding from eastern Europe to eastern Russia, northern China and northern India; wintering south to northern Africa, India and South-East Asia to the Malay Peninsula. Its small population appears to be declining due to persecution and habitat loss. **VU. Status and distribution:** Uncommon and localized winter visitor at low elevations, south to Singapore, where it is rare. Particularly regular at rice-growing schemes, especially at Sekinchan (Selangor) and Seberang Perak (Perak). The most frequently seen *Aquila* eagle. **Habitat:** Open country, ricefields. **Habits:** Usually solitary, sometimes in small groups. Soars very high over ricefields and open country. Often walks in rice stubble, feeding on frogs and carrion. Has been observed harrying Black Kite (86), forcing it to drop its catch. **Note:** *Aquila* eagles take up to 6 years to acquire full adult plumage. The transitional stages between juvenile to full adult can be very confusing.

111. Steppe Eagle, *Aquila nipalensis* — Plate 18

Helang Gurun

Description: 76–86 cm (30–34 in). Large, broad-winged dark eagle. From Greater Spotted Eagle (110) by larger size, more robust build, shaggy 'trousers' and gape extending behind eye. In flight, wings are held flat and horizontal with primaries widely splayed. Plumage uniform dark brown with hindcrown and nape often yellow. In upperwing flight pattern, shows small pale patch at base of primaries which are darker than wing coverts. In underwing flight pattern, wing coverts slightly darker than flight feathers. **Juvenile** (not illustrated): Light brown plumage with pale crescentic patch on rump together with blackish primaries and tail. From juvenile Eastern Imperial Eagle (112) by blackish bill contrasting sharply with pale yellowish cere. In upperwing flight pattern of juveniles and subadults (not illustrated), shows broad whitish wingbar between brown wing coverts and darker flight feathers as well as dark tail with very narrow pale terminal band. **Voice:** Generally silent in wintering quarters. **Range:** Breeding in Africa and South-East Europe east through India and China to Hainan; wintering south to Myanmar and the Malay Peninsula. **Status and distribution:** Rare and localized winter visitor at low elevations, south to Singapore. The first record, a juvenile at Sekinchan (Selangor) in February 1987 with subsequent sightings there through March 1987. In Singapore, first known from a juvenile at Punggal in December 1987 with subsequent sightings there between November 1989 and January 1990. **Habitat:** Open country, ricefields. **Habits:** Like Greater Spotted Eagle, habitually soars over open country. Takes carrion. Other habits little known.

112. Eastern Imperial Eagle, *Aquila heliaca* — Plate 18

Helang Tengkuk Kuning

Description: 79–84 cm (32–34 in). Large long-necked eagle. When soaring, wings are held flat with tail usually closed in adult (appearing narrow) but spread in juvenile. Blackish brown plumage with tawny buff crown and hindneck together with huge greenish yellow bill diagnostic. Tail silvery grey with broad black subterminal band. In upperwing flight pattern, dark upperparts contrast with tawny buff hindneck. In underwing flight pattern, silvery grey flight feathers and basal part of tail contrast slightly with darker wing coverts and distal part of tail respectively. **Juvenile:** Paler colouration and proportionately larger bill separate it from juvenile Greater Spotted (110) and Steppe Eagles (111). Cere and bill whitish, showing very little contrast (see Steppe Eagle (111)). Pale yellowish brown head and brownish upperparts marked with buff streaks on neck and back with buff bars on wings distinctive. In upperwing flight pattern, yellowish brown head and wing coverts contrast with blackish flight feathers which are marked with narrow pale buff bars and very narrow pale trailing edge. Also shows whitish rump which contrasts with dark tail with very narrow pale terminal band. In underwing flight pattern, shows distinct whitish panel on inner primaries. In subadult plumage (not illustrated), pale body develops dark blotches with the appearance of dark feathers. **Voice:** A series of well spaced deep goose-like *gahk* notes (extralimital data) seldom heard in wintering range. **Range:** Breeding throughout southern Europe, temperate Asia, the Indian subcontinent east to China and Taiwan; wintering south to North Africa, the Middle-East, southern India, South-East Asia and the Malay Peninsula. Its small global population is declining largely due to habitat degradation, persecution, nest robbing, collision with power lines and prey depletion. **VU. Status and distribution:** Rare and localized winter visitor; a juvenile at Sekinchan (Selangor) in February 1986. Also single subadults at Sekinchan and Punggol (Singapore) respectively in February and March 1987 with subsequent sightings of individuals at Punggol in January and December 1989 and November 1990. **Habitat:** Open country, ricefields. **Habits:** Usually soars high on thermals. Often perches on trees, displacing other smaller raptors present. Takes carrion. Other habits little known.

113. Booted Eagle, *Hieraaetus pennatus* Plate 18

Helang Junam

Description: 48–56 cm (19–22 in). Small crestless eagle with relatively long tail. When gliding, wings are usually held level with tips slightly bowed downwards. In flight, resembles Black Kite (86), but tail square-cut. **Dark morph:** Uniform dark brown plumage with white neck patches. In upperwing flight pattern, blackish flight feathers and tail contrast with pale brown wing coverts and narrow wingbar. In underwing flight pattern, wings all dark, tail cinnamon buff **Pale morph:** White underparts and underwing coverts, black primaries and secondaries. Tail grey with black teminal band Shows dark ear coverts. **Voice:** A thin, high-pitched *kik-kik,* when harassed by Large-billed Crow (W1). **Range:** Breeding from North Africa and southern Europe throughout southern temperate Asia, east to north India and China; wintering south to tropical Africa, South and South-East Asia with stragglers reaching Bali. **Status and distribution:** Rare and localized winter visitor at low elevations, south to Singapore. The dark morph is more common. **Habitat:** Open country, wooded areas. **Habits:** Usually solitary, circling and soaring on thermals. Dives steeply when pursuing prey in flight. Has been observed taking shorebirds on the ground.

114. Rufous-bellied Eagle, *Lophptrorchis kienerii* Plate 18

Helang Perut Merah

Description: 51–61 cm (20–24 in). Small eagle with distinct short crest when perched. In flight, wing and tail shape pattern closely resemble Common Buzzard

(108). Black head sharply demarcated from white throat and upper breast. Lower breast, flanks, belly, thighs and undertail coverts pale to dark rufous, perceivable under good lighting conditions. Underparts marked with dark streaks. Upperparts blackish with pale fringes. Tail brown with blackish bands. Bill blackish with yellow cere. Legs yellow. In underwing flight pattern, rufous underparts and wing coverts (which has narrow blackish border) contrast with pale wings. **Juvenile:** Blackish brown upperparts with conspicuous blackish eye patch and whitish supercilium distinctive. Underparts whitish. **Voice:** Several ascending high-pitched notes terminating in a scream. Also a high-pitched *kliu* (M and W). **Range:** South-West India and Sri Lanka, North-East India through South Asia to the Greater and Lesser Sundas east to Sulawesi and the Philippines. **Status and distribution:** Locally common resident, passage migrant and winter visitor from low elevations up to 1280 m, south to Singapore where it is a rare migrant. **Habitat:** Forests, the forest edge. **Habits:** Usually solitary, gliding or circling low over the forest canopy.

115. Changeable Hawk-eagle, *Nisaetus cirrhatus* Plate 15

Helang-Rajawali Hindik/Helang Hindik

Description: 56–75 cm (22–30 in). Slender and broad-winged eagle with long, rounded tail. Wings usually held level when gliding and soaring. Short crest separates it from other hawk-eagles. **Dark morph:** Blackish brown plumage with brownish tail slightly contrasting with darker terminal band. From Black Eagle (109) by smaller and narrower wings which show pale band on flight feathers in underwing flight pattern. **Pale morph:** Upperparts dark brown with whitish fringes; underparts whitish with prominent dark mesial stripe and bold streaks on breast and belly. Shows pale supercilium. Tail brownish, marked with fine narrow dark bands. In underwing flight pattern, shows largely whitish underparts with blackish wingtips, dark streaks on breast and belly and dark subterminal tail band. **Juvenile:** White head and underparts and dark brown upperparts **Voice:** A series of loud and shrill ascending whistles, *kri-kri-kri-kri-kri-kri-krik*. Also a loud *kreee-kit*. **Range:** The Indian subcontinent and South-East Asia to the Greater and Lesser Sundas east to the Philippines. **Status and distribution:** Fairly common resident primarily at low elevations up to 800 m, occasionally 1200 m, (W1) south to Singapore, where it is uncommon. The dark morph is more common. Fairly common large eagle but less common than Crested Serpent-eagle (97). **Habitat:** Forests, the forest edge, wooded country, plantations. **Habits:** Usually solitary or in pairs, gliding over the forest canopy and forest edge. Hunts from both the air and from perches. Will take poultry from rural settlements. Very vocal especially during courtship when a pair may engage in display flight with both wings and tail tipped upwards.

116. Mountain Hawk-eagle, *Nisaetus nipalensis* Plate 15

Helang-Rajawali Besar

Description: 66–75 cm (26–30). Tall erectile crest (frequently carried up cocked) is distinctive in both adult and juvenile and is blackish tipped white. At rest from pale morph Changeable Hawk-eagle (115) by length of crest and larger size and from

Blyth's (117) and Wallace's Hawk-eagle (118) by much larger size and stockier build. In flight shows broader wings than the other hawk-eagles with more distinct finger projection. **Adult:** Dark mesial stripe distinctive. Upperparts largely rufous-brown and black with pale buff fringes. Tail has four dark bands which are much broader than pale morph Changeable Hawk-eagle. Breast and underwing coverts tawny with blackish streaks limited to upper breast with rest of underparts, including flanks, marked with rufous-brown bars. **Juvenile:** Like adult but tawny buff head and underparts unmarked with dark tail bars narrower and less distinct. **Voice:** A shrill trisyllabic *tlueet-weet-weet* (L and R). **Range:** South-west India and Sri Lanka, the eastern Himalayas and north-east India to southern China (including Hainan), Taiwan, Japan, and hilly regions of continental South-East Asia to the Malay Peninsula. **Status and distribution:** Rare and localized resident only on Gunung Raya, Pulau Langkawi (Kedah). **Habitat:** Forests. **Habits:** Like its smaller congeners, still hunts from perches. Predation on a juvenile Great Hornbill (333) was observed; one adult dived on a Great Hornbill pair escorting a juvenile, knocking it down, and as it tumbled downwards, was caught in mid air in the talons of another Mountain-Hawk-eagle from below, which flew off with the hornbill firmly in its grip (Indra and Irshad, 2009). Frequently soars over the forest.

117. Blyth's Hawk-eagle, *Nisaetus alboniger* — Plate 15

Helang-Rajawali Gunung

Description: 51–58 cm (20–23 in). Blackish head and upperparts with long erectile crest distinctive. Underparts whitish with prominent dark mesial stripe. Has black streaks on breast and bars on belly and flanks. In underwing flight pattern, shows diagnostic single broad central white tail band, sandwiched between two darker bands with very narrow white terminal band. **Juvenile:** Brownish upperparts and unmarked buff underparts (light streaking gradually acquired through age) with tail narrowly marked with alternate dark and pale bands. **Voice:** A series of shrill whistles, *kew-kew-kew*, with the initial syllable stressed. Also a disyllabic *kreee-kit* (L and R). **Range:** The Malay Peninsula, Sumatra, and Borneo. **Status and distribution** Locally common resident principally in hilly country up to 1980 m, south to Singapore where it is a rare dispersant with sightings of an adult and subadult from 27 December 2000 through 28 January 2001 at Kent Ridge Park and an adult at Mount Faber on 3 February 2003 (W and H). **Habitat:** Forests. **Habits:** Frequents the canopy where it often perches in tall trees. Often soars over the forest.

118. Wallace's Hawk-eagle, *Nisaetus nanus* — Plate 15

Helang-Rajawali Rimba

Description: 46 cm (18 in). Small eagle with long erect crest, which is blackish, tipped white. Larger than Jerdon's Baza (81) and at rest differs from it by the wing length which does not reach tail tip. From adult Blyth's Hawk-eagle (117) by smaller size and distinctive three dark bands and two broader pale bands on tail. In underwing flight pattern, shows rufous wing coverts and black trailing edge of wings. **Juvenile:** Similar to juvenile Blyth's Hawk-eagle. **Voice:** A high-pitched musical whistle, *tui-tuik*, with an upward inflection on the second note (M and W). **Range:** Tenasserim (Myanmar) and the Malay Peninsula to Sumatra and Borneo. Undergoing a rapid population decline owing to widespread and ongoing loss of lowland forest. **VU. Status and distribution:** Uncommon and localized resident at low elevations, south to Johor. **Habitat:** Forests. **Habits:** Frequents the canopy. Principally still hunts from perches.

RAILS AND CRAKES

ORDER *Gruiformes* **FAMILY** *Rallidae*

World: *128 species* **Peninsular Malaysia and Singapore:** *12 species*

Terrestrial swamp-dwelling wading birds with long legs and toes. Walk with typical horizontal carriage, often cocking short tails. Run well. Most species are capable of swimming. Flight weak and direct with much flapping on short wings (although known to migrate long distances). Most are shy and secretive, skulking in thick cover. Similar looking species are often difficult to identify especially in thick cover. Calls are loud and distinct in many species. Solitary or in small loose groups. Sexes similar in most species.

119. Red-legged Crake, *Rallina fasciata* Plate 19
Sintar Api

Description: 25 cm (9 in). Reddish brown head, neck and breast, together with short, thin bill and red legs diagnostic. Mantle and wing coverts olive-brown, heavily marked with white barring on primaries. Belly, flanks and undertail coverts marked with black and white bars. **Voice:** A series of loud and nasal monosyllabic *ehh* followed by an ascending trill; most frequent at dawn and dusk. **Range:** Breeding in North-East India, Myanmar, Thailand, the Malay Peninsula and the Greater Sundas; northern populations are migratory, wintering south to the Malay Peninsula and the Greater and Lesser Sundas with stragglers reaching North-West Australia. **Status and distribution:** Uncommon and localized resident, passage migrant and winter visitor at low elevations, south to Singapore. Occurs above 900 m, only during autumn and spring passage. **Habitat:** Inland freshwater swamps, edges of settled mining pools, flooded ricefields, wooded areas, plantations. **Habits:** Solitary or in small groups. Has crepuscular habits. Calls at night. Skulks in dense vegetation.

120. Slaty-legged Crake, *Rallina eurizonoides* Plate 19
Sintar Merah

Description: 25 cm (9 in). From Red-legged Crake (119) by duller upperparts, more robust bill, narrower white barring on flanks, dark greenish legs and lacking the white markings on wing coverts. Throat whitish. **Voice:** A disyllabic *kek-kek,* frequently uttered at night. **Range:** The Indian subcontinent and Myanmar discontinuously through South-East Asia. Sumatra, Java and the Philippines. **Status and distribution:** Rare and localized passage migrant and winter visitor from low elevations up to 900 m, south to Singapore. **Habitat:** Banks of forested streams and rivers. **Habits:** Generally shy, keeping to thick cover.

121. Slaty-breasted Rail, *Gallirallus striatus* Plate 19

Sintar

Description: 30 cm (12 in). Long reddish bill, chestnut crown and nape together with slaty grey breast and fine white barring on entire upperparts diagnostic. Throat pale grey; flanks and undertail coverts blackish with white barrings. **Juvenile** (not illustrated): Underparts browner with less distinct streaking; crown streaked blackish. **Voice:** A sharp and distinct whistle, *cirrk*. Also short jerking screams when alarmed. **Range:** The Indian subcontinent and southern China through South-East Asia, east to the Greater and Lesser Sundas, Sulawesi and the Philippines. **Status and distribution:** Locally common resident at low elevations, south to Singapore. **Habitat:** Inland freshwater swamps, flooded ricefields, reedbeds, banks of small streams, overgrown drains and ditches. **Habits:** Usually solitary. Skulks in thick vegetation, appearing briefly at the edge of reedbeds or grassy verges of roads.

122. White-breasted Waterhen, *Amaurornis phoenicurus* Plate 19

Ruak-Ruak

Description: 33 cm (13 in). White face and underparts contrasting with dark slaty grey upperparts together with yellow bill with red base diagnostic. Undertail coverts cinnamon. Legs greenish. **Voice:** Typical call is a harsh disyllabic *krruwak, krruwak*... persistently uttered. Also grunts, chuckles and croaks. Alarm call is a high-pitched *wheet* (M). **Range:** The Indian subcontinent and southern China through South-East Asia to the Greater and Lesser Sundas and the Philippines. Northern populations are migratory. **Status and distribution:** Common and widespread resident and migrant principally at low elevations, south to Singapore. Also in the hills up to 1500 m, especially during autumn and spring passage. The most frequently seen rail. **Habitat:** Inland freshwater swamps, flooded ricefields, swampy vegetation, reedbeds fringing settled mining pools, banks of forested streams, gardens, short and wet grassy areas. **Habits:** Usually solitary or in small loose groups. Skulks in thick vegetation but often feeds on short grassy lawns. Sometimes clambers onto small bushes and trees. Shy and wary, quickly scurrying to cover at the slightest sign of danger. Very vocal.

123. Baillon's Crake, *Porzana pusilla* Plate 19

Sintar Kecil

Description: 19 cm (7.5 in). Small size and small greenish bill distinctive. Upperparts brownish flecked with white; breast and head slaty grey with indistinct brownish eyeline; flanks and undertail coverts barred black and white. **Voice:** A sharp whinnying trill, similar to Little Grebe (33). **Range:** Breeding in Africa, Europe, temperate Asia, northern India discontinuously to Australia and New Zealand Northern populations are migratory, wintering south to tropical Africa, South and South-East Asia to the Greater Sundas. **Status and distribution:** Uncommon and localized winter visitor at low elevations, south to Singapore. **Habitat:** Inland freshwater swamps, reedbeds. **Habits:** Usually solitary. Feeds at the edge of reedbeds and swamps, moving quickly through the vegetation. Rarely seen in flight.

124. Ruddy-breasted Crake, *Porzana fusca* Plate 19
Sintar Belacan

Description: 22 cm (8.5 in). Dull reddish chestnut head and underparts together with dark olive-brown upperparts diagnostic. Hind flanks and undertail coverts marked with black and white stripes. Legs red. **Voice:** A soft *kek*, uttered infrequently. Also a sharp whinnying trill like Little Grebe (33) but with a slightly descending pitch. **Range:** The Indian subcontinent to China and Japan, south through South-East Asia to the Greater and Lesser Sundas, Sulawesi, and the Philippines. **Status and distribution:** Locally common resident at low elevations, south to Singapore, where it is an uncommon resident. **Habitat:** Inland freshwater swamps, reedbeds, plantations. **Habits:** Solitary and shy. Skulks in thick cover, occasionally appearing to feed at the edge. Very vocal especially at dawn and dusk.

125. Band-bellied Crake, *Porzana paykullii* Plate 19
Sintar Jalur

Description: 23 cm (9 in). From Red-legged (119) and Slaty-legged Crakes (120) by dark olive brown crown, nape and mantle with white flecking on wing coverts. Throat and supercilium whitish. Flanks and undertail coverts marked with black and white bars. Legs red. **Voice:** Unknown in review area. **Range:** Breeding in eastern Russia and northern China; wintering south to South-East Asia and the Greater Sundas. Agricultural intensification within its range is suspected to be causing a population decline at an unknown rate. **NT.** Status **and distribution:** Very rare passage migrant and winter visitor at low elevations at least south to Johor. Occurs above 900 m only during autumn and spring passage. **Habitat:** Inland freshwater swamps. **Habits:** Keeps to thick cover. Very secretive and other habits poorly known.

126. White-browed Crake, *Porzana cinerea* Plate 19
Sintar Dahi Putih

Description: 20 cm (8 in). Greenish bill with red base to maxilla, black eyeline together with white supercilium and stripe below eye diagnostic. Head and breast grey; throat whitish; upperparts dark brown mottled with buff; flanks buff brown; undertail coverts cinnamon buff; legs greenish yellow. **Juvenile** (not illustrated): Upperparts brownish; face markings indistinct; lacks red base on bill. **Voice:** A high pitched nasal piping, *cutchi-cutchi-cutchi*, usually uttered from thick cover. **Range:** The Malay Peninsula to the Greater Sundas, east to New Guinea, Australia, the western Pacific and Polynesia. **Status and distribution:** Fairly common resident at low elevations, south to Singapore. **Habitat:** Inland freshwater swamps, flooded ricefields, overgrown drains and ditches. **Habits:** Usually solitary or in pairs. Generally keeps to thick cover but sometimes walks onto floating vegetation, picking up food material. Shy and quickly scurries for cover at the slightest disturbance.

127. Watercock, *Gallicrex cinerea* Plate 19

Ayam-Ayam

Description: 43 cm (17 in). Large size distinctive. **Breeding male:** Plumage blackish grey mottled buff brown on upperparts; bill, frontal shield and legs red; undertail coverts buff. **Female and non-breeding male:** Buff brown with dark streaks above with narrow barrings on flanks. Bill (thicker in male) and legs yellowish green. Male lacks frontal shield. **Voice:** A series of deep booming notes, *gok-gok-gok-gok*... uttered rapidly. **Range:** Breeding from India, east to eastern Russia, China, Korea, and Japan, south to South-East Asia to the Philippines; wintering south to the Greater Sundas and Sulawesi. Northern populations migratory. **Status and distribution:** Uncommon and localized resident at low elevations in Perlis and Kedah. Elsewhere locally common non-breeding visitor at low elevations, south to Singapore, where it is uncommon. Occurs above 900 m only during autumn and spring passage. **Habitat:** Inland freshwater swamps, flooded ricefields, wet grasslands. **Habits:** Usually solitary. Skulks in thick cover, occasionally coming out to feed in the open. Partly nocturnal.

128. Purple Swamphen, *Porphyrio porphyrio* Plate 20

Pangling Ungu

Description: 43 cm (17 in). Large size, dark purplish blue plumage with bright red bill and frontal shield diagnostic. Head pale grey; legs red; undertail coverts white. **Juvenile** (not illustrated): Paler with duller frontal shield. **Voice:** Harsh cackles, grunts, hoots and rippling notes. **Range:** Africa through southern Europe and the Middle-East through the Indian subcontinent and South-East Asia through Australia, New Zealand and the western Pacific. **Status and distribution:** Fairly common resident at low elevations, south to Singapore, where it is a rare resident. Distribution much reduced due to the filling in of disused mining pools. **Habitat:** Disused mining pools, flooded ricefields, overgrown irrigation ditches. **Habits:** Usually in small loose groups, keeping to reedbeds and waterside vegetation. Often feeds in the open. Walks on floating vegetation, picking up food material. Constantly flicks tail. Readily takes to flight when disturbed

129. Common Moorhen, *Gallinula chloropus* Plate 20

Pangling Kecil/Tiong Air

Description: 33 cm (13 in). Blackish plumage with white line along flanks and white sides of undertail coverts together with bright red bill (tipped yellow) and frontal shield diagnostic. Upperparts tinged brownish; underparts tinged greyish. When tail is cocked, shows two large white oval patches on undertail coverts. Legs greenish. **Juvenile:** Overall brownish (including bill) with less distinct line along flanks. **Voice:** A loud and abrupt *prruk*. **Range:** Tropical and temperate latitudes worldwide except Australia and New Zealand. **Status and distribution:** Fairly common resident at low elevations west of the Titiwangsa Range, south to Singapore. Very rare east of the Titiwangsa Range. Distribution much reduced due to filling in of disused mining pools. **Habitat:** Disused mining pools, flooded ricefields, overgrown drains and ditches. **Habits:** Largely aquatic. Usually in small loose groups, dabbling largely on water surface vegetation. Sometimes feeds on land near water edge. Dives well. When alarmed, patters along water surface in a short, sustained flight. Constantly flicks tail while swimming or on land.

130. Common Coot, *Fulica atra* Plate 20
Pangling Hitam

Description: 41 cm (16 in). From Common Moorhen (129) by more robust and hunched appearance and all glossy black plumage with white bill and frontal shield. Trailing edge of wings whitish, conspicuous only in flight. Legs greenish grey with long lobed toes. **Voice:** A high pitched *kyow*, repeated persistently. **Range:** Africa and Europe through temperate Asia, the Indian subcontinent, China discontinuously to the Greater Sundas, Sulawesi, New Guinea, Australia, and New Zealand. **Status and distribution:** Vagrant, known from four records, one each from Poyan (Singapore) in December 1983 and January 1988, and one each from Malim Nawar (Perak) in July 1998 and Timah Tasoh reservoir (Perlis) from 5–8 February 1999. **Habitat:** Disused mining pools, inland freshwater swamps. **Habits:** Strictly aquatic, often diving to feed on underwater vegetation. Patters along the water surface before taking off.

FINFOOTS

ORDER *Gruiformes* **FAMILY** *Heliornithidae*

World: *3 species* **Peninsular Malaysia and Singapore:** *1 species*

Tropical aquatic birds with long thin necks and heavy tapered bills. Feet lobed. Swim partly submerged with neck bobbing back and forth. Dive well. Inhabit dense vegetation fringing mangroves, freshwater swamps and forested waterways. Usually solitary, rarely in small groups.

131. Masked Finfoot, *Heliopais personatus* Plate 20
Pedendang

Description: 53 cm (21 in). Yellow pointed bill and black mask together with olive brown upperparts diagnostic. Legs and feet green. **Male:** Forecrown, throat and front of neck black with narrow white border running down to front of neck. Develops small knob at base of bill during breeding. **Female:** Like male but throat and front of neck white. **Juvenile (not illustrated):** Like female but black on crown replaced with dark brown. **Voice:** Deep bubbling notes. Like a large bubble being blown under water, only the sound is deeper, followed by a few more before changing to clucks with increasing tempo (Mah Yoong Loong). **Range:** North-East India through South-East Asia and the Malay Peninsula to Sumatra and Java. Declining throughout due to loss and degradation of wetlands, especially lowland riverine forest in Asia. EN. **Status and distribution:** In Peninsular Malaysia, rare and localized non-breeding visitor in small numbers at low elevations, south to Johor, but one was mist-netted at 1400 m at night at Fraser's Hill (Pahang) on 27 December 1976, suggesting the possibility of nocturnal movements (Wells, 1983). Regularly seen along the Sungai Tahan, above Kuala Tahan (Pahang) where the evidence of a juvenile and a male in breeding condition suggest the possibility of a breeding population. In Singapore, vagrant with three confirmed records; a male at Sungei Buloh Wetland Reserve on 12–13 January 1999 and another (sex unknown) at Upper Seletar Reservoir on 12 April 2002 (W and H). A male at Upper Seletar Reservoir on 9–11 January 2010 (SINAV 24 (1). **Habitat:** Mangroves, forested rivers, disused mining pools, fish ponds. **Habits:** Swims with neck moving back and forth, keeping close to the pond or riverside vegetation. Shy, rapidly skitting across the water surface before disappearing into dense vegetation. Sometimes tame. Perched birds are known to occasionally jerk head and neck up simultaneously in what appears to be low-level alarm behaviour (W2). Secretive and other habits little known.

CRANES

ORDER *Gruiformes* **FAMILY** *Gruidae*

World: *15 species* **Peninsular Malaysia:** *1 species*

Very large wading birds with long necks, long legs and long pointed bills. From storks and herons by elongate tertiaries which droop over short tails when at rest. In flight, necks are fully outstretched with legs trailing beyond tails. Calls are loud distinctive trumpeting notes. Feed on both small animals and vegetable matter. Usually in pairs or small loose groups. Sexes alike. Not recorded from Singapore.

132. Sarus Crane, *Grus antigone* Plate 5

Keria

Description: 152 cm (60 in). Large size with bare red head and upperneck together with grey plumage diagnostic. Primaries black, concealed at rest; crown and bill whitish; legs pinkish. In upperwing and underwing flight patterns, black primaries contrast with grey secondaries and wing coverts. **Juvenile** (not illustrated): Grey plumage tinged brownish; head and upperneck buff and feathered. **Voice:** A sequence of sonorous trumpeting notes, usually uttered antiphonally by pairs (extralimital data). **Range:** India through mainland South-East Asia to the Philippines and northern Australia. Declining due to widespread exploitation and reduction in the extent and quality of its wetland habitats together with the effect of pollutants. **VU. Status and distribution:** A former presumed resident, now locally extinct; said to have ranged south to Kuala Kangsar (Perak) (W1). **Habitat:** Open country, ricefields. **Habits:** Where common, usually frequents open country singly or in pairs. Forages mainly for small vertebrates and invertebrates in grassy areas. Usually noisy.

BUTTONQUAILS

ORDER *Gruiformes* **FAMILY** *Turnicidae*

World: *16 species* **Peninsular Malaysia and Singapore:** *2 species*

Small and rounded ground-dwelling birds with short tails. Resemble true quails (Phasianidae) but differ by having all three toes pointing forward with no hind toe, and in flight by sandy-buff upperwing coverts contrasting with darker flight feathers. Bills are proportionately longer and more slender. Polyandrous, the sex roles are reversed; the females are brighter coloured than males in most species and mate with several males, laying their eggs, and leaving the task of egg incubation and rearing of young to the male. Usually solitary or in pairs. Inhabit largely open country.

133. Small Buttonquail, *Turnix sylvaticus* Plate 1
Puyuh Kecil

Description: 14 cm (5.5 in). From Barred Buttonquail (134) by much smaller size, a characteristic upright stance, and pink legs. Has distinct yellowish-buff supercilium which curves around dark grey ear coverts. Shows broad dark brown lateral crown stripes and a yellowish-brown central crown stripe. Hind neck and sides of neck chestnut-brown with some vertical streaking on sides of breast, with blackish spots on foreneck which extend from the ear coverts to the carpal joint. Mantle grey-brown with blackish terminal mark, edged with a clear yellowish-buff line. Underparts plain yellowish-buff with rear flanks and vent washed rufous. Iris whitish and bill pale greyish. Female is slightly larger and more rufescent than male. **Voice:** Unknown in review area. **Range:** North Africa and south-west Europe east to the Indian subcontinent, south China, Taiwan, and continental south-east Asia south to the Philippines, Java, and Bali. **Status and distribution:** Rare and localized resident near Kulim (Kedah). First seen on 12 July 2006 (Bakewell, 2006), and subsequently found breeding there (Bakewell, 2010a). **Habitat:** Open scrub. **Habits:** Unobtrusive like Barred Buttonquail. Other habits little known.

134. Barred Buttonquail, *Turnix suscitator* Plate 1
Puyuh

Description: 17 cm (6.5 in). **Female:** Black throat and upper breast diagnostic. Upperparts mottled brown; underparts bright rufous buff from flanks down to belly and undertail coverts. Shows conspicuous blackish bars on sides of breast. **Male:** Like female but lacks black throat and upper breast. Bill and legs grey in both sexes. **Voice:** A prolonged series of soft booming notes which increase in volume before terminating abruptly. Female utters a purring *krrrr* for several seconds at a time during courtship. **Range:** The Indian subcontinent and southern China east through South-East Asia to Sumatra, Java and Bali. **Status and distribution:** Fairly common resident at low elevations up to 1600 m at Cameron Highlands (Pahang) south to Singapore. **Habitat:** Open grassland and cultivation. **Habits:** Usually solitary, sometimes in pairs. Generally unobtrusive, keeping to thick cover.

When flushed, takes off and flies low with fast whirring wingbeats before dropping back to cover.

THICK-KNEES

ORDER *Charadriiformes* **FAMILY** *Burhinidae*

World: *9 species* **Peninsular Malaysia and Singapore:** *1 species*

Medium to large size shorebirds with compact bodies, long legs, large heads and bills. All have distinctive yellow eyes. Plumage mainly combination of brown, grey, black and white. Nocturnal and crepuscular. Inactive during day. Usually in pairs or in small parties; sometimes in larger groups. Calls, eerie wailing notes.

135. Beach Thick-knee, *Esacus neglectus* Plate 20
Kedidi Malam

Description: 53 cm (21 in). Blackish brown crown, broad blackish eyeline, white supercilium and throat together with massive straight dagger-like black bill with yellow base diagnostic. Upperparts greenish sandy; breast brownish; belly and underwing white; legs greenish yellow; wings greyish with broad black band near shoulder at rest. In flight, shows distinct arched profile of back with greyish upper wing coverts contrasting with black flight feathers; underwing all white with black flight feathers and trailing edge of wings. **Voice:** Territorial call is a mournful and harsh disyllabic *wee-loo*. **Range:** The Andamans and Nicobars through South-West China (including Hainan) to South-East Asia and the Greater Sundas, the Philippines south to Australia and New Caledonia. Declining due to human disturbance and predation by invasive mammals. **NT. Status and distribution:** In Peninsular Malaysia, presumably very rare resident known from two records; one off rocky islets off Jeram (Selangor) in June 1878 (W1) and two on a coastal ricefield on 27 October 2003 on Pulau Langkawi (Kedah) (W2). An individual on Pulau Busing (Singapore) from March 1978 to March 1983 when the island was reclaimed. Subsequently sighted on Pulau Hantu, Pulau Salu and Pulau Sudong (all islands adjacent to each other). Last seen on Pulau Salu in 1993. All sightings possibly refer to the same individual. **Habitat:** Sea coasts, river banks. **Habits:** Where common, feeds in small parties on the coast. Calls sporadically at night. Shy.

OYSTERCATCHERS

ORDER *Charadriiformes* **FAMILY** *Haematopodidae*

World: *6 species* **Peninsular Malaysia:** *1 species*

Large and chunky shorebirds with black or boldly patterned black and white plumage and brightly coloured bills and legs. Bills of adults long and heavy with a blunted point, being specially adapted to prise open intertidal bivalve molluscs on which oystercatchers feed. Sexes alike but bills of females longer and thinner. Solitary or gregarious.

136. Eurasian Oystercatcher, *Haematopus ostralegus* — Plate 20
Kedidi Tiram

Description: 45 cm (18 in). Head, breast, mantle and broad terminal tail band black; rest of underparts, rump and wingbar white. Juvenile and non-breeding adult have white throat patch. In flight white rump, wingbars and base of tail contrast sharply with black head and mantle. Bill, iris and narrow orbital skin red; feet pinkish. **Juvenile (not illustrated):** Like adult but black parts greyer with faint white scalings on mantle and wings. Tip of red bill blackish and pointed. **Voice:** A clear piping *kleep* in flight or when disturbed (extralimital data). **Range:** Africa, Europe and temperate Asia through India and China. **Status and distribution:** Vagrant, a juvenile at Kapar (Selangor) from 29 August 1992 to the first week of April, 1993 (W1). **Habitat:** In review area, ash-ponds of a coastal power-station (as a high tide roost habitat, otherwise inter-tidal mudflats). **Habits:** Where common, gathers in large noisy flocks, feeding mainly on molluscs at intertidal estuaries or rocky beaches.

CRAB PLOVER

ORDER *Charadriiformes* **FAMILY** *Dromadidae*

World: *1 species*

Thickset black and white shorebird with large head, long legs and heavy dagger-like bill. Superficially resembles thick-knees (135). Strictly coastal. Sexes alike. Gregarious.

137. Crab Plover, *Dromas ardeola* — Plate 20
Kedidi Ketam

Description: 38 cm (15 in). White plumage with black mantle, inner scapulars and flight feathers. Bill black; legs grey. **Juvenile** (not illustrated): Mantle pale grey with blackish streaks on rear crown and nape. In upperwing flight pattern, neck is extended with legs trailing beyond tail; shows white triangular rump patch with white wing coverts contrasting with black mantle and flight feathers. **Voice:** Contact call consists of harsh chattering notes. **Range:** Shores of the Indian Ocean and adjacent seas from the east coast of Africa and the Arabian Peninsula to the Indian subcontinent, the Andaman and Nicobar Islands. **Status and distribution:** Vagrant, recorded off the coast of Selangor; six on Pulau Pintu Gedung on 24 September 1912 and a juvenile on Pulau Tengah on 24 January 1990. **Habitat:** Mudflats, coral reefs, estuaries. **Habits:** Where common, gathers in noisy flocks. Feeds largely on crabs which it dismembers with powerful stabs of its bill. Often rests on tarsi. **Note:** The Andaman Sea population has experienced a significant decline over the years and are no longer regular visitors to Ko Libong (Trang Province, Peninsular Thailand) and therefore likelihood of future occurrence in review area is much reduced (John Howes, pers. comm.).

STILTS

ORDER *Charadriiformes* **FAMILY** *Recurvirostridae*

World: *7 species* **Peninsular Malaysia and Singapore** *1 species*

(This family also includes avocets which do not occur in the review area)

Shorebirds with long slender legs and straight bills. Plumage mainly black and white. Legs trail far beyond tail in flight. Gregarious and noisy.

138. Black-winged Stilt, *Himantopus himantopus* Plate 21
Kedidi Kaki Panjang

Description: 38cm (15 in). Head, neck and entire underparts white. **Male:** Mantle, scapulars and tertials black. **Female:** Black parts replaced with dark brown with white head and neck suffused with pale grey. Legs pink in both sexes. In flight white triangular rump patch and uppertail coverts distinctive. **Voice:** A shrill, high pitched *keek,* uttered repeatedly. **Range:** Breeding throughout much of southern Europe and temperate Asia, tropical Africa and Asia; northern populations are migratory, wintering south to tropical Africa, South and South-East Asia to Borneo, Java and the Philippines. **Status and distribution:** Fairly common resident, passage migrant, and winter visitor at low elevations, south to Singapore. Nesting has been observed at the Ulu Dedap ricefields (Perak) while copulating individuals have been observed at Kuala Selangor (Selangor). **Habitat:** Flooded ricefields, mudflats, fringes of freshwater ponds. **Habits:** Moves in small groups, feeding by gleaning small insects off the surface of mud, water and vegetation. Nesting birds have been observed mobbing Eastern Marsh-harriers (98) at Ulu Dedap. Generally shy.

LAPWINGS AND PLOVERS

ORDER *Charadriiformes* **FAMILY** *Charadriidae*

World: *67 species* **Peninsular Malaysia and Singapore:** *13 species*

Small to medium sized shorebirds with compact bodies, short and thick necks, short tails and long legs with three forward toes. Bills short and stubby, limiting them to food material on the surface only. Inhabit a variety of habitats from coastal mudflats and sandy shores to muddy riverbanks and wet and dry grassland. Run quickly in an upright stance. Feeding technique unique; consisting of short runs interspersed with pauses before tilting to pick up food material. Fly strongly in compact flocks. Most species are migratory and are highly gregarious in winter, often feeding and roosting in mixed species flocks. Sexes alike in most species. Represented by the following groups and genera. **Lapwings,** genus *Vanellus* (139–141). Generally larger and plumper than plovers. Some species have conspicuous facial wattles and wing spurs. Flight rather dippy with continuous moderate wingbeats on broad rounded wings, usually marked with black and white. Have similar plumage during breeding and non-breeding. **Plovers,** genera *Pluvialis* (142, 143) and *Charadrius* (144–151). Smaller than lapwings with longer, narrower and more pointed wings. Lack facial wattles and wing spurs. Have different plumage patterns during breeding and non-breeding. *Pluvialis* are larger and have distinct black flanks and underparts during breeding while some of the breeding *Charadrius* are beautifully marked with shades of bright rufous on the sides of neck and breast. All are rather soberly clad in shades of brown and grey when not breeding. Many of the *Charadrius* are difficult to identify especially in non-breeding plumage and so attention should be paid to relative size, length of bill as well as facial and wing patterns. Leg colour is also a valuable criterion but this is often obscured by mud. Calls are diagnostic in many.

139. Yellow-wattled Lapwing, *Vanellus malabaricus* — Plate 21

Rapang India

Description: 27 cm (10 in). Small lapwing. Bright yellow wattle hanging down from base of bill in front of eye, narrow white stripe between black cap and brown mantle and upper breast diagnostic. White underparts separated from brown upper breast by narrow blackish breast band. Bill blackish; legs yellow. In upperwing flight pattern, white wing bar contrasts with brown mantle and black primaries and secondaries. Tail white with black subterminal band. In underwing flight pattern (not illustrated), wings all white except for black primaries and secondaries. **Voice:** A loud and plaintive *ti-ee, ti-ee*. **Range:** The Indian subcontinent. **Status and distribution:** Vagrant, an individual on the playing fields of the Agriculture University of Malaysia in Serdang (Selangor) from December 1979 to April 1980. Largely sedentary in its indigenous range, the appearance of this lapwing in the review area is a point that needs further investigation. **Habitat:** Open grassland. **Habits:** Where common, usually feeds in pairs or in small loose groups outside the breeding season. Prefers drier habitat than Red-wattled Lapwing (140). Noisy and conspicuous in flight.

140. Grey-headed Lapwing, *Vanellus cinereus* — Plate 21

Rapang Kepala Kelabu

Description: 35 cm (14 in). Grey head and breast, greyish brown upperparts, blackish breast band, white underparts, yellow legs and yellow bill with dark tip diagnostic. Small wattle in front of eye ye low. In upperwing flight pattern, black primaries and white secondaries contrast with greyish brown mantle and wing coverts which extend down to base of tail in a triangle. Tail white with broad black subterminal band. In underwing flight pattern (not illustrated), black primaries contrast with white secondaries and wing coverts. **Juvenile** (not illustrated): Lacks breast band; head brownish; upperparts scaled whitish. **Voice:** A loud and sharp *kirrit-kirrit*. **Range:** Breeding in Mongolia, northern China and Japan; wintering south through southern China, India, South-East Asia to the Philippines. **Status and distribution:** An uncommon and localized passage migrant and winter visitor at low elevations, south to Selangor.

Flocks regularly winter in disused tin mines around Bidor (Perak). **Habitat:** Muddy riverbanks, flooded ricefields, mudflats. **Habits:** Usually solitary or in small groups when on passage. Feeds in typical plover fashion. Shy, taking to flight at the slightest sign of danger. Very vocal.

141. Red-wattled Lapwing, *Vanellus indicus* Plate 21

Rapang Minta Duit/Burung Minta Duit

Description: 33 cm (13 in). Black head and breast, white ear coverts, bronze-brown upperparts, narrow white collar, white underparts, yellow legs, red bill with dark tip and small red wattle in front of eye diagnostic. In upperwing flight pattern, narrow white collar separates black head from bronze brown mantle. Tail white with broad black subterminal band. In underwing flight pattern (not illustrated), black throat and breast together with black primaries and secondaries contrast with white wing coverts and underparts. **Juvenile** (not illustrated): Like adult but duller with whitish throat. **Voice:** A loud and high-pitched plaintive *di-di-du-it*, uttered in flight and on the ground. **Range:** South-West Asia through the Indian subcontinent and southern China to South-East Asia. **Status and distribution:** Locally common resident at low elevations, south to Johor. Also on on Pulau Langkawi (Kedah). In Singapore, a rare resident which seems to becoming more common (W and H) especially at Tuas, Poyan and Kranji. Also on Pulau Tekong and Pulau Ubin. **Habitat:** Pasture land, flooded ricefields, sandbars in rivers, newly planted oil palm fields. **Habits:** Usually in pairs, feeding in typical plover fashion, making short runs and tilting to pick up food material. During courtship, noisy and conspicuous and when nesting, will engage in an alarm display flight to distract intruders. Often active at night, calling intermittently.

142. Pacific Golden Plover, *Pluvialis fulva* Plate 21

Rapang Kerinyut

Description: 25 cm (10 in). Large golden yellow plover. **Breeding:** Mantle and wings mottled with golden yellow and black. Face, throat and underparts black; narrow band from forehead and supercilium down to flanks white. **Non-breeding:** Upperparts dark brown mottled with golden yellow; underparts pale brown; narrow supercilium yellowish white. In upperwing flight pattern, black outer wings show less contrast with faint whitish wingbar; rump concolorous with mottled brown and golden yellow mantle. In underwing flight pattern, wings, including axillaries all grey. Autumn migrants show traces of breeding plumage on arrival and many spring migrants moult into full breeding plumage before departure. **Voice:** A clear disyllabic whistle, *chu-wit*, with a rise on the second syllable. **Range:** Breeding in the Arctic of North-East Asia and Alaska; wintering south to South-East Asia, Australia and the western Pacific. **Status and distribution:** Common and widespread passage migrant and winter visitor at low elevations, south to Singapore. Common on playing fields and the grassy verges of airports in coastal districts; rarer inland. **Habitat:** Mudflats, wet and dry grassland, sandy shores, coastal power-station ash-ponds. **Habits:** Usually in small loose groups, exclusively on its own or in the company of other smaller plovers. Flies in compact flocks from one feeding area to another. Often feeds and calls at night.

143. Grey Plover, *Pluvialis squatarola* Plate 21
Rapang Kelabu

Description: 28 cm (11 in). Large and heavy-bodied greyish plover. From Pacific Golden Plover (142) by larger size, heavier bill and largely greyish plumage. In underwing flight pattern, black axillaries contrasting with all white wing diagnostic. **Breeding:** Grey upperparts (including crown) mottled white, black face, throat and belly together with broad white band from forehead and supercilium down to flanks diagnostic. Undertail coverts all white. **Non-breeding:** Upperparts brownish grey; underparts white with faint dark streaks on breast. Broad white band reduced to pale supercilium. In upperwing flight pattern, white wingbars contrast with black flight feathers and grey wing coverts. Rump white. Autumn migrants show traces of breeding plumage on arrival and many spring migrants moult into full breeding plumage before departure. **Voice:** A clear trisyllabic *tiee-e-ee,* with an upward inflection on the last syllable. **Range:** Breeding in the Arctic throughout temperate Asia and North America; wintering south to temperate and tropical latitudes worldwide. **Status and distribution:** Locally common passage migrant and winter visitor at low elevations, south to Singapore. Fairly common in coastal districts but very rare inland. Not as numerous as Pacific Golden Plover. **Habitat:** Sandy shores, mudflats, coastal power-station ash-ponds. **Habits:** Usually in small loose groups, feeding in coastal habitats during low tide. Congregates with other shorebirds at high tide roosts. Shy.

144. Common Ringed Plover, *Charadrius hiaticula* Plate 22
Rapang Gelang Besar

Description: 19 cm (7.5 in). From Little Ringed Plover (146) by larger size and conspicuous white wingbars in upperwing flight pattern. **Breeding:** Broad black breast separates white throat from entire white underparts; broad white forehead separated by broad black midcrown from dark brown hindcrown; narrow white collar separates crown from dark brown upperparts. Facial pattern distinct; broad black eyeline, yellow bill with dark tip, small white patch and small white patch above eye. Legs bright orange. **Non-breeding:** Generally duller with black parts replaced by dark brown; bill all black; breast band narrower and usually broken in the middle; legs dull orange-yellow. **Voice:** A diagnostic soft and fluty *tuu-it* with an upward inflection on the second syllable. **Range:** Breeding in Greenland and North-East Canada through northern Europe and temperate Asia; wintering south to the coasts of Europe, North and tropical Africa, the Middle East, South Asia, and Australia. **Status and distribution:** Very rare non-breeding visitor recorded only from Georgetown (Pulau Pinang) and Jurong and Serangoon (Singapore). **Habitat:** The fringes of freshwater ponds, mudflats. **Habits:** Usually in small loose groups where common, feeding in typical plover fashion.

145. Long-billed Plover, *Charadrius placidus* Plate 22
Rapang Paruh Panjang

Description: 21 cm (8.5 in). Slender build, long yellow legs with long thin bill distinctive. **Breeding:** Blackish facial markings, black midcrown above white forehead and narrow black breast band. **Non-breeding:** Black facial markings replaced by brown and is concolorous with crown and nape, black midcrown duller, black breast band much narrower. In upperwing flight pattern white collar contrasts with narrow black hindneck collar especially at close range. White wing bars thin and indistinct but greyish black flight feathers contrast slightly with sandy brown upperparts. **Voice:** A loud and distinct disyllabic whistle, *pe-wee*. **Range:** Breeding in eastern Russia, northern China and Japan; wintering south to North-East India, southern China and northern continental South-East Asia with stragglers reaching the Malay Peninsula and the Greater Sundas. **Status and distribution:** Vagrant, an individual at Tanjong Ru, Pulau Langkawi (Kedah) on 19 March 1968. **Habitat:** Sandy coasts, stony river beds and banks. **Habits:** Where common, congregates in small groups on shingle river beds, foraging in typical plover fashion. Other habits here little known.

146. Little Ringed Plover, *Charadrius dubius* Plate 22

Rapang Gelang Kecil

Description: 18 cm (7 in). Small and compact. **Breeding:** Like Common Ringed Plover (144) but proportionately longer legged with conspicuous thin yellow orbital skin, longer white patch above eye, narrower breast band, pinkish legs and dark brown mantle and blackish flight feathers without white wingbars. **Non-breeding:** Lacks broad black eyeline. Breast band indistinct and usually broken in the middle or reduced to dark smudges on sides of breast. **Juvenile** (not illustrated): Like non-breeding but upperparts faintly scaled whitish; crown, throat and sides of neck tinged yellow. **Voice:** A diagnostic disyllabic whistle, *pee-oo,* on a falling tone. **Range:** Breeding throughout Europe and temperate Asia discontinuously through South and South-East Asia to New Guinea. Populations from temperate latitudes are migratory; wintering south to tropical Africa and Asia. **Status and distribution:** Fairly common and widespread winter visitor at low elevations, south to Singapore. Fairly common on playing fields and the grassy verges of airports in coastal districts. Also occurs inland. **Habitat:** Open grassland, sandy shores, fringes of freshwater ponds, mudflats, brackish ponds, coastal power-station ash-ponds (in drier areas only). **Habits:** Usually in small loose groups, feeding in typical plover fashion. Sometimes unobtrusive, standing motionless. Flies in small compact flocks from one feeding ground to another, often calling in flight.

147. Kentish Plover, *Charadrius alexandrinus* Plate 22

Rapang Pantai

Description: 15 cm (6 in). *C. a. alexandrinus*: Small size with sandy brown upperparts, white underparts together with distinctive dark lateral breast patches and narrow white nuchal collar distinguish it from other smaller *Charadrius* plovers. Legs rather longish and bluish-grey. **Male breeding:** Chestnut brown crown and hindneck, black eyeline, white forehead (with black frontal bar on mid crown) and supercilium distinctive. Bill blackish. **Female and non-breeding male:** Crown, eyeline and broken breast band greyish brown. In upperwing flight pattern, narrow white collar and wing bars across primaries contrast with sandy-brown mantle. Also shows narrow white trailing edge to secondaries. **Voice:** A soft *chit* while taking off. Also a short trill, *prriit*. **Range:** Breeding throughout temperate and tropical latitudes worldwide with northern populations wintering south to tropical latitudes. **Status and distribution:** Locally common passage migrant and winter visitor at low elevations, south to Singapore. **Habitat:** Sandy shores, mudflats, estuaries, coastal power-station ash-ponds. **Habits:** Usually solitary or in small groups, often in association with other small shorebirds. Roosts in large flocks.

147.1. Kentish Plover (White-faced Plover), *Charadrius a. dealbatus* Plate 22
Rapang Cina

Description: 16 cm (6.25 in). **Adult (non-breeding):** In review area, like nominate *Charadrius a. alexandrinus* but noticeably larger, flatter-backed, heavier breasted, larger headed with heavier and proportionately longer bill. White forehead, lores, broad supercilium, and lower ear coverts distinctive, creating a 'white-faced' appearance and contrasting greatly with both dark eye and black bill. Legs proportionately longer than those of other *Charadrius* plovers in review area and are pinkish or flesh-coloured. Also shows shorter and narrower lateral breast patches than nominate Kentish (very obvious when seen side by side) Upperparts pale sandy brown compared with darker sandy brown of nominate Kentish. In upperwing flight pattern **(not illustrated)**, white wing bar across inner primaries broader and more extensive than in nominate Kentish. White trailing edge to the secondaries broader and more conspicuous. **Male breeding:** Entire crown bright orange, paler and brighter than in male nominate Kentish. Black frontal bar on mid crown variable in width but most of the time wider than those of male nominate Kentish. Lateral breast patches dark. Legs turn pale lead grey, which are still paler than those of nominate Kentish. **Female breeding:** White forehead and supercilium, sandy brown crown, upper ear coverts and lores, with strong rufous to orange tone. Lateral breast patches sandy brown and slightly more extensive than male. Legs remain pale pinkish or flesh coloured. **Voice:** Similar to nominate Kentish Plover. **Range:** Breeding in south and south-west China, wintering south to Indo China, the Malay Peninsula, Sumatra, and Borneo. **Status and distribution:** Very rare passage migrant and winter visitor at low elevations, south to Singapore. **Habitat:** Sandy shores, mudflats, estuaries, coastal power-station ash-ponds. **Habits:** Similar to nominate Kentish Plover with which it associates but has been observed to be more active and skittish. When feeding, has a more upright posture than nominate Kentish Would often hunt by lowering head and pursuing prey, usually small crabs. Nominate Kentish is less inclined to do this. When roosting, lies on the sand with legs folded beneath with head held low whereas nominate Kentish would be more inclined to stand up (Bakewell and Kennerley, 2008). **Note:** Included as a full separate description due to structural differences, plumage patterns and distinct behavioural patterns from nominate Kentish. Comparisons are not made with other subspecies of Kentish Plover. Kennerley *et al.* (2008) have described this subspecies in great detail, proposing it be elevated to species rank, White-faced Plover *Charadrius dealbatus*. Further research, including DNA analysis would be very crucial in establishing this.

148. Malaysian Plover, *Charadrius peronii*

Plate 22

Rapang Pasir

Description: 15 cm (6 in). From Kentish Plover (147) by proportionately shorter legs and whiter underparts. Greyish brown upperparts faintly scaled with white. **Male:** From breeding male Kentish Plover by black hindneck collar and more extensive black breast band. In upperwing flight pattern, shows white wingbars and all black tail with white edges. Black hindneck collar contrasts with white forecollar, especially at close range. **Female:** Pale rufous crown, ear coverts and broken breast band. **Juvenile:** Like female but lacks rufous crown and sides of neck. **Voice:** Soft *chit* notes, similar to Kentish Plover. **Range:** The Malay Peninsula to the Greater and Lesser Sundas. Declining owing to development on the coastal areas it inhabits. **NT. Status and distribution:** Uncommon and localized coastal resident, south to Singapore. Also on Pulau Langkawi (Kedah) and in the Tioman Archipelago (Pahang). Favours sandy shores and is therefore more frequent along the east coast. Especially common at Tanjong Resang, near Mersing and Telok Mahkota (both on the east coast of Johor). In Singapore, rare resident at Changi. **Habitat:** Sandy shores, usually in sheltered bays, sandy fringes of mudflats. **Habits:** Usually solitary or in pairs and sometimes in very small groups. Never gathers in large flocks. Feeds exclusively on its own most of the time, seldom associating with other shorebirds. Feeds by running swiftly, before pausing to pick up food material, a feature shared by most of the smaller *Charadrius* plovers. Sometimes tame and confiding.

149. Lesser Sand Plover, *Charadrius mongolus*

Plate 22

Rapang Mongolia

Description: 20 cm (8 in). **Breeding:** White throat (bordered by a thin black line in *C. m. mongolus* only) sharply demarcated from orange rufous breast which extends to nape and flanks with prominent black eyeline and white supercilium. Underparts white. Lores white in nominate subspecies *C. m. mongolus* **(not illustrated)** and all black in *C. m. atrifrons*. **Non-breeding:** *C. m. mongolus*. Upperparts brownish grey, underparts whitish with brownish breast band, facial markings sandy buff with pale buff supercilium. *C. m. atrifrons* **(not illustrated):** Has distinct dark brown and white facial pattern. In upperwing flight pattern, shows conspicuous white wingbars. Greyish brown crown concolorous with mantle and wings. Legs greyish in all seasons. **Juvenile (not illustrated):** Like non-breeding adult but duller with whitish scales on crown and upperparts. Many assume full breeding plumage during spring passage. **Voice:** A series of short, hard trills, *krri-krri...*, constantly uttered especially in flight. **Range:** Breeding in the highlands of central Asia and eastern Russia; wintering south to the coasts of southern Africa, South and South-East Asia to Australia and the South-West Pacific. **Status and distribution:** Common and widespread passage migrant and winter visitor at low elevations, south to Singapore. Prefers largely coastal habitats and very rarely occurs inland. **Habitat:** Mudflats, sandy shores, the fringes of settled mining pools, brackish ponds, coastal power-station ash-ponds. **Habits:** Usually congregates in flocks of several hundred at low tide feeding grounds or at high tide roosts. Associates with other feeding shorebirds. Shy.

150. Greater Sand Plover, *Charadrius leschenaultii* Plate 22
Rapang Besar

Description: 24 cm (9 in). Similar to Lesser Sand Plover (149) in both breeding and non-breeding plumages but is larger with broader head, longer and thicker bill and proportionately longer and paler yellower legs (often stained dark with mud). Breeding plumage birds always show white face patch in black mask and orange breast band doesn't extend as much. In upperwing flight pattern, shows conspicuous white wing bars and legs project more beyond tail. Many assume full breeding plumage during spring passage. **Voice:** Short and hard trilling notes, *krri-krri…*, similar to but deeper than Lesser Sand Plover. **Range:** Breeding in central Asia; wintering south to coasts of the Indian Ocean and adjacent seas from southern Africa through South and South-East Asia to Australasia and the South-West Pacific. **Status and distribution:** Fairly common passage migrant and winter visitor at low elevations, south to Singapore. **Habitat:** Mudflats, sandy shores, brackish ponds, coastal power-station ash-ponds. **Habits:** Usually in small loose groups. Associates with other feeding shorebirds, especially Lesser Sand Plover.

151. Oriental Plover, *Charadrius veredus* Plate 22
Rapang Timur

Description: 24 cm (9 in). Large *Charadrius* plover with long legs. In breeding plumage, lacks the dark eyeline so characteristic of most *Charadrius* plovers; dark iris stands out in white head in breeding male. In underwing flight pattern, shows all dark underwing contrasting with paler underparts, an important field character at all times. **Male breeding:** Greyish brown crown and nape concolorous with mantle and uppertail coverts, white throat and sides of neck contrasting with orange rufous breast, which is separated from white underparts by black breast band. **Female breeding:** Head and breast tinged orange rufous with faint whitish scales on breast. Throat white. **Non-breeding (both sexes):** (not illustrated) Shows diffused brownish eyeline, white supercilium and throat with brownish mottles on sides of breast. Bill black and legs yellow in both sexes in all seasons. In upperwing flight pattern, legs project beyond tail. Greyish black flight feathers and wing tips show little contrast with greyish brown upperparts. In underwing flight pattern **(not illustrated)** shows all dark underwing contrasting with paler underparts. **Voice:** A loud and clear piping *kwink*. Also a plaintive *chip-chip-chip* in flight. **Range:** Breeding in eastern Russia, Mongolia and northern China; wintering south through South-East Asia to Australia and the South-West Pacific. **Status and distribution:** Very rare passage migrant and winter visitor at low elevations, south to Singapore. **Habitat:** Wet grasslands, the fringes of settled mining pools. **Habits:** Largely solitary. Feeds mostly exclusively on its own, sometimes in association with Pacific Golden Plover (142).

PAINTEDSNIPE

ORDER *Charadriiformes* **FAMILY** *Rostratulidae*

World: *2 species* **Peninsular Malaysia and Singapore:** *1 species*

Superficially resemble true snipe *Gallinago* sp (156–158) but flight slower on broader and more rounded wings with legs dangling. Very secretive with crepuscular habits.

Sexually dimorphic; the female is larger and more brightly coloured. Polyandrous with sex roles reversed; the male incubates the eggs and raises the young while the female consorts and mates with different males and lays the eggs. Solitary or in pairs.

152. Greater Paintedsnipe, *Rostratula benghalensis* Plate 20
Meragi

Description: 25 cm (10 in) male; 28 cm (11 in) female. Slightly drooping pale yellow bill distinctive. Yellowish central crown stripe and scapular V together with broad white stripe curving along shoulder from white underparts diagnostic. **Male:** Eyering and stripe behind eye yellowish white with brownish upperparts marked with buff spots. **Female:** Eyering and stripe behind eye white. Sides of neck rich chestnut. Upperparts dark brown with faint buff streaks. **Voice:** Breeding female utters a long *koh-koh-koh*, usually at night. Otherwise usually silent. **Range:** Africa, the Indian subcontinent east to Japan and south through South-East Asia to Australia. **Status and distribution:** Locally common resident at low elevations, south to Singapore. Also recorded as high as 900 m on Gunung Jerai (Kedah). **Habitat:** Open freshwater marshes, flooded rice-fields. **Habits:** Keeps to thick cover, emerging at clearings only to feed. When it walks, bobs tail up and down like Common Sandpiper (175).

JACANAS

ORDER *Charadriiformes* **FAMILY** *Jacanidae*

World: *7 species* **Peninsular Malaysia and Singapore:** *2 species*

Swamp birds with extremely long toes which allow them to walk on floating aquatic vegetation in ponds and lakes where they feed. Also called lily-trotters. Flight, weak and direct on short rounded wings with legs trailing beyond tail Swim and dive well. Sexes alike. Polyandrous like Paintedsnipe (152); the female mates with a number of males which incubate the eggs and raise the young.

153. Pheasant-tailed Jacana, *Hydrophasianus chirurgus* Plate 20
Burung Teratai Ekor Panjang

Description: 30 cm (12 in). In flight, white upper and lower wings contrasting with black outer primaries diagnostic. **Breeding:** Sports long black tail. Lower breast, belly and undertail coverts blackish brown. Upperparts glossy brown. Face, throat and upper breast white separated from yellow hindneck by narrow black line running down side of neck. **Non-breeding:** Tail much shorter, entire underparts white with blackish brown reduced to band around breast. Yellow hindneck duller, crown brownish, supercilium whitish with narrow black line from eye running down side of neck into breast band. **Voice:** Nasal mewing notes, *teun, teun*... **Range:** India and China through South-East Asia to the Greater Sundas and the Philippines. **Status and distribution:** Uncommon and localized non-breeding visitor in small numbers, south to Singapore. **Habitat:** Open ponds, disused mining pools, fish ponds, freshwater marshes, slow flowing rivers and streams. **Habits:** Walks slowly on floating vegetation, picking up food material. Makes short flights from one feeding spot to another. Shy. **Note:** Birds seen in Peninsular Malaysia and Singapore are almost always in non-breeding plumage.

154. Bronze-winged Jacana, *Metopidius indicus* Plate 20
Burung Teratai Tembaga

Description: 28 cm (11 in). From Pheasant-tailed Jacana (153) by wholly darker plumage without extensive areas of white. Bronze olive mantle and wings, broad white supercilium together with black neck, breast and lower belly diagnostic. Undertail and uppertail coverts dark chestnut. Bill yellow. In flight, blackish primaries and secondaries contrast with olive bronze upperwing coverts. **Voice:** A repertoire of high-pitched piping notes. Also low guttural notes. **Range:** India, discontinuously through South-East Asia to Java and Sulawesi. **Status and distribution:** Vagrant, an individual collected in Taiping (Perak) in December 1911; two adults at Kampung Tasoh, Timah-Tasoh (Perlis) from 27 December 2009–23 January 2010 (BiW). **Habitat:** Open ponds, fish ponds. **Habits:** Like Pheasant-tailed Jacana.

SNIPE, GODWITS, CURLEWS, SANDPIPERS, AND PHALAROPES

ORDER *Charadriiformes* **FAMILY** *Scolopacidae*

World: *91 species* **Peninsular Malaysia and Singapore:** *37 species*

A diverse group of small to large long-legged shorebirds, inhabiting mostly coastal mudflats, intertidal estuaries and freshwater wetlands. Have generally long and slender bills, down-curved in some species and straight or upturned in others; all used for probing into soft mud and shallow water during low tide to feed on an assortment of small aquatic vertebrates and invertebrates, including crabs. Often feed together in mixed species flocks with usually no threat of competition as different sizes, shapes and lengths of bills ensure specialized exploitation of particular food sources. Length of legs vary from short to long. Flight strong and direct with rapid wingbeats on long, pointed wings. Migrate long distances from northern temperate breeding grounds to winter in tropical latitudes. Sexes alike in most species. Breeding plumage is brightly coloured but cryptic in most species and is usually acquired during spring passage northwards. During non-breeding, most are clad in varying shades of dull brown or grey (usually mottled on upperparts). Intermediate plumage patterns during autumn passage southwards can be sometimes confusing. Best identification clues are general shape and size of body and bill, tail and wing patterns (distinct in many species) and colour of bill, rump and legs; the last is often covered with mud and is therefore of limited aid. Calls are high-pitched plaintive whistles or shrill piping notes and are diagnostic in many species. Highly gregarious during winter, flying in large compact flocks. Also roost together in mixed species flocks during high tide. Roosting flocks are frequently put to flight by the presence of birds of prey overhead. Feeding and roosting behaviour of many species are continuous throughout day and night, being largely governed by tide movements. Represented by the following groups and genera: **Snipe and Woodcock:** Medium size and short legged with plump bodies, long and straight bills and densely streaked cryptic plumage. Very secretive, staying in thick cover. Snipe, genus *Gallinago* (156–158), are all very similar and very difficult to identify in the field. They have distinct tail structures (apparent only when examined in hand or when fanned out in alarm), a sound knowledge of which is essential in positive identification. Colour of legs vary slightly in each species but these are usually concealed in thick cover. Flight pattern erratic with zig-zag movements. Woodcock, genus *Scolopax* (155) is larger and chunkier and has broad black transverse bars on hindcrown and nape **Dowitchers,** genus *Limnodromus* (159). Large size with long, straight and relatively thick bill with slightly swollen tip. **Godwits,** genus *Limosa* (160, 161). Smaller than curlews with long straight or slightly up-curved bills **Curlews and Whimbrel,** genus *Numenius* (162–165). Large with characteristic long down-curved bills. **Sandpipers:** Genus *Tringa* (166–173). Medium sized with long bills and distinctive white rumps and uppertail coverts. Genus *Xenus* (174). Small size and short legged with thin, slightly up-curved bill. Genus *Actitis* (175). Small size and short straight bill with characteristic tail bobbing. **Turnstones,** genus *Arenaria* (176). Small size and short legged with short, thick bill. **Knots, stints, and other small sandpipers,** genus *Calidris* (177–187). Medium to very small size with comparatively short bills which are straight or down-curved. Probe and pick food material off surface. Genus *Eurynorhynchus* (188). Very small with short spatulate bill. Genus *Limicola* (189). Small with longish, slightly down-curved bill. **Ruff,** genus *Philomachus* (190). Medium sized with short straight bill and largely buff plumage. Sexually dimorphic. **Phalaropes,** genus *Phalaropus* (191). Small and slim shorebirds with straight narrow bills. Have lobed toes and thick downy feathers which enable them to swim efficiently and maintain buoyancy. Sexually dimorphic; the females are larger than males and have brighter and more colourful breeding plumage

155. Eurasian Woodcock, *Scolopax rusticola* — Plate 23

Berkek Besar

Description: 36 cm (14 in). Chunky appearance, triangular shaped head, proportionately heavier bill, diagnostic transverse black bars on hindcrown and nape together with the lack of pale buff longitudinal stripes separate it from snipe (156–158) at rest. Flight slower on broader and more rounded wings. Upperparts marked with varying shades of rusty brown and buff. Underparts pale buff with narrow blackish bars on flanks. **Voice:** Usually silent in wintering quarters. **Range:** Breeding throughout Europe and temperate Asia and the Himalayan region. Northern populations are migratory, wintering south to North Africa, South and South-East Asia to the Philippines. **Status and distribution:** Very rare winter visitor principally at low elevations, south to Singapore, where it is vagrant with no recent records. Recorded as high as 1200 m at Fraser's Hill (Pahang). **Habitat:** Dense cover along streams, moist forest undergrowth. **Habits:** Usually solitary, often sitting in a crouched position in thick cover. Active at dawn and dusk, feeding at the forest edge, probing into the moist earth for grubs.

156. Pintail Snipe, *Gallinago stenura* Plate 23
Berkek Berbintik

Description: 25 cm (10 in). Brown upperparts with fine black and white streaks, pale buff longtitudinal stripes from mantle to rump and broad cinnamon-rufous subterminal tail band are common features in all three *Gallinago* snipe wintering here. All three species also have blackish median stripe, eyeline and line below eye as well as pale buff supercilium. Underparts are whitish with blackish-brown bars on flanks. In Pintail, bill brown with dark tip and is proportionately shorter than Swinhoe's (157) and Common Snipe (158) (only apparent when seen together or examined in hand). Tail consists of 24–28 feathers, of which 6–9 pairs of outer tail feathers are pin-shaped. In underwing flight pattern, shows white barred wing coverts which contrast slightly with blackish grey flight feathers. Legs yellowish and in flight extend more beyond tail than other snipe. **Voice:** A brief rasping *squak*. Also a rasping *ketk* when flushed. **Range:** Breeding throughout much of northern temperate Asia; wintering south to East Africa, South and South-East Asia east to Sulawesi and Lombok. **Status and distribution:** Locally common passage migrant and winter visitor principally at low elevations, south to Singapore. Has been mist-netted at 1200 m at Fraser's Hill (Pahang) on passage. The most common snipe in review area. **Habitat:** Flooded rice-fields, the margins of open freshwater swamp, wet grassland. **Habits:** Usually solitary or in small scattered groups. When approached, often runs away swiftly.Generally unobtrusive and is usually flushed. Flight fast and erratic with much zig-zagging.

157. Swinhoe's Snipe, *Gallinago megala* Plate 23
Berkek Tiruk

Description: 28 cm (11 in). At rest, generally indistinguishable in field from Pintail Snipe (156) but is generally larger with proportionately longer bill. Flight is more lumbering with greyish legs just extending beyond tail. In the hand, tail consisting of 20–22 feathers is diagnostic. In underwing flight pattern, greyish flight feathers contrast slightly with heavily barred white wing coverts and is almost identical with Pintail Snipe. **Voice:** A short and coarse *shrek* when flushed. **Range:** Breeding in northern temperate Asia; wintering south to South and South-East Asia to Australia. **Status and distribution:** Uncommon and localized winter visitor in small numbers, south to Singapore. **Habitat:** Margins of open freshwater swamps, wet grassland. **Habits:** General habits similar to Pintail Snipe, with which it sometimes associates, but flight slower with less manouverability.

158. Common Snipe, *Gallinago gallinago* Plate 23
Berkek Ekor Kapas

Description: 28 cm (11 in). At rest, usually indistinguishable from both Pintail (156) and Swinhoe's Snipe (157). In flight differs by narrow white trailing edge to secondaries and more white on underwing coverts. Tail consists only of 14 feathers. Legs just extend beyond tail in flight. **Voice:** A long nasal *jeyht* when flushed, distinct from both Pintail and Swinhoe's Snipe. **Range:** Breeding throughout temperate North America east through Europe and temperate Asia; wintering south to tropical latitudes worldwide. **Status and distribution:** Uncommon and localized winter visitor in small numbers principally at low elevations, south to Singapore. Has been recorded at 1500 m in Cameron Highlands (Pahang). The least common snipe in review area. **Habitat:** Wet and marshy grassland. **Habits:** Usually solitary. General habits like other snipe. Flight fast and swift with much zig-zagging like Pintail Snipe.

159. Asian Dowitcher, *Limnodromus semipalmatus* Plate 24

Kedidi Paruh Tegak

Description: 35 cm (14 in). From Black-tailed (160) and Bar-tailed Godwits (161) by smaller size and straight all black bill with slightly swollen tip. **Breeding:** Bright rufous-chestnut from head down to breast and flanks with whitish bars on sides of breast and dark brownish bars on flanks. Centre of belly whitish; upperparts blackish with pale brown fringes. **Non-breeding:** Head and entire underparts whitish with greyish wash on breast and neck with brownish streaks on sides of breast and flanks. Crown blackish brown with pale supercilium and indistinct dark eyeline. Upperparts blackish grey with whitish fringes. In upperwing flight pattern, shows whitish triangular rump patch and tail marked with dark bars near tip. Legs extend beyond tail. **Voice:** Usually silent but also utters a soft moaning *kiaow*. Also a disyllabic *chep-chep*. **Range:** Breeding in Mongolia, eastern Russia and north-eastern China; wintering south through South and South-East Asia to northern Australia. Declining possibly due to pollution and development on wintering grounds. **NT. Status and distribution:** Uncommon and localized passage migrant and winter visitor in small numbers at low elevations in coastal districts, south to Singapore. **Habitat:** Mudflats, estuaries, brackish ponds, coastal power-station ash-ponds. **Habits:** Usually solitary or in twos or threes, feeding usually exclusively on its own. Feeding behaviour distinct, plodding forward stiffly with an upright stance before lowering head and making rapid 'pumping' probes into mud for food material. Associates with other shorebirds at high-tide roosts. Shy.

160. Black-tailed Godwit, *Limosa limosa* Plate 24

Kedidi Ekor Hitam

Description: 38 cm (15 in). Upright stance and long and straight pink flesh coloured bill with dark tip distinctive (see Asian Dowitcher (159)). Knee joint almost half way down leg, giving a long-legged appearance. Has conspicuous white supercilium and indistinct black eyeline in both breeding and non-breeding plumages. Legs dark grey in all seasons. **Breeding:** Bright rufous-chestnut head and breast with blackish bars on flanks and blackish mottles on upperparts. The male is brighter than the female. **Non-breeding:** Upperparts uniform grey-brown with pale fringe and darkish streaks; head to breast pale grey-brown; rest of underparts whitish. In upperwing flight pattern, shows conspicuous white wing bars which contrast with blackish flight feathers and grey brown wing coverts. Also shows characteristic white rump and uppertail coverts contrasting with black terminal tail band. Legs extend well beyond tail. **Voice:** A sharp *kit-kit-kit*, frequently uttered in flight. **Range:** Breeding throughout northern Europe and temperate Asia; wintering south to southern Europe, Africa, South and South-East Asia and Australia. Although widespread with a large global population, its numbers have declined rapidly in parts of its range owing to agricultural practices. NT. **Status and distribution:** Locally common passage migrant and winter visitor at low elevations principally in coastal districts, south to Singapore. **Habitat:** Mudflats, estuaries, flooded ricefields, inland freshwater swamps, brackish ponds, coastal power-station ash-ponds. **Habits:** Usually in small loose groups on feeding grounds. Probes into mud and shallow water to catch small vertebrates and invertebrates. Shy.

161. Bar-tailed Godwit, *Limosa lapponica* Plate 24

Kedidi Ekor Berjalur

Description: 42 cm (16 in). From Black-tailed Godwit (160) by slightly up-curved bill, less upright stance and proportionately shorter legs with knee joint closer to body. Lacks white wing bars in flight. Also shows conspicuous white supercilium and indistinct blackish eyeline. Legs dark grey in all seasons. **Breeding:** Bright rufous chestnut from head down to breast and flanks. **Non-breeding:** Greyish brown underparts with pale brownish markings on breast and flanks. Upperparts heavily marked with whitish fringes. In upperwing flight pattern, the nominate subspecies, *L. l. lapponica* shows conspicuous white triangular rump patch and uppertail coverts with terminal end of tail finely barred with black and white while in *L. l. baueri* **(not illustrated)**, the white rump and uppertail coverts are much reduced. Legs only just extend beyond tail. **Voice:** A sharp *kek-kek*, frequently uttered in flight. **Range:** Breeding from North-East Europe east through northern temperate Asia to Alaska (North America); wintering south to Africa, South and South-East Asia to Australia and the western Pacific. **Status and distribution:** Fairly common passage migrant and winter visitor in fair numbers at low elevations, south to Singapore. More numerous than Black-tailed Godwit. **Habitat:** Mudflats, estuaries, sandy shores, brackish ponds, coastal power-station ash-ponds. **Habits:** Usually in small loose groups, feeding in muddy estuaries and shallow water. Often roosts with other shorebirds during high tide. Shy.

162. Little Curlew, *Numenius minutus* Plate 24

Kendi Kerdil

Description: 30 cm (12 in). From Whimbrel (163) by much smaller size and shorter and straighter bill which is black with pinkish brown basal half of mandible. Shows pale buff supercilium and blackish coronal stripe with rear part of blackish eyeline extending behind from below level of eye. Upperparts strongly tinged buff with white belly and undertail coverts. In upperwing flight pattern, pale brown rump and uppertail coverts with blackish brown barrings do not contrast with rest of upperparts. In underwing flight pattern, shows buff brown wing coverts and greyish flight feathers. **Voice:** A sharp rising trisyllabic *kwit-kwit-kwit* in flight. **Range:** Breeding in North-East Russia; wintering through the Philippines and the Indonesian Archipelago south to Australia. **Status and distribution:** Vagrant, an individual each on a golf course at Tuas from 25 October to 2 November 1997, Sungei Buloh Wetland Reserve on 24 April 1998 (both in Singapore) (W and H) and in a ploughed ricefield at Sungai Balang, Muar District (Johor) from 11–25 October 2008 (BiW) **Habitat:** Open grassland, mudflats (during passage only). **Habits:** Feeds largely in open grassy areas.

Whimbrel, *Numenius phaeopus variegatus* (163)
Shows much reduced white rump patch, marked with fine blackish bars

163. Whimbrel, *Numenius phaeopus* Plate 24

Kendi Pisau Raut

Description: 43 cm (17 in). From Eurasian (164) and Far-eastern Curlews (165) by smaller size, proportionately shorter bill and legs and conspicuous black lateral coronal stripes and white median coronal stripe. Plumage darker and browner than Eurasian Curlew, lacking the buff tinge. In upperwing flight pattern, shows pale triangular white patch which varies from being pure white in *N. p. phaeopus* to much reduced white with fine blackish bars in *N. p. variegatus* (see figure). In underwing flight pattern **(not illustrated)**, entire wing white, heavily marked with dusky bars. **Voice:** A series of distinct shrill plaintive tittering whistles, *pi-pi-pi-pi-* of about 6–7 notes. **Range:** Breeding in the Arctic from North America throughout Europe and temperate Asia; wintering south to South America, Africa, South and South-East Asia, Australia, New Zealand, and the South-West Pacific. **Status and distribution:** Fairly common passage migrant and winter visitor at low elevations, south to Singapore. Some birds overwinter in this region and are seen throughout the year. More common on mudflats than sandy shores. **Habitat:** Mudflats, estuaries, sandy shores, mangroves, brackish ponds, rocky shores, coastal power-station ash-ponds. **Habits:** Feeds in large numbers on low tide feeding grounds, often in association with other shorebirds. Often takes crabs. Flies in compact and noisy flocks. Frequently roosts in mangrove trees during high tide. Shy.

164. Eurasian Curlew, *Numenius arquata* Plate 24

Kendi Besar

Description: 58 cm (23 in). Large size and long down-curved bill obvious. Upperparts brownish-buff (except for white rump) with bold black streaks; underparts whitish.

In upperwing flight pattern, shows conspicuous white triangular rump patch. In underwing flight pattern, shows all white underwings. **Voice:** A loud plaintive, *courlee,* with a rising inflection on the second syllable; uttered frequently in flight and on the ground. Alarm call is a shrill *kwi-kwi-kwi-kwi*... **Range:** Breeding throughout much of northern Europe and northern temperate Asia; wintering south through Africa, South and South-East Asia to the Greater Sundas and the Philippines. **Status and distribution:** Fairly common passage migrant and winter visitor at low elevations, south to Singapore. Some birds overwinter in this region and are seen throughout the year. **Habitat:** Mudflats, estuaries, sandy shores, brackish ponds, coastal power-station ash-ponds. **Habits:** Often in small groups during low tide, foraging together with other smaller shorebirds. Feeds largely on an assortment of crabs, other smaller crustaceans and fish. Flies in compact and noisy flocks between feeding and roosting grounds. Shy.

165. Far Eastern Curlew, *Numenius madagascariensis* Plate 24
Kendi Timur

Description: 60 cm (24 in). Large size and long down-curved bill obvious. From Eurasian Curlew (164) by darker brown plumage and longer bill (apparent when seen together). In upperwing flight pattern, rump concolorous with rest of upperparts. In underwing flight pattern, shows heavily barred underwing. **Voice:** Like Eurasian Curlew. **Range:** Breeding in far eastern Russia and northern China; wintering south through South-East Asia to the south-west Pacific, Australia, and New Zealand. Rapidly declining possibly due to habitat loss and deterioration **VU. Status and distribution:** Locally common passage migrant and winter visitor at low elevations, south to Singapore. **Habitat:** Mudflats, estuaries, sandy shores, brackish pond. **Habits:** Usually feeds singly among larger flocks of other shorebirds. General habits similar to Eurasian Curlew.

166. Spotted Redshank, *Tringa erythropus* Plate 25
Kedidi Berbintik

Description: 30 cm (12 in). **Breeding:** Slaty black plumage with white mottles and spots on upperparts diagnostic. Shows faint white bars on flanks. Legs reddish brown. Bill has red maxilla and black mandible in all seasons. **Non-breeding:** From Common Redshank (167) by longer and more slender bill (with slight drooped appearance), longer legs and more upright stance. Upperparts grey, mottled white on mantle. Conspicuous white supercilium extends beyond eye. Underparts white with greyish bars on sides of neck and upper flanks. Legs bright orange-red. From Common Redshank in upperwing flight pattern by the lack of broad white trailing edge of wing. Shows white triangular rump patch. Legs extend well beyond tail. **Voice:** A sharp explosive *chu-it,* uttered both in flight and at rest. **Range:** Breeding from northern Europe east throughout northern temperate Asia; wintering south to Africa, South and South-East Asia to the Malay Peninsula with stragglers reaching Borneo. **Status and distribution:** Rare and localized passage migrant and winter visitor at low elevations in coastal districts, south to Singapore where it is very rare. **Habitat:** The fringes of freshwater ponds, mudflats. **Habits:** Usually in small loose parties on low tide feeding grounds, often with other shorebirds.

167. Common Redshank, *Tringa totanus* Plate 25

Kedidi Kaki Merah

Description: 28 cm (11 in). Orange-red legs and red base of bill with dark tip readily separate it from all *Tringa* sandpipers except non-breeding Spotted Redshank (166). **Breeding:** Grey-brown head and upperparts with blackish streaks. Underparts white with bold brown streaks on breast and dark brown spots on lower flanks and belly. **Non-breeding:** Plumage lighter with less distinct dark streakings on underparts. From non-breeding Spotted Redshank by proportionately shorter, stouter bill and shorter legs and less prominent white supercilium. In upperwing flight pattern, shows triangular rump patch and broad white trailing edge of wing. Legs just extend beyond tail. **Voice:** A loud ringing cadence of 3–4 notes, *teu-heu,heu,heu*. **Range:** Breeding throughout northern Europe east through northern temperate Asia; wintering south to Africa, South and South-East Asia to the Greater Sundas and the Philippines. **Status and distribution:** Common and widespread passage migrant and winter visitor at low elevations in coastal districts, south to Singapore. **Habitat:** Mudflats, mangroves, sandy beaches, fringes of freshwater ponds, brackish ponds, coastal power station-ash-ponds. **Habits:** Feeds in small loose parties, often in association with other shorebirds. Often roosts in mangrove trees during high tide. Very vocal in flight.

168. Marsh Sandpiper, *Tringa stagnatilis* Plate 25

Kedidi Paya

Description: 25 cm (10 in). From Common Greenshank (169) by slimmer appearance as well as slender and straight black bill. Has long greenish legs. **Breeding:** Upperparts greyish brown mottled buff grey and black; underparts white, spotted blackish brown on breast. Shows indistinct white supercilium. **Non-breeding:** Upperparts pale grey; face and underparts white. Shows prominent white supercilium. In upperwing flight pattern, shows a long triangular rump patch, extending up back in a long narrow wedge with terminal end of tail finely barred blackish brown. Legs extend well beyond tail. **Voice:** A plaintive *peuk*, uttered while foraging. Also loud piping notes, *teu-heu-heu*... almost similar to Common Redshank (167). **Range:** Breeding from eastern Europe through temperate Asia; wintering south to Africa, South and South-East Asia to Australia and New Zealand. **Status and distribution:** Fairly common passage migrant and winter visitor at low elevations in coastal districts, south to Singapore. **Habitat:** Fringes of inland freshwater ponds, flooded ricefields, mudflats, brackish ponds, sandy shores, coastal power-station ash-ponds. **Habits:** Usually in very small loose groups, often with other shorebirds. Strides with an upright gait, sometimes running in shallow water, frequently pausing to probe for food material. More common in freshwater habitats. Shy.

169. Common Greenshank, *Tringa nebularia* Plate 25

Kedidi Kaki Hijau

Description: 35 cm (14 in). Large and chunky appearance with slightly up-curved pale greenish bill with dark tip. **Breeding:** Head, neck, breast and flanks boldly streaked black; upperparts greyish brown with blackish streakings and warm brown markings.

Non-breeding: Head and neck white; upperparts grey, finely flecked with black and white. Underparts white with fine blackish streaks on sides of neck, breast and flanks. Legs greyish green in all seasons. In upperwing flight pattern, white triangular rump patch extends up back in a long narrow wedge. In underwing flight pattern **(not illustrated)**, shows dark barring on white part of wing, including axillaries. Legs extend well beyond tail. **Voice:** A diagnostic loud ringing *teu-teu-teu*. **Range:** Breeding throughout northern Europe and northern temperate Asia; wintering south to Africa, South and South-East Asia, the Greater and Lesser Sundas to Australia and New Zealand. **Status and distribution:** Fairly common passage migrant and winter visitor at low elevations in coastal districts, south to Singapore. **Habitat:** Fringes of freshwater ponds, mudflats, sandy shores, brackish ponds, coastal power-station ash-ponds. **Habits:** Usually feeds solitarily or in small loose groups. Probes into soft mud as well as side sweeps bill in shallow water to obtain food material. Very vocal in flight. Shy.

170. Spotted Greenshank, *Tringa guttifer* Plate 25

Kedidi Kaki Hijau Berbintik

Description: 32 cm (13 in). Like Common Greenshank (169) but appears more compact and chunky with proportionately thicker black bill with yellowish basal half and proportionately shorter ochreous yellow legs. **Breeding:** Upperparts grey with whitish triangular spots on scapulars; underparts white with dense blackish spots on breast and flanks. **Non-breeding:** Pale grey upperparts with unstreaked underparts. Legs yellow in all seasons. In upperwing flight pattern, white triangular rump patch extends up back in a long narrow wedge. In underwing flight pattern (not illustrated), shows all white axillaries. Legs just extend beyond tail. **Voice:** Generally silent but occasionally utters a harsh *gwaak,* lacking the shrill piping quality characteristic of *Tringa* sandpipers. **Range:** Breeding in far eastern Russia only on the island of Sakhalin and coasts of Sea of Okhotsk; wintering south to East and South-East Asia to Borneo and the Philippines. Peninsular Malaysia is probably the main wintering ground for this species. Its small population is declining due to development of coastal wetlands throughout range, for industry, infrastructure projects, and aquaculture. **EN. Status and distribution:** Uncommon and localized passage migrant and winter visitor at low elevations in coastal districts, south to Singapore where it is very rare with no recent records. **Habitat:** Mudflats, fringes of freshwater ponds, brackish ponds, coastal power-station ash-ponds. **Habits:** Usually solitary or in twos and threes, feeding among other shorebirds in the inter-tidal zone. Appears to specialize by feeding on crabs.

171. Green Sandpiper, *Tringa ochropus* Plate 25

Kedidi Pasir Hijau

Description: 24 cm (9.5 in). Dark green-brown upperparts contrasting with white underparts and base of underwing distinctive. Has narrow white eyering and faint supercilium. From Wood Sandpiper (172) by darker upperparts at rest, dark greenish legs, less distinct supercilium, and indistinct white spotting on upperparts (see figure). From Common Sandpiper (175) by larger size, colour of legs, and lack of white wingbars in upperwing flight pattern. **Breeding (not illustrated):** Sides of neck and breast marked with bold dark streaks. **Non-breeding:** Streaks on neck and breast reduced to greenish brown smudges. In upperwing flight pattern, lacks the characteristic triangular rump patch of most *Tringa* species; showing all dark upperparts contrasting with white square rump patch and tail which is marked with broad dark bars at the terminal end. All dark underwing readily distinguishes it from other sandpipers. **Voice:** A shrill and distinct *tweet-weet-weet*. **Range:** Breeding throughout northern Europe through northern temperate Asia; wintering south to Africa, South and South-East Asia to Borneo and the Philippines with stragglers reaching Australasia. **Status and distribution:** Rare and localized passage migrant and winter visitor at low elevations in coastal districts, south to Singapore. **Habitat:** Fringes of freshwater ponds, brackish ponds. **Habits:** Usually solitary, feeding in typical sandpiper fashion in freshwater localities. Bobs tail up and down like Common Sandpiper. When alarmed, takes off in a zig-zag flight pattern. Shy.

172. Wood Sandpiper, *Tringa glareola* — Plate 25

Kedidi Bendang

Description: 23 cm (9 in). From Green Sandpiper (171) by slimmer appearance and more upright stance, yellow legs, prominent white supercilium and more densely white spotted brownish grey upperparts (see figure). **Breeding:** Breast buff white with bold dark streaks and barrings on flanks. **Non-breeding:** Upperparts greyer; breast darker with faint brown streaks, lacking the barrings on flanks. In upperwing flight pattern, shows white square rump patch and tail, marked with narrow dark bars. In underwing flight pattern **(not illustrated)** whitish basal part of underwing and less contrasting tail markings separate it from Green Sandpiper. Legs extend further beyond tail than Green Sandpiper. **Voice:** A shrill and high pitched *chip-chip-chip*. **Range:** Breeding throughout northern Europe through northern temperate Asia; wintering south to Africa, South and South-East Asia to Australasia and the western

Wood Sandpiper, *Tringa glareola* (172, left) and **Green Sandpiper**, *Tringa ochropus* (171, right)

Pacific. **Status and distribution:** Fairly common passage migrant and winter visitor at low elevations in coastal districts, south to Singapore. The most common *Tringa* sandpiper in freshwater habitats. **Habitat:** Fringes of freshwater ponds, flooded rice-fields and grasslands, rarely mudflats. **Habits:** Generally in small loose groups, feeding exclusively on its own or in association with other shorebirds. Also associates with Pacific Golden Plover (142) in wet grasslands. Shy, taking off at the slightest sign of danger, uttering its shrill cry.

173. Grey-tailed Tattler, *Tringa brevipes* Plate 25

Kedidi Kelabu

Description: 25 cm (10 in). Pale grey plumage, short yellow legs and the lack of white curve in front of the bend of the wing readily distinguish it from Common Sandpiper (175). Upperparts uniform slaty grey with rump faintly marked with white bars, perceivable only at close range. Has distinct white supercilium and blackish lores and thin eyeline. **Breeding:** Dark grey barrings on breast and flanks. **Non-breeding:** Barring on breast and flanks replaced with uniform pale grey. In upperwing flight pattern, shows all grey upperparts with no distinct white markings on wings or tail. In underwing flight pattern **(not illustrated)**, greyish wings show slight contrast with pale underparts. Legs do not extend beyond tail. **Voice:** A disyllabic *tu-whit*. **Range:** Breeding in eastern Russia; wintering south to southern China through South-East Asia to Australasia and the western Pacific. **Status and distribution:** Scarce passage migrant and winter visitor at low elevations, recorded only from the east coast, south to Singapore. No confirmed records from the west coast but possibilities exist. **Habitat:** Sandy shores, mudflats, brackish ponds. **Habits:** Usually solitary or in twos and threes, actively running about and feeding at the water's edge, sometimes associating with other shorebirds. Habitually bobs tail like Common Sandpiper.

174. Terek Sandpiper, *Xenus cinereus* Plate 25

Kedidi Sereng

Description: 23 cm (9 in). Small and short-legged. Greyish brown upperparts with long and thin up-curved black bill with yellow basal half and yellow to orange legs diagnostic. White supercilium prominent in front of eye but indistinct behind. **Breeding:** Shows black scapular lines on wings and blackish streaks on breast and sides of neck. **Non-breeding:** Lacks black scapular lines; streaks on breast indistinct. In upperwing flight pattern, shows narrow trailing edge of secondaries; tail greyish brown and unmarked. Legs do not extend beyond tail. **Voice:** A fluty disyllabic *twit-twit*. **Range:** Breeding from north-eastern Europe throughout temperate Asia; wintering south to the coasts of Africa, South and South-East Asia to Australasia. **Status and distribution:** Fairly common passage migrant and winter visitor at low elevations in coastal districts, south to Singapore. **Habitat:** Mudflats, estuaries, sandy shores, brackish ponds, rocky shores, coastal power-station ash-ponds. **Habits:** Usually in small loose groups, feeding with other shorebirds. Very active, running rapidly, pausing to feed before moving on. Strictly an inter-tidal species, feeding mostly on crabs. Flies in compact flocks, usually exclusively on its own between feeding and roosting grounds. Shy.

175. Common Sandpiper, *Actitis hypoleucos* Plate 25
Kedidi Pasir

Description: 20 cm (8 in). Small sandpiper with straight dark bill Crown and upperparts dark olive brown; breast and belly white with conspicuous white curve up bend of wing. Has conspicuous white supercilium and thin black eyeline. **Breeding** (not illustrated): Sides of neck and area across breast finely streaked olive brown, forming a full breast band. **Non-breeding:** Upperparts paler, streaks on sides of neck and breast become indistinct, disappearing from the middle of the breast, forming a broken breast band. Legs yellow in all seasons. In upperwing flight pattern, shows conspicuous white wing bars which contrast with black flight feathers and olive brown upperparts. Tail unmarked and concolorous with mantle. Legs do not extend beyond tail. **Voice:** A shrill piercing *twee-wee-wee,* frequently uttered in flight. **Range:** Breeding across Europe through temperate Asia and also in the Himalayas; wintering south to Africa, South and South-East Asia to Australasia. **Status and distribution:** Abundant and widespread passage migrant and winter visitor principally at low elevations, south to Singapore. Also occurs far inland and has been recorded at 1000 m in the Cameron Highlands (Pahang). Individuals probably overwinter here because of its presence all year round. Especially numerous during autumn and spring passage. **Habitat:** Mudflats, estuaries, sandy shores, rocky shores, fringes of freshwater ponds, flooded ricefields, riverbanks, wet grasslands, brackish ponds, canals, sewage oxidation ponds, coastal power-station ash-ponds. **Habits:** Usually solitary, sometimes in twos and threes. Very active while feeding, moving about in short and quick strides. Also skims low over large rivers from one bank to another. Flight pattern distinctive with shallow wing beats. Frequently bobs tail. When alarmed, quickly takes to flight, uttering its shrill cry. Roosts at dusk irrespective of state of tide. Generally shy but sometimes tame and confiding.

176. Ruddy Turnstone, *Arenaria interpres* Plate 23
Kedidi Batu

Description: 23 cm (9 in). Stocky short legged shorebird with short, dark wedge-shaped bill, mandible slightly angled upwards giving an upturned appearance. Legs orange-red in all seasons. **Male breeding:** Head white streaked lightly with black on crown with black facial stripes and broad black breast band. Upperparts bright chestnut with black and white markings. Underparts white. **Female breeding** (not illustrated): Like breeding male but has less white on head with more black streaks on crown. Chestnut upperparts much duller. **Non-breeding:** Both sexes lose white head and chestnut upperparts which are replaced with dark brown, paler on head. In upperwing flight pattern (all seasons), shows white wingbars, back, rump, uppertail coverts and broad black subterminal tail band. **Voice:** A soft *kiuk*. Also a rapid *kikekikekike* when alarmed. **Range:** Breeding in the arctic from North America, across northern Europe and arctic Asia; wintering south to temperate and tropical latitudes worldwide. **Status and distribution:** Fairly common passage migrant and winter visitor at low elevations in coastal districts, south to Singapore. **Habitat:** Mudflats, rocky shores, sandy shores, brackish ponds, coastal power-station ash-ponds. **Habits:** Usually in small groups, running and feeding close to the shore. Feeds exclusively on its own but joins other shorebirds at high tide roosts. True to it names, flips over small stones and shells with bill to locate hidden crustacea and molluscs.

177. Great Knot, *Calidris tenuirostris* Plate 23

Kedidi Dian Besar

Description: 30 cm (12 in). Largest *Calidris* shorebird, has small-headed and plump-bodied appearance. From Red Knot (178) by larger and more robust build. Bill is thicker and proportionately longer, slightly down-curved at tip. **Breeding:** Upperparts brownish black with rufous chestnut patches on scapulars. Develops blackish bars and spots on breast, forming a breast band. Head and neck marked with fine black streaks; underparts white. **Non-breeding:** Upperparts greyish with pale fringes; sides of breast and flanks marked with black spots, more prominent than in Red Knot. Legs greenish in all seasons. In upperwing flight pattern, shows faint white wing-bar and all white rump. In underwing flight pattern, shows all white underwings. **Voice:** *Chak-chuka-chak or chaka-ruk-chak* (W1). **Range:** Breeding at high altitudes in North-East Russia; wintering south through South and South-East Asia to Australia. **Status and distribution:** Fairly common passage migrant and winter visitor in small numbers at low elevations in coastal districts, south to Singapore. **Habitat:** Mudflats, estuaries, sandy beaches, coastal power-station ash-ponds. **Habits:** Very sociable and usually in small loose groups. Will associate with other shorebirds at feeding and roosting grounds.

178. Red Knot, *Calidris canutus* Plate 23

Kedidi Dian Kecil

Description: 25 cm (10 in). Like Great Knot (177) but with shorter bill and less pot-bellied appearance. **Breeding:** Sides of head, breast and upper belly bright rufous chestnut; upperparts scaled blackish and rufous with narrow white fringes. **Non-breeding:** Upperparts pale greyish with faint whitish scales and faint pale supercilium; underparts whitish with faint dark spots on breast and upper belly. Legs yellowish-green in all seasons. In upperwing flight pattern, shows narrow white wing-bars and pale greyish rump with fine darkish bars. In underwing flight pattern, shows pale greyish wings with broad white markings. **Voice:** Normally silent in wintering quarters but sometimes utters a short disyllabic *chut-chut*. **Range:** Breeding in the arctic from North America east through Europe and temperate Asia; wintering south to temperate and tropical latitudes worldwide. **Status and distribution:** Uncommon and localized winter visitor at low elevations in coastal districts, south to Singapore. **Habitat:** Mudflats, estuaries, brackish ponds, coastal power-station ash-ponds. **Habits:** Usually solitary or in small loose groups. Often associates with other shorebirds. Walks with an upright gait. Probes rapidly into soft mud while feeding.

179. Sanderling, *Calidris alba* Plate 26

Kedidi Kapak

Description: 20 cm (8 in). Short, thick black bill and black legs distinctive. **Breeding:** Deep chestnut head, neck and breast with small dark brown streaks. Belly and rest of underparts white. **Non-breeding:** White head and underparts contrasting with pale grey upperparts with diagnostic black patch at bend of closed wing. In upperwing flight pattern, white wingbars contrast sharply with blackish flight feathers and grey upperparts. Shows white sides of rump contrasting with darker central tail and terminal tail band. **Voice:** A sharp, high pitched *tvik* or *plit*. **Range:** Breeding in the arctic from North America east through Europe and temperate Asia; wintering south to temperate and tropical latitudes worldwide. **Status and distribution:** Uncommon and localized winter visitor in small numbers at low elevations in coastal districts, south to Singapore. More common on the east coast because of its preference for sandy substrates. **Habitat:** Sandy shores, brackish ponds, coastal power-station ash-ponds. **Habits:** Usually solitary or in small loose groups. Runs rapidly in a crouched manner along the water's edge, following receding waves to pick up food material. Usually exclusively on its own.

180. Red-necked Stint, *Calidris ruficollis* Plate 26

Kedidi Luris Leher

Description: 15 cm (6 in). Small size with short pointed blackish bill Very similar to Little Stint (181). **Breeding:** Distinctive rufous crown, throat, sides of head and upper breast with fine dark streaks on crown only. Black streaks on sides of breast restricted more to white underparts and extends little into rufous sides of neck. Chestnut on mantle restricted only to upper scapulars. From Spoon-billed Sandpiper (188) by bill shape and all grey lower scapulars. **Non-breeding:** Upperparts pale grey brown marked with narrow dark shafts. In upperwing flight pattern, shows thin white wing bars and white sides to rump. **Voice:** A distinctive cadence, *chui-chui-chiriri-chiriri*. **Range:** Breeding in north central Russia east to Alaska (North America): wintering south to southern China and South-East Asia to the western Pacific, Australia and New Zealand. **Status and distribution:** Fairly common passage migrant and winter visitor at low elevations in coastal districts, south to Singapore. **Habitat:** Mudflats, estuaries, sandy shores, fringes of open freshwater pools, brackish ponds, coastal power-station ash-ponds. **Habits:** Gathers in very large flocks on feeding grounds, often associating with other shorebirds. Dainty and nimble in appearance, it makes brisk strides or dashes about, picking up food material. Flies in compact flocks between feeding and roosting grounds.

181. Little Stint, *Calidris minuta* Plate 26

Kedidi Kerdil

Description: 13 cm (5 in). From Red-necked Stint (180) by more pointed black bill and proportionately longer legs. Also appears rounder and shorter in length. **Breeding:** Entire upperparts blackish with rufous fringes together with bright orange-brown sides of face and breast, marked with fine blackish streaks separate it from Red-necked Stint. Throat and stripe behind eye white; underparts whitish. **Non-breeding:** Like non-breeding Red-necked Stint but has pale grey-brown breast band which contrasts slightly with white throat and underparts. In upperwing flight pattern, resembles Red-necked Stint but white wingbars thinner (not apparent in field).

Voice: A short and distinctive *tit-tit*. **Range:** Breeding in northern Europe and western temperate Asia; wintering south to Africa and South Asia. **Status and distribution:** Uncommon and localized passage migrant and winter visitor, recorded from Pulau Pinang and Perak, south to Selangor. **Habitat:** Mudflats, estuaries, coastal power-station ash-ponds. **Habits:** Solitary but may gather in loose flocks where common. Very active, making short dashes to pick up food material. Often associates with other *Calidris* shorebirds.

182. Temminck's Stint, *Calidris temminckii* Plate 26
Kedidi Temminck

Description: 15 cm (6 in). Small nondescript shorebird with yellow legs. Looks like a miniature Common Sandpiper (175). **Breeding:** Upperparts dark grey-brown marked with greyish and buff fringes; underparts white, contrasting with uniform grey-brown breast and sides of neck. Shows short white supercilium. **Non-breeding:** Uniform grey-brown upperparts with dark grey-brown breast band which is sometimes broken in the middle. Supercilium faint and pale. In upperwing flight pattern, shows thin white wingbar and white side to rump. **Voice:** A high pitched metallic trilling *tirrrrr*. Also a soft *kililik*. **Range:** Breeding from North-East Europe throughout southern temperate Asia; wintering south to Africa, South and South-East Asia to Borneo and the Philippines. **Status and distribution:** Uncommon and localized passage migrant and winter visitor at low elevations in coastal districts, south to Singapore. Found almost exclusively in freshwater habitats. **Habitat:** Fringes of freshwater pools, flooded ricefields, brackish ponds, riverbanks. **Habits:** Usually in small loose groups feeding in the same manner as other small *Calidris* shorebirds. Seldom associates with other shorebirds.

183. Long-toed Stint, *Calidris subminuta* Plate 26
Kedidi Jari Panjang

Description: 16 cm (6.25 in). Upright stance and long necked appearance together with thin black bill and rather longish green legs distinctive. Has distinct white supercilium and streaks on sides of neck and breast. **Breeding:** Crown rufous with rufous fringes on black and grey upperparts with heavy streaks on sides of breast. **Non-breeding:** Lacks rufous crown and rufous fringes on upperparts as well as dark streaks on sides of neck. In upperwing flight pattern, shows indistinct thin pale wing bars and white sides to rump. **Voice:** A trilling and musical *trrrrrp*. **Range:** Breeding in North-East temperate Asia; wintering south to East Africa, South and South-East Asia to Australia and New Zealand. **Status and distribution:** Locally common passage migrant and winter visitor at low elevations in coastal districts, south to Singapore. **Habitat:** Freshwater pools, flooded ricefields, rarely on mudflats, in estuaries, or brackish ponds. **Habits:** Usually solitary or in small loose groups. Often associates with other shorebirds while feeding. Feeds at a leisurely pace; seldom dashing about. When alarmed, assumes a more upright stance than other *Calidris* shorebirds.

184. Pectoral Sandpiper, *Calidris melanotos* Plate 24

Kedidi Dada Hitam

Description: 23 cm (9 in). Like Sharp-tailed Sandpiper (185) but with less distinct supercilium. Foreneck and breast more heavily streaked and sharply demarcated from white belly. Male is larger. Legs yellowish. **Breeding:** Neck and breast washed buff with dark streaks forming a pectoral band, more distinct in male, and appearing blackish with white mottles when viewed from a distance. Rest of underparts white, with a few dark streaks on flanks. Dull chestnut crown marked with dark brown streaks with blackish brown feathers of upperparts fringed with chestnut brown and buffish-white, the white on some outer mantle and scapulars forming Vs. **Non-breeding:** Like breeding but duller without the chestnut brown and buff tones. In upperwing flight pattern, shows dark centre of rump and basal half of tail with white edges. **Voice:** No information from review area. **Range:** Breeding throughout much of the north-east Palaearctic and the Nearctic; wintering south mainly to the Neotropics, and also Australia and New Zealand. **Status and distribution:** Very rare winter visitor with four confirmed records from Peninsular Malaysia; one at Georgetown (Pulau Pinang) on 25 March 1984 (Speight, 2007), an adult in breeding plumage at Nibong Tebal, Seberang Prai, (Pulau Pinang) on 12 September 2007, another adult in non-breeding plumage (probably the same individual) together with a juvenile at the same place on 25 October 2007 (Bakewell, 2007a), and an adult at Sungai Balang, Muar District (Johor), on 30 August 2008 (BiW). In Singapore known from an individual at Changi on 26 November 1999 and again on 1–3 January 2000 (W and H). **Habitat:** In review area, reclaimed land, landfill sites, and flooded ricefields. **Habits:** In review area, solitary. When alarmed, would assume an upright posture with neck stretched up vertically (Bakewell, 2007a). Like other *Calidris* sandpipers, feeds by probing into soft earth near pools of water. Other habits unknown.

185. Sharp-tailed Sandpiper, *Calidris acuminata* Plate 26

Kedidi Ekor Tajam

Description: 22 cm (8.5 in). Slightly down-curved black bill with yellowish tinge at base distinctive. Shows sharp contrast between dark rufous crown and long white supercilium. Has a narrow but distinct pale eyering. **Breeding:** Rufous buff tinge on neck and breast heavily marked with dark brown spots. Lower breast and flanks develop brownish V-shaped markings. Upperparts strongly washed rufescent. **Non-breeding:** Upperparts dull grey-brown with faint streaks on sides of breast and flanks. Legs yellowish green in all seasons. In upperwing flight pattern, shows very thin and faint white wingbars with faint dark streaks on white sides to rump. **Voice:** A soft liquid *tweep-tweep-tweep*. Also a low *preeet*. **Range:** Breeding in North-East Russia; wintering south to Australasia and the western Pacific. **Status and distribution:** Vagrant, with sporadic records of individuals from Selangor and Singapore. **Habitat:** Fringes of freshwater pools, mudflats, coastal power-station ash-ponds. **Habits:** Usually solitary, probing for food material in soft mud. Sometimes associates with other smaller shorebirds.

186. Curlew Sandpiper, *Calidris ferruginea* — Plate 26
Kedidi Pasir Kendi

Description: 22 cm (8.5 in). From Dunlin (187) in non-breeding plumage by longer and more prominent down-curved black bill. **Breeding:** Head, neck, breast, and belly bright chestnut with whitish base of bill. Shows whitish supercilium and fine dark streaks on crown and sides of neck. Flanks faintly barred whitish; mantle and scapulars blackish with chestnut fringes; wing coverts grey. **Non-breeding:** Upperparts uniform grey-brown with pale fringes on wing coverts. Shows whitish supercilium. Underparts white with fine pale greyish streaks on sides of neck and breast. Legs black in all seasons. In upperwing flight pattern, white rump and uppertail coverts contrasting with all grey terminal tail band separates it from other *Calidris* shorebirds. Legs extend slightly beyond tail. **Voice:** A soft *tirrit*. **Range:** Breeding in North-East Europe and western temperate Asia; wintering south to Africa, South and South-East Asia to Australia and New Zealand. **Status and distribution:** Fairly common passage migrant and winter visitor at low elevations in coastal districts, south to Singapore. **Habitat:** Mudflats, estuaries, freshwater pools, sandy shores, brackish ponds, coastal power-station ash-ponds. **Habits:** Normally in large flocks, flying in tight formation. Usually associates with other shorebirds on both feeding and roosting grounds. Makes rapid and short runs before stopping to probe for food.

187. Dunlin, *Calidris alpina* — Plate 26
Kedidi Dunlin

Description: 19 cm (7.5 in). Longish black bill, slightly down-curved at tip distinctive (see Curlew Sandpiper (186) and Broad-billed Sandpiper (189)). **Breeding:** Black patch on centre of belly diagnostic. Mantle and scapulars blackish with broad chestnut fringes; wing coverts grey; breast and sides of neck marked with fine dark brown streaks. White supercilium contrasts with streaked crown and dark lores. **Non-breeding:** Upperparts uniform pale grey-brown; underparts white with pale greyish streaks on sides of breast. Legs black in all seasons. In upperwing flight pattern, shows thin white wingbars and conspicuous white sides to rump and tail with black centre of tail. **Voice:** A soft *treeer* with a metallic quality, distinct from Curlew Sandpiper. **Range:** Breeding in the Arctic from North America east through Europe and temperate Asia; wintering south to temperate and tropical latitudes worldwide. **Status and distribution:** Very rare winter visitor at low elevations in coastal districts, south to Singapore. **Habitat:** Mudflats, estuaries, sandy shores, brackish ponds, coastal power-station ash-ponds. **Habits:** Usually solitary or in small groups, feeding together with other shorebirds.

188. Spoon-billed Sandpiper, *Eurynorhynchus pygmeus* — Plate 26
Kedidi Paruh Sudu

Description: 15 cm (6 in). From Red-necked Stint (180) by black spatulate bill (often difficult to discern when viewed from side) and in breeding plumage, by the lack of grey on lower scapulars. **Breeding:** Face, neck and upper breast reddish rufous with entire upperparts black with all rufous chestnut fringes. Underparts white with blackish streaks on sides of breast. **Non-breeding:** Upperparts greyish brown; wing coverts grey with whitish fringes. Shows conspicuous white supercilium. Legs black in all seasons. In upperwing flight pattern, shows thin white wingbars and white sides to rump. **Voice:** Usually silent in wintering quarters. Sometimes utters a soft *pii-pii*. **Range:** Breeding in far eastern Russia; wintering south to North-East India and the coasts of southern China and South-East Asia with stragglers reaching the Greater Sundas. Its extremely small population is declining rapidly due to habitat loss in breeding, passage, and wintering grounds, which are compounded by human disturbances, hunting, and climatic change, as well as poor fledging success. **CR. Status and distribution:** Very rare and localized passage migrant and winter visitor at low elevations in coastal districts, south to Singapore. **Habitat:** Mudflats, fringes of freshwater pools, brackish ponds, coastal power-station ash-ponds. **Habits:** Usually solitary, sometimes in twos or threes, often associating with other shorebirds. Has unique feeding habit, sweeping bill from side to side in mud or shallow water; an important field characteristic to distinguish it from the Red-necked Stint.

189. Broad-billed Sandpiper, *Limicola falcinellus* Plate 26
Kedidi Paruh Lebar

Description: 18 cm (7 in). Small and compact size with longish black bill, slightly down-curved at tip. Whitish supercilium split by darker coronal stripe diagnostic. From Dunlin (187) and Curlew Sandpiper (186) by smaller size and proportionately shorter legs. **Breeding:** Crown blackish; upperparts blackish with chestnut and white fringes; underparts white with breast, sides of neck and flanks spotted black. **Non-breeding:** Coronal stripe paler, making supercilium more indistinct; upperparts grey-brown with dark streaks and white fringes, sometimes showing blackish patch at bend of wing (see Sanderling (179)). White underparts streaked light grey on breast. In upperwing flight pattern, shows white wingbars and white sides to rump; blackish flight feathers and carpal joint contrast with grey-brown upperparts. **Voice:** A low pitched trill *chreeek* with a metallic quality. **Range:** Breeding in northern Europe through temperate Asia; wintering south to Africa, South and South-East Asia to Australia. **Status and distribution:** Locally common passage migrant and winter visitor at low elevations in coastal districts, south to Singapore. **Habitat:** Sandy shores, mudflats, estuaries, fringes of freshwater pools, brackish ponds, coastal power-station ash-ponds. **Habits:** Usually solitary, seldom associating with other shorebirds. Probes vertically into soft mud while feeding. Generally shy.

190. Ruff, *Philomachus pugnax* Plate 23
Kedidi Ropol

Description: Male: 30 cm (12 in). Female: 23–28 cm (9–11 in). Generally recognized by upright stance, relatively long necked and small headed appearance with distinctive long legs. Short straight bill is slightly down-curved at tip. **Male breeding:** Has distinct neck ruff and ear tufts ranging from all white to black, cinnamon or dark rufous. Upperparts generally blackish with pale grey or buff fringes. Bill colour usually pink, orange or blackish (highly variable in individuals). Full breeding plumage is not seen in wintering quarters but northward bound spring migrants usually show traces of black or cinnamon on face or breast. **Non-breeding:** Upperparts dark blackish brown with narrow pale fringes; head, neck and underparts vary from buff brown to whitish. **Female:** Like non-breeding male but shorter with proportionately shorter bill. Colour of legs in both sexes vary from yellowish to greenish or orange to reddish. In upperwing flight pattern, shows narrow pale wingbar and white sides to rump contrasting with narrow grey inner centre of tail and black tail tip. Legs extend beyond tail. **Voice:** Usually silent in wintering quarters. **Range:** Breeding throughout Europe and temperate Asia; wintering south to the Mediteranean throughout Africa to South and South-East Asia; the Greater Sundas and the Philippines with stragglers reaching Australia. **Status and distribution:** Uncommon and localized winter visitor at low elevations, south to Singapore. **Habitat:** Flooded ricefields, inland freshwater pools, mudflats, brackish ponds. **Habits:** Usually solitary or in twos and threes, rarely in larger numbers. Feeds by leisurely probing into soft mud. Seldom associates with other shorebirds.

191. Red-necked Phalarope, *Phalaropus lobatus* — Plate 21
Kedidi Paruh Jarum

Description: 21 cm (7.5 in). Slender needle-like black bill, black eyeline and black legs with distinctive lobed feet separate it from small *Calidris* shorebirds (177–187). **Female breeding:** Crown, face and hindneck dark grey contrasting with bright chestnut upper breast and sides of neck. Throat and small patch above eye white; upperparts dark grey mottled with golden buff horizontal bars. Sides of lower breast and flanks greyish, paler on underparts. **Male breeding (not illustrated):** Like breeding female but duller. Sometimes shows a distinct white supercilium. Underparts pale grey with darker grey streaks on flanks. **Non-breeding:** Shows distinct black crown and eyeline. Rest of plumage pale grey above and white below. In upperwing flight pattern **(non-breeding)**, shows white wing bars and broad white base of outer tail. **Voice:** A high-pitched *chek*, uttered in flight. Also a series of rapidly uttered *chek* notes when alarmed. **Range:** Breeding throughout temperate North America through Europe and temperate Asia; wintering south to temperate and tropical latitudes worldwide. **Status and distribution:** Rare and localized non-breeding visitor in coastal waters, first recorded south of the Terengganu oil field in the South China Sea on 7 October 1982 (Wells, 1990) In the Straits of Melaka, 14 (including a female in breeding plumage) off Pulau Tengah (Selangor) on 8 April 1986 (Wells, 1990) with more sporadic occurances on the Selangor coast., the latest being one at Sungai Besar (Selangor) on 9 January, 2005 (BiW). In Singapore, known from a single record at Tuas from 16–25 November 1994 (W and H). **Habitat:** Open sea, sea-coasts, and freshwater pools (on passage stops only). **Habits:** Highly pelagic outside breeding, swimming and feeding in flocks on the sea. Feeding behaviour distinctive, consisting of head pumping and spinning in circles in shallow water to stir up minute surface plankton and small aquatic insects. Tame and confiding.

PRATINCOLES

ORDER *Charadriiformes* **FAMILY** *Glareolidae*

World: *18 species* **Peninsular Malaysia and Singapore:** *2 species*

Medium to small sized shorebirds with short slightly down-curved bills and wide gapes. Wings long and pointed, the tips projecting beyond tails at rest. Tails deeply forked. Flight buoyant like terns (198–211) with slow wingbeats. Aerial feeders, hawking flying insects. Very active towards dusk. Gregarious, nesting in loose colonies on the ground. Sexes alike.

192. Oriental Pratincole, *Glareola maldivarum* Plate 21

Kedidi Padang

Description: 25 cm (10 in). Earthy brown crown and mantle contrasting with white rump and deeply forked black tail diagnostic (conspicuous in upperwing flight pattern). Belly and undertail coverts whitish. In underwing flight pattern, deep chestnut wing coverts contrast with black flight feathers and trailing edge of wings. **Breeding:** Rufescent wash on breast and flanks; throat creamy with black outline; base of lower mandible red. **Non-breeding** (not illustrated): Duller with black throat outline reduced to dark brownish streaks. **Juvenile (not illustrated):** Brownish with whitish scales on upperparts. Throat, belly, and undertail coverts whitish. **Voice:** A grating *krit-krit-krit,* often uttered in flight. **Range:** Breeding in the Indian subcontinent through South-East Asia to the Malay Peninsula; wintering south to the Greater and Lesser Sundas to northern Australia. **Status and distribution:** In Peninsular Malaysia along the west coast, locally common resident in Perlis and Kedah with breeding recorded south at Batang Berjuntai (Selangor) in May 1996. Along the east coast, breeds from Kelantan south to Kuantan (Pahang). Elsewhere locally common passage migrant and winter visitor, south to Singapore. Particularly common on the grassy verges of airports and on freshly harvested ricefields. **Habitat:** Open grassy areas, harvested ricefields. **Habits:** Usually in small loose groups. Makes swooping flights as it pursues flying insects. Flight buoyant and graceful.

193. Small Pratincole, *Glareola lactea* Plate 21

Kedidi Padang Kecil

Description: 18 cm (7 in). From Oriental Pratincole (192) by much smaller size, paler colouration and slightly forked tail. Mantle pale sandy grey; crown grey; breast pale buff grey; belly white. In upperwing flight pattern, black flight feathers contrast with pale grey wing coverts and white base of secondaries. White rump contrasts with black tail with very narrow white terminal band. In underwing flight pattern, black wing coverts contrast with white base of secondaries and pale underparts. **Voice:** A rather shrill *tiririt, tiririt, tiririt,* when disturbed. **Range:** The Indian subcontinent and southern China through Myanmar, Thailand and Indo-China. **Status and distribution:** Vagrant in Singapore with 4 records of individuals from Jurong in November 1964 through January 1965, at Changi from December 1987 through January 1988, at Serangoon in January 1989, and another at Changi on 23 December 1999 (W and H). **Habitat:** Sand bars in rivers; margins of freshwater swamps, freshwater reservoirs, sandy shores. **Habits:** Where common, gathers in loose flocks. Prefers wetter areas than Oriental Pratincole. When hawking insects, flight swifter than Oriental Pratincole with faster wingbeats and frequent upward swoops.

GULLS AND TERNS

ORDER *Charadriiformes* **FAMILY** *Laridae*

World: *97 species* **Peninsular Malaysia and Singapore:** *18 species*

Small to large sized seabirds with mostly white and grey plumage. Feet webbed. Sexes alike. Gregarious, feeding in groups and nesting in large colonies. Represented by two distinct groups and the following genera:

Gulls: Genus *Larus* (194–197). Large and robust. Wings broad based, tails squarish to rounded and bills thick and slightly hooked at tip. Some have prominent dark hoods during breeding. Inhabit coastal waters, feeding largely on offal and small fish picked up from the water surface. Often sit on water. Non-breeding visitors only. **Terns:** Generally smaller, slimmer and more streamlined than gulls. Some species large. Wings long, narrow and pointed, tails proportionately longer and usually deeply forked with elongated outer tail feathers. Bills slim and pointed. Inhabit both coastal waters and the open sea and to a lesser extent, freshwater habitats inland. Colonial breeders on rocky stacks offshore or coastal sandbanks. Some species nest on floating vegetation in inland wetlands. Most species develop dark caps during breeding. Juveniles of many species are dependent on parents for several months during post birth are may be often seen begging for food outside breeding season. Feeding methods vary. **Marsh Terns** genus *Chlidonias* (209, 210), **Gull-billed tern,** genus *Gelochelidon* (198) and **Noddies,** genus *Anous* (211) pick up food material from the water surface while **Caspian Tern** genus *Hydroprogne* (199), **Crested Terns,** genus *Thalasseus* (200, 201), and **Typical Terns,** genus *Sterna* (202–208) largely plunge-dive into the water to feed on small fish and squid. Some species follow fishing boats.

194. Heuglin's Gull, *Larus heuglini* Plate 27

Camar Cina.

Description: 65 cm (25.6 in). **Adult:** Large white headed gull with dark grey upperparts, pale iris with narrow red eyering, and yellow legs. Bill larger and more thickset than other gulls in review area, and is yellow with orange-red mark on tip of mandible. **Breeding:** All white head and neck. **Non-breeding (not illustrated):** Head and hindneck heavily streaked. **Juvenile:** White of head and underparts, including flanks, marked with dull brownish streaks and mottles. Upperparts dark brownish with whitish to brownish-white fringes. Bill blackish, legs dull pinkish, both gradually turning yellow from first through second winter plumages. **First winter (not illustrated):** Moulted new scapulars and tertiaries greyer with darker and more diffuse darker central markings with less defined pale borders. Head noticeably whiter. **Second winter (not illustrated):** Like first winter but mantle, back and scapulars predominantly grey. **Voice:** Unknown from review area. **Range:** Breeding throughout the north-east, west, and north central Palaearctic, wintering south to the coasts of East Africa, the Arabian Peninsula, coastal Iran and Pakistan, to India, Sri Lanka, and China **Status and distribution:** Vagrant, with one confirmed sighting of two large pale juvenile gulls on Pulau Ketam (Selangor), on 27 February 2006 (David Li, RCS). **Habitat:** In review area, coastal mudflats fringing mangroves. **Habits:** Little known. **Note:** The larger pale headed gulls, especially in juvenile or transition plumage types to full adult plumage can be very confusing and careful notes together with digital images should be taken to avoid misidentification.

195. Brown-headed Gull, *Larus brunnicephalus* Plate 27
Camar Topeng Coklat

Description: 46 cm (18 in). At rest from Common Black-headed Gull (196) by larger size, stocky build with proportionately thicker neck and bill, and in flight by black wingtips with conspicuous white spot. **Breeding** Dark chocolate brown hood. **Non-breeding:** Brown hood reduced to small crescent on ear coverts. Bill and legs red. **Juvenile (not illustrated):** Like non-breeding adult but entire trailing edge of wing dark brownish black. Lacks white spot on black wingtips. Brownish mottling on wing coverts. Black terminal tail band. **Voice:** A harsh *gek, gek*. **Range:** Breeding in the central Asian Highlands from the Himalayas to China; wintering south to South and South-East Asia. **Status and distribution:** Uncommon and localized winter visitor in small numbers on the west coast; fairly regular from Kedah to Perak, especially off the Kerian coast. South of Perak, it is vagrant south to Singapore with two records from Kranji in 1967 and 1986. More numerous than Common Black-headed Gull. **Habitat:** Sea-coasts, mudflats. **Habits:** Moves and feeds in small loose groups; larger flocks may form during passage. Rests on mudflats together with Common Black-headed Gull as well as terns and shorebirds. Often rests buoyantly on the sea.

196. Common Black-headed Gull, *Larus ridibundus* Plate 27
Camar Topeng Hitam

Description: 40 cm (15 in). From Brown-headed Gull (195) by smaller size and more slender build with all white leading edge of outer upperwing. Mantle and wings pale grey; underparts white; legs red. **Breeding:** Chocolate brown hood with narrow crescents above and below eye. Bill all red. **Non-breeding:** Brown hood reduced to small crescent on ear coverts; bill red with dark tip. **Juvenile (not illustrated):** Like non-breeding adult but wing coverts marked brownish. Secondaries and subterminal tail band blackish. Bill dark orange with dark tip. **Voice:** A raucous *kwaar* **Range:** Breeding throughout Europe and temperate Asia; wintering south to Africa, South and South-East Asia. **Status and distribution:** Rare and localized winter visitor on the west coast, south to Singapore. Very rare on the east coast. **Habitat:** Sea-coasts, mudflats, coastal power-station ash-ponds. **Habits:** Similar to Brown-headed Gull.

197. Laughing Gull, *Larus atricilla*

Plate 27

Camar Topeng Hitam Amerika

Description: 41 cm (16.1 in). Longish gull with rather long, drooping bill, and long and slender wingtips which extend beyond tail. **Breeding:** Dark slate hood extends over nape to hindneck like Common Black-headed Gull (196) but with more pronounced and conspicuous white crescents above and below eye, bill dark red with black ring near orange tip. Mantle and greater wing coverts slate grey, with tertiaries very narrowly tipped white, and primaries very dark with the wingtips noticeably darker than the inner upperwing. Neck, breast, and rest of underparts white with a pinkish suffusion. **Non-breeding:** Lacks hood. Head white with dark greyish smudged ear coverts and hindcrown, without the characteristic ear crescents of Brown-headed (195) and Common Black-headed Gulls (196) Bill dark with reddish tip. **Underwing flight pattern (not illustrated):** Dark primaries contrast with paler secondaries, with a white trailing edge across the latter. Rump and tail white, and legs black in all seasons. **Voice:** In review area, a single yelp was noted (Tebb *et al.*, 2003). **Range:** Breeding mainly in eastern North America along the Atlantic coast from New Brunswick and Nova Scotia south to Florida, around the Gulf of Mexico through the Caribbean islands to the northern coast of Venezuela and on the Pacific coast from southern California to western Mexico. Outside breeding, it is widespread on the Atlantic coast of the Americas from Florida to the estuary of the Amazon river in Brazil while on the Pacific coast, from southern Mexico to southern Peru, with vagrants reported in western Europe, North Africa, West Africa, and southern Africa (Namibia and South Africa) and also in the Hawaiian islands, Japan, eastern Australia (Queensland), and the Malay Peninsula. **Status and distribution:** Vagrant, an adult in breeding plumage at Kapar (Selangor), on 1 April 2000 (Tebb *et al.*, 2003). **Habitat:** In review area, roosting at coastal power-station ash-ponds during high tide. **Habits:** The individual threw back its head and uttered a yelping laugh, with bill pointed up. Was observed chasing White-winged terns (209) at roost site. Other habits unknown. **Note:** The appearance of this new world species in South-East Asia is something that requires more observation on the species' movement patterns.

198. Gull-billed Tern, *Gelochelidon nilotica*

Plate 27

Camar Tiram

Description: 38 cm (15 in). From breeding Common Tern (204) by heavier build, thick black bill and slightly forked tail. Upperwing, rump and tail pale grey; neck and underparts whitish. **Breeding:** Entire cap black. **Non-breeding:** Black on head reduced to spot on ear coverts with some individuals having dark streaks on hindneck. **Voice:** A soft, sharp *chewik,* uttered persistently in flight. **Range:** Breeding in temperate latitudes worldwide; wintering in tropical latitudes. **Status and distribution:** Locally common passage migrant and winter visitor, south to Singapore where it is rare. **Habitat:** Sea-coasts, mudflats, coastal power-station ash-ponds. **Habits:** Usually solitary or in small loose groups, skimming low over the water surface or inter tidal mudflats to feed on crabs. Also feeds on dry land, picking up food material from the ground. Often rests on mudflats together with shorebirds and other terns.

199. Caspian Tern, *Hydroprogne caspia* Plate 27

Camar Caspian

Description: 53 cm (21 in). Very large size with massive dusky tipped bright red bill diagnostic. Tail slightly forked. Mantle pale greyish; underparts whitish; legs black. **Breeding:** All black cap. **Non-breeding:** White streaks on blackish cap. **Voice:** A loud raucous *kraaah* **Range:** Temperate and tropical latitudes worldwide. **Status and distribution:** Locally common non-breeding visitor in small numbers, south to Singapore, where it is rare. **Habitat:** Sea-coasts, mudflats, coastal power-station ash-ponds. **Habits:** Flies with slow and leisurely wingbeats. Plunge-dives vertically with force to fish. Often rests on mudflats with other terns and shorebirds.

200. Lesser Crested Tern, *Thalasseus bengalensis* Plate 28

Camar Kecil Berjambul

Description: 40 cm (15 in). From Great Crested Tern (201) by slimmer build, bluish grey upperparts and more slender bright orange bill. In flight, neck is held forward. **Breeding (not illustrated):** Forehead, crown and shaggy nuchal crest black. **Non-breeding:** Entire forehead and crown white with black reduced to crest and area behind eye. Legs black. **Voice:** A loud *chirruk*, less strident than Great Crested Tern. **Range:** Breeding along the coasts of the Mediterranean and the Red Seas through the Indian Ocean to Australia. Migrates to the coasts of Africa, South and South-East Asia. **Status and distribution:** Uncommon and localized non-breeding visitor in small numbers, south to Singapore. **Habitat:** Sea-coasts, mudflats. **Habits:** Gregarious, gathering in small numbers on mudflats, often with Great Crested Tern. Frequently perches on *kelong* poles and floating drift wood. Often fishes far out at sea, making deep vertical dives. Does not follow fishing boats.

201. Great Crested Tern, *Thalasseus bergii* Plate 28

Camar Besar Berjambul

Description: 46 cm (18 in). Large and robust with shaggy nuchal crest and thick slightly down-curved yellow bill diagnostic. Mantle dark grey with neck and entire underparts white. Legs black. In flight, neck is usually tucked back into shoulders, giving a rather heavy breasted appearance. **Breeding (not illustrated):** Forehead white; crown and crest black. **Non-breeding:** Forecrown white with blackish streaks with black reduced to crest and area behind eye. **Voice:** A series of loud strident notes, *ke-eek*. Also a clear *chew*. **Range:** Breeding along the coasts of the Red Sea and the Indian Ocean to South and South-East Asia, Australia and the

South Pacific. **Status and distribution:** Locally common non-breeding visitor, south to Singapore. Nearest known breeding stations are the Mergui Archipelago (southern Myanmar) in the Andaman Sea and Terumbu Layang-Layang (Swallow reef) in the South China Sea. **Habitat:** Sea-coasts, mudflats. **Habits:** Gregarious, gathering in fairly large numbers. Often perches on *kelong* poles and floating drift wood in coastal waters. When hunting, dives clumsily into the water, often submerging completely. Also feeds far out at sea. Sometimes rests on mudflats with shorebirds and other terns. Flight is leisurely with slow wingbeats. Does not follow fishing boats but will however hang around stationary fishing boats, waiting for fish to be discarded. **Note:** The highly endangered Chinese Crested Tern *Thalasseus bernsteini* (ranked **CR**) which breeds off the coasts of east and south-east China and north-east Taiwan (Matsu archipelago), is very rare and its current wintering range is unknown. Has occurred as a vagrant on the east coast of Peninsular Thailand (Nakhon Si Thammarat) and north-west Borneo (Bako-Buntal Bay, Sarawak) and could turn up in review area. Measuring 42 cm (16.5 in) it has much paler upperparts than both 200 and 201 with a black-tipped yellow bill. During breeding, entire crown is black.

202. Roseate Tern, *Sterna dougallii* — Plate 28

Camar Jambu

Description: 39 cm (15 in). At rest, long deeply forked tail extends beyond tips of folded wings. From Common Tern (205) by much lighter mantle and the lack of black trailing edge to underwing. **Breeding:** All black cap. Bill bright red with dark tip. Legs bright red. **Non-breeding:** Forehead and crown white, lightly streaked black on the latter with black reduced to nape and stripe behind eye. Legs yellowish. **Juvenile:** Crown black with distinctive 'U' shaped blackish marks on pale grey back. In underwing flight, shows pale grey trailing edge to primaries. **Voice:** A soft disyllabic *che-weet*. **Range:** Temperate and tropical latitudes in the Atlantic Ocean, the Mediterranean Sea and the Indian Ocean through South and South-East Asia east to China and south to Australia. **Status and distribution:** Uncommon and localized offshore resident in Peninsular Malaysia with breeding colonies recorded on Pulau Tenggol (Terengganu) and Pulau Yu (Johor) in the South China Sea; these have been reduced by egg collecting fishermen. No breeding colonies noted in the Straits of Melaka although birds in breeding plumage have been observed there. Outside the breeding season, may disperse over wider areas. No recent records from Singapore. **Habitat:** Sea-coasts, rocky islets, coral reefs. **Habits:** Flies in small loose groups. Plunges steeply to fish. Flight buoyant and very graceful. Seldom comes close to mainland shores.

203. Black-naped Tern, *Sterna sumatrana* — Plate 28

Camar Tengkok Hitam

Description: 31 cm (12 in). From all other terns by wholly white plumage, black eye-line and nape band and deeply forked tail. Bill and legs black. Develops a rosy flush to plumage when breeding. **Juvenile:** Like adult but tail unforked and rounded. Crown spotted with black and upperparts scaled black. **Voice:** A sharp *tsii-chee-chi-chi* **Range:**

Breeding in the eastern Indian Ocean from South and South-East Asia eastward to Australia and the tropical western Pacific. **Status and distribution:** Locally common offshore resident, south to Singapore where it is uncommon. Breeds on rocky stacks and islets in both the Straits of Melaka and South China Sea. Egg collecting fishermen have reduced breeding colonies. **Habitat:** Sea-coasts, rocky islets. **Habits:** Gregarious, gathering in small flocks. Feeds further offshore than Common (204), Great (201), and Lesser Crested Terns (200). Does not perch on drift wood or *kelong* poles. Flight buoyant and graceful.

204. Common Tern, *Sterna hirundo*

Plate 28

Camar Siput

Description: 35 cm (14 in). Slender build with long deeply forked tail distinctive. At rest, tail same length with tips of folded wings. Mantle, wings, and upperback blue-grey; rump, tail and underparts white. In underwing flight, shows slight black trailing edge to primaries. **Breeding:** All black cap. In *S. h longipennis*, bill and legs black with pale greyish wash to white underparts. In *S. h. minussensis*, bill bright red with dark tip. Legs bright red. **Non-breeding:** Forehead and underparts white. Legs and bill black. **Juvenile (not illustrated):** Like non-breeding adult but upperparts scaled black and white with black bar on upperwing coverts. **Voice:** A harsh descending disyllabic *kirri, kirri*. **Range:** Breeding in West Africa through Europe and temperate Asia and North and Central America; wintering south to parts of tropical Africa, South and South-East Asia, Australia, and South America. **Status and distribution:** Locally common passage migrant and winter visitor, more common on the west coast, south to Singapore, where it is uncommon. **Habitat:** Sea-coasts, mudflats, coastal power-station ash-ponds. **Habits:** Gathers in small loose flocks, keeping mainly to coastal waters. Fishes by plunging steeply into water. Frequently rests on *kelong* poles and floating drift wood. Flight buoyant and graceful. Often follows fishing boats. **Note:** In non-breeding plumage, both *S. l. longipennis* and *S. l. minussensis* are currently indistinguishable in the field.

205. Little Tern, *Sterna albifrons*

Plate 28

Camar Kecil

Description: 23 cm (9 in). Small size with moderately forked tail distinctive. Upperparts greyish contrasting with white rump and uppertail coverts. Entire underparts white. Black of forehead tapers to a point with white of crown usually reaching eye. **Breeding:** Forehead white, crown, lores and eyeline black; bill yellow with dark tip; legs yellow. **Non-breeding:** Lores white; bill and legs black. **Juvenile:** Upperparts grey with buff scalings; bill black with yellow base. **Voice:** A high pitched *krii-krii*. **Range:** Breeding throughout temperate and tropical latitudes worldwide; temperate populations winter in tropical latitudes. **Status and distribution:** Fairly common resident and winter visitor, south to Singapore. **Habitat:** Sea-coasts, mudflats, estuaries, large rivers, reservoirs, sewage oxidation ponds, coastal power-station ash-ponds. **Habits:** Usually in pairs or small loose groups but never in large numbers. Flies with fairly rapid wingbeats with bill pointing vertically down. Hovers in mid-air before plunging deep into the water to fish.

206. Aleutian Tern, *Sterna aleutica* Plate 28

Camar Aleutia

Description: 33 cm (13 in). From breeding Common Tern (204) at rest by white forehead. Otherwise by slaty grey upperparts and upperwing (apparent when seen together) and in flight by shorter tail. In underwing flight pattern, shows dark, white-edged bar on secondaries. When perched on floating drift wood, holds tail up at approximately 45 degrees angle (Common Tern usually holds tail straight or sometimes slightly lifted). **Breeding:** Whitish grey breast, belly and flanks. **Non-breeding (not illustrated):** Entire underparts and flanks white. Legs and bill black in all seasons. **Voice:** A distinct and squeaky *twee-ee-ee* **Range:** Breeding in North-East Russia, the Aleutian Islands (North-West Pacific) and Alaska. Wintering range not well known, birds on passage have been seen off Hong Kong and Singapore in October. **Status and distribution:** Probably rare and localized passage migrant, south to Singapore, where it is rare. Has perhaps been frequently overlooked as Common Tern. Birds seen have been in non-breeding plumage. May acquire breeding plumage during spring passage. **Habitat:** Oceans, sea-coasts. **Habits:** Like other *Sterna* terns, plunge-dives when feeding. Sometimes perches on floating drift wood. Other habits little known.

207. Bridled Tern, *Sterna anaethetus* Plate 28

Camar Batu

Description: 37 cm (14.5 in). Brownish black upperparts separated from black nape and hindneck by narrow pale grey collar. Crown black with white forehead extending above eye as thin white supercilium. Underparts white. Tail deeply forked. **Juvenile (not illustrated):** Like adult but duller with upperparts scaled pale grey. **Voice:** A sharp *wep-wep*. Also harsh grating notes. **Range:** Pantropical oceans. **Status and distribution:** Locally common offshore resident south to Johor, with non-breeding dispersants off Singapore, where it is uncommon. Breeds on rocky stacks and islets in both the Straits of Melaka and South China Sea, often together with Black-naped Tern (203). Like other resident terns, egg collecting fishermen have reduced breeding populations. **Habitat:** Oceans, oceanic islands, sea-coasts. **Habits:** Usually in small loose groups. Feeds in the open sea, rarely coming to coastal waters. Seldom dives, preferring to pick up food material from the water surface. Flight buoyant and graceful.

208. Sooty Tern, *Sterna fuscata* Plate 28

Camar Angin

Description: 41 cm (17 in). From Bridled Tern (207) by sooty black upperparts. White of forehead broader and does not extend above eye. Black cap and hindneck concolorous with mantle. Tail deeply forked. **Juvenile (not illustrated):** Head, throat and breast dark brown. Upperparts brownish with white scales. Belly whitish. **Voice:** A harsh *kur-wak*, persistently uttered at breeding colonies. **Range:** Pantropical oceans. **Status and distribution:** Vagrant, with individuals recorded only on Pulau Perak (Kedah).

No breeding recorded. Nearest known breeding station is Terumbu Layang-Layang (Swallow Reef) in the South China Sea. **Habitat** Oceans, oceanic islands. **Habits:** Where common, highly gregarious, breeding in dense and noisy colonies. Feeds exclusively in the open sea and never comes to coastal waters. Plunge dives like most other *Sterna* terns.

209. Whiskered Tern, *Chlidonias hybridus* Plate 27
Camar Bermisai

Description: 26 cm (11 in). From non-breeding Common Tern (204) by shallow fork in tail. **Breeding:** Broad white cheek stripe and throat contrasting sharply with black cap and dark grey upperparts and underparts diagnostic Vent and undertail coverts white. Bill and legs red. **Non-breeding:** White crown finely streaked with black. Black patch on ear coverts concolorous with rear crown and hindneck. Mantle, wings, rump, and uppertail coverts pale grey. Bill and legs black. **Juvenile (not illustrated):** Like non-breeding adult but mantle and back brownish boldly mottled black. **Voice:** A sharp *ki-kitt*. **Range:** Breeding in Africa, southern Europe, temperate Asia to South and South-East Asia discontinuously to Australia and New Zealand. Populations from temperate latitudes winter in tropical latitudes. **Status and distribution:** Fairly common passage migrant and winter visitor, south to Singapore, where it is rare. **Habitat:** Sea-coasts, mudflats, flooded ricefields, freshwater ponds, coastal power-station ash-ponds. **Habits:** Usually in small loose flocks, sometimes larger groups with White-winged Tern (210). Feeds largely by skimming low over water surface. Sometimes makes shallow plunges.

210. White-winged Tern, *Chlidonias leucopterus* Plate 27
Camar Hitam Sayap Putih

Description: 23 cm (9 in). From Whiskered Tern (209) by smaller size and rounder head. White rump contrasts with pale grey tail. **Breeding:** Entire head, back, throat, breast upper belly and underside of wing coverts black. Upperwing coverts, lower belly and undertail coverts white. Rest of upperwing pale grey. Bill and legs red. **Non-breeding:** White crown and hindneck finely streaked black with black patch on ear coverts extending below level of eye (giving a "dog-eared" appearance). Bill and legs black. **Juvenile (not illustrated):** Crown blackish; mantle and back brownish black. Many assume full breeding plumage during spring passage. **Voice:** A sharp *kweek*, uttered persistently. Also a sharp *kwek-kwek*. **Range:** Breeding throughout Europe and temperate Asia; wintering south to Africa and South-East Asia and Australia. **Status and distribution:** Fairly common passage migrant and winter visitor, south to Singapore. More numerous than Whiskered Tern. **Habitat:** Sea-coasts, mudflats, rivers, disused mining pools, flooded and dry ricefields, freshwater ponds and reservoirs, coastal power-station ash-ponds. **Habits:** Gathers in small loose flocks, feeding by wheeling and skimming low over water surface. In rice growing areas, often follows harvesting machinery, taking insects disturbed in small swoops. Frequently perches on *kelong* poles and also electricity power lines (in rice growing areas). Follows fishing boats.

211. Brown Noddy, *Anous stolidus* Plate 28
Camar Angguk

Description: 42 cm (17 in). From other terns by wholly dark chocolate brown plumage (upperwing coverts and wing lining paler) and heavy wedge shaped tail with slight notch in centre when fanned. Appears black from a distance. Thick bill, lores and legs black; narrow crescent below eye white; crown and nape greyish white. **Juvenile (not illustrated):** Crown brown with slight whitish supercilium. **Voice:** A harsh *kaaark*. Also a *kwok-kwok*. **Range:** Pantropical oceans. **Status and distribution:** Rare and localized resident in small numbers only on Pulau Perak (Kedah) where numbers continue to decline due to poaching of eggs by fishermen. The other nearest known breeding station is Terumbu Layang-Layang (Swallow Reef) in the South China Sea. **Habitat:** Open sea, oceanic islands. **Habits:** Strictly marine in habits, it feeds exclusively in the open sea, picking up small fish from the water surface. Seldom plunge-dives like other sea terns. Flies with slow, leisurely wingbeats. During courtship, pairs will often engage in nodding to one another; hence the name.

JAEGERS

ORDER *Charadriiformes* **FAMILY** *Stercorariidae*

World: *7 species* **Peninsular Malaysia:** *3 species*

Medium to large sized predatory seabirds characterized by heavy bodies and distinctive projecting tail feathers. Resemble gulls (194–197) but plumage predominantly dark brown with conspicuous white markings on primaries. Bill strong and hooked at the tip. Flight is swift and direct on long angular wings. Largely kleptoparasitic, relentlessly pursuing other smaller feeding seabirds in flight, harassing them until they disgorge their food. Often rest on water surface. Pelagic outside breeding season. Sexes alike.

212. Pomarine Jaeger, *Stercorarius pomarinus* Plate 27

Description: 50 cm (20 in). Larger, heavier and broader-winged than Parasitic (213) and Long-tailed Jaegers (214). Twisted spoon-shaped elongated central tail feathers diagnostic. **Pale morph:** Black crown, yellowish cheek and nape; sooty brown upperparts; white throat and belly separated by broad blackish breast band; vent and flanks barred brownish; undertail coverts and tail streamers blackish. **Dark morph (not illustrated):** Uniformly dark sooty brown. Both morphs show white flashes on primaries in both upperwing and underwing flight patterns. **Juvenile (not illustrated):** Uniformly dark brown plumage with whitish bars on underparts and undertail coverts. Central tail feathers rudimentary and rounded. Heavier build than juvenile Parasitic and Long-tailed Jaegers. **Voice:** Usually silent at sea **Range:** Breeding throughout the Arctic region; wintering south to temperate and tropical oceans. **Status and distribution:** Rare and localized winter visitor with sporadic records from the east and west coasts of Peninsular Malaysia. All sightings have been of light morph individuals. **Habitat:**

Oceans, coastal waters. **Habits:** Skims over the sea in swift hawk-like flight, harassing feeding seabirds, forcing them to regurgitate their food. Often rests on the sea.

213. Parasitic Jaeger, *Stercorarius parasiticus* Plate 27

Description: 45 cm (18 in). From Pomarine (212) and Long-tailed Jaegers (214) by short pointed central tail feathers. **Pale morph:** Breast band paler showing less contrast. **Dark morph (not illustrated):** Uniformly dark sooty brown. Both morphs show white flashes in primaries in both upperwing and underwing flight patterns **Juvenile (not illustrated):** Dark brown with narrower whitish bars than Pomarine Jaeger on lower breast, belly and undertail coverts. Central feathers rudimentary but pointed. **Voice:** Usually silent at sea **Range:** Breeding throughout the Arctic region; wintering south to temperate and tropical oceans. **Status and distribution:** Vagrant, a pale morph individual off Tanjung Agas, Pekan (Pahang) on 29 April 1986 **Habitat:** Oceans, coastal waters **Habits:** Like Pomarine and Long-tailed Jaegers.

214. Long-tailed Jaeger, *Stercorarius longicaudus* Plate 27

Description: 50 cm (20 in). Slimmest of all jaegers. Long pointed central tail feathers diagnostic. **Pale morph:** Throat, breast and belly white. **Dark morph** (not illustrated): Uniformly dark sooty brown. In flight both morphs show little white markings on primaries. **Juvenile (not illustrated):** Like juvenile Pomarine (212) and Parasitic Jaegers (213) but plumage greyer with whitish centre of belly. Central tail feathers rudimentary and pointed. **Voice:** Usually silent at sea **Range:** Breeding throughout the Arctic region; wintering south to temperate and tropical oceans. **Status and distribution:** Rare passage migrant, observed in the Straits of Melaka in April from 2008–2010. On the east coast, a pale morph individual off Pulau Tioman (Pahang) on 5 May 1986 (Wells, 1990). The only inland record is of a juvenile on Tasik Temengor (Perak) on 7 June 2006 (Lim and Lim, 2006). **Habitat:** Oceans, coastal waters, inland freshwater lake (very rare). **Habits:** The most pelagic of the jaegers, keeping more to open sea. Flight and manouverability swifter than Pomarine Jaeger. Other habits similar to other jaegers.

PIGEONS

ORDER *Columbiformes* **FAMILY** *Columbidae*

World: *306 species* **Peninsular Malaysia and Singapore:** *20 species*

Small to large sized birds with plump bodies, small heads, short bills and legs. Wings vary from narrow to broad and pointed to rounded. Flight strong and direct with moderate to fast continuous wingbeats. Feed largely on fruit, seeds and buds. Some species are nomadic, making erratic movements over long distances in search of food. Most species are gregarious. Nests are flimsy platforms of loosely placed sticks, usually constructed in trees. Represented by the following groups and genera: **Wood-pigeons,** genus *Columba* (215). Large but smaller and more slender with more pointed wings than *Ducula*. A largely arboreal group but the sole representative here has terrestrial habits. Sexes alike. **Doves:** Smaller and more slender than pigeons with more pointed wings and usually proportionately longer tails. Have distinct wing and tail patterns in flight. Forage mainly on the ground. Sexes alike or nearly so. Genus *Streptopelia* (216, 217). Medium size and slender with tail of moderate length. Genus *Macropygia* (218, 219). Distinctive long and graduated tails with largely rufescent brown plumage. Genus *Chalcophaps* (220). Plump and short-tailed with broad, rounded wings. Genus *Geopelia* (221). Small and slender. **Nicobar Pigeon,** genus *Caloenas* (222). A monotypic genus. Large and robust short-tailed terrestrial pigeon with hackles on the neck. Sexes alike. **Green-pigeons,** genus *Treron* (223–230). Small to fairly large size with predominantly light yellowish green plumage and conspicuous yellowish wingbars. Sexes differ, the males are usually more colourful with maroon mantles and shoulder patches or bright orange patches on breast. All except one of the local species have red legs. Highly gregarious, often gathering in mixed species flocks in fruiting fig trees. Strictly arboreal but will sometimes descend at salt-licks to feed on minerals. Most utter soft musical whistling notes. **Fruit-doves,** genus *Ptilinopus* (231). A distinct group with the centre of distribution in the Australasian and South Pacific regions. Much darker green than *Treron* and lack wingbars. Solitary or in pairs. Sexes differ, the males are brighter. **Imperial-pigeons,** genus *Ducula* (232–234). Very large and robust with broad rounded wings. Less gregarious than *Treron*, gathering in small flocks at fruiting trees. Strictly arboreal, usually keeping to the canopy. Flight slower than other smaller pigeons. Sexes alike.

215. Rock Pigeon, *Columba livia* — Plate 29
Merpati

Description: 33 cm (13 in). The genetically pure wild stock is the evolutionary ancestor of all domestic breeds of town pigeons. Strongly pointed wings distinctive. The genetically pure wild individual (which many domestic varieties closely resemble) is overall blue-grey, paler on the mantle and wings and much darker on the head and neck which is glossed green and purplish; broad black wingbars; black tail tip and whitish underwing diagnostic. Bill black, legs red. Plumage of domesticated forms varies from all white and black to a combination of white, black, or brown. **Voice:** A rolling trisyllabic *oo-roo-coo*, the notes almost running together. **Range:** Europe, North Africa, the Middle East and the Indian subcontinent. Worldwide populations are the results of widespread domestication and introduction. **Status and distribution:** Introduced. Feral populations throughout Peninsular Malaysia and Singapore are of domesticated forms and are abundant and widespread from low elevations up to 1500 m at the Cameron Highlands (Pahang). **Habitat:** Limestone cliffs, open country, buildings in towns and cities. **Habits:** Gregarious, often gathering in large flocks. A commensal of man, it feeds largely on the ground close to human habitations, and is to an extent dependent on man for food. as well as for nesting sites. Perches freely on trees and buildings. Nests mainly under the eaves of buildings as well as on limestone cliffs and ledges. Very vocal day and night. Flocks have a characteristic wheeling flight with much gliding. Displaying males will inflate necks and perform bowing behavior to prospective females.

216. Red Collared-dove, *Streptopelia tranquebarica* Plate 29

Tekukur Merah

Description: 23 cm (9 in). Short tail and plump appearance distinctive. Narrow black collar around hindneck, broad white tip on blackish tail separated by blackish central tail feathers (very conspicuous in flight) diagnostic in both sexes. **Male:** Blue-grey head contrasting with vinaceous red body plumage (darker on upperparts) and blackish flight feathers diagnostic. **Female:** Like male but head greyish brown and body plumage dull brown. Bill greyish and legs brownish grey in both sexes. **Voice:** A deep and distinct *cru-u-u-u*, repeated at a rate of 4–5 times. **Range:** India (including the Andamans), southern China, South-East Asia east to the Philippines. Some northern populations are known to migrate or undertake long-distance dispersal movements (W1). **Status and distribution:** Uncommon and localized resident, known by very small populations with records from the Perak coast (where breeding was recorded in March 1933), and Melaka, probably a range expansion by unaided dispersants from continental breeding populations (W1). Four on Pulau Langkawi on 8 November 2008 (BiW). In Singapore, common resident, probably established from released captives (W1). Relatively common along the north coast between Changi and Seletar, and also in the Yishun-Senoko area (W1) Also on Pulau Ubin where it is uncommon (W and H). **Habitat:** Scrub, open country. **Habits:** Gathers in small to large parties, foraging on the ground. Fairly tame.

217. Spotted Dove, *Streptopelia chinensis* Plate 29

Tekukur Leher Berbintik/Tekukur

Description: 30 cm (12 in). Long and slender wedge-shaped tail distinctive. Pale greyish head, broad black hindneck collar with fine white spots together with pale vinaceous brown throat, breast and upper belly, merging into whitish belly and undertail coverts diagnostic. Upperparts brown with blackish streaks. In flight, shows conspicuous broad whitish tips to outer tail feathers and pale greyish crescentic patch on outer greater wing coverts and bend of wing. Iris orange, bill black and legs red. **Voice:** A soft trisyllabic *ku-kroo-krooo* **Range:** The Indian subcontinent and China through South-East Asia east to the Moluccas. Widely introduced eleswhere. **Status and distribution:** Abundant and widespread resident from low elevations, south to Singapore. Recorded at 1500 m in the habited areas of Cameron Highlands (Pahang). A conspicuous garden bird. **Habitat:** Open country, scrub, cleared banks of large rivers, plantations, gardens. **Habits:** Generally gathers in small loose parties. A commensal of man, it lives close to human habitations especially in rural areas. Forages on the ground, frequenting open roads, grassy verges of roads and paths in plantations. When disturbed, takes off with noisy wingbeats, keeping close to the ground. Sometimes glides with tail feathers spread out, resembling a small raptor. Usually delivers its call from a low perch.

218. Barred Cuckoo-dove, *Macropygia unchall* Plate 29

Tekukur-Sewah Berjalur

Description: 38 cm (15 in). Slender build with long graduated barred tail distinctive. **Male:** Upperparts distinctly marked with narrow dark rufous and black bars; head rufous, contrasting with buff throat; breast and flanks vinaceous with narrow black barrings. Crown, nape, upper back and breast have iridescent green and purplish wash. Belly and undertail coverts buff. **Female:** Like male but entire underparts dark buff, distinctly marked with narrow blackish bars; darker and bolder on breast. Bill black and legs red in both sexes. **Voice:** A soft but deep disyllabic *u-ooo* (the first note audible only at close range), uttered on a falling tone at an interval of 1–2 seconds. **Range:** The Himalayas and southern China discontinuously through South-East Asia to Sumatra, Java and Lombok. **Status and distribution:** Uncommon and localized resident above 1200 m in the Larut Hills (Perak) and the Titiwangsa Range south to Genting Highlands (Pahang). Has not been known to disperse to feed at lower elevations. More common at the higher elevations of the Cameron Highlands (Pahang). **Habitat:** Forests. **Habits:** Usually in pairs or in small parties, flying fast and straight over tree tops. Feeds largely on the ground. Has the habit of holding tail up when on the ground. Shy.

219. Little Cuckoo-dove, *Macropygia ruficeps* Plate 29
Tekukur-Sewah Kecil/Tekukur Api

Description: 30 cm (12 in). From Barred Cuckoo-dove (218) by smaller size, less distinctly barred plumage, bright orange rufous head, broad rufous scalings on upperparts and pale orange rufous underparts. **Male:** Whitish scalings on upper breast; greenish gloss on neck. **Female:** Duller than male with blackish scalings on upper breast. Bill pinkish brown and legs red in both sexes. **Voice:** A high pitched and rapid *kru-wuck*, the first note audible only at close range, persistently repeated over a brief period. **Range:** South-East Asia to the Greater and Lesser Sundas. **Status and distribution:** Fairly common resident principally above 800 m. Frequently disperses to feed at much lower elevations but nests in the mountains, generally above 900 m. Recorded in the Larut Hills (Perak) and the Titiwangsa Range south to Negeri Sembilan. Common and conspicuous at the hill-stations. **Habitat:** Forests. **Habits:** Usually in pairs or in small loose parties. Flies great distances, usually above tree tops to sources of food. Feeds largely on the ground but delivers call from a high perch. Courting pairs have been noted engaged in a nodding display with throats puffed out. Shy.

220. Emerald Dove, *Chalcophaps indica* Plate 29
Punai Zamrud/Punai Tanah

Description: 25 cm (10 in). Dark iridescent green mantle together with vinaceous red sides of head and underparts diagnostic. **Male:** Crown and nape blue-grey; forehead and supercilium whitish; lower back black with two whitish bars. **Female:** Like male but duller with grey forehead and supercilium and brownish crown and nape. Bill and red bright red in both sexes. **Voice:** A low mournful rising *ti-coooo,* the preliminary note audible only at close range. **Range:** The Indian subcontinent and southern China through South-East Asia and the Indonesian Archipelago to New Guinea, Australia and the western Pacific. **Status and distribution:** Common and widespread resident from low elevations up to 1400 m, south to Singapore, where it is uncommon. **Habitat:** Forests, the forest edge, mangroves, plantations. **Habits:** Frequents the ground usually solitarily, feeding largely on fallen seeds and grub. Flies low with rapid wingbeats, skilfully dodging past trees. Generally keeps to thick cover. Drinks from pools of water. Nomadic, flying over great distances.

221. Zebra Dove, *Geopelia striata* Plate 29
Merbuk

Description: 21 cm (8.5 in). Small and slender build with proportionately long slender tail distinctive. Grey head and forecrown and brown hindcrown together with pale brown upperparts, marked with fine black and white barrings on neck and flanks diagnostic. Breast pale vinaceous pink; belly and undertail coverts buff white. Orbital skin blue-grey, bill pale horn and legs dark red Blackish tail has narrow whitish edges, conspicuous in flight. **Voice:** A soft trilling cadence with a metallic quality, ending in a series of short coos. Variable but has a definite rhythmic pattern. **Range:** Tenasserim (Myanmar) through South-East Asia, the Greater and Lesser Sundas, the Philippines, New Guinea and Australia. Widely introduced eleswhere. **Status and distribution:** Abundant and widespread resident at low elevations, south to Singapore. Specimens seen at the hill-stations are most likely to be caged escapees. **Habitat:** Open country, plantations, gardens. **Habits:** Usually solitary or in small loose groups. Forages largely on the ground but will also perch in low trees. Sometimes nomadic, flying over great distances. Very vocal. Often tame and confiding.

222. Nicobar Pigeon, *Caloenas nicobarica* Plate 29
Merpati Emas

Description: 41 cm (16 in). Large and plump short-tailed pigeon. Dark metallic green plumage, long iridescent bronze hackles around neck and back together with white tail diagnostic; the latter is conspicuous both at rest and in flight. Head, neck and breast slaty black with a purplish blue sheen. Legs dark red. **Juvenile** (not illustrated): Like adult but duller and lacks hackles on neck and back. Tail blackish with a greenish sheen. **Voice:** Usually silent but sometimes utters a deep croak. **Range:** Small forested islands from the Andamans and Nicobars in the Indian Ocean east throughout the Indonesian Archipelago and the Philippines to New Guinea and the Solomons in the western Pacific. Declining as a result of capture for food, the pet trade, habitat destruction and predation by introduced mammals **NT. Status and distribution:** Rare and localized resident on offshore islands, recorded in the Langkawi archipelago (Kedah), Pulau Jarak (Perak) and Pulau Pisang (Johor) in the Straits of Melaka and Pulau Redang, Pulau Perhentian (Terengganu), and the Tioman archipelago (Pahang) in the South China Sea. **Habitat:** Forested islands. **Habits:** Forages on the forest floor solitarily or in pairs. Has crepuscular habits. Generally unobtrusive, spending the day resting in low perches. Rather nomadic, flying great distances from island to island in search of food resources. Sometimes occurs in forested mainland coasts. Other habits little known.

223. Cinnamon-headed Green-pigeon, *Treron fulvicollis* Plate 30
Punai Bakau

Description: 26 cm (10 in). Both sexes have red at base of bill. **Male:** Rusty chestnut head merging into maroon mantle and ochraceous orange breast diagnostic.

Undertail coverts cinnamon. **Female:** From all female green pigeons, except Thick-billed Green-pigeon (227) by red base of bill. From female Thick-billed Pigeon by thinner bill, greenish (not bluish grey) crown, yellowish thighs and the lack of broad bluish green orbital skin. **Voice:** Unknown. **Range:** Southern Vietnam and the Malay Peninsula to Sumatra and Java. Undergoing a moderately rapid population decline due to habitat destruction. **NT. Status and distribution:** Uncommon and localized resident at low elevations, distributed mostly in coastal districts south to Singapore where it is presently a non-breeding dispersant, recorded from Pulau Ubin, Sungei Buloh Nature Park, and the Sime Forest. **Habitat:** Mangroves, forests. **Habits:** Usually inhabits the canopy in small flocks, most often exclusively on its own. Often gathers to feed in fruiting trees. General habits similar to other *Treron* pigeons.

224. Little Green-pigeon, *Treron olax* Plate 30
Punai Kecil/Punai Siul

Description: 20 cm (8 in). Small size and dark appearance separate it from its larger congeners. Blackish tail with broad pale grey terminal band distinctive. **Male:** Dark grey head, throat and neck, dark maroon mantle, broad orange breast band and chestnut undertail coverts diagnostic. **Female:** From female Pink-necked (225) and Orange-breasted Green-pigeons (226) by size and tail pattern. Upperparts dull dark green, underparts paler without any yellow suffusion. Throat whitish. Greyish crown gradually merges into greenish sides of head. **Voice:** A series of soft whistling notes, distinctly higher pitched than other *Treron* pigeons. **Range:** The Malay Peninsula to the Greater Sundas. **Status and distribution:** Locally common resident at low elevations, south to Singapore where it is a rare resident in the Central Catchment Nature Reserve and a non-breeding dispersant to Pulau Tekong and Pulau Ubin. Also occurs in wooded suburban gardens close to forests. **Habitat:** Forests, the forest edge. **Habits:** Usually in pairs or in small parties, keeping to the canopy and middle storey. Like other *Treron* pigeons, will gather to feed in fruiting fig trees, often in mixed species flocks.

225. Pink-necked Green-pigeon, *Treron vernans* Plate 30
Punai Leher Jambu/Punai Gading

Description: 25 cm (10 in). Grey tail with broad subterminal black band and very narrow grey tip separate it from respective sexes of Orange-breasted Green-pigeon (226). **Male:** Grey head, green mantle, purplish pink neck and upper breast together with orange lower breast patch diagnostic. Undertail coverts dark chestnut. **Female:** Dull green with mild yellowish suffusion on underparts. Undertail coverts buff. **Voice:** A series of pleasing chuckling *coos*, uttered up and down the scale. **Range:** Tenasserim (Myanmar) and the Malay Peninsula to the Greater Sundas, Sumbawa, Sulawesi and the Philippines. **Status and distribution:** Common and widespread resident at low elevations, south to Singapore. The only common *Treron* pigeon in patches of wooded or open country within large cities, e.g. Kuala Lumpur and Singapore. **Habitat:** Mangroves, the forest edge, plantations, wooded gardens, open country. **Habits:** Frequents the canopy and the crowns of smaller trees in pairs or in moderate to large flocks. Gathers in large numbers to feed in fruiting trees and shrubs and also on the fruits of the Mac Arthur palm in cities. Forms large communal roosts in trees.

226. Orange-breasted Green-pigeon, *Treron bicinctus* Plate 30

Punai Dada Jingga/Punai Siam

Description: 26 cm (10 in). Blackish tail with broad greyish terminal band distinctive, a feature more obvious in flight (folded tail often obscures this feature). Bill pale grey with whitish tip. **Male:** Orange lower breast band distinctive. From male Pink-necked Green-pigeon (225) by tail pattern, all green head with greyish suffusion on hindcrown and nape. Purplish pink upper breast band does not extend around neck. **Female:** From female Pink-necked Green-pigeon by tail pattern, greyish suffusion on hindcrown and nape and denser yellowish suffusion on underparts. **Voice:** A distinct series of gurgling notes introduced by a pleasing whistle. Also utters rapid chuckling notes. **Range:** The Indian subcontinent and southern China, discontinuously through South-East Asia to the Malay Peninsula and Java. **Status and distribution:** Uncommon and localized resident at low elevations, south to Melaka. Shares similar habitat preference as Pink-necked Green-pigeon but is far less common. **Habitat:** Open country, scrub. **Habits:** Usually in pairs or in small parties. Generally keeps to the canopy or the crowns of smaller trees, but will also descend to feed in low fruiting bushes. Sometimes occurs in mixed species flocks, usually with the Pink-necked Green-pigeon.

227. Thick-billed Green-pigeon, *Treron curvirostra* Plate 30

Punai Paruh Tebal/Punai Daun

Description: 27 cm (11 in). Broad bluish green orbital skin and red at base of bill diagnostic in both sexes. Has bluish grey crown and nape. **Male:** Maroon mantle and cinnamon undertail coverts. **Female:** Lacks maroon mantle, undertail coverts greenish yellow marked with bold dark green streaks. **Voice:** A series of low chucking guttaral *coos*, similar to that of Pink-necked Green-pigeon (225) but softer and higher pitched with less variety. **Range:** The eastern Himalayas through South-East Asia to Sumatra, Borneo and the Philippines. **Status and distribution:** Fairly common resident from low elevations up to 900 m, south to Singapore, where it is uncommon (W and H). Common at the lower elevations of the hill-stations especially in fruiting fig trees. In Singapore, uncommon resident in the Central Catchment Nature Reserve and scarce non-breeding dispersant elsewhere. **Habitat:** Mangroves, forests, the forest edge, wooded gardens. **Habits:** Frequents the canopy and middle storeys in small parties. Feeds in fruiting fig trees often with other *Treron* pigeons. Flies in compact flocks between roosting and feeding grounds.

228. Large Green-pigeon, *Treron capellei* Plate 30

Punai Besar/Bakok/Lengguak

Description: 37 cm (14 in). From other *Treron* pigeons by very large size, thick and robust pale greenish bill and yellow legs. Plumage pale green with broad and conspicuous yellow wingbars. Tail blackish with whitish terminal band on underside;

on upperside, separated by dull green central tail feathers. **Male:** Has distinct bright orange breast band and brown undertail coverts. **Female:** Orange breast band much duller and lighter, appearing as a suffusion. Undertail coverts buff with dark green streaks. Both sexes show narrow yellowish orbital skin. **Voice:** A series of deep and hoarse growling notes. **Range:** Tenasserim (Myanmar) to the Greater Sundas. Populations are suspected to be rapidly declining due to loss and degradation of lowland rainforests. **VU. Status and distribution:** Locally common resident principally at low elevations, south to Johor. Has also been recorded up to 1200 m at Fraser's Hill (Pahang). **Habitat:** Forests, the forest edge. **Habits:** Frequents the canopy, usually in pairs or small parties of up to 5 but never in large flocks like its smaller congeners. Feeds principally on large figs, usually exclusively on its own.

229. Yellow-vented Green-pigeon, *Treron seimundi* — Plate 30

Punai Ekor Runcing

Description: 32 cm (12 in) of which 2 cm (1 in) is length of elongate central tail feathers. From other *Treron* pigeons by elongate pointed central tail feathers protruding from wedge shaped tail, bright cobalt blue bill, lores and orbital skin. Belly white, vent and undertail coverts bright yellow, marked with dark green streaks. **Male:** Maroon patch on scapulars. **Female:** Lacks maroon scapular patch, dark green streaks on undertail coverts bolder. **Voice:** A series of soft and low mournful but pleasing whistling notes, uttered up and down the scale. **Range:** Thailand, Indo-China, and the Malay Peninsula. **Status and distribution:** Uncommon and localized resident at low elevations, recorded in the Manjong District (Perak) and on the Sala River (Kedah) (M and W), and Kuala Selangor (Selangor) (BiW). More common in the Titiwangsa Range between 800–1500 m. Fairly common at The Gap (Selangor) below Fraser's Hill, especially in fruiting fig trees. Presumably, this pigeon undertakes erratic movements over great distances and should therefore occur throughout the Titiwangsa Range, though Genting Highlands (Pahang) is its southernmost known range (D.R. Wells, pers. comm). **Habitat:** Forests, mangroves. **Habits:** Frequents the canopy and middle storeys, usually in small parties. Frequently gathers to feed at fruiting fig trees, often with Thick-billed Green-pigeon (227). Generally unobtrusive, sometimes sitting puffed up on a perch for long periods. Other habits little known.

230. Wedge-tailed Green-pigeon, *Treron sphenurus* — Plate 30

Punai Ekor Baji

Description: 32 cm (12 in). From Yellow-vented Green-pigeon (229) by the lack of elongate central tail feathers on wedge-shaped tail. Bill purplish mauve with horn tip, more slender and less robust than most *Treron* pigeons. Yellowish wingbars thinner and less conspicuous than other *Treron* pigeons. Head and underparts yellowish green, contrasting with darker olive-green upperparts. **Male:** Maroon patch on scapulars, greyish tinge on mantle. **Female:** Lacks both maroon scapular patch and greyish tinge on neck. **Voice:** A deep and distinct whistling *coo* with a grunting note. **Range:** The Himalayas and southern China through South-East Asia, south to Sumatra, Java and Lombok. **Status and distribution:** Rare and localized resident from 760–1600m, recorded in the Larut Hills (Perak) and the Titiwangsa Range from Gunung Batu Brinchang in the Cameron Highlands (Pahang) south to Genting Highlands (Pahang). Also on Gunung Tahan (Pahang). The most common *Treron* pigeon above 1200 m and may sometimes be seen at Bukit Larut (Perak) and Fraser's Hill (Pahang). **Habitat:** Forests. **Habits:** Frequents the canopy and middle storeys in small parties. Gathers to feed in fruiting fig trees, usually exclusively on its own. Often unobtrusive. Has been observed nesting in an exotic conifer *Pinus* sp in a hill-station garden adjacent to forest. Other habits little known.

231. Jambu Fruit-dove, *Ptilinopus jambu* Plate 30
Punai Jambu

Description: 27 cm (11 in). Bright iridescent dark green upperparts, white underparts, whitish orbital skin, orange-yellow bill and red legs diagnostic. **Male:** Crimson face, black throat, white breast with small rosy patch and dark brown undertail coverts. **Female:** Purplish face, greenish grey breast and cinnamon-buff undertail coverts. **Juvenile** (not illustrated): Duller with greenish face and pale yellow throat. Lacks white orbital skin. **Voice:** A soft *hoo* uttered at short intervals, similar to Emerald Dove (220), but lacking the preliminary clicking note. **Range:** The Malay Peninsula to the Greater Sundas. Has suspected to have undergone a moderately rapid population decline due to deforestation. **NT. Status and distribution:** Locally common resident from low elevations up to 1300 m, south to Singapore, where it is an uncommon non-breeding dispersant principally to Bukit Timah and the Central Catchment Nature Reserve. **Habitat:** Forests, the forest edge, mangroves. **Habits:** Frequents the canopy and middle storey solitarily or in pairs. Very retiring and unobtrusive, often sitting motionless on a high perch. Often feeds in fruiting trees with other frugivores.

232. Green Imperial-pigeon, *Ducula aenea* Plate 29
Pergam Hijau

Description: 43 cm (17 in). Pale grey head, neck and underparts, tinged pinkish, contrasting with dull bronze green upperparts. Undertail coverts dark brown; entire tail blackish. Bill grey with red base of maxilla. Iris brown. Legs dark red. **Voice:** A deep disyllabic *click-hroooo*, with the second syllable stressed. Also utters chuckling coos. **Range:** The Indian subcontinent east to South-East Asia to the Greater and Lesser Sundas, Sulawesi and the Philippines. **Status and distribution:** Uncommon and localized resident at low elevations, south to Singapore, where it is currently a rare non-breeding dispersant to Pulau Ubin and the Sime Road Forest. Sometimes overlaps with Mountain Imperial-pigeon (233) in the lowlands. **Habitat:** Mangroves, forests, the forest edge. **Habits:** Frequents the canopy in pairs or in small loose parties, often gathering to feed in fruiting fig trees.

233. Mountain Imperial-pigeon, *Ducula badia* Plate 29
Pergam Gunung/Pergam Bukit

Description: 47 cm (18.5 in). From Green Imperial-pigeon (232) by broad dull greyish terminal band on blackish tail and buff undertail coverts which contrast with pinkish grey underparts. Head and neck grey with pinkish tinge; wings and mantle brownish maroon; throat and chin white. Iris creamy white. Orbital skin red. Bill whitish with red base. Legs bright red. **Voice:** A deep resonant disyllabic *whoo-OO* (with a one second interval between both notes), the second note louder and deeper. **Range:** India and southern China discontinuously through South-East Asia to the Greater Sundas. **Status and distribution:** Fairly common resident principally above 970 m in the mountains but regularly disperses over great distances to feed at low elevations. Breeding has only been recorded in the mountains. Common and conspicuous at the hill-stations. **Habitat:** Forests, mangroves. **Habits:** Frequents the canopy in small parties of up to 5 or more. Often gathers to feed in fruiting fig trees with other frugivores. When feeding, is generally unobtrusive but is very vocal especially during breeding. Sometimes tame and confiding.

234. Pied Imperial-pigeon, *Ducula bicolor* Plate 29
Pergam Rawa/Rawa

Description: 40 cm (16 in). Overall creamy white with broad black terminal tail band and flight feathers diagnostic. Bill and narrow orbital skin grey-blue; legs purplish grey. **Voice:** A series of chuckling notes, *hu-hu-hu-hu...* **Range:** The Andamans and Nicobars east to the coasts of Myanmar, South-East Asia, the Indonesian Archipelago to western New Guinea. **Status and distribution:** Locally common resident at on small offshore islands off the east and west coasts, south to Singapore. Recorded on the Langkawi group (Kedah), Pulau Jarak and the Sembilan group (Perak), Pulau Perhentian, Pulau Serenggeh, and the Redang group (Terengganu) and the Tioman group (Pahang). More common on the east coast. In Singapore, a scarce non-breeding dispersant, recorded from St Johns Island, Pulau Semakau and Pulau Sentosa. **Habitat:** Forests, mangroves. **Habits:** Frequents the canopy in small loose parties. Undertakes long flights from island to island in search of new food resources. Roosts communally in tall trees. **Note:** The similar looking Silvery Wood-pigeon *Columba argentina* 40 cm (16 in) of the smaller islands in the Greater Sundas could occur in Peninsular Malaysia and Singapore. Best distinguished by having the white parts tinged pale grey except for lower belly. Bill pale green with red base; narrow orbital skin red. In flight, shows more pointed wings.

PARROTS, HANGING-PARROTS, PARAKEETS, AND COCKATOOS

ORDER *Psittaciformes* **FAMILY** *Psittacidae*

World: *359 species* **Peninsular Malaysia and Singapore:** *7 species*

A large and diverse family of brightly coloured birds with large heads, short powerful hooked bills and strong zygodactylous feet. Have swift and direct flight on narrow and pointed wings. Flight silhouette distinctive with blunt head and long pointed tails in parakeets, genus *Psittacula* (238–240) and short rounded and slightly protruding tails in Blue-rumped Parrot, genus *Psittinus* (241) and hanging parrots, genus *Loriculus* (235). The three indigenous species are largely greenish besides the array of other colours. Feral populations of non-indigenous species, derived largely from escaped caged specimens have established themselves in certain localities in review area. This includes two parakeets and two cockatoos, genus *Cacatua* (236, 237); the latter are largely white. Arboreal, feeding largely on fruit, seeds, flower buds, pollen and also grain. Very versatile, using bills as climbing aid and feet to manipulate food. Calls high-pitched and screeching notes in most species. Gregarious, with some species forming moderate to large communal roosts. Nest in loose colonies in tree holes. Sexes differ in most species in review area. Many species are much sought after for the caged bird trade and have experienced population declines due to excessive trapping.

235. Blue-crowned Hanging-parrot, *Loriculus galgulus* Plate 31

Serindit

Description: 14 cm (5.5 in). Tiny green parrot with short tail distinctive. **Male:** Small blue crown patch, red breast patch, golden yellow patch on mantle together with yellow band on rump diagnostic. Rump and uppertail coverts red **Female:** Like male but lacks blue crown patch, red breast patch and yellow band on rump. Bill blackish in both sexes. **Voice:** A shrill high-pitched *dzi*. **Range:** The Malay Peninsula, Sumatra and Borneo. A very popular caged bird it is frequently trapped for the caged bird trade. **Status and distribution:** Common and widespread resident from low elevations up to 1400 m, south to Singapore, where it is uncommon. **Habitat:** Mangroves, forests, the forest edge, plantations, wooded gardens. **Habits:** Usually frequents the canopy and crowns of tall trees, especially flowering coral trees, *Erythrina* spp. In flight, wingbeats produce a soft whirring sound. Flocks often call in flight. Will sometimes descend low to feed on small fruit and flower buds. Also feeds on the mesocarp of oil palm fruit. When roosting, hangs upside down like a bat.

236. Tanimbar Corella, *Cacatua goffini* Plate 31

Kakatua Tanimbar

Description: 32 cm (12.5 in). From Yellow-crested Cockatoo (237) by greyish white bill and the lack of long erectile yellow crest. Has pinkish lores and greyish legs. In flight, shows yellowish tinge on underwing and tail. **Voice:** Loud and raucous screeching notes like Yellow-crested Cockatoo. **Range:** Endemic to the Tanimbar Islands (Indonesia). Heavy trapping for the caged bird trade as well as habitat destruction have resulted in population declines. **NT. Status and distribution:** Introduced circa 1980 (W and H). Common with breeding populations well established in Singapore, especially in the Botanic Gardens, Sentosa and St. Johns Islands and Changi. Far more common and widespread than Yellow-crested Cockatoo. **Habitat:** Parks, wooded areas. **Habits:** Gregarious, feeding mainly at tree-top level. Flies noisily from place to place.

237. Yellow Crested Cockatoo, *Cacatua Sulphurea* Plate 31

Kakatua Jambul Kuning

Description: 33 cm (13 in). Large and robust with white plumage, yellow tinge on ear coverts and long erectile yellow crest diagnostic. Bill blackish; legs black. In flight, shows yellowish tinge on underwing and tail. **Voice:** A series of loud raucous screeching notes, typical of the genus. Also a variety of whistled notes. **Range:** Sulawesi and the Lesser Sundas. **Status and distribution:** Introduced circa 1927 (BAC). Small feral populations were established in Singapore but have rapidly declined and is uncommon today with no recent breeding records. **Habitat:** Wooded areas, parks, gardens. **Habits:** Usually in pairs or in very small groups, keeping to the tree-tops. Very noisy and conspicuous in flight which consists of rapid wingbeats interspersed by glides. Erects and lowers crest when calling.

238. Rose-ringed Parakeet, *Psittacula krameri* Plate 31
Bayan Lepas

Description: 36 cm (14 in). From respective sexes of Reddish-cheeked Parakeet (240) by lack of both reddish sides of head and broad black malar stripe. Upperparts green with bluish grey wash on nape and sides of head. Underparts yellowish green. Elongate tail blue with green outer feathers. **Male:** Narrow pink collar on hindneck together with black facial stripe running from base of bill to pink collar diagnostic. **Female:** Like male but lacks both pink collar and black facial stripe. Maxilla red and mandible black in both sexes. **Voice:** Harsh and shrill screaming notes, typical of the genus. A distinct squealing note on a falling tone, repeatedly uttered in flight. **Range:** Africa east to the Indian subcontinent and Myanmar. **Status and distribution:** Introduced. Small feral populations only on Pulau Pinang and in Singapore, where breeding has been recorded. Has greatly declined in Pulau Pinang and is uncommon in Singapore. **Habitat:** Wooded areas, parks, scrub. **Habits:** Like other parakeets.

239. Red-breasted Parakeet, *Psittacula alexandrii* Plate 31
Bayan Dada Merah

Description: 36 cm (14 in). From Reddish-cheeked (240) and Rose-ringed Parakeets (238) by diagnostic pinkish breast and blue head. Has broad black malar stripe and narrow black stripe across forehead. **Male:** Pink breast tinged lilac; maxilla red; mandible blackish. **Female:** Pink breast tinged orange; bill all black. **Voice:** A series of distinct high-pitched disyllabic chattering screams, typical of the genus. **Range:** The Himalayas and North-East India to southern China discontinuously through South-East Asia and the islands of Simeulue, the Banyak group and Nias (off west Sumatra), south to Borneo, Java and Bali. **Status and distribution:** Introduced. Feral population in Singapore which was discovered in 1943 (W1) is locally common today. Indigenous range probably extended south to northern Kelantan in the form of probable dispersants from southern Thailand with sightings of a flock near Kota Bharu in August 1889 (W1). Currently not known to occur naturally in the Malay Peninsula. **Habitat:** Parks, wooded areas, scrub. **Habits:** Where common, highly gregarious. Usually in very small loose noisy flocks, feeding in the tree-tops.

240. Reddish-cheeked Parakeet, *Psittacula longicauda* Plate 31
Bayan Nuri

Description: 41 cm (16 in). **Male:** Green crown, reddish sides of head and nape together with broad black malar stripe diagnostic. Wings blue, mantle tinged bluish, tail blue with yellow tips. Maxilla red with horn tip, mandible brown. **Female:** Like male but has shorter tail and lacks bluish tinge on mantle. Reddish sides of head duller, malar stripe dark green and bill dull brown. **Juvenile (not illustrated):** Like female but has greenish sides of head. In underwing flight, both sexes show yellow wing lining which contrast with dark flight feathers. **Voice:** A high-pitched strident screech, *kiak*, like a toy trumpet. Uttered while perched or in flight. **Range:** The Andaman and Nicobars, the Malay Peninsula, Sumatra and Borneo. Is experiencing moderately rapid declines due to deforestation but is not considered more threatened because of its ability to utilize human modified habitats. **NT. Status and distribution:** Fairly common resident at low elevations, south to Singapore. Has adapted well to a diet of the mesocarp of oil palm fruit, establishing colonies in or near plantations. **Habitat:** Mangroves, forests, the forest edge, open country, wooded areas, oil palm plantations. **Habits:** Usually frequents the canopy and crowns of tall trees in noisy flocks while foraging but forms larger flocks while roosting. Frequently perches on tall dead trees. Flies swiftly in huge compact flocks between feeding and roosting sites.

241. Blue-rumped Parrot, *Psittinus cyanurus* — Plate 31

Puling/Serindit Gajah

Description: 19 cm (7.5 in). Stocky and short-tailed appearance distinctive. **Male:** Blue-grey head, neck, lower back, rump and uppertail coverts diagnostic. Back blackish, shoulder patch red, maxilla red, mandible dark grey, iris red. **Female:** Brown head and dull brown bill with blue tinge confined to back. Iris creamy white. In flight, both sexes show blackish underwing contrasting with diagnostic broad red wing lining. **Voice:** A series of loud, melodious musical whistling notes and trills. **Range:** Tenasserim (Myanmar) through the Malay Peninsula, Sumatra, and Borneo. Has experienced population declines due to deforestation and trapping for the caged bird trade but is not considered more threatened because of its ability to utilize secondary habitats and also because it occurs in lower montane forest. **NT. Status and distribution:** Locally common resident principally at low elevations, south to Singapore, where it is rare and found mainly in the Bukit Timah Nature Reserve and the Central Catchment Forest (W and H). Has been recorded at 1300 m at Fraser's Hill (Pahang). Has also adapted to feeding on the mesocarp of oil palm fruit and established itself on some plantations. **Habitat:** Mangroves, forests, the forest edge, oil palm plantations. **Habits:** Frequents the canopy and crowns of tall trees. Often feeds in flowering trees. Flies swiftly in small flocks.

CUCKOOS

ORDER *Cuculiformes* **FAMILY** *Cuculidae*

World: *144 species* **Peninsular Malaysia and Singapore:** *26 species*

A highly varied family of small to large sized birds characterized by slender bodies with soft plumage and long graduated tails. Bills slightly down-curved. Have zygodactylous feet. Solitary. Sexes similar or nearly so in most species. Represented by the following groups and genera:

Parasitic Cuckoos: Brood parasites, laying their eggs in the nests of other birds, leaving foster parents to incubate eggs and raise the young. Largely arboreal and shy, keeping to the forest canopy or the crowns of tall trees, where they are difficult to observe. Their monotonous repetitive calls are distinctive and often the best guide to positive identification since many species share similar plumage patterns. May be divided into the following subgroups. **Crested cuckoos,** genus *Clamator* (242, 243). Rather slender and long-tailed with tall and flexible crests and attractive plumage. Have rather loud calls. **Hawk-cuckoos,** genus *Hierococcyx* (244–246). Characterized by short, broad wings and greyish or brown tails, marked with alternate dark and pale bands above and below; in this respect they resemble *Accipter* hawks (101–106) but have shallow wingbeats in flight as well as proportionately longer and slender bills Underparts usually marked with dark streaks. **Typical cuckoos,** genus *Cuculus* (247, 248), Have longer and more pointed wings than Hawk-cuckoos. Tails are dark and plain (without alternate dark and pale bands) and have whitish tips and markings on the sides. Head, throat and breast are usually uniform slate grey (without any markings) and are sharply demarcated from whitish lower breast and underparts which are marked with blackish bars. **Brush cuckoos,** genus *Cacomantis* (249–251). Closely allied to Hawk-cuckoos but are smaller with largely buff or rusty unbarred underparts in most species. Have rather pleasing plaintive calls. **Bronze-cuckoos,** genus *Chrysococcyx* (252–255). A distinct group of small cuckoos with largely iridescent upperparts, usually greenish or violet with a bronze sheen. Underparts are predominantly white and broadly barred. Territorial calls are rather weak and are hence overlooked. **Drongo Cuckoo,** genus *Surniculus* (256). All black with a drongo-like shaped tail and gait. Inhabits the canopy where its call is the only clue of its presence. **Koels,** genus *Eudynamys* (257). Large size with long tail. Sexually dimorphic, the male is black while the female is brownish marked with buff streaks and mottles. Have loud piercing calls. **Non-parasitic cuckoos:** Larger and heavier-built than parasitic cuckoos. Build nests, lay eggs and raise young. In review area, represented by the following two groups. **Malkohas,** genera *Rhopodytes*, *Rhinortha*, and *Zanclostomus* (258–263). Large and extremely long-tailed with broad brightly coloured orbital skins and pale green robust bills in most species. All except one species have white tips to tail feathers. Usually in pairs. Arboreal. Clamber about in dense foliage like squirrels and are often difficult to observe. Feed largely on insects gleaned from the foliage. Flight slow with much gliding on short, broad and rounded wings. Usually silent but also utter low croaks and soft mewing notes. Nests are cup-shaped structures of sticks. **Coucals,** genus *Centropus* (264–266). Largely terrestrial with a walking gait. Flight weak and direct, consisting of moderately fast flapping interspersed with short glides, usually low over the ground. Have fairly long, broad ended tails with largely black and chestnut plumage. Inhabit scrub, grassy and dense vegetation, feeding on insects and small vertebrates. Calls are resonant booming notes. Nests are globular structures of dead vegetation with lateral openings, usually placed in dense bushes.

242. Pied Cuckoo, *Clamator jacobinus* — Plate 32

Sewah Berjambul Hitam-Putih

Description: 33 cm (13 in). Like Chestnut-winged Cuckoo (243) but smaller with shorter crest and glossy black hindneck and wings, contrasting with white underparts. Has a conspicuous white patch across base of primaries, which is very conspicuous in flight. Outer tail feathers broadly tipped white. Bill black. **Juvenile:** Like adult but crest shorter and upperparts and wings drab brown, with white underparts tinged buff. **Voice:** Unknown in review area. **Range:** Breeding in Sub-Saharan Africa, the Indian subcontinent (including Sri Lanka) South-East Tibet, and central Myanmar. In Africa, a transequatorial migrant. Absent or very rare in Asia during the northern hemisphere winter, it is presumed that Asian breeding populations also winter in Africa (W1). **Status and distribution:** Vagrant, a juvenile at Penanti, Seberang Prai (Pulau Pinang) on 6 December, 2009 (Choo, 2009). **Habitat:** In review area, open scrubland. **Habits:** Observed clumsily hopping from branch to branch in the manner of a Greater Coucal (265). On the ground was seen catching insects and pulling out worms from the soft earth (Choo, 2009).

243. Chestnut-winged Cuckoo, *Clamator coromandus* Plate 32

Sewah Berjambul Sayap Perang

Description: 46 cm (18 in). Large cuckoo with diagnostic black head and tall erect crest and long broad-ended tail together with chestnut wings and white nuchal collar. Mantle black; throat and upper breast buff rufous; lower breast and belly whitish; thighs greyish; undertail coverts blackish. In flight, crest is held flat against crown. **Voice:** Raucous screams, *kreech-kreech-kreech*... Also a hoarse whistle. **Range:** Breeding in the Himalayas and southern China through South-East Asia to Tenasserim (Myanmar). Northern populations are migratory, wintering south to the Indian subcontinent and South-East Asia to the Malay Peninsula, the Greater Sundas and the Philippines. **Status and distribution:** Uncommon and localized passage migrant and winter visitor principally at low elevations, south to Singapore, where it is uncommon. Occurs above 900 m only during autumn and spring passage. **Habitat:** Mangroves, the forest edge, scrub, wooded gardens. **Habits:** Clambers about in dense vegetation in the manner of malkohas (258–263), hunting for insects. Generally shy.

244. Large Hawk-cuckoo, *Hierococcyx sparveriodes* Plate 32

Sewah-Rajawali Besar

Description: *H. s. bocki* 33 cm (13 in). *H. s. sparveriodes* 44 cm (16 in). Large size separates it from other hawk-cuckoos. Head, nape and cheeks grey, showing little contrast with brownish grey upperparts (browner in *H. s. bocki*). Chin black; breast rufous with bold dark grey-brown streaks; belly whitish with rufous brown bars. White tipped tail marked with broad greyish brown and blackish bars. *H. s. bocki* is darker and more richly coloured. Only *H. s. sparveriodes* is illustrated. **Juvenile:** Upperparts marked with broad rufous brown bars; underparts whitish with brownish streaks. **Voice:** The territorial call of *H. s. bocki* is a typical loud disyllabic *pi-pi*, repeatedly uttered on a rising pitch before culminating in a shrill climax. Also a more subdued disyllabic *pi-pi*, gradually rising on an accelerating pace before falling on a slower pace (This utterance usually precedes the typical one). *H. s. sparveriodes* is silent in wintering quarters. **Range:** Breeding in the Himalayas and southern China discontinuously through South-East Asia and the Malay

Peninsula to Sumatra and Borneo. Northern populations winter south through South and South-East Asia to the Greater Sundas, Sulawesi and the Philippines. **Status and distribution:** *H. s. bocki* is a locally common resident, from 1100–1800 m in the Larut Hills (Perak) and the Titiwangsa Range from Cameron Highlands, south to Genting Highlands (Pahang). Also on Gunung Tahan and Gunung Benom (Pahang). Fairly common at the hill-stations. *H. s. sparveriodes* is an uncommon and localized passage migrant and winter visitor generally at low elevations, south to Singapore. Occurs above 900 m only during autumn and spring passage. **Habitat:** Forests, the forest edge, wooded gardens and plantations (migrants only). **Habits:** Frequents the canopy solitarily, keeping to thick foliage. *H. s. bocki* calls intensely especially from mid September to late May (W1). The migrant subspecies has been reported feeding on the ground in a suburban lowland garden. The brood host is unknown, but W1 suggests keeping a close watch on laughingthrushes *Garrulax* sp. (535, 537). **Note:** Payne and Klitz (2005), split *bocki* from *sparveriodes*, citing differences in size, plumage colour and vocalizations, naming the new taxon Dark Hawk-cuckoo, which is confined to the Malay Peninsula, Sumatra, and Borneo. A DNA analysis will settle the issue.

245. Moustached Hawk-cuckoo, *Hierococcyx vagans* Plate 32

Sewah-Rajawali Bermisai

Description: 30 cm (12 in). Broad slaty malar stripe contrasting with pale greyish cheek and white throat diagnostic. Brown upperparts contrast with whitish underparts, marked with dark streaks on breast and upper belly. Crown and nape dark slate blue. White tipped tail marked with pale brownish and blackish bands. **Juvenile** (not illustrated): Like adult but crown and nape brown. **Voice:** Territorial call of male is a disyllabic *kan-ko*, repeated for long periods. Also a series of rising disyllabic whistles which rises to a shrill climax before ceasing abruptly. **Range:** Tenasserim (Myanmar) and the Malay Peninsula south to Borneo and Java. **Status and distribution:** Rare and localized resident but distributed from low elevations up to 900 m, south to Segamat district (Johor) and the Endau-Rompin State Park (Johor/Pahang). **Habitat:** Forests, the forest edge. **Habits:** Frequents the canopy, usually keeping to thick foliage, where presence is betrayed by call. Calls from Late October to May (W1). Brood host unknown.

246. Hodson's Hawk-cuckoo, *Hierococcyx fugax* Plate 32

Sewah-Rajawali Rimba

Description: 29 cm (11.5 in). *H. f. fugax*: From Moustached Hawk-cuckoo (245) by stockier build, proportionately shorter tail, the lack of slaty malar stripe and by narrow rufous tail tip. Upperparts dark slaty grey contrasting with whitish underparts, marked with dark rufous brown streaks. Throat whitish with faint streaks. Has narrow white patch on side of neck. *H. f. nisicolor*: Upperparts paler grey; unmarked whitish underparts, tinged pale rufous on breast and upper belly. **Voice:** Territorial call of *H. f. fugax* is a cadence of rapid disyllabic notes, *pi-kwik*... Also a rapid *pi-pi*, which increases in tempo before tapering off briefly. **Range:** Breeding in the Himalayas and North-East India, eastern Russia, China and Japan through South-East

Asia to the Greater Sundas and the Philippines. Northern populations are migratory, wintering south to the Greater Sundas, northern Sulawesi and the Philippines. **Status and distribution:** *H. f. fugax* is an uncommon and localized resident from low elevations up to 300 m, south to Johor. Also on Pulau Tioman (Pahang). *H. f. nisicolor* occurs as an uncommon and localized passage migrant and winter visitor from low elevations up to 1200 m, south to Singapore. **Habitat:** Forests, the forest edge, wooded suburban gardens (migrants only). **Habits** Frequents the lower and middle storeys. Generally shy and secretive. Vocal from February to August (W1), including at night. Brood hosts unknown. **Note:** *nisicolor* was split from *fugax* by King (2002) who assigned Hodgson's Hawk-cuckoo for the former and named the latter Malaysian Hawk-cuckoo. The songs of both however remain the same (W2).

247. Indian Cuckoo, *Cuculus micropterus* — Plate 32

Sewah India

Description: 33 cm (13 in). Broad blackish subterminal band on white tipped brownish tail and dark brown upperparts together with greyish green orbital skin separate it from Oriental Cuckoo (248) at close range or in the hand. Otherwise separated by call. Slate grey head, throat and upper breast contrast sharply with whitish underparts which are marked with blackish bars; female has rufous tinge on breast. **Voice:** A loud, clear and fluty medium paced *ko-ka-ka-ko*, with the final syllable on a falling tone. This cadence is sometimes accelerated. **Range:** The Indian subcontinent, far eastern Russia and China through South-East Asia and the Greater Sundas. Northern populations are migratory, wintering south through India, South-East Asia and the Greater Sundas. **Status and distribution:** Fairly common resident, passage migrant and winter visitor from low elevations up to 800 m, south to Singapore. Also occurs above 900 m only during autumn and spring passage. **Habitat:** Forests, the forest edge. **Habits:** Forages in the canopy or crowns of tall trees. Calls persistently for long periods, most intensely between March to June (W1), often at night. A known brood parasite of the Black-and-Yellow Broadbill (383).

248. Oriental Cuckoo, *Cuculus saturatus* — Plate 32

Sewah Gunung Asia

Description: 33 cm (13 in). Grey head, throat, upper breast and upperparts contrast with blackish, white-tipped tail. Lower breast and rest of underparts whitish with blackish bars. Orbital skin yellow. **Hepatic morph female:** Rufous-bay head, upper breast and entire upperparts (including tail) marked with blackish bars. Throat whitish with blackish markings. **Voice:** A deep and hollow *pup, pup-pup-pup*, with a slight pause after the first note. **Range:** Central Asia to eastern Russia and Japan, south through China, the Himalayas, South-East Asia, and the Malay Peninsula through the Greater Sundas to Timor. Northern populations are migratory, wintering south throughout South-East Asia and the Indonesian Archipelago to New Guinea, Australia, and the Solomons with stragglers reaching New Zealand. **Status and distribution:** Uncommon and localized resident and rare passage migrant and winter visitor from low elevations to above 1400 m, south to Singapore. The resident population occurs principally above 900 m in the Titiwangsa Range south to Genting Highlands

(Pahang) but has also been recorded on rare occasions at low elevations. Fairly common at Fraser's Hill. Migratory populations tend to frequent low elevations. The hepatic morph female is very rare. **Habits:** Frequents the canopy where it forages in thick foliage. The resident subspecies calls intensely during the breeding season from January-July. Known brood parasite of the Chestnut-crowned (489) and Yellow-breasted Warblers (490). **Note:** King (2004) proposed that *saturatus* be split into three different taxons, *optatus* (Oriental Cuckoo), *saturatus* (Himalayan Cuckoo), and *lepidus* (Sunda Cuckoo) based on size, plumage, and new studies on vocalizations. Only DNA analysis of the different subspecies will confirm whether or not they are distinct species.

249. Banded Bay Cuckoo, *Cacomantis sonnerattii* — Plate 33
Sewah Berjalur/Burung Takuweh

Description: 23 cm (9 in). From *Cuculus* cuckoos, especially hepatic morph female Oriental Cuckoo (248) by smaller size and very fine barrings on entire underparts as well as an upright stance while perched. From juvenile *Cacomantis* cuckoos by proportionately longer and down-curved bill. Dark rufous upperparts, finely barred blackish; dark patch below eye; whitish supercilium and sides of neck together with whitish underparts, finely barred blackish diagnostic. **Voice:** Typical territorial call is a frequently uttered whistling sequence, *pi-pi-piu-piu*, the last two notes on a falling tone. Also an ascending sequence of four whistling notes which gradually rise in pitch and tempo before ceasing abruptly. **Range:** The Indian subcontinent through South-East Asia and the Malay Peninsula to the Greater Sundas and the Philippines. **Status and distribution:** Locally common resident from low elevations up to at least 820 m at The Gap (Selangor), south to Singapore, where it is uncommon. Very often heard near rural settlements and heavily wooded areas around towns. **Habitat:** Forests, the forest edge, plantations, heavily wooded areas. **Habits:** Frequents the canopy and crowns of tall trees, usually keeping to dense foliage but sometimes perching on open branches. Very vocal. A known brood parasite of the Common Iora (399) (W and H) and the Green Iora (400) (Khoo, 2001).

250. Plaintive Cuckoo, *Cacomantis merulinus* — Plate 33
Sewah Perut Jingga/Burung Mati Anak

Description: 22 cm (8.5 in). Pale grey head, throat and upper breast contrasting sharply with rufous buff lower breast and belly diagnostic. Mantle and wings dark grey-brown. Tail has whitish markings on the side as well as an incomplete narrow white terminal band. **Juvenile and hepatic morph:** Rufous bay head and upperparts marked with blackish barrings on head, mantle, wings and entire uppertail. Underparts whitish with fine blackish barrings. Has faint whitish supercilium and small white patch below eye. **Voice:** Typical territorial call is a three note whistle, followed by a rapidly descending sequence of slurred whistling notes. Also a three note ascending whistle, *pi-piu-piu*, gradually rising in pitch and tempo before ceasing abruptly. **Range:** The Indian subcontinent and southern China through South-East

Asia and the Malay Peninsula to the Greater Sundas, Sulawesi and the Philippines. **Status and distribution:** Fairly common resident principally at low elevations, south to Singapore, where it is uncommon. Very often heard near rural settlements and heavily wooded areas near towns. The hepatic morph is very rare. **Habitat:** The forest edge, scrub, plantations. **Habits:** Frequents the crowns of tall trees. Sometimes descends to catch insects on the ground. Very vocal, sometimes calling at night. Known brood parasite of the Dark-necked Tailorbird (548) and yellow-bellied Prinia (545).

251. Rusty-breasted Cuckoo, *Cacomantis sepulcralis* Plate 33

Sewah Dada Jingga

Description: 24 cm (9.5 in). From Plaintive Cuckoo (250) by narrow bright yellow orbital skin. Rufous buff underparts extends to upper breast and throat, gradually merging with greyish sides of head. Crown and nape dark grey, showing little contrast with dark grey-brown mantle and wings. **Juvenile:** Head and entire upperparts blackish with narrow rufous bars on head, mantle, wings and sides of tail. Underparts whitish with fine blackish barrings. **Voice:** Typical territorial call is a long cadence of whistling notes, *piu,...piu...piu...*, uttered on an asecending scale of about 10–15 times at a one second interval before gradually descending on the last few notes. Also a three note ascending whistle *pi-piu-piu*, similar to Plaintive Cuckoo. **Range:** The Malay Peninsula through the Greater and Lesser Sundas to Australia and the Solomons. **Status and distribution:** Locally common resident from low elevations up to 600 m, south to Singapore, where it is uncommon. **Habitat:** Mangroves, forests, the forest edge, scrub, plantations, wooded gardens. **Habits:** Frequents the canopy and crowns of trees. Prefers more forested and heavily wooded areas than Plaintive Cuckoo. Known brood parasites of the Pied Fantail (432), Tickell's Blue-flycatcher (604), and Olive-backed Sunbird (636).

252. Little Bronze-cuckoo, *Chrysococcyx minutillus* Plate 33

Sewah Daun

Description: 16 cm (6.25 in). From female Violet Cuckoo (255) by all black bill, whitish forehead and the lack of rufous outer tail feathers. Crown and nape brown with strong bronzy green tinge and are concolorous with mantle, wings, lower back and tail. From Horsfield's Bronze-cuckoo (253) by lack of brown ear patch and whitish supercilium. Entire whitish underparts barred blackish, narrower on throat and breast, broader on belly and flanks. Orbital skin red in male, green in female. **Voice:** A descending whistling cadence of 3–5 notes, *teu, teu, teu...* Also a high-pitched trill on a falling tone. **Range:** The Malay Peninsula to the Greater and Lesser Sundas to the Philippines, New Guinea and Australia. **Status and distribution:** Locally common resident principally at low elevations and foothills, south to Singapore. Has been recorded at 1200 m at Fraser's Hill (Pahang). **Habitat:** Mangroves, the forest edge, scrub, wooded gardens. **Habits:** Frequents the crowns of trees where it is often unobtrusive as it forages. Usually calls from a high perch. Sometimes descends to lower branches. A known brood parasite of the Golden-bellied Gerygone (392).

253. Horsfield's Bronze-cuckoo, *Chrysococcyx basalis* Plate 33
Sewah Australia

Description: 16 cm (6.25 in). From Little Bronze-Cuckoo (252) by dark brown ear coverts, pale supercilium, unmarked whitish belly and rufous base to sides of tail. Upperparts brown with a bronzy green tinge with fine white fringing to mantle and scapulars. Underparts whitish with faint streaks on throat and thin wavy barrings on breast (both discernable only at close range). Bolder barrings on underparts confined to flanks only. **Voice:** Unknown in review area. **Range:** Breeding in Australia; wintering northwards to New Guinea and the Indonesian Archipelago during the southern hemisphere winter. **Status and distribution:** In Peninsular Malaysia, vagrant, known from an individual at Kampung Likir, Sitiawan (Perak) on 24 July 2008 (Tan Ah Lai, RCS). In Singapore, rare winter visitor, with sporadic records in from June–August during the southern hemisphere winter. **Habitat:** Open country. **Habits:** In review area, known to occur solitarily. Feeds principally on caterpillars and insects, gleaned from branches and foliage. Will also forage on the ground.

254. Asian Emerald Cuckoo, *Chrysococcyx maculatus* Plate 33
Sewah Hijau

Description: 18 cm (7 in). **Adult male:** Iridescent dark green head, upperparts and upper breast together with white lower breast and belly marked with dark green barrings diagnostic. Orbital skin orange; bill orange-yellow with dark tip. **Subadult male (not illustrated):** Like adult male but entire underparts (including throat and breast) marked with white and dark green bars. **Female:** Pale coppery green upperparts together with chestnut crown and nape diagnostic. Entire underparts barred. In flight, both sexes show broad white band on underwing, at base of flight feathers. **Voice:** A sharp *chweek*, usually uttered in flight. **Range:** The known breeding range extends from the Himalayas east to Myanmar, northern Thailand and China; south of this range, sporadic occurences in India, Sri Lanka, South-East Asia, the Malay Peninsula and Sumatra fall mainly during the wintering period and are therefore assumed to be either non-breeding visitors or vagrants due to the lack of evidence of breeding within this range. **Status and distribution:** Rare and localized migrant at low elevations, south to Johor. **Habitat:** Forests, the forest edge. **Habits:** Frequents the canopy and crowns of tall trees. Other habits little known.

255. Violet Cuckoo, *Chrysococcyx xanthorhynchus* Plate 33
Sewah Ungu

Description: 17 cm (6.5 in). **Male:** Iridescent dark purple head, upperparts and upper breast together with bright orange bill with red base diagnostic. Belly white, marked with dark purple bars. **Female:** From Little Bronze-cuckoo (252) by orange bill (duller than male), reddish bronze (with slight greenish tinge) forehead, crown, nape and upperparts and rufous outer tail feathers. Underparts whitish marked with dark bars; finer on throat and breast and bolder on belly. Orbital skin reddish in both sexes. In flight both sexes show white band on underwing which is narrower and shorter than Asian Emerald Cuckoo (254). **Voice:** A disyllabic *ke-vick*, usually uttered in flight. Also a shrill and rapid trill. **Range:** The Himalayas and southern China through South-East Asia and the Malay Peninsula to the Greater Sundas and the Philippines. Northern populations migrate to winter in southern latitudes. **Status and distribution:** Locally common resident, passage migrant and winter vistor from low elevations up to 500 m, south to Singapore, where it is uncommon. Occurs above 900 m only during autumn and spring passage. **Habitat:** Forests, the forest edge, plantations. **Habits:** Frequents the canopy and crowns of tall trees. Forages mainly in dense foliage, sometimes coming out to call on exposed perches. Has dippy flight. Known brood parasite of the Brown-throated (632) (W1) and Purple-throated Sunbirds (634) (W and H) and Little Spiderhunter (640).

256. Drongo Cuckoo, *Surniculus lugubris* Plate 33

Sewah Cecawi/Burung Hamba Kera/Burung Sawai

Description: 25 cm (9.25 in). Overall glossy black. Small white nuchal patch and white thighs are difficult to observe in the field. From drongos (425–430) by thinner slightly down-curved bill, slightly forked tail with white barrings on vent and undertail coverts. **Juvenile (not illustrated):** Duller with white spots. **Voice:** Territorial call is a cadence of high pitched ascending clear whistles of 5–6 notes, *pi-pi-pi-pi-pi-pi*, uttered persistently. Also a rapid and rising trilled whistle, *kree-kree-kree-kree*., falling on the last two or three notes. **Range:** The Indian subcontinent and southern China through South-East Asia to the Greater Sundas, Bali, Sulawesi and the Philippines. Northern populations are migratory. **Status and distribution:** Common and widespread resident, passage migrant and winter visitor from low elevations up to 800 m, south to Singapore where residents are uncommon and confined to remaining forests; migrants often turn up in wooded areas. Occurs above 900 m only during autumn and spring passage. **Habitat:** Forests, the forest edge. **Habits:** Keeps to the canopy and the crowns of tall trees. Known to capture termites on aerial sallies (W1). Very vocal. A known brood parasite of the Chestnut-winged Babbler (504), Rufous-fronted Babbler (502), Striped Tit-Babbler (510), Sooty-capped Babbler (516) (W1), Horsfield's Babbler (523) and Olive-winged Bulbul (464) (Lim, in prep.).

257. Asian Koel, *Eudynamys scolopaceus* Plate 32

Sewah Tuwu/Tahu/Burung tuwu

Description: 43 cm (17 in). Rather large cuckoo with long broad ended tail, heavy pale greenish bill and red iris diagnostic. **Male:** Overall glossy black. From crows (441) by length and shape of tail and rapid flapping in flight. **Female:** Dark brown (paler below) with bold buff spots and streaks. Tail marked with narrow rufous bars. **Voice:** Territorial call of male is a loud piercing *ko-el*, with the second syllable stressed. Also a harsh rising *kwok-kwok-kwok-kwok*... Also a high-pitched *kik-kik-kik-kik*... usually uttered in flight. **Range:** The Indian subcontinent and China discontinuously through South-East Asia and the Malay Peninsula to the Indonesian Archipelago, New Guinea, Australia and the western Pacific. Northern populations are migratory. **Status and distribution:** Common and widespread resident, passage migrant and winter visitor principally at low elevations, south to Singapore. Occurs above 900 m only during autumn and spring passage. A known brood parasite of the House (441) and

Large-billed Crows (443). Very common around Kuala Lumpur and along the coast of Selangor, presumably because of expanding range of the House Crow. **Habitat:** Mangroves, scrub, plantations, reed beds, wooded gardens in towns and cities. **Habits:** Usually solitary, foraging in the crowns of trees. Keeps to dense foliage and is seldom observed. Very vocal, especially during breeding.

258. Black-bellied Malkoha, *Rhopodytes diardii* Plate 34

Cenuk Perut Hitam

Description: 37 cm (14.5 in). Small size and dark appearance distinctive. Throat and belly dark grey, merging into black belly and undertail coverts. Orbital skin red. **Voice:** A variety of vocalizations is given in W1, from a frog-like *gwaup* and a hurried disyllabic *gwagaup* to a louder, more carrying *pauk* that is hard to separate from the advertising call of the Banded Pitta (387). **Range:** Tenasserim (Myanmar) through the Malay Peninsula to Sumatra and Borneo. Declining due to extensive loss and fragmentation of lowland rainforest. **NT. Status and distribution:** Locally common resident principally from low elevations up to 200 m, south to Johor but also recorded as high as 1200 m on Gunung Rabong (Kelantan). Locally extinct in Singapore since 1950 (L and G). **Habitat:** Forests, the forest edge. **Habits:** Frequents the middle storey, keeping to thick cover where it forages for insects. Generally unobtrusive.

259. Chestnut-bellied Malkoha, *Rhopodytes sumatranus* Plate 34

Cenuk Perut Coklat

Description: 40 cm (16 in). From Black-bellied Malkoha (258) by heavier bill and chestnut belly and undertail coverts (usually inconspicuous in poor light). Orbital skin pinkish orange. **Voice:** Very high pitched mewing notes. **Range:** Tenasserim (Myanmar) through the Malay Peninsula to Sumatra and Borneo. Declining due to extensive loss and fragmentation of lowland rainforest, but can still use secondary growth. **NT. Status and distribution:** Uncommon and localized resident principally at low elevations south to Singapore where it is the only malkoha. Has also been recorded as high as 1000 m on Gunung Tahan (Pahang). The only malkoha in the mangroves (see Chestnut-breasted Malkoha (263)). **Habitat:** Mangroves, forests, the forest edge, plantations. **Habits:** Forages mainly in the middle storey. Generally unobtrusive.

260. Green-billed Malkoha, *Rhopodytes tristis.* Plate 34

Cenuk Paruh Hijau/Burung Kera

Description: 56 cm (22 in). From all malkohas by large size, robust bill and very long tail. Crown, sides of head and underparts greyish (much paler than Black-bellied Malkoha (258)). Has narrow border of whitish feathers around broad red orbital skin. **Voice:** A soft descending *poou*. Also deep guttural grunts and croaks. **Range:** The Himalayas and southern China through South-East Asia to the Malay Peninsula and Sumatra. **Status and distribution:** Fairly common resident south to Selangor. Occupies varying altitudinal zones; from Kedah and Kelantan south to coastal Perak in the west to Kuala Terengganu (Terengganu) in the east, occurs from about sea level up to 1500 m. In Pahang and Selangor, occurs between 820–1500 m, principally on the

Titiwangsa Range and its foothills. Fairly regular at The Gap (Selangor) and Fraser's Hill (Pahang). **Habitat:** Forests (including coastal forests with stands of *Melaleuca*), the forest edge, scrub, plantations, wooded gardens. **Habits:** Frequents the middle storey but in cleared areas will also forage low in thick vegetation. Sometimes perches on bare branches especially at the forest edge. General habits like other malkohas.

261. Raffles's Malkoha, *Rhinortha chlorophaea* — Plate 34
Cenuk Kecil

Description: 33 cm (13 in). Smallest malkoha in the review area with pale greenish orbital skin and distinctive chestnut mantle and wings. **Male:** Buff rufous throat and breast together with blackish tail (faintly marked with very narrow dark rufous bars) diagnostic. **Female:** Pale grey head, neck, throat and upper breast together with chestnut tail (concolorous with mantle) diagnostic. **Voice:** 3–4 soft descending mewing notes, *kiau-kiau-kiau-kiau*. **Range:** Tenasserim (Myanmar) through the Malay Peninsula to Sumatra and Borneo. **Status and distribution:** Fairly common resident from low elevations up to 1000 m, south to Johor. Locally extinct in Singapore since 1895 (L and G). **Habitat:** Forests, the forest edge, plantations, wooded gardens. **Habits:** Frequents the middle storey, usually in pairs. Makes short glides from tree to tree in search of insect prey.

262. Red-billed Malkoha, *Zanclostomus javanicus* — Plate 34
Cenuk Paruh Merah/Cenuk Api

Description: 45 cm (18 in). From all malkohas by diagnostic red bill, blue orbital skin, pale cinnamon throat and upper breast merging into rufescent grey belly. Undertail coverts cinnamon. Upperparts (including tail) greyish with slight gloss. **Voice:** A soft monosyllabic *kuk*, repeated many times. **Range:** Tenasserim (Myanmar) through the Malay Peninsula south to the Greater Sundas. **Status and distribution:** Locally common resident from low elevations up to 1200 m, south to Johor. Locally extinct in Singapore since 1895 (L and G). **Habitat:** Forests, the forest edge. **Habits:** Frequents the middle storey. Generally unobtrusive. General habits like other malkohas.

263. Chestnut-breasted Malkoha, *Zanclostomus curvirostris* — Plate 34
Cenuk Dada Perang

Description: 46 cm (18 in). From all malkohas by entire chestnut underparts and the lack of white tips to tail feathers. Upperparts dark iridescent green, including most of uppertail. Has broad chestnut terminal band on upper tail. Sides of entire tail narrowly marked chestnut. Orbital skin red. **Voice:** A cadence of rapid *tok* notes, *tok-tok-tok-tok-tok*... **Range:** Tenasserim (Myanmar) and the Malay Peninsula south to the Greater Sundas, Bali and Palawan (the Philippines). **Status and distribution:** Fairly common resident from low elevations up to 1000 m, south to Johor. Also on Pulau Pinang, Pulau Langkawi (Kedah), and Pulau Tioman (Pahang). **Habitat:** Mangroves (on Pulau Langkawi only), forests, the forest edge, plantations, wooded gardens. **Habits:** Frequents the middle storey, usually in pairs. Sometimes forages low in dense scrub. Generally unobtrusive, perching for long periods in the foliage.

264. Short-toed Coucal, *Centropus rectunguis* Plate 34

Bubut Rimba

Description: 37 cm (14.5 in). A miniature Greater Coucal (265) but tail proportionately shorter with breast, nape, and tail washed violet. From Lesser Coucal (266) by darker and more glossy black parts and brighter chestnut mantle and wings. Has different call and habitat preference. **Juvenile:** Rufous head and breast with pale buff streaks and pale buff supercilium. Wings chestnut, lower breast and belly blackish marked with rufous bars on flanks. **Voice:** A series of 4–5 deep resonant booming notes, *boop-boop-boop-boop-boop*, distinctly deeper and slower than Greater Coucal. Also a series of more rapid *boop* notes on an ascending scale, frequently uttered at dusk. **Range:** The Malay Peninsula, Sumatra, and Borneo. Being restricted to lowland forest, populations are declining due to extensive loss of this type of habitat. **VU. Status and distribution:** Rare and localized resident from low elevations up to 600 m, south to Johor. **Habitat:** Forests. **Habits:** General habits like Greater Coucal but keeps to deep forest; rarely associating with it at the forest edge.

265. Greater Coucal, *Centropus sinensis* Plate 34

Bubut Besar

Description: 53 cm (21 in). Large size with glossy black head, long broad-ended tail and underparts together with bright chestnut mantle and wings diagnostic. Has black underwing coverts. Iris red. Bill and legs black. **Juvenile:** Chestnut parts duller and marked with blackish bars; black parts faintly marked with greyish barrings or spots. **Voice:** A series of deep and level *boop* notes which tapers of briefly on a descending scale, *boop-boop-boop-boop*... **Range:** The Indian subcontinent and southern China through South-East Asia and the Malay Peninsula to the Greater Sundas, Bali and the Philippines. **Status and distribution:** Common and widespread resident from low elevations up to 700 m, south to Singapore, where it is uncommon (W and H). **Habitat:** The forest edge, scrub, riverine vegetation, plantations. **Habits:** Usually solitary or in pairs, foraging on the ground in dense vegetation. Small groups may sometimes gather in open grassland to feed on insects, especially after rains. Sometimes hops about in small trees and low bushes. Often perches on bare branches, sunning itself with partly outstretched wings. Does not associate with Lesser Coucal (266).

266. Lesser Coucal, *Centropus bengalensis.* Plate 34

Bubut Kecil

Description: 38 cm (15 in). Plumage pattern similar to Short-toed (264) and Greater Coucals (265) but duller without gloss. Head, neck and mantle marked with long, fine pale buff streaks. Bill and legs black. **Juvenile:** Brown head and mantle marked with long fine buff streaks; underparts buff with fine streaks on throat and breast. Blackish tail finely marked with buff bars on uppertail coverts. Bill pale buff. **Voice:** A distinct series of booming notes, terminating in a series of sharp stacatto hoots, *hup-hup-hup-hup...tokato, tokato, tokato*... Also a level staccato *tuk-tuk-tuk-tuk-tuk-tuk*... descending on the last few notes. **Range:** The Indian subcontinent and southern China through South-East Asia and the Greater Sundas east to the Moluccas. **Status and distribution:** Fairly common resident from low elevations up to 1500 m in the cleared areas of Cameron Highlands (Pahang) south to Singapore. Also recorded at the cleared areas of The Gap (Selangor) and Fraser's Hill (Pahang). The most commonly seen coucal in open country habitats. **Habitat:** Wet and dry grassland, scrub, reedbeds. **Habits:** Forages on the ground, preferring more open vegetation than Greater Coucal. Often makes short flights with much flapping and gliding, low over the ground. Frequently suns itself on bare branches or tall reeds, with wings partly outstretched. Sometimes calls in the dead of night.

BARN OWLS AND BAY-OWLS

ORDER *Strigiformes* **FAMILY** *Tytonidae*

World: *15 species* **Peninsular Malaysia and Singapore** *2 species*

Nocturnal birds of prey characterized by heart-shaped faces, dark eyes and broad feathered facial discs which amplify sounds to the ears. Barn Owls, genus *Tyto* are long, slender and long legged with varying amounts of white particularly on the underparts. Inhabit mainly open and wooded country. Locate prey by hearing. Usually solitary or in pairs, sometimes in larger groups. Bay-owls, genus *Phodilus* resemble Barn Owls in facial shape but are shorter-legged and more compact with general habits closer to typical owls (269–282). Inhabit forests. Both genera have soft plumage, hooked bills, powerful talons and feathered tarsi. Flight moderate, leisurely and silent on broad rounded wings. Sexes alike.

267. Barn Owl, *Tyto alba* Plate 35

Jampuk Putih

Description: 34 cm (13.5 in). Broad white heart-shaped facial disc contrasting with dark brown eyes distinctive. Upperparts golden buff and grey, lightly marked with white spots. Underparts pure white. Tail marked with narrow darker bars. Shows all white underwing in flight. **Voice:** A loud screech. Also harsh rasping and hissing notes. **Range:** Worldwide except the Arctic region in the north and New Zealand in the south. **Status and distribution:** Fairly common resident at low elevations, south to Singapore, where it is uncommon. Distribution and numbers are more or less stable due to widespread introduction and breeding programmes in oil palm plantations and rice planting schemes to combat rodent problems. **Habitat:** Open country, inland freshwater swamps, ricefields, plantations, cultivation, wooded areas and buildings in towns and cities. **Habits:** Usually solitary but congregates in larger groups where common. Typically seen flying low over open country at dusk. Feeds principally on rodents which it swoops down to catch from open perches. Also catches fish in its talons from flooded ricefields and freshwater swamps. Roosts in trees and uninhabited buildings by day.

268. Oriental Bay-owl, *Phodilus badius* Plate 35
Jampuk Rimba

Description: 29 cm (11.5 in). From Barn Owl (267) by smaller size, shorter legs and tail and broader forecrown and facial disc, which protrude slightly above eye in short ear-like projections. Facial disc is separated from underparts by a thin white 'necklace'. Upperparts dark reddish chestnut marked with small black and white spots. Forecrown, facial disc and underparts vinous with small dark spots on breast and belly. Dark smudges below eye give it a rather 'big eyed' appearance. **Voice:** A series of musical and rather mournful whistles, starting on a level note and rising on the rest. Usually uttered for long periods. **Range:** South India and Sri Lanka discontinuously north to the Himalayas, east through Myanmar and South-East Asia to the Greater Sundas, Bali and Samar (the Philippines). **Status and distribution:** Uncommon and localized resident from low elevations up to 1200 m, south to Johor. Also on Pulau Pinang. Locally extinct in Singapore since 1928 (L and G). **Habitat:** Forests, the forest edge, plantations, landward fringes of mangroves. **Habits:** Frequents the lower storey solitarily or in pairs. Unobtrusive, perching vertically on tree trunks, vines and steep sloping branches, often with one foot high above the other. Generally shy and other habits little known.

TYPICAL OWLS

ORDER *Strigiformes* **FAMILY** *Strigidae*

World: *187 species* **Peninsular Malaysia and Singapore:** *14 species*

Tiny to large nocturnal birds of prey with large heads, short tails, strong hooked bills, powerful talons and soft plumage. Differ from Barn and Bay-owls (268, 269) by more robust build and puffed up appearance, rounder and proportionately smaller facial discs with conspicuous erect tufts of feathers on the sides of heads of many species. Legs are also proportionately shorter. Eyes are large and directed forward, giving excellent binocular vision, essential in locating prey with great accuracy. Facial discs greatly aid in focusing sounds to the ears. Fly silently with fast (in smaller species) to moderate (in larger species) wingbeats on broad, rounded wings; flight silhoutte distinctive with short tail and large blunt head. Most species assume a typical upright carriage while perched. During the day, roost inconspicuously in trees. Often difficult to identify by sight at night; their calls which vary from soft monotonous hoots to harsh screams and loud maniacal laughing hoots are distinct in many species, often providing the key to positive identification (see Appendix I). Solitary. Sexes alike. Represented by the following groups and genera: **Scops-owls**, genus *Otus* (269–273). Small with conspicuous ear tufts. Usually inhabit the lower storeys of forests. **Eagle-owls**, genus *Bubo* (274, 275). Very large with long and erect ear tufts set in the corners of the head, giving a 'horned' appearance. The bills of all local species are pale yellowish. Tarsi feathered. **Fish-owls**, genus *Ketupa* (276, 277). Very large. Ear tufts less erect than *Bubo*. Bills greyish. Tarsi unfeathered. Have narrow buff bars on dark tail. Particularly attached to waterways and swampy localities. **Wood-owls**, genus *Strix*

(278, 279). Large to very large. Lack ear-tufts, giving facial discs a more rounded appearance. **Owlet**, genus *Glaucidium* (280). Tiny and lack ear tufts. Assumes a more horizontal position when perched. **Boobook**, genus *Ninox* (281). Medium sized. A largely Australasian genus with a dumpy hawk-like appearance and lacking both facial disc and ear tufts. **Eared Owl**, genus *Asio* (282). Medium sized with ear-tufts set closer to centre of head.

269. White-fronted Scops-owl, *Otus sagittatus* Plate 35

Jampuk Kening Putih

Description: 27 cm (10.5 in). Largest Scops-owl. In shape resembles Collared Scops-owl (272) but is larger and proportionately longer tailed and lacks the buff nuchal collar. Broad whitish forehead and eyebrows together with rufescent plumage and whitish bill diagnostic. Iris brown. **Voice:** A hollow *hoo* like Reddish Scops-owl (270) but starts and ends abruptly. Also soft moans. **Range:** Tenasserim (Myanmar) and the Malay Peninsula to northern Sumatra. Being highly dependent on lowland rainforest, much of which has been lost and fragmented, its highly fragmented population has declined greatly. **VU. Status and distribution:** Very rare resident from low elevations up to 600 m, south to Negeri Sembilan. **Habitat:** Forests. **Habits:** Frequents the lower storey. Feeds mainly on small vertebrates and invertebrates caught on the ground. Other habits little known.

270. Reddish Scops-owl, *Otus rufescens* Plate 35

Jampuk Merah

Description: 19 cm (7.5 in). Small Scops-owl with rufescent plumage marked with white or buff spots on back which are tipped with blackish V-shaped markings. Underparts paler and marked with fine blackish spots. Iris brown but on rare occasions, yellow (see Mountain Scops-owl (271)). **Voice:** A hollow and distinct high-pitched *hooo* without any inflection, given at regular intervals. **Range:** The Malay Peninsula through the Greater Sundas. Being forest dependent, is declining due to loss and fragmentation of lowland rainforest. **NT. Status and distribution:** Uncommon and localized resident principally at low elevations up to 200 m, in central Perak and northern Pahang, south to Johor. **Habitat:** Forests. **Habits:** Usually frequents the lower storey, sometimes higher. General habits little known.

271. Mountain Scops-owl, *Otus spilocephalus* Plate 35

Jampuk Gunung

Description: 20 cm (8 in). From Reddish-Scops-owl (270) by yellow iris and uniform rufescent plumage with no blackish spots on underparts which are marked with very fine whitish barrings (altitudinal ranges of both species do not overlap). Scapulars have white spots. **Voice:** A soft and fluty disyllabic, *who-who*, on a level tone with a one second interval between both notes. **Range:** The Himalayas, southern China (including Taiwan) and Myanmar through South-East Asia to the Greater Sundas. **Status and distribution:** Locally common resident above 800 m in the Larut Hills (Perak) and the Titiwangsa Range, south to Genting Highlands (Pahang). Also on Gunung

Tahan and Gunung Benom (Pahang). Frequently heard at the hill-stations. **Habitat:** Forests. **Habits:** Frequents the lower and middle storeys. Feeds mainly on small prey caught on the ground. Calls persistently especially after dusk, often throughout the night. Sometimes tame.

272. Collared Scops-owl, *Otus bakkamoena* Plate 35
Jampuk Kubur

Description: 23 cm (9 in). Dull brownish plumage with pale buff nuchal collar and brown iris diagnostic. Dark brown inverted triangular patch on forehead contrasts sharply with whitish buff eyebrows. Underparts marked with faint blackish streaks. **Voice:** A soft disyllabic *bo-ou*, both notes almost running together with the second note slightly lower. Uttered persistently with a 10–15 second interval between each call. **Range:** South-East Asia to the Greater Sundas and the Philippines. **Status and distribution:** Fairly common resident principally at low elevations, south to Singapore, where it is common (W and H). Has also been recorded at 850 m along The Gap-Raub road (Pahang). Probably the most commonly seen and heard small owl. **Habitat:** Forests, the forest edge, landward fringes of mangroves, plantations, wooded surburbs of towns and cities. **Habits:** Frequents the lower and middle storeys. Watches for small vetebrate prey from a low perch before pouncing to catch it. Sometimes hawks for insects attracted by street lights, especially in wooded surburban areas. When alarmed, will raise ear tufts in an erect position. Often comes close to human habitation. Very vocal and will sometimes call during the day.

273. Oriental Scops-owl, *Otus sunia* Plate 35
Jampuk Kecil

Description: 19 cm (7.5 in). Small Scops-owl with diagnostic bold blackish vertical streaks on underparts, pale eyebrows and yellow iris. **Grey morph:** Greyish plumage. **Rufous morph:** Brownish with strong rufescent tinge especially on face and underparts. **Voice:** Usually silent in wintering quarters. Sometimes utters a purring trisyllabic *toik-to-toik*. **Range:** Breeding throughout much of East Asia through the Indian subcontinent and South-East Asia discontinuously to the Philippines. Northern populations are migratory, wintering south to tropical latitudes. **Status and distribution:** Locally common passage migrant and winter visitor principally at low elevations, south to Singapore, where it is very rare. Occurs above 900 m only during autumn and spring passage. Grey morph birds are more common. **Habitat:** Forests, mangroves, plantations. **Habits:** Frequents the lower storey, often sitting on exposed perches. Like other Scops-owls, feeds largely on small prey.

274. Barred Eagle-owl, *Bubo sumatranus* Plate 36
Burung Hantu Jalur

Description: 48 cm (19 in). Large size. Dark brown upperparts with narrow buff bars and whitish underparts with fine brownish scaly bars diagnostic. Ear tufts and eyebrows brown with fine whitish vertical bars, contrasting with dark brown inverted triangular patch on forehead. Iris dark brown. **Voice:** A deep disyllabic *hu-hu*. Also a harsh gecko-like *gagagagogo*. Also a hooting scream. **Range:** Tenasserim (Myanmar) through the Malay Peninsula to the Greater Sundas and Bali. **Status and distribution:** Locally common resident from low elevations up to 1400 m at Fraser's Hill (Pahang), south to Johor. Has bred on Bukit Tunku, a wooded suburb of Kuala Lumpur in March 1989. Locally extinct in Singapore since 1925 (L and G); an individual at Bukit Timah Nature Reserve in October 1996 could be probably a stray from Johor. **Habitat:** Forests, the forest edge, overgrown plantations, heavily wooded suburbs of towns and cities. **Habits:** Usually frequents the middle storey. Emerges from daytime roost at dusk, flying low to locate prey which includes a variety of vertebrates. Usually hunts from perches. Frequently descends to the ground, often bathing in forest streams.

275. Dusky Eagle-owl, *Bubo coromandus* Plate 36

Burung Hantu Kelabu

Description: 53 cm (21 in). Large size. From Fish-owls (276, 277) by highly erect ear tufts when alert, feathered tarsi, darker and greyer colouration with less distinct dark streaks on underparts. Facial disc grey with fine shaft lines, brow and outer margin blackish, Bill greenish horn to blue-grey with blackish base. Upperparts uniform drab greyish-brown marked with narrow dark streaks; underparts paler greyish buff evenly marked with regularly spaced thin blackish vertical shaft streaks from base of chin to vent with finer horizontal blackish vermiculations, discernible at close range. Iris bright orange-yellow and eyelids pale pinkish, conspicuous when drooped; in Fish-owls, iris is bright pale yellow and eyelids are concolorous with facial disc) Legs lead grey with black claws. **Voice:** Typical territorial call is a series of deep *ko* notes, starting with a *kok* and gradually accelerating before tapering off abruptly, *kok, ko-ko-ko-kokokoko*…, recalling a locomotive puffing out of a station. A loud screech (Sharon Chan) and a wheezy throaty *ee-ee* discontinuous scream (Sharon Chan). **Range:** The Indian subcontinent and South-East China through western and central Myanmar to Tenasserim (Myanmar) to south-western Thailand and the Malay Peninsula. **Status and distribution:** Rare and localized resident from Ulu Muda (Kedah) south to Sekinchan (Selangor). Originally known from three skins, the last collected in May 1910 on Gunung Semanggol, Kerian District (Perak), this species was never seen for 88 years until a juvenile was sighted at the edge of peat swamp forest in Sekinchan (Selangor) on 6 December 1998. Subsequently four were sighted from 5 October–29 November 2001 in forest near Tasik Bukit Merah, Kerian District (Perak), not far from where the last known specimen was collected. Two were seen on 15 July 2004 in the Ulu Muda Forest Reserve (Kedah), bridging gaps in its distribution in Peninsular Malaysia. **Habitat:** Forests, including peat swamp. **Habits:** Usually solitary or in small groups. Has been observed using pylons as perches. Sometimes active during the day. Other habits little known.

276. Brown Fish-owl, *Ketupa zeylonensis* Plate 36

Ketuk Ketampi Rimba

Description: 53 cm (21 in). Large size. Brownish buff upperparts with narrow black shaft streaks on back. Has prominent whitish patch below throat. Underparts pale buff marked with narrow blackish vertical streaks and very fine indistinct brownish buff horizontal bars (usually indistinct in field). Iris yellow. **Voice:** A series of laughing hoots in a definite rhythm. Also a low hoarse scream. **Range:** Turkey through the Middle-East, the Indian subcontinent and southern China to South-East Asia and the Malay Peninsula. **Status and distribution:** Very rare and localized resident known only from Kedah and Perlis where it bred at least up to the 1940s. The last authentic sighting was of an individual in a forest patch in Chuping Estate (Perlis) in November 1983. With extensive forest clearance in both Kedah and Perlis, this species has probably greatly declined in the review area with no recent records. **Habitat:** Forests. **Habits:** Usually frequents the middle storey, often in the vicinity of small streams or watercourses. Feeds on a variety of small animal prey captured on the ground. Also catches fish from ponds and streams. Very vocal.

277. Buffy Fish-owl, *Ketupa ketupu* — Plate 36

Ketuk Ketampi

Description: 50 cm (18 in). Large size. From Brown Fish-owl (276) by darker brown upperparts with bolder black streaks on back and whitish bars on scapulars. Underparts warmer and richer buff with bolder vertical streaks and lacking the fine horizontal bars. Whitish patch below throat less conspicuous. Has prominent white patch on forehead above base of bill. Iris yellow. **Voice:** A repertoire of long drawn out shrill and wheezy notes. Also a soft monotonous and rapid *pup-pup-pup-pup-pup*..., uttered by a displaying pair. **Range:** North-East India east through Myanmar, southern Thailand, Indo-China south to the Malay Peninsula, the Greater Sundas and Bali. **Status and distribution:** Common and widespread resident at low elevations, south to Singapore where it is very rare. The most common large owl in open country habitats and rural settlements throughout lowlands in Peninsular Malaysia. **Habitat:** Mangroves, forested waterways, freshwater swamps, flooded ricefields, plantations, wooded suburban gardens. **Habits:** Usually frequents the middle storey or tall trees. Frequently seen flying low at dusk with moderate and regular wingbeats. Hunts from perches and often descends to the ground. Will also perch on the roofs of houses in rural settlements. Also catches fish from streams, ponds and flooded ricefields.

278. Spotted Wood-owl, *Strix seloputo* — Plate 36

Burung Hantu Berbintik

Description: 48 cm (19 in). Large size. Dark rufous brown upperparts boldly marked with white spots, pale rufous facial disc and the lack of ear tufts diagnostic. Shows pale line below chin. Underparts pale buff with bold blackish brown horizontal bars. Tail brown with narrow buff bars. **Voice:** A rising mellow *hoo*, uttered at an interval of about 5 seconds. Also a series of rapid *hoo* notes in a definite sequence. **Range:** Tenasserim (Myanmar), Thailand, Vietnam, and the Malay Peninsula discontinuously to Java and Palawan (the Philippines). **Status and distribution:** Common and widespread resident from low elevations up to 300 m, south to Singapore where it is the common large owl. Particularly frequent on oil-palm plantations throughout Peninsular Malaysia (not as common as Barn Owl (267)) where rodents form a great proportion of its diet. **Habitat:** Forests, the forest edge, plantations, wooded suburban gardens.

Habits: Frequents the middle storey or tall trees in suburban woodlands. Calls frequently after dusk, often in duets. In plantations, utilizes lamp posts or telegraph posts as vantage points to look for prey.

279. Brown Wood-owl, *Strix leptogrammica* Plate 36

Burung Hantu Rimba

Description: 55 cm (22 in). From Spotted Wood-Owl (278) by larger size, reddish brown upperparts and rufous facial disc encircled by reddish brown edge of face, giving a broad 'spectacled' effect. Dark smudges surrounding brown eyes create a 'big-eyed' effect. Underparts pale buff with finer and more compact horizontal brownish bars. Tail reddish brown with narrow buff bars. **Voice:** A distinct and deep disyllabic *hoo-hoo*. Sometimes utters a rapid four note *go-ko-kaloo*. **Range:** The Indian subcontinent and southern China through South-East Asia to the Greater Sundas. **Status and distribution:** Locally common resident from low elevations up to 1700 m, south to Johor. The common large owl at the hill-stations. **Habitat:** Forests. **Habits:** Frequents the middle storey where it often sits unobtrusively, its cryptic colour blending with the surroundings. Hunts for a variety of small animals, usually captured on the forest floor. Very vocal especially at dusk.

280. Collared Owlet, *Glaucidium brodiei* Plate 35

Pungguk Kerdil Bercekak

Description: 16 cm (6.5 in). Tiny size and round head together with buff hind collar and two black and buff patches on hindneck and neck, creating a false facial pattern diagnostic. Head greyish marked with fine white spots; neck collar white; upperparts brown with rufous buff bars; underparts pale buff with rufous buff bars. Iris yellow. **Voice:** A monotonous four note *ponk, ponk-ponk-ponk* with a brief pause after the first note. Uttered indefinitely over a long period. **Range:** The Himalayas and southern China through South-East Asia to Sumatra and Borneo. **Status and distribution:** Locally common resident from 700–1800 m in the Larut Hills (Perak) and the Titiwangsa Range south to Negeri Sembilan. Also on Gunung Tahan and Gunung Benom (Pahang) and Gunung Rabong (Kelantan). Frequently heard at the hill-stations. **Habitat:** Forests, the forest edge. **Habits:** Frequents the lower storey. Largely diurnal in habits and often calls during the day. Feeds principally on insects as well as small vertebrates and invertebrates. Has been observed capturing and devouring an adult tailorbird. Frequently cocks tail.

281. Brown Boobook, *Ninox scutulata* Plate 35

Pungguk

Description: 30 cm (12 in). Slim build and rounded head lacking facial disc together with proportionately long tail distinctive. Upperparts dark brown with small pale patch on base of bill between eyes. Underparts whitish with broad brown streaks. Tail dark brown with broad buff bars. Iris yellow. **Voice:** The territorial song of the resident, *N. s. scutulata* is a pleasing and persistently uttered disyllabic *hoo-wup*, the second note briefer and rising. The migrant *N. s. japonica* is usually silent in wintering quarters. **Range:** The Indian subcontinent and eastern Russia, China and Japan south to South-East Asia to the Greater and Lesser Sundas and the Philippines. Northern populations winter south to tropical latitudes. **Status and distribution:** Fairly common resident, passage migrant and winter visitor at low elevations, south to Singapore. Occurs above 900 m only during autumn and spring passage. **Habitat:** Mangroves, forests, the forest edge, mature plantations, wooded suburban gardens. **Habits:** Frequents the lower and middle storeys. Largely insectivorous, emerging at dusk from roost to feed on flying insects, often captured in mid-air. In wooded suburban areas, also feeds on insects attracted to street lights. Very vocal. **Note:** King (2002a) split *japonica* from *scutulata* based on differences in vocalizations in respective breeding areas, naming the new taxon Northern Hawk-owl or Northern Boobook. In the field both subspecies are virtually indistinguishable and DNA analysis will confirm if *japonica* is a full species.

282. Short-eared Owl, *Asio flammeus* Plate 35
Burung Hantu Sayap Panjang

Description: 38 cm (15 in). Long-winged with very short ear tufts, usually inconspicuous in the field. Upperparts rich tawny buff mottled with shades of buff and blackish. Underparts buff with blackish streaks, bolder on breast and sides of neck. Facial disc pale greyish white with conspicuous white eyebrows. Yellow eyes surrounded by blackish feathers, creating a 'black-eyed' effect. In upperwing flight pattern, shows conspicuous black patch on carpals. In underwing flight pattern, shows dark crescentic patch on carpals as well as blackish markings on trailing edge of otherwise all white underwing. **Voice:** Usually silent in wintering quarters. **Range:** Breeding in temperate North America, Europe and temperate Asia, Central and South America and islands of the Pacific; wintering south to Central America, Africa, South and South-East Asia to the Malay Peninsula. **Status and distribution:** Rare and localized winter visitor at low elevations, south to Singapore. **Habitat:** Scrub, open country, plantations. **Habits:** Assumes a more horizontal raptor like stance than other owls when perched. Feeds on a variety of small animals caught on the ground. Flight is wavering with rather floppy wingbeats, usually low over the ground when hunting. Sometimes flies about during the day.

FROGMOUTHS

ORDER *Caprimulgiformes* **FAMILY** *Podargidae*

World: *13 species* **Peninsular Malaysia:** *3 species*

A distinct group of nocturnal birds with soft and mottled cryptic plumage. Differ from nightjars (286–290) by heavier and broader bills, shorter and more rounded wings and an upright posture when perched. Have tuft of bristles on base of bill. Arboreal, inhabiting mainly the middle and lower storeys. Feed mainly on insects gleaned from branches. Do not hawk for insects on the wing. Very vocal during breeding; calls range from soft wheezy whistles to harsh tremulous rattles. Under normal viewing conditions, they are difficult to identify; the call notes which are distinct in many species are often the key to positive identification (see Appendix 1). Build small and soft padded nests on horizontal branches or in tree forks. Sexes alike or almost so in most species. Solitary. Two species are declining due to loss and fragmentation of lowland rainforest, but because of the ability to also use secondary forest, are ranked as **NT.** These are stated appropriately in each species account below.

283. Large Frogmouth, *Batrachostomus auritus* Plate 36
Burung Segan Besar

Description: 41 cm (16 in). Large size and heavy growth of compact bristles on sides of cheek, resembling flaps, separate it from the other two species. Upperparts chestnut with white double collar around hindneck; wing coverts spotted white; throat and breast rufous brown with white spots merging into buff belly. Tail buff with rufous brown bars; iris dark brown. When viewed directly from below, appears all puffed up with features of upperparts not visible. **Voice:** A loud, deep and hollow tremulous rattle, *kworrrrrr*, repeated up to 7–8 times. **Range:** The Malay Peninsula to Sumatra and Borneo. **NT. Status and distribution:** Uncommon and localized resident at low elevations, south to Johor. **Habitat:** Forests. **Habits:** Usually frequents the middle storey, gleaning insects from branches. Very vocal during breeding. Other habits little known.

284. Gould's Frogmouth, *Batrachostomus stellatus* Plate 36
Burung Segan Bintik Emas

Description: 25 cm (9.75 in). Sexes almost alike, differentiated by variation in colour. **Male:** Pale chestnut brown upperparts with white hind collar; throat, upper breast tawny with rest of underparts creamy buff, all marked with tawny rufous scales. Tail barred brown and buff. Iris yellow or brown. **Female:** Darker and more chestnut. **Voice:** A distinct but far carrying *whoo-wheeow*, with the second syllable longer and on a falling tone. **Range:** The Malay Peninsula, Sumatra and Java. **NT. Status and distribution:** Locally common resident at low elevations, south to Johor. Locally extinct in Singapore since 1895 (L and G) **Habitat:** Forests, the forest edge. **Habits:** Frequents the lower storey but will also ascend to middle storey level. Very territorial and vocal, often calling from a low perch.

285. Javan Frogmouth, *Batrachostomus javensis* Plate 36
Burung Segan Jawa

Description: 25 cm (9.75 in). From Gould's Frogmouth (284) by heavier growth of bristles above base of bill and behind eyes, forming superficial ear tufts. **Male:** Plumage sandy grey-brown above, more rufous below with white hind collar and markings on scapulars, breast and belly. Barrings on tail less conspicuous than Gould's Frogmouth. **Female:** Upperparts and underparts uniform rufous. From Gould's Frogmouth by white markings on breast and very fine inconspicuous barrings on tail. **Voice:** Utters a repertoire of tremulous whistled notes. Among the most frequently uttered is a short whistle, *wheeow*, or an ascending scale. Also utters a descending harsh and shrill *kerr-ker-ker* and also falling *gwaa* notes. **Range:** Myanmar and Thailand south through the Malay Peninsula to Sumatra, Java, and Borneo east to Palawan (the Philippines). **Status and distribution:** Locally common resident from low elevations up to 800m on the lower slopes of Genting Highlands (Pahang) (BiW), south to Johor. Probably the most commonly seen and heard frogmouth in the review area. **Habitat:** Forests, the forest edge, plantations. **Habits:** Frequents the middle and lower storeys. Habits like other frogmouths. Very vocal especially during breeding.

NIGHTJARS

ORDER *Caprimulgiformes* **FAMILY** *Caprimulgidae*

World: *88 species* **Peninsular Malaysia and Singapore:** *5 species*

Nocturnal insectivorous birds with flattened heads, short necks, long tails and long narrow wings. Plumage soft and beautifully mottled brown with shades of white, buff and grey. This cryptic colouring together with their typically crouched posture offer excellent camouflage amongst dead leaves and vegetation on the ground. Have large eyes. Bills small with large gapes and growth of rictal bristles. Tarsi short with small feet. Feed principally on insects caught on the wing. Flight hawk-like with moderate and erratic wingbeats interspersed with short glides. Very active at dawn and dusk. Calls diagnostic in each species (see Appendix 1). Largely terrestrial, roosting on the ground during the day with eyes closed to a slit. Do not build nests, laying eggs on the bare ground or in leaf litter. Will also perch lengthwise on horizontal branches. Most species are resident but one which breeds in northern temperate Asia migrates to the review area during winter. Solitary. Represented by the following groups and genera: **Eared-nightjars,** genus *Eurostopodus* (286, 287). Medium to large size with feathers of rear sides of crown elongated into conspicuous ear tufts on sides of head. These are seldom held erect and are usually held flat against head. Lack conspicuous rictal bristles around gape. Tarsi thickly feathered and feet stronger than typical nightjars. Show no white in wings and tail. Feed at great heights, usually above tall trees. Sexes alike. **Typical Nightjars,** genus *Caprimulgus* (288–290). Medium sized. Lack ear tufts. Have pronounced rictal bristles around gape. Legs weak, limiting movement on the ground; birds immediately take to flight when flushed. Wings and tail usually marked with white. All local species show some white markings on sides of throat. Show minor sexual dimorphism. Feed at moderate heights, usually below the level of tall trees.

286. Malaysian Eared-nightjar, *Eurostopodus temminckii* Plate 37

Taptibau Kecil

Description: 27 cm (10.5 in). At rest, medium size with erect ear tufts and conspicuous white collar diagnostic. Underparts whitish with narrow blackish-brown bars. Tail buff marked with indistinct dark bands. **Voice:** A diagnostic rich trisyllabic whistle, *tip-ti-tiew*, uttered in flight; the first note loudest followed by a brief pause before terminating in two closely spaced notes with the last note descending; a characteristic sound of the rainforest at dusk. **Range:** The Malay Peninsula, Sumatra and Borneo. **Status and distribution:** Fairly common resident from low elevations up to 800 m, south to Singapore where it uncommon, and confined to the central catchment forest, where population may not be self-sustaining in the long term (W and H). Fairly common at the lower elevations of the hill-stations, especially at The Gap (Selangor) below Fraser's Hill (Pahang). **Habitat:** Forests, the forest edge. **Habits:** Usually solitary or in pairs. Roosts on the forest floor during the day, emerging at dusk to feed over the forest. Engages in a rapid and erratic flight, constantly calling especially at dawn and dusk. Feeds very high in the air. Perches lengthwise on branches at night in between dusk and dawn feeding periods.

287. Great Eared-nightjar, *Eurostopodus macrotis* Plate 37
Taptibau Besar

Description: 41 cm (16 in). From Malaysian Eared-nightjar (286) at rest by much larger size and bulkier build. Ear tufts more erect and conspicuous with white neck collar and buff nuchal collar diagnostic. On the wing by buoyant harrier-like flight with much slower wingbeats. Underparts whitish with blackish-brown bars. Tail rich buff marked with conspicuous dark bands. **Voice:** A loud and distinct trisyllabic whistle *pit, whee-wheeeow*, uttered in flight; the first note sharp and low in volume (sometimes inaudible) followed by a brief pause, then a louder and longer second note before terminating in a much longer and slightly descending third note; the whole sequence louder and more prolonged than Malaysian Eared-nightjar. **Range:** The Indian subcontinent east to southern China and Myanmar through South-East Asia to the Malay Peninsula, Simeulue Island (Sumatra), Sulawesi and the Philippines. **Status and distribution:** Locally common resident at low elevations, recorded on the west coast from Chuping and Tasik Timah Tasoh (Perlis), Sungai Sedim Forest Reserve and Kota Kuala Muda (Kedah), Air Itam Forest Reserve and Juru, south to Bukit Panchor Forest Reserve, Seberang Prai (Pulau Pinang) (BiW). On the east coast, known from a single sighting at Kota Bharu (Kelantan). Fairly regular in the rice growing areas of northern Kedah and Perlis. **Habitat:** Mangroves, the forest edge, open country. **Habits:** Usually solitary or in pairs, sometimes in larger groups of three to four especially when feeding at dusk. Feeds high in the air, but usually lower than Malaysian Eared-Nightjar. Has also been observed quartering over dry ricefields (W1). Like the preceding species, calls persistently at dawn and dusk.

288. Grey Nightjar, *Caprimulgus indicus* Plate 37
Burung Tukang Kelabu

Description: 28 cm (11 in). Plumage darker and greyer than other nightjars. **Male:** Has small white wing patch and white subterminal spots on outer four pairs of tail feathers, conspicuous in flight when flushed during the day. **Female:** Like male but wing patches buff. **Voice:** A long series of rapidly uttered notes, *chook-chook-chook-chook*... given at rest. Seldom calls in wintering quarters. **Range:** Breeding in the Indian subcontinent and northern Myanmar, China, eastern Russia and Japan; populations from northern latitudes are migratory, wintering south through South-East Asia to the Greater Sundas, Sulawesi, and New Guinea. **Status and distribution:** Locally common passage migrant and winter visitor at low elevations up to 1300 m, south to Singapore, where it is rare. Fairly common at the higher elevations of the hill-stations during winter. **Habitat:** The forest edge, open country. **Habits:** Usually solitary, sometimes in pairs. Emerges at dusk to feed. Usually hunts higher than Large-tailed Nightjar (289). Seldom roosts on the ground, preferring to roost lengthwise on branches of medium height. Sometimes calls at dusk and dawn from a perch.

289. Large-tailed Nightjar, *Caprimulgus macrurus* Plate 37
Burung Tukang Ekor Besar/Burung Malas

Description: 30 cm (12 in). Plumage greyish brown marked with bold black streaks on scapulars. Has rusty buff nuchal collar. **Male:** Prominent white wing patches and broad white corners to tail. **Female:** Wing patch and corners to tail buff. **Voice:** A loud and resonant *chonk* delivered at rest and sometimes in flight, uttered at an interval of one second, usually preceded by a rapid purring *kru-kru-kru-kru*. Calling is seasonal and usually begins in September and stops in June. **Range:** The Indian subcontinent, Myanmar and southern China through South-East Asia, the Greater and Lesser Sundas, Sulawesi, and the Philippines, south to New Guinea and northern Australia. **Status and distribution:** Abundant and widespread resident at low elevations, south to Singapore. The most frequently seen and heard nightjar. **Habitat:** The forest edge, mangroves, open country, plantations, wooded gardens. **Habits:** Usually solitary or in pairs. Commonly seen hawking for insects at dusk and dawn. Often gathers to feed on the many insects gathering around street lights; its slow fluttery flight gives the impression of a huge moth rather than a bird. Very vocal, delivering its call from a perch; in many cases a fence post or an electric cable. When calling, puffs out throat and cocks tail. Often sits on roads where its slow take off at the approach of a vehicle sometimes results in a 'road kill'.

290. Savanna Nightjar, *Caprimulgus affinis* Plate 37
Burung Tukang Belukar

Description: 25 cm (10 in). Plumage greyish brown with fine blackish mottles on upperparts. From Large-tailed Nightjar (289) by slimmer appearance and proportionately shorter tail and lacking the bold black markings on the upperparts of the former. **Male:** From all nightjars by all white outer tail feathers. Has white patches on wing. **Female:** Entire tail greyish brown with buff wing patch. **Voice:** A harsh rasping high-pitched *chweeez*, uttered both in flight and at rest. **Range:** The Indian subcontinent east to southern China, South-East Asia discontinuously to the Greater and Lesser Sundas, Sulawesi and the Philippines. **Status and distribution:** Locally common resident from Singapore, north to Peninsular Malaysia. First seen in grassland in Tuas in May 1988, presumably colonists from the Riau Archipelago. Common in the coastal scrub belt of Teluk Mahkota (Johor), especially around Sedili Kecil. With the construction of the new North-South Highway and the subsequent creation of open grassy areas which this species favours, it has rapidly expanded its range north into Peninsular Malaysia along both coastal plains. Seen at Pagoh (Johor) in May 1991 and at Ampang Pecah, near Kuala Kubu Bahru (Selangor) in September 1996. The northernmost limit is currently Chuping (Perlis) (BiW). **Habitat:** Scrub, open country, grassland with open stoney patches (favourite nesting sites). **Habits:** Usually solitary but small parties of threes and fours may hawk for insects together. Very vocal in flight, which is swifter and more erratic than Large-tailed Nightjar. Often sits on tracks at dusk.

TREESWIFTS

ORDER *Apodiformes* **FAMILY** *Hemiprocnidae*

World: *4 species* **Peninsular Malaysia and Singapore:** *2 species*

Closely related to true swifts (293–304) but differ in being brighter coloured and less strictly aerial in habits. Often perch on exposed branches and wires, from which they sally forth after flying insects. When perched, slender and longish appearance distinctive; the long wingtips crossed and the tails deeply forked. In flight, blunt head, long crescentic shaped wings together with long tail feathers held closely together like a spine distinctive. Fly with rapid wingbeats, gliding intermittently. Gregarious. Sexes differ slightly.

291. Grey-rumped Treeswift, *Hemiprocne longipennis* Plate 38

Lelayang Pokok Berjambul

Description: 23 cm (9 in). Crest prominent when erect. **Male:** Dark blue-green crown, nape and mantle contrasting with greyish rump diagnostic. Ear coverts chestnut. Throat and breast greyish merging into whitish centre of belly. Patch on tertials pale grey. Flight feathers grey-brown; underwing coverts dark blue-green. **Female:** Like male but lacks chestnut ear coverts. **Voice:** A shrill high pitched chatter, *pi-pi-peet, pi-pi-peet,* uttered in flight. **Range:** Tenasserim (Myanmar) to the Greater Sundas, Philippines, and Sulawesi. **Status and distribution:** Locally common resident from low elevations up to 1200 m, south to Singapore. **Habitat:** Mangroves, the forest edge, plantations, wooded gardens. **Habits:** Frequents the tree tops in small noisy parties. Sometimes gathers in larger parties of up to 50. Often perches on electricity cables. Swoops low over clearings in pursuit of insects.

292. Whiskered Treeswift, *Hemiprocne comata* Plate 38

Lelayang Pokok Kecil

Description: 15 cm (6 in). From Grey-rumped Treeswift (291) by smaller and more compact size. Crest indistinct, just above nape. **Male:** Bronzy brown plumage, glossy blue-black head, wings and tail together with two narrow horizontal white facial stripes diagnostic. Patch on tertials white. Ear coverts chestnut. **Female:** Like male but lacks chestnut ear coverts. **Voice:** Similar to Grey-rumped but thinner and higher pitched. **Range:** Tenasserim (Myanmar) to Sumatra, Borneo and the Philippines. **Status and distribution:** Fairly common resident from low elevations up to 1110 m, south to Johor. Scarce non-breeding dispersant to Singapore. **Habitat:** Forests, the forest edge. **Habits:** Frequents the canopy in small parties. Makes longer aerial swoops than Grey-rumped Treeswift Keeps more to the forest and shuns open habited areas. Often perches on the tips of high bamboo stands at the forest edge.

SWIFTS

ORDER *Apodiformes* **FAMILY** *Apodidae*

World: *80 species* **Peninsular Malaysia and Singapore:** *12 species*

Aerial feeding insectivorous birds with distinct cigar-shaped bodies. Superficially resemble swallows and martins (446–452) but differ by longer and proportionately narrower sickle-shaped wings. Flight consists of rapid and stiff wingbeats interspersed by long glides. Differ from treeswifts (291, 292) by inability to perch; their short claws are only adapted to cling onto vertical surfaces. Calls vary from chitters and burbling notes to loud and high-pitched screaming trills. Highly gregarious, nesting in loose or dense colonies on cliffs and buildings. Most species are clad in combinations of black, brown and white and are especially difficult to identify when gliding at great heights. Best identified by distinct wing and tail shapes as well as flight pattern especially when they descend during bad weather or skim low over water bodies to drink. Sexes alike. Represented by the following groups and genera: **Swiftlets,** genera *Hydrochous* (293), *Collocalia* (294), and *Aerodramus* (295–297). Medium to small size with rapid fluttery flight on slender wings. Tails slightly notched. Nest in large colonies in caves and sometimes buildings. Nests are attached to vertical cave walls and are small cup-shaped structures of largely hardened saliva. The genus *Aerodramus* utilizes echolocation, uttering high pitched clicks and rattles (audible to human ears) to orientate in dark caves and in some species, the nest material provides ingredients for birds' nest soup. **Spinetail,** genus *Rhaphidura* (298). Smaller than Needletails with a slower and more fluttery flight on long broad wings which taper at tip. **Needletails,** genus *Hirundapus* (299–301) Heavy-bodied with short squarish tails with needle-like projections (usually inconspicuous in the field). Larger and broader winged than other swifts. Underparts generally dark brown with conspicuous broad whitish 'V'-shaped patch on vent and flanks. Among the fastest flying birds, flight is very swift and powerful on broad scythe-shaped wings. Nest in tree hollows. **Typical Swifts,** genus *Apus* (303, 304). Medium sized and heavy-bodied with deeply forked or notched tails. Flight fast with swift beats of long, tapered wings. Nest on cliff faces or on buildings. **Palm-swifts,** genus *Cypsiurus* (302). Resemble swiftlets but tail narrower and deeply forked, wings generally narrower and more sickle-shaped and flight more direct and less fluttery. Roost and nest in fan-palms.

293. Waterfall Swift, *Hydrochous gigas* Plate 38

Lelayang Hitam

Description: 16 cm (6.5 in). From other smaller swiftlets by larger size and notched tail. Upperparts black; underparts dark brown. **Voice:** Sharp chittering notes. **Range:** The Malay Peninsula and the Greater Sundas. Its small and disjunct population may be declining due to loss and degradation of forests within its range. **Status and distribution:** Very rare resident, originally known from a specimen collected at The Gap (Selangor) in 1900. In recent times, confirmed sightings are known only from the Titiwangsa Range between approximately 800–1500 m (W1), from Cameron Highlands and Fraser's Hill, south to Genting Sempah (Pahang-Selangor border). Sight records from the lowlands must be treated with caution. **Habitat:** Forests. **Habits:** Generally flies higher than other swiftlets. Relatively large eyes suggests it probably has acute vision in poor light and is to an extent crepuscular. Nests behind waterfalls. Other habits little known.

294. Glossy Swiftlet, *Collocalia esculenta* Plate 38

Lelayang Perut Putih

Description: 10 cm (4 in). The smallest swiftlet. Glossy blue-black upperparts, dark greyish throat and breast together with white belly diagnostic. **Voice:** Burbling calls and twittering chirrups at nesting colonies. **Range:** The Andamans and Tenasserim (Myanmar) through South-East Asia and the Greater Sundas east to the Philippines and south to Christmas Island, New Guinea, northern Australia and the South-West Pacific to New Hebrides. **Status and distribution:** Fairly common resident from sea level to the summit of the highest mountains, south to Singapore where it is very rare. Very common at the hill-stations. **Habitat:** Feeds over forests, the forest edge, open country, towns and cities. **Habits:** Feeds in small flocks, generally lower than the larger swifts. Often skims over water bodies to drink. Nests in small colonies under the eaves of occupied buildings, in uninhabited buildings, under culverts and in cave mouths.

295. Himalayan Swiftlet, *Aerodramus brevirostris* Plate 38

Lelayang Himalaya

Description: 13 cm (5 in). From Edible Nest Swiftlet (297) by greyish to all blackish brown rump (like rest of upperparts). From Black-nest Swiftlet (296) by more slender body and wings and in hand, by unfeathered to lightly feathered tarsi. In flight from both Edible Nest and Black-nest Swiftlets by deeper fork in tail and longer wings which are slightly swept back. **Voice:** Clicking and rattling echolocation calls similar to Edible and Black-nest Swiftlets. **Range:** Breeding in the Himalayas and southern China discontinuously through South-East Asia to Sumatra, Java and the Philippines; some populations are migratory. **Status and distribution:** Uncommon and localized passage migrant and winter visitor from low elevations to the summit of mountains, south to Johor, with unconfirmed records from Singapore. **Habitat:** Feeds over forests, the forest edge and open country. **Habits:** Feeds in small flocks. General habits similar to other swiftlets.

296. Black-nest Swiftlet, *Aerodramus maximus* Plate 38

Lelayang Sarang Hitam

Description: 14 cm (5.5 in). Inseparable in field from Edible-nest Swiftlet (297) but is slightly stockier with longer and broader wings. This is apparent when both species share a common nesting site. Tail slightly notched. In hand from both Edible-nest Swiftlet and Himalayan Swiftlets (295) by densely feathered tarsi. Colour of rump ranges from greyish to all blackish brown **(not illustrated).** Nest is white but mixed with feathers, giving it a blackish appearance; another good field characteristic in identification when both species nest together. The nest is edible but of lower commercial value than Edible-nest Swiftlet. **Voice:** Shrill chirrups and burbling notes at nesting colonies. Echolocation call is similar to Edible-nest Swiftlet but is louder (apparent when heard together). **Range:** The Himalayas through Myanmar, Thailand and the Malay Peninsula to the Greater Sundas. **Status and distribution:** Fairly common resident from sea level to the summit of mountains, south to Singapore, where it nests in disused bunkers on Sentosa Island. Large breeding colonies exist on Pulau Redang (Terengganu) south to the Tioman Archipelago (Pahang) where breeding is often disrupted by the periodic harvesting of nests. **Habitat:** Feeds over forests, the forest edge and open country. **Habits:** Generally similar to Edible-nest Swiftlet.

297. Edible-nest Swiftlet, *Aerodramus fuciphaga* Plate 38
Layang-Layang Gua

Description: 14 cm (5.5 in). Small swiftlet with blackish brown upperparts with colour of rump ranging from whitish to all blackish brown **(not illustrated).** In the field, virtually inseparable from Black-nest (296) and Himalayan Swiftlets (295) but is generally shorter winged with a more rapid flight. Tail slightly notched. In hand, tarsi unfeathered or lightly feathered. The nest is opaque and white in colour and is edible (see Black-nest Swiftlet). **Voice:** High-pitched chirrups and burbling notes at breeding colonies. When echolocating, utters sharp clicking notes in succession, forming a rattle-like call (M and W). **Range:** The Andamans and the coasts of South-East Asia and the Indonesian Archipelago. **Status and distribution:** Fairly common resident at low elevations up to 1280 m, south to Singapore, where it nests in the city and in disused bunkers on Sentosa Island. Breeds principally in coastal districts especially in limestone caves and rock crevices. The only known inland nesting colony is the Kuala Lumpur clock tower, which it shares with Black-nest Swiftlet. **Habitat:** Feeds above forests, the forest edge and open country. **Habits:** Feeds in small flocks. Has a less fluttery flight and flies higher than Glossy Swiftlet (294). Wingbeats more powerful and stiffer. Skims low over freshwater bodies to drink.

298. Silver-rumped Spinetail, *Rhaphidura leucopygialis* Plate 38
Lelayang Ekor Pendek

Description: 11 cm (4.5 in). Glossy black plumage with white rump and entire uppertail coverts (giving a 'tailless' impression) together with long, broad wings which gradually taper at tips (resembling a butter knife) diagnostic. Has a distinct flight silhouette with fluttery wingbeats like a bat. The smaller swiftlets (294–297) have more slender wings and darker rumps. **Voice:** A series of high-pitched notes, *tirrr, tirrr*, uttered by feeding flocks. **Range:** Tenasserim (Myanmar) through the Malay Peninsula to the Greater Sundas. **Status and distribution:** Fairly common resident from low elevations up to 1200 m, south to Johor. Not seen in Singapore since 1950. **Habitat:** Feeds over forests and the forest edge. **Habits:** Flies low over forests in small flocks, often within the vicinity of rivers and streams.

299. White-throated Needletail, *Hirundapus caudacutus* Plate 38
Lelayang Leher Putih

Description: 20 cm (8 in). Conspicuous white throat, sharply demarcated from brown breast and belly diagnostic. Has small white patch on side of neck. In upperwing flight pattern, shows conspicuous silvery white patch on back together with small white patch on tertials. Upperparts glossy-black. **Voice:** High-pitched chittering notes. **Range:** Breeding in the Himalayas, far eastern Russia, northern China, Korea and Japan; wintering south through South-East Asia and the Indonesian Archipelago to Australia. **Status and distribution:** Scarce passage migrant from low elevations up to 1600 m, south to Singapore with a single sighting from Bukit Timah. **Habitat:** Feeds over forests, the forest edge and open country. **Habits:** Similar to other Needletails.

300. Silver-backed Needletail, *Hirundapus cochinchinensis* Plate 38
Lelayang Rengkong Putih

Description: 20 cm (8 in). From White-throated Needletail (299) by greyish throat and the lack of small white patch on tertials. From Brown-backed Needletail (301) by smaller size and silvery white patch on back. Upperparts glossy-black. **Voice:** Unknown. **Range:** The Himalayas and North-East India through South-East Asia to Sumatra and Java. **Status and distribution:** Uncommon and localized passage migrant and winter visitor from low elevations to at least 1500 m, south to Singapore. **Habitat:** Feeds over forests, the forest edge and open country **Habits:** Flies fast in small flocks. Other habits similar to Brown-backed Needletail (301).

301. Brown-backed Needletail, *Hirundapus giganteus* Plate 38
Lelayang Besar

Description: 25 cm (10 in). From both White-throated (299) and Silver-backed Needletails (300) by larger size, white lores (in migrant subspecies *H. g. indica* **(not illustrated)**) and bronzy brown patch on back, conspicuous in upperwing flight pattern. Upperparts glossy blue-black. The resident subspecies *H. g. giganteus* has dark lores. **Voice:** Brittle chatters and squeaking notes **Range:** Breeding in the Indian subcontinent through South-East Asia to the Greater Sundas, Sulawesi and the Philippines. Northern populations are migratory; wintering south to the Indonesian Archipelago. **Status and distribution:** Fairly common resident, passage migrant and winter visitor from low elevations to the summit of mountains, south to Singapore. **Habitat:** Feeds over forests and the forest edge. **Habits:** Usually in small flocks, flying very fast with audible swooshing of wingbeats. Often flies low over forest streams to drink and bathe.

302. Asian Palm-swift, *Cypsiurus balasiensis* Plate 38
Lelayang Pokok Palma

Description: 13 cm (5 in). Small swift with dark sooty brown plumage (appears all black in field). Tail deeply forked when open but when closed, appears pointed and spine shaped. From treeswifts (291, 292) by much smaller size, darker plumage and proportionately shorter tail. **Voice:** Shrill high-pitched screaming chatters, often in roosting and nesting sites. **Range:** The Indian subcontinent and southern China through South-East Asia and the Malay Peninsula to the Greater Sundas, Bali and the Philippines. **Status and distribution:** Fairly common resident from low elevations up to 1200 m, south to Singapore. Distribution largely governed by availability of fan-palms on which it is highly dependent on for roosting and nesting. **Habitat:** Open country, forests, the forest edge, plantations. **Habits:** Usually in small groups, feeding over a variety of habitats. Flies in and out of fan-palms, often near human habitations. Builds tiny cup-shaped nests on surface of drooping palm fronds.

303. Fork-tailed Swift, *Apus pacificus* — Plate 38
Lelayang Ekor Cabang

Description: 19 cm (7.5 in). Deeply forked tail (also visible when folded) and long narrow crescentic shaped wings distinctive. Plumage blackish with narrow curved white band on rump (sometimes inconspicuous). Throat whitish with fine blackish streaks. Breast and belly have whitish scalings and barrings (discernible only at close range). **Voice:** A series of distinct long and high-pitched squeaking notes. Also twittering notes. **Range:** Breeding in North-East Asia, the Himalayas and northern Myanmar; wintering throughout South-East Asia through the Indonesian Archipelago to New Guinea and Australia. **Status and distribution:** Locally common passage migrant from low elevations to summits of high mountains, south to Singapore. Main autumn migration route follows the western coastal plain from Kedah to Negeri Sembilan, where migrating flocks converge at Tanjung Tuan before crossing the Straits of Melaka. **Habitat:** Feeds over forests, the forest edge and open country **Habits:** Hunts in flocks high in the air, often in association with other swifts. Engages in flutters and turns when pursuing insects.

304. House Swift, *Apus affinis* — Plate 38
Lelayang Rumah

Description: 15 cm (6 in). From Fork-tailed Swift (303) by smaller size, slightly notched tail, white throat and larger square white rump patch. **Voice:** A high-pitched screaming stacatto in flight and at roost. Also a high-pitched *kik-kik-kik-kik*...when flying in to roost. **Range** Africa, southern part of the Iberian Peninsula through the Middle-East, the Indian subcontinent, southern China and South-East Asia to the Greater Sundas and the Philippines. **Status and distribution:** Abundant resident from sea-level to the summit of the highest mountains, south to Singapore. The most common swift. Distribution has expanded greatly in the twentieth century due to the increased number of buildings in urban and suburban settlements but in recent years have declined sharply (W2). **Habitat:** Feeds over open country, forests, the forest edge, towns and cities. **Habits:** Highly gregarious, feeding in small noisy flocks high in the air over a variety of habitats. A known commensal of man, it breeds in fairly large colonies under eaves of buildings and under bridges in busy towns and cities. Under natural conditions, nests in cave mouths or in crevices on cliff faces. Very noisy and conspicuous especially in the evenings when flocks come in to roost. Often calls intermittently at night.

TROGONS

ORDER *Trogoniformes* **FAMILY** *Trogonidae*

World: *42 species* **Peninsular Malaysia:** *6 species*

Brightly coloured, characterized by soft fluffy plumage, long square-cut tails and short rounded wings. Broad, stubby bills and bluish or purplish orbital skins conspicuous in field. Forest inhabitants, preferring close-canopy forest, they are often unobtrusive, perching on branches and vines with a characteristic upright posture. Presence often betrayed by territorial calls which are distinctive, consisting of soft plaintive hoots or rattling trills. All species also utter similar whirring notes. Sexually dimorphic, males are brighter than females. In addition the vermiculations on the wings are whitish in males and buff in females. Principally insectivores, catching prey on short flights from perches or gleaning them from foliage. May also take fruit. Solitary or in pairs. Locally extinct in Singapore (L and G). Most species in the review area have declined greatly over the years due to loss and fragmentation of close-canopy lowland rainforest, and in global terms rank as **NT.** This is stated appropriately in each species account below.

305. Red-naped Trogon, *Harpactes kasumba* Plate 39

Kesumba Tengkuk Merah

Description: 32 cm (12.5 in). **Male:** Head, throat and upper breast black; orbital skin deep blue; nuchal patch and underparts bright red; crescent across breast white. **Female:** Throat and breast dark brownish grey contrasting sharply with tawny brown underparts. **Juvenile (not illustrated):** Both sexes like female, the male soon attaining adult plumage. **Voice:** Territorial call of male is a series of 5–6 evenly spaced *kau* notes on an even pitch, *kau-kau-kau-kau...* Call of female slightly higher pitched and more reedy (L and R). **Range:** The Malay Peninsula, Sumatra and Borneo. **NT. Status and distribution:** Locally common resident from low elevations up to 600 m, south to Johor. Locally extinct in Singapore since 1921 (L and G). **Habitat:** Forests, including peatswamp. **Habits:** Frequents the middle storey, often perching motionless for long periods. Presence most often detected by unmistakable call. Flies off to another perch when disturbed.

306. Diard's Trogon, *Harpactes diardii* Plate 39

Kesumba Bukit

Description: 30 cm (12 in) **Male:** Similar to male Red-naped Trogon (305) but distinguished by purple (not deep blue) orbital skin; fine black vermiculations on greyish white underside of retrices and less contrasting pinkish crescent across breast. Hindcrown brownish maroon; nuchal patch and underparts red. **Female:** Head and breast brownish, contrasting with pinkish red underparts. **Juvenile (not illustrated):** Both sexes like female, the male soon attaining adult plumage. **Voice:** A series of 10–12 *kau* notes on a descending scale, faster and higher pitched than Red-naped Trogon. Uttered by both sexes. **Range:** The Malay Peninsula, Sumatra and Borneo. **NT. Status and distribution:** Fairly common resident from low elevations up to 900 m, south to

Johor. Locally extinct in Singapore since 1921 (L and G). **Habitat:** Forests **Habits:** Occupies very much the same niche as Red-naped Trogon and shares similar habits.

307. Cinnamon-rumped Trogon, *Harpactes orrhophaeus* Plate 39
Kesumba Pinggul Perang

Description: 25 cm (10 in). **Male:** From Scarlet-rumped Trogon (308) by more thick-set bill and cinnamon rump which is concolorous with mantle and uppertail coverts. Underparts red. Black of head does not extend down to breast. **Female:** Sides of head and underparts rusty buff. Orbital skin above eye pale blue in both sexes. **Juvenile (not illustrated):** Like female but pale vermiculations on wing much broader. **Voice:** Territorial call of male is a cadence of 3–4 soft *taup* notes on a descending scale. **Range:** The Malay Peninsula, Sumatra and Borneo. **NT. Status and distribution:** Uncommon and localized resident at low elevations up to 180 m, south to Johor. **Habitat:** Forests. **Habits:** Very shy and seen less frequently than the others, it frequents the middle and lower storeys where it often perches lower than other trogons.

308. Scarlet-rumped Trogon, *Harpactes duvaucelii* Plate 39
Kesumba Pinggul Merah

Description: 25 cm (10 in). **Male:** Black of head does not extend down to breast. Rump bright red, contrasting sharply with rusty buff mantle and uppertail coverts. **Female:** Head brownish; rusty buff breast grading into reddish pink underparts; rump reddish pink. Orbital skin above eye brilliant pale blue in both sexes. **Juvenile (not illustrated):** Both sexes like female but lack the characteristic pink in the plumage. Males soon acquire adult plumage. **Voice:** Territorial call of male is an accelerating high pitched rattling cadence of 12 notes; the first two level and the rest rapidly descending. **Range:** The Malay Peninsula, Sumatra and Borneo. **NT. Status and distribution:** Probably the most common trogon in review area. Fairly common resident from low elevations up to 1000 m, south to Johor. **Habitat:** Forests. **Habits:** Frequents the middle storey where it is most often detected by its characteristic call. Sometimes seen in pairs.

309. Orange-breasted Trogon, *Harpactes oreskios* Plate 39
Kesumba Dada Jingga/Kesumba Harimau

Description: 30 cm (12 in). From all other trogons by orange-yellow underparts and the lack of any red in the plumage. **Male:** Head and throat olive green; mantle chestnut; breast bright orange grading into orange-yellow underparts. **Female:** Mantle olive-green; olive-green head and throat duller; orange-yellow underparts paler. Orbital skin bluish in both sexes. **Juvenile (not illustrated):** Male resembles juvenile female but has chestnut mantle, scapulars, and back. Juvenile female rufous-chestnut tinge to head, upperparts, and breast. **Voice:** Territorial call of male is a series of 3–5 rapidly uttered *tiu* notes on a level pitch; *tiu-tiu-tiu-tiu-tiu*. **Range:** South-West China through South-East Asia to Sumatra, Java and Borneo. **Status and distribution:** Locally common resident in hilly country from 60–1200 m, south to Endau-Rompin (Pahang/Johor). Fairly common at the lower elevations of the hill-stations. **Habitat:**

Forests. **Habits:** Frequents the middle storey, singly or in pairs, usually perched motionless, waiting for insect prey. Sometimes seen along roads where it is particularly conspicuous.

310. Red-headed Trogon, *Harpactes erythrocephalus* Plate 39
Kesumba Kepala Merah

Description: 34 cm (13.5 in). Large and robust. **Male:** From all other trogons by red head. **Female:** Head and breast brownish cinnamon which is concolorous with mantle. Orbital skin bluish in both sexes. Both sexes have broad white crescent across breast. **Juvenile (not illustrated):** Both sexes resemble female but head, breast, and upperparts rufescent, the male soon attaining the red head and upper breast. **Voice:** Territorial call of male is a cadence of 5–6 widely spaced mellow notes on a descending scale; *tiaup…tiaup…tiaup…* Also a low rattling, *twirrrr*. **Range:** The eastern Himalayas and southern China through South-East Asia to Sumatra. **Status and distribution:** Locally common resident above 700 m in the Larut Hills (Perak) and the Titiwangsa Range south to Negeri Sembilan. Also on Gunung Benom (Pahang). Fairly common at the lower and upper elevations of the hill-stations. **Habitat:** Forests. **Habits:** Frequents the middle storey. Most often seen in typical trogon fashion, quietly perched and peering over its back at an intruder. Very conspicuous when it appears along roads. Sometimes participates in bird waves.

ROLLERS

ORDER *Coraciiformes* **FAMILY** *Coraciidae*

World: *12 species* **Peninsular Malaysia and Singapore:** *2 species*

Thickset with large heads, robust bills, long broad wings and relatively short tails. Flight strong but leisurely. Indulge in acrobatic rolling display flights during courtship; hence their name. Utilize exposed perches as vantage points to hunt insects and small vertebrates by either dropping to the ground or by making aerial sallies. Calls mainly harsh croaking notes. Solitary or in pairs. Sexes alike.

311. Indian Roller, *Coracias benghalensis* Plate 41
Tiong Gajah

Description: 33 cm (13 in). Crown turquoise; nape and mantle greenish-olive tinged brown, sides of head, breast and upper belly vinous brown with pale streaks on the former; iris dark with narrow yellow orbital skin; bare skin around eye yellow; bill black with thick bristles at base; patch of deep purple-blue on scapulars of folded wings; lower belly and undertail coverts pale blue; rump pale blue, very conspicuous in flight. Appears all brown at rest in poor light. **Juvenile (not illustrated)** Like adult but browner above including forehead and crown; the turquoise being confined to the supercilium; sides of head and underparts paler and browner. **Voice:** A single harsh retching *kyak*. **Range:** The Persian Gulf region through the Indian subcontinent, Sri Lanka and islands south to the Maldives, southern China and continental South-East Asia to the Malay Peninsula. **Status and distribution:** In Peninsular Malaysia, locally common to uncommon resident at low elevations on the east coast from Kelantan and Terengganu, south to Beserah, Kuantan District (Pahang) (BiW). Particularly common at the Setiu estuary (Terengganu) and in ricefields with stands of coconut palm between Tok Bali (Kelantan) and Besut (Terengganu). On the west coast, distribution more sporadic, occurring at Chuping (Perlis), Sungai Petani (Kedah), Balik Pulau, Seberang Prai (Pulau Pinang), and in Perak, from Kamunting, Taiping, and along the open scrub and oil palm plantations between Kuala Gula and Kuala Kurau, south to Tanjung Buloh, Teluk Intan, with a single sighting there on 29 January 2009. Also an inland record along the Sungai Tembeling between Kuala Tahan and Kuala Kenyam (Pahang) on 27 July, 1996 (BiW). **Habitat:** Beach scrub, casuarinas, margins of ricefields, coconut groves, plantations. **Habits:** Frequently utilizes telephone or electricity wires, the tops of dead coconut palms, or low branches as vantage points to survey the ground for prey. Drops to the ground to grab large grasshoppers and small vertebrates. Will also take winged insects on aerial sallies.

312. Dollarbird, *Eurystomus orientalis* Plate 41
Tiong Batu

Description: 30 cm (12 in). At rest from Indian Roller (311) (where ranges overlap) by shorter and thicker orange-red bill together with dark greenish-blue plumage. Head and mantle dark brown, throat and flight feathers purplish-blue, iris hazel with narrow red eyering, legs red. Broad silvery turquoise patch on base of primaries conspicuous only in flight. Appears longer-winged than Indian Roller with more leisurely wingbeats. At rest appears all black in poor light. **Juvenile (not illustrated):** Duller with flesh coloured bill. **Voice:** Harsh, scratchy notes, *kiak-kiak* uttered especially in flight, often in rapid sequence. **Range:** The Indian subcontinent, temperate eastern Russia, China and Japan through South-East Asia to Australia and the Solomons; northern populations migrate to tropical latitudes during winter. **Status and distribution:** Fairly common resident, passage migrant and winter visitor throughout Peninsular Malaysia. Occurs above 900 m only during spring and autumn passage. In Singapore, rare resident and common winter visitor. **Habitat:** Mangroves, forest edge, beach scrub, plantations, open country. **Habits:** Particularly favours tall dead trees on which it often perches for long periods. Does not drop to the ground to catch prey, preferring to catch winged insects on aerial sallies. Very active when ant and termite swarms are present, sallying after them well into dusk. Usually solitary or in pairs but small groups may be encountered on passage.

KINGFISHERS

ORDER *Coraciiformes* **FAMILY** *Alcedinidae*

World: *95 species* **Peninsular Malaysia and Singapore:** *12 species*

A distinct group with proportionately large heads, long pointed bills, short legs and medium length to very short tails. Many are brilliantly coloured. Sit with a characteristic upright stance on bare perches overlooking water or open ground. Nest in horizontal burrows in steep earth banks or termite nests. Solitary or in pairs. Sexes alike in most species. Represented by the following groups and genera: **Wood Kingfishers:** Medium to large size. Most are rather stocky with long and massive pointed bills. Generally found close to water bodies but some are forest inhabitants, living far from water. Most feed on an assortment of vertebrates and invertebrates such as frogs, lizards and insects which are mostly captured on the ground while those which inhabit mangroves may also take fish and crabs. Flight swift and direct. Very vocal, calls ranging from plaintive cadences to harsh discordant shrieks or cackles. May be further divided into the following subgroups. **Forest Kingfishers**, genera *Actenoides* (313) and the monotypic *Lacedo* (314), inhabit forests and are not particularly attached to water. **Storkbills**, genus *Pelargopsis*, (315, 316) are large with orange-rufous underparts and massive red bills, inhabiting both freshwater and brackish habitats. **Typical Woodland Kingfishers**, genus *Halcyon* (317–319) are smaller than storkbills and inhabit forest, scrub and mangrove edges. All are colourful with red bills. Not particularly attached to water. Mangrove Kingfishers, genus *Todiramphus* (320) have their centre of distribution in Australasia and the western Pacific. Mostly blue upperparts, contrasting with white underparts. Inhabits mangroves and beach scrub. **Dwarf and River Kingfishers,** differ from wood kingfishers by smaller size, slender bodies, much shorter tails, and proportionately longer and more slender bills. Calls are thin, shrill squeaks. **Dwarf kingfishers**, genus *Ceyx* (321) are tiny and predominantly rufous above and yellow below, feeding principally on insects and other small animals, often far away from water. **River kingfishers**, genus *Alcedo* (322–324) are larger than the former and are mostly brilliant blue above with rufous underparts. Flight is swift and direct, low over the water surface. Typically sit upright on waterside perches and catch fish and other aquatic animals by plunge-diving.

313. Rufous-collared Kingfisher, *Actenoides concretus* Plate 40
Pekaka Leher Perang

Description: 24 cm (9.5 in). Rufous collar and underparts, green crown, dark blue malar stripe and yellow bill diagnostic. **Male:** Mantle dark blue. **Female:** Mantle green with buff spots. Both sexes show a bright blue line down lower back and rump, conspicuous especially in flight. **Juvenile (not illustrated):** Duller, male wing coverts and scapulars buff-spotted, female upperparts darker. **Voice:** A rising whistle, *kwi-i, kwi-i*...repeated at a rate of approximately one note every second. **Range:** Tenasserim (Myanmar) through the Malay Peninsula to Sumatra and Borneo. Being dependent on close canopy forests, it is declining owing to deforestation. **NT. Status and distribution:** Uncommon and localized resident at low elevations up to 1300 m, south to Johor. Locally extinct in Singapore circa 1938 (W and H). **Habitat:** Forests **Habits:** Generally unobtrusive, sitting motionless for long periods on a shaded perch. Usually frequents the middle and lower storeys, often far from water. Feeds largely on insects and small invertebrates.

314. Banded Kingfisher, *Lacedo pulchella* Plate 40
Pekaka Berjalur/Kaing-Kaing Bukit

Description: 23 cm (9 in). Has distinctive silhouette with relatively long tail and short broad slightly uptilted orange-red bill. **Male:** Upperparts black with blue and white bars; throat white; bill red; forehead and sides of head chestnut; underparts white with pale rufous buff wash on breast and flanks. **Female:** Upperparts rufous with black bars; underparts white with fine black bars on breast and flanks. **Voice:** Territorial call of male is a long cadence of up to 15 *chi-wiu* notes, with increasing stress on the first syllable until it gradually dies away. Also a sharp *wiak wiak* (W1). **Range:** Myanmar through Indo-China and the Malay Peninsula to the Greater Sundas. **Status and distribution:** Fairly common resident in the lowlands and hills up to 900 m, south to Johor. Also on Pulau Langkawi (Kedah) Locally extinct in Singapore since 1950 (L and G). **Habitat:** Forests. **Habits:** Frequents the middle storey where it hunts for small vertebrates and invertebrates. Will also probe on the forest floor and forage for grubs among fallen wood. Often perches motionless for long periods. When alarmed, long crown and nape feathers are slowly erected and retracted. Very vocal especially at dawn and dusk. Shy but sometimes allows close approach. Not particularly attached to water.

315. Brown-winged Storkbill, *Pelargopsis amauroptera* Plate 40

Pekaka Emas Bakau/Burung Buaya

Description: 36 cm (14 in). Large size with huge coral-red bill with dark tip distinctive. From Common Storkbill (316) by diagnostic dark chocolate brown mantle, wings and tail contrasting with wholly bright orange-yellow head, concolorous with underparts. Rump bright blue, conspicuous in flight. Legs red. **Juvenile (not illustrated):** Orange-yellow parts marked with fine dark fringing. **Voice:** A series of low whistles on a descending scale; *pree preew-preew*. Also harsh cackling notes, *kek-ek-ek-ek*. **Range:** The north eastern coast of India through Bangladesh and Myanmar to the Malay Peninsula Declining owing to loss and degradation of mangroves. **NT. Status and distribution:** Uncommon and localized resident only in the Langkawi archipelago (Kedah), recorded at Kuah, Sungai Kuala Melaka, Teluk Burau, Kisap-Kilim Forest Reserves, Kubang Badak and Teluk Datai on the main island. Also on Pulau Dayang Bunting and Pulau Singa Besar (Yeap, 2005). **Habitat:** Mangroves. **Habits:** In review area, a strict salt and brackish water specialist, avoiding even freshwater habitats close to mangroves. More secretive than Common Storkbill but fairly regular where it occurs. Perches on low branches overlooking mangrove creeks. Feeds largely on crabs and other crustacea.

316. Common Storkbill, *Pelargopsis capensis* Plate 40

Pekaka Emas Biasa/Burung Buaya

Description: 37 cm (14.5 in). The largest kingfisher in the review area. Huge coral-red bill with dark tip and pale brown head together with dull blue to greenish-blue mantle, wings and tail which contrast sharply with bright blue rump diagnostic. Collar, throat and entire underparts orange-yellow. Legs red. Juvenile: (not illustrated) Collar, lower throat and breast finely marked with dark fringing. In flight from White-throated (318) and Black-capped Kingfishers (319) by lack of white wing patches. **Voice:** A disyllabic whistle, *pree–preew*, usually uttered in flight. Also harsh cackling notes similar to Brown-winged Storkbill (304). **Range:** The Indian subcontinent through South-East Asia and the Greater and Lesser Sundas to the Philippines. **Status and distribution:** Fairly common resident at low elevations, south to Singapore (including offshore islands), where it is uncommon. Also on Pulau Pinang and Pulau

Langkawi (Kedah), where in the latter, may overlap with Brown-winged Storkbill in saline and brackish habitats. **Habitat:** Mangroves, water bodies in open country, forested river courses. **Habits:** Often sits on open perches. Feeds largely on frogs and reptiles captured on the ground. Will take crabs in the mangroves.

317. Ruddy Kingfisher, *Halcyon coromanda* Plate 40
Pekaka Ungu

Description: 25 cm (10 in). Bright violet tinged rufous upperparts contrasting with bright rufous buff underparts diagnostic. Patch on rump bluish white (conspicuous in flight); bill and legs bright red. **Voice:** A series of 4 or 5 mournful whistles on a descending scale. **Range:** India (including the Andamans) through China, Korea and Japan through South-East Asia to the Greater Sundas, the Philippines and Sulawesi. Temperate populations winter south to tropical latitudes. **Status and distribution:** Uncommon and localized resident, passage migrant and winter visitor at low elevations up to 450 m, south to Singapore where it is a rare resident on Pulau Tekong Besar and Pulau Tekong Kecil and also an uncommon passage migrant and winter visitor. Occurs above 900 m only during autumn and spring passage. **Habitat:** The resident population inhabits mangroves; winter visitors also inhabit forests in the interior. **Habits:** Generally unobtrusive. Perches low in dense mangroves, feeding largely on crabs and other crustacea. Shy.

318. White-throated Kingfisher, *Halcyon smyrnensis* Plate 40
Pekaka Dada Putih

Description: 28 cm (11 in). Unmistakeable. Red bill, white throat and breast with bright chocolate brown head, flanks and belly diagnostic. Wings bright blue, legs red. **Juvenile (not illustrated):** Very young birds have dark bills which turn dull red with age. Breast feathers have fringes with lesser wing coverts mottled black. White patch at base of primaries is mostly concealed at rest but conspicuous in flight. **Voice:** Territorial call is a long shrill musical rattle on a descending scale *kiri-kiri-kiri*..., uttered when perched. Also a shrill *kiri-ki, kiri-ki*... in flight. The alarm call is a harsh *kreek-kreek-kreek-kreek kreek-kreek-kreek-kreek*. Also a percing *tsik* (W1). **Range:** Turkey through the Middle East, southern China, South and South-East Asia to Sumatra, Java and the Philippines. **Status and distribution:** Abundant and widespread resident at principally low elevations, south to Singapore. Has also been observed at 830 m below Genting Highlands (Pahang). The most common kingfisher inland throughout Peninsular Malaysia **Habitat:** Scrub, mangrove and forest edges, ricefields, cultivated land, plantations, gardens, large forested river courses. **Habits:** Frequently utilizes bare branches, fence posts and telephone wires as vantage points to catch large insects, reptiles and amphibians. Very vocal. Not particularly attached to water.

319. Black-capped Kingfisher, *Halcyon pileata* Plate 40
Pekaka Ubun Hitam

Description: 30 cm (12 in). From White-throated Kingfisher (318) at rest and in flight by black head and rufous belly and flanks separated by white throat and collar. Bill brighter red, mantle, wings and uppertail coverts deeper royal blue. Wing coverts black, legs red. White patch at base of primaries is mostly concealed at rest but conspicuous in flight. **Voice:** A piercing *tsik* and a harsh *kreek-kreek-kreek* in alarm, both like White-throated Kingfisher. **Range:** Breeds in the Indian subcontinent, Myanmar, China and Korea. Winters throughout South-East Asia to Sumatra, Borneo, Sulawesi and the Philippines. **Status and distribution:** Fairly common passage migrant and winter visitor from low elevations up to 1200 m, south to Singapore, where it is uncommon. **Habitat:** Mangroves, ricefields, the forest edge, forested river courses, reservoirs. **Habits:** Territorial in its wintering quarters, it is aggressive and often displaces the resident White-throated Kingfisher. Not as vocal as White-throated Kingfisher. Shy.

320. Collared Kingfisher, *Todiramphus chloris* Plate 40
Pekaka Bakau

Description: 24 cm (9.5 in). Blue-green upperparts with white collar and underparts diagnostic. Wings and tail deeper blue; bill black with flesh coloured mandible. **Juvenile (not illustrated):** Upperparts duller, collar, breast, and flanks tinged buff with fine blackish vermiculations in the latter. **Voice:** A shrill *kiank-kiank, kiank-kiank,* uttered both in flight and while perched. **Range:** Coasts of the Red Sea, South and South-East Asia, the Indonesian Archipelago, northern Australia, the western Pacific to Fiji and Samoa. **Status and distribution:** Common and widespread resident, passage migrant and winter visitor along the coastal plains, south to Singapore. South of Negeri Sembilan, frequently occurs inland and in Singapore, is more common than the White-throated Kingfisher (318). The most common kingfisher in coastal districts. **Habitat:** Mangroves, beach scrub, coastal gardens, plantations. **Habits:** Conspicuous in coastal towns where it often perches on telephone or electricity wires. Very noisy and conspicuous especially during courtship. On the coast, feeds largely on crabs and crustacea during low tide. Inland it will take insects, small reptiles and amphibians.

321. Oriental Dwarf Kingfisher, *Ceyx erithaca* Plate 40
Pekaka Kerdil

Description: 14 cm (5.5 in). Tiny size, coral red bill and bright yellow underparts diagnostic. **Black-backed form:** Mantle and wings black tinged dark blue on scapulars, dark blue spot on forehead (usually) and side of neck. **Rufous-backed form:** Mantle and wings dark rufous tinged lilac, lacks dark spot on head and sides of neck. Line running down rump and uppertail coverts bright lilac in both forms, conspicuous only in flight. Intermediate forms (not illustrated) also occur. **Voice:** A weak high pitched acidic *dzeeep*, usually uttered in flight. **Range:** The Indian subcontinent and southern China through South-East Asia, the Greater and Lesser Sundas to the Philippines. **Status and distribution:** Locally common resident, and fairly common passage migrant and winter visitor from low elevations up to 450 m, south to Singapore where it uncommon. Occurs above 900 m only during autumn and spring passage. The resident population comprises largely the Rufous-backed form while migrants are mostly the Black-backed form. **Habitat:** Mangroves, forests (including stream courses), plantations, wooded areas. **Habits:** Keeps to the lower storey, often far from water. Feeds principally on insects captured on aerial sallies from open perches. Shy.

322. Blue-banded Kingfisher, *Alcedo euryzona* Plate 40

Raja Udang Rimba

Description: 20 cm (8 in). From Blue-eared Kingfisher (323) by larger size and blackish-blue upperparts with fine blue speckles on wing coverts. Bluish white line running down back and rump conspicuous in flight. Ear spot rufous above, white below. Throat white; bill black with mandible sometimes reddish. **Male:** Scaly blue breast band on white underparts diagnostic. Flanks pale rufous. Bill largely blackish. **Female:** Pale rufous underparts, blackish maxilla and dull reddish mandible. From Common Kingfisher (324) by larger size, duller upperparts and the lack of rufous ear coverts. **Voice:** Like Common Kingfisher but less shrill. **Range:** Myanmar through the Malay Peninsula to the Greater Sundas. Being restricted to lowland forested rivers, it is suspected to have declined rapidly due to significant losses of this habitat type. **VU. Status and distribution:** Uncommon and localized resident from low elevations up to 900 m, south to Johor. Favours forested streams to small and larger forested river courses Generally absent along the larger river courses whose banks today are devoid of natural vegetation. Populations may still be self sustaining in foothill river courses, making it vital to restrict development in these areas. **Habitat:** Mangrove edge, small forested rivers and streams. **Habits:** Typically seen on a low riverside perch or in flight low over the water surface. Fishing and feeding techniques similar to Common Kingfisher.

323. Blue-eared Kingfisher, *Alcedo meninting* Plate 40

Raja Udang Telinga Biru/Binti-Binti

Description: 16 cm (6 in). From Common Kingfisher (324) by dark blue upperparts and ear coverts, darker rufous underparts and all blackish bill. **Female:** Base of bill reddish. **Juvenile** (not illustrated): Orange ear coverts. **Voice:** Like Common Kingfisher but higher pitched and usually uttered on one note. **Range:** The Indian subcontinent through South-East Asia and the Greater Sundas to Bali, Lombok and the Philippines. **Status and distribution:** Locally common resident principally at low elevations, south to Singapore where it is rare, confined to the central catchment forest. Has also been recorded at Pulau Tekong Besar, Poyan, and the Sungei Buloh Wetland Reserve (W and H). **Habitat:** Mangroves, forested streams. **Habits:** Sits on low perches overlooking forest streams. Fishing and feeding techniques like Common Kingfisher.

324. Common Kingfisher, *Alcedo atthis* Plate 40

Raja Udang Biasa

Description: 17 cm (7 in). Upperparts blue-green with bright pale blue line running down back and rump (conspicuous in flight). Underparts and ear coverts rufous; throat and neck spot white; legs red. **Male:** Bill all black with extreme base of mandible red. **Female:** All black maxilla with basal half of mandible orange-reddish. **Voice:**

A shrill high pitched trill, uttered twice or thrice usually in flight. **Range:** North Africa through Europe, temperate Asia, South and South-East Asia to New Guinea and the Solomons. Populations from temperate latitudes migrate to the tropics in winter. **Status and distribution:** Fairly common resident and winter visitor at low elevations, south to Singapore. Especially common during winter. **Habitat:** Mangroves, river and stream banks, water bodies in open country, fish ponds. **Habits:** Typically seen on low perch overlooking water. Catches small fish by plunge-diving; victim is firmly secured in bill and stunned with blows on perch before swallowed whole, head first. Flies low over water surface.

BEE-EATERS

ORDER *Coraciiformes* **FAMILY** *Meropidae*

World: *26 species* **Peninsular Malaysia and Singapore:** *4 species*

Small to medium sized slender-bodied, upright perching birds with long narrow down-curved bills, longish square-cut tails, and long and pointed triangular wings. Plumage predominantly green with a contrasting coloured head in most species. Feed mainly on bees and other insects caught on the wing on swift aerial sallies from exposed perches. Bees and wasps are swiftly immobilized and devenomized by repeated rubbing and beating on a perch before being swallowed. Sexes virtually alike, juveniles duller. Represented by the following groups and genera: **Forest Bee-eaters,** genus *Nyctyornis* (325). Larger and more robust than typical bee-eaters with broader wings and more prominently decurved bills. Inhabits forests. Does not undertake long distance flights. Solitary or in pairs. Calls deep harsh guttural croaks. **Typical Bee-eaters,** genus *Merops* (326–328). Slender built inhabitants of open country. Flight is undulating and rather erratic with rapid flapping, interspersed with glides, often accompanied by repeated calling. All species in review area display a narrow black eyeline and coppery underwing in flight. Two species have elongated central tail feathers; these are not diagnostic features as they are sometimes abraded or lost. Gregarious, moving about in groups over long distances, sometimes at night, where incessant calls, which are pleasing liquid burring trills give the clue.

325. Red-bearded Bee-eater, *Nyctyornis amictus* Plate 41

Beberek Janggut Merah/Beberek Tunggal

Description: 32 cm (13 in). Robust appearance with strongly decurved bill distinctive. **Male:** Largely bright green plumage, throat and shaggy breast crimson; forehead lilac. **Female** (not illustrated): Like male but forehead red. Underside of tail pale orange-yellow with broad blackish terminal band in both sexes. **Juvenile:** Entire head and underparts mostly green including underside of tail. **Voice:** A single (sometimes two) deep harsh guttural croak *arrk*, uttered from a perch. Also a harsh chatter *ka-ka-ka-ka-ka-ka,* uttered repeatedly on a descending scale, tailing off abruptly. **Range:** Tenasserim (Myanmar), southern Thailand through the Malay Peninsula to Sumatra and Borneo. **Status and distribution:** Locally common resident from low elevations up to 1300 m, south to Johor. **Habitat:** Forests, the forest edge. **Habits:** Typically seen hawking for insects at the forest edge. Will also probe into wood for grubs in the manner of woodpeckers. Usually keeps to the lower canopy and middle storey. When vocalizing, puffs out throat feathers and raises crown feathers. Nests in earth banks near forest streams.

326. Blue-tailed Bee-eater, *Merops philippinus* Plate 41

Beberek Ekor Biru

Description: 30 cm (12 in). At rest and in flight, noticeably heavier built and broader winged than Blue-throated Bee-eater (327). Crown, nape and upper back greenish; upper throat yellowish; lower throat brownish. Shows conspicuous pale blue streak below black eyeline. In flight, blue rump concolorous with uppertail coverts and elongate central tail feathers. **Juvenile** (not illustrated): Chestnut patch on lower throat duller, lacks elongate central tail feathers. **Voice:** A ringing trill, *be-rek - be-rek,* very similar to that of Blue-throated Bee-eater. Also a sharp *pit* while perched. **Range:** Breeding in the Indian subcontinent and southern China discontinuously through South-East Asia to the Malay Peninsula, and in New Guinea. Winters through South and South-East Asia to Sulawesi and the Lesser Sundas. **Status and distribution:** Common and passage migrant and winter visitor, and sparse breeder, principally at low elevations, south to Singapore. **Habitat:** Beach scrub, ricefields, open lightly wooded country, tin mines, mangroves. **Habits:** Often seen on open perches (including telephone and electricity wires) waiting for insect prey. Skims low over fields in pursuit of insects. Roosts communally in mangroves or in tall trees.

327. Blue-throated Bee-eater, *Merops viridis* Plate 41

Beberek Leher Biru

Description: 28 cm (11 in). Dark chestnut crown, nape and upper back together with blue throat diagnostic. Rump iridescent pale blue, conspicuous in flight. **Juvenile** (not illustrated): Crown, nape and upper back green, concolorous with mantle; lacks elongate central tail feathers. **Voice:** A metallic ringing trill, *be-rek - be-rek,* richer than similar calls of Blue-tailed Bee-eater (326), safely distinguishable only when heard together. **Range:** Southern China through South-East Asia to the Philippines and the Greater Sundas; northen populations are migratory. **Status and distribution:** Fairly common breeding migrant and non-breeding visitor from low elevations up to 670 m (sometimes higher) south to Singapore. Breeding population disperses to Sumatra during non-breeding season, then replaced in its niche by the Blue-tailed Bee-eater. **Habitat:** Breeding migrants frequent beach scrub, open lightly wooded country, gardens and tin mines while non-breeding migrants frequent forest canopies, the forest edge and mangroves. **Habits:** Frequently utilizes open perches (including telephone and electricity wires) in open country as vantage points. Travels in small noisy groups. Forms large communal roosts in tall trees.

328. Chestnut-headed Bee-eater, *Merops leschenaulti* Plate 41

Beberek Kepala Perang

Description: 23 cm (9 in). From Blue-throated Bee-eater (327) by all green tail, creamy yellow sides of neck and narrow chestnut and blackish gorget. Crown, nape, and upper back bright chestnut; rump iridescent pale blue, conspicuous in flight.

Lacks elongate central tail feathers of its other congeners. **Juvenile** (not illustrated): Crown, nape and upperback green. **Voice:** A soft and melodious burring *pruip-pruip* in flight. Also a soft, airy *chewy-chewey-chewy* (W1). **Range:** India to southern China discontinuously through South-East Asia to Java and Bali. **Status and distribution:** In Peninsular Malaysia, locally common resident south to Chemor (Perak) on the west coast and Setiu (Terengganu) on the east, with post post-breeding dispersants seen further south at Kuala Lumpur and Tekam (Pahang) (W1). Also on Pulau Langkawi and Pulau Pinang. **Habitat:** Open lightly wooded country, beach scrub, plantations. **Habits:** Travels in small groups. Hawks for insects from exposed perches, sometimes utilizing telephone and electricity wires.

HOOPOES

ORDER *Coraciiformes* **FAMILY** *Upupidae*

World: *3 species* **Peninsular Malaysia and Singapore:** *1 species*

Slender build with long erectile fan-like crest and slender slightly down-curved bill. Wings broad and rounded. Flight leisurely and undulating. Largely terrestrial. Sexes alike. Juveniles are duller with shorter bills and crests. Solitary.

329. Eurasian Hoopoe, *Upupa epops* Plate 41
Burung Huphup

Description: 30 cm (12 in). Plumage pinkish buff with tip of crest black; wings black with white bars; tail black with white band. Long crest usually held flat on head while feeding and in flight. Black and white wing pattern stands out boldly in flight. **Voice:** A soft and mellow far carrying *hoop-hoop-hoop*. **Range:** Europe and temperate Asia east to the Indian subcontinent, China, and the Korean peninsula, wintering south through continental South-East Asia, north Borneo, and the Phillippines. **Status and distribution:** Rare and localized non-breeding visitor; a road kill near Kuala Lumpur on 16 September 2001 and individuals at Balik Pulau (Pulau Pinang) on 28 October 1997, 30 November 2003 and 17 March 2004 (W2), the latest, an individual at Juru, Seberang Perai (Pulau Pinang) on 24–25 October 2006 (Tan, 2006). In Singapore known from two sightings, individuals each at Upper Pierce Reservoir on 25 November 1983 and Seletar on 24 October 1987 (W2). In the 1930s was locally common resident from Perlis to southern Kedah. **Habitat:** Open country, gardens, cultivated land. **Habits:** Feeds on insect larvae and minute invertebrates obtained by probing into soft earth and decaying organic matter. When alarmed, holds up crest.

HORNBILLS

ORDER *Coraciiformes* **FAMILY** *Bucerotidae*

World: *50 species* **Peninsular Malaysia and Singapore:** *10 species*

Large sized with broad wings, long tails and mainly brown, black and white plumage. Bills huge and down-curved, adorned above with gaudily coloured casques. In all species, these casques are hollow except for that of the Helmeted Hornbill (335) which is a solid lump of ivory. Casques are rudimentary in juveniles and over the years gradually enlarge to adult size; varying in size with each species. Most are forest inhabitants. Sexually dimorphic in many species; the males are usually larger and have larger casques and some sport drooping crests. Juveniles more or less resemble adults but are usually duller with shorter tails. White of plumage often stained yellow with oil from preen gland. Flight, moderately fast and laboured with noisy wingbeats in many species. Principally frugivorous, gathering to feed in fruiting fig trees; forest species favour primary and selectively logged forest with a high density of figs and other fruits. Also take small mammals, birds, reptiles and insects. Nest in natural crevices high up in big trees. The female is sealed in the crevice which is regularly plastered by mud and droppings until reduced to a slit, through which it receives food and expels droppings. It remains sealed from the time the eggs are laid until the young hatch and is totally dependent on food from the male or in some species, other members of the flock called helpers. Calls loud and harsh, varying from deep roaring barks and growls to high-pitched yelps, cackles and squeals. Gregarious outside breeding season, gathering in flocks and flying vast distances in search of food. Being highly dependent on forest fruiting trees as well as the availability of large tree crevices for nesting, their presence is an indicator of the carrying capacity of the forest. Throughout their global ranges, most species in the review area have declined greatly due to loss and fragmentation of lowland rainforest and in global terms rank between **NT** and **VU**. These are stated appropriately in each species account below.

330. Bushy-crested Hornbill, *Anorrhinus galeritus* Plate 42

Enggang Berjambul/Burung Kawan/Enggang Buluh

Description: 89 cm (35 in). Blackish plumage and bill together with drooping nuchal crest diagnostic. Tip of tail black with basal two-thirds brownish grey. Bare skin around eye and patch at base of bill bluish white. In underwing flight pattern, dark plumage contrasts sharply with brownish grey base of tail. **Voice:** A series of high-pitched squealing laughs, persistently uttered by foraging flocks. **Range:** Tenasserim (Myanmar) through the Malay Peninsula to Sumatra and Borneo. **Status and distribution:** Locally common resident from low elevations up to 1200 m, south to Johor. **Habitat:** Forests. **Habits:** Frequents the canopy. Highly gregarious, gathering up to 15. Often gathers to feed in fruiting fig trees. During nesting, the female is not only fed by the male but by other members of the flock as well.

331. Oriental Pied Hornbill, *Anthracoceros albirostris* Plate 42

Enggang Kelingking/Burung Kelingking/Burung Lilin

Description: 70 cm (27 in). Black upperparts and breast contrast sharply with white belly and small patch below eye. **Male:** Bill and casque ivory white. **Female:** Casque smaller with more blackish markings. In *A. a. albirostris* (not illustrated), outer tail feathers black with white tips. Basal two thirds of undertail coverts black with the rest white. In *A. a. convexus*, outer tail feathers together with undertail coverts white. In underwing flight pattern, black breast and wings with white trailing edge contrasting with white belly and tail distinctive. Intermediate forms (not illustrated) occur. **Voice:** A high-pitched cackling bray, *kek-kek-kek-kek*... **Range:** The Indian subcontinent and southern China through South-East Asia to the Greater Sundas and Bali. **Status and distribution:** Fairly common resident principally at low elevations but recorded up to up to 150m, south to Singapore where it is an uncommon visitor to Pulau Tekong and has bred on Pulau Ubin from 1997, now spreading on mainland Singapore, where a few are escaped captive individuals. *A. a. albirostris* occurs in Kedah (including Pulau Langkawi) while *A. a. convexus* replaces it south of Kedah to Singapore. Also on Pulau Pangkor (Perak). **Habitat:** Forests (near streams and rivers), the forest edge, plantations, beach scrub. **Habits:** Frequents the canopy and middle storeys, usually in pairs or small loose groups. Usually noisy. Feeds largely on insects and small vertebrates which are sometimes obtained from the ground. Also gathers to feed in fruiting trees. Flight consists of rapid flaps followed by short glides, wingbeats producing a soft chuffing sound. Tame.

332. Black Hornbill, *Anthracoceros malayanus* Plate 42

Enggang Hitam/Burung Kekek/Burung Gatal Birah

Description: 76 cm (30 in). Entire plumage black with white outer edges of tail tip. **Male:** Bill, casque, and facial skin whitish. Some individuals have broad white supercilium. **Female:** Casque blackish; facial skin dull pink. In underwing flight pattern, all black wings and underparts with white outer edges of tail tip distinctive. **Voice:** Harsh, resonant retching growls. **Range:** The Malay Peninsula to Sumatra and Borneo. **NT. Status and distribution:** Locally common resident principally at low elevations up to 200 m, south to Johor. **Habitat:** Forests, the forest edge. **Habits:** Frequents the canopy singly or in small loose groups of up to 5; sometimes larger groups may form. Will sometimes descend to feed on ground. **Note:** Occasional birds 1986–2003 at Seletar and Nee Soon (Singapore) were probably escaped captive individuals.

333. Great Hornbill, *Buceros bicornis* Plate 42

Enggang Papan

Description: 122 cm (48 in). Pale yellow bill, casque and neck contrasting with black mask and body plumage diagnostic. Broad white bar on trailing edge and centre of wing. Tail white with black subterminal band. **Male:** Front edge of casque black; iris red. **Female** (not illustrated): Iris white. In underwing flight pattern, black wings with white bar and trailing edge together with white tail and black subterminal tail band distinctive. **Voice:** A series of deep resonant *honk* notes while perched; lower pitched than Rhinoceros Hornbill (334). Also loud barking and roaring notes. In flight, a disyllabic *ho-honk,* uttered antiphonally by pairs. **Range:** The Indian subcontinent through South-East Asia and the Malay Peninsula to Sumatra. **NT. Status and distribution:** Locally common resident from low elevations up to 1200 m, south to the Panti Forest Reserve (Johor). Also on Pulau Langkawi (Kedah) and Pulau Pangkor (Perak). Sightings in Singapore are escaped captive individuals. **Habitat:** Forests. **Habits:** Frequents the canopy singly or in pairs, sometimes in larger loose groups. Feeds in fruiting trees together with other hornbills. In the wild, has been observed associating with Rhinoceros Hornbill; an individual crossing bills with one of the latter in an unaggressive manner. Sound of wingbeats like Rhinoceros Hornbill.

334. Rhinoceros Hornbill, *Buceros rhinoceros* Plate 42
Enggang Sumbu

Description: 122 cm (48 in). Up-turned reddish and yellow casque on ivory-white bill diagnostic. Upperparts and breast black; belly white; tail white with black subterminal band. **Male** (not illustrated): Casque larger with black line running down sides; iris red. **Female:** Iris white. In underwing flight pattern, all black wings contrasting with white belly and tail with black subterminal band distinctive. **Voice:** A deep resonant *honk* while perched. A loud disyllabic *ho-honk* in flight, uttered antiphonally by pairs. **Range:** The Malay Peninsula to the Greater Sundas. **NT. Status and distribution:** Locally common resident from low elevations up to 1200 m, south to Johor. Locally extinct in Singapore since 1898 (L and G). **Habitat:** Forests. **Habits:** Frequents the canopy singly or in pairs, sometimes larger groups may form especially when food is in abundance. Feeds largely on fruit but also takes small vertebrates. Noisy. Wingbeats produce a chuffing sound (not as loud or sharp and swishing like *Aceros* and *Rhyticeros* hornbills (337–339)) like a locomotive labouring uphill.

335. Helmeted Hornbill, *Rhinoplax vigil* Plate 42
Enggang Bertopi/Burung Tebang Mentua/Burung Torak

Description: 127 cm (50 in). From other hornbills by shorter conical bill, deeply truncated casque and elongated central tail feathers, which are 50 cm (20 in) more in length Dark brown upperparts and breast contrast sharply with white belly and tail. Bill and casque red at base with tip yellow. **Male:** Bare skin around neck red and coarsely wrinkled. **Female (not illustrated):** Skin around neck also coarsely wrinkled but sky-blue. In underwing flight pattern, elongate central tail feathers, all dark wings with white trailing edge together with white belly and tail with dark subterminal band distinctive. **Voice:** Territorial call consists of a series of hollow *poop* notes, separated by long intervals before accelerating on a disyllabic *poop-poop* on a descending scale and ending in a harsh cackling laugh. Also a high pitched *ka-hank-ka-hank* in flight. **Range:** Tenasserim (Myanmar) through the Malay Peninsula, Sumatra and Borneo. **NT. Status and distribution:** Uncommon and localized resident from low elevations up to 1500 m, south to Johor. Locally extinct in Singapore since 1895 (L and G). **Habitat:** Forests. **Habits:** Frequents the canopy. Usually solitary or in pairs but never in larger flocks. Often gathers in fruiting fig trees to feed together with other hornbills and frugivorous birds. Flight consists of slow flaps with long intermittent glides.

336. White-crowned Hornbill, *Berenicornis comatus* Plate 42
Enggang Ubun Putih

Description: 90 cm (36 in). Blackish bill and bushy white crest covering small casque diagnostic. Mantle black, tail white. **Male:** Neck and underparts white. **Female:** Neck and underparts black. In underwing flight pattern, white trailing to edge of black wings together with all white tail distinctive. **Voice:** A series of deep resonant pigeon-like hooting notes, *huu…hu-hu-hu, hu-hu-hu…* **Range:**

Tenasserim (Myanmar) through the Malay Peninsula to Sumatra and Borneo. **NT**. **Status and distribution**: Uncommon and localized resident from low elevations up to 1680 m, south to Johor. More common in hilly elevations between 120–900 m. **Habitat:** Forests, the forest edge, plantations fringing forests. **Habits:** Frequents the canopy, usually singly or in pairs; sometimes in larger groups. Other habits little known.

337. Wrinkled Hornbill, *Aceros corrugatus* — Plate 42

Enggang Berkedut

Description: 81 cm (32 in). From Wreathed Hornbill (339) by smaller size, larger and more prominent narrow and high standing reddish casque (in male) and black basal half of tail with the distal half white, usually strongly stained yellow to buffish. Bill yellowish; orbital skin blue. **Male:** White neck and gular pouch. **Female:** Black head and neck, pale blue gular pouch. In underwing flight pattern, all black wings, belly and basal half of tail together with pale buffish stained tail distinctive. **Voice:** Contact call is a short barking *kak-kak*. **Range:** The Malay Peninsula to Sumatra and Borneo. **NT**. **Status and distribution:** Uncommon and localized resident confined to low elevations, south to Johor, but seen up to 800 m at the Temengor Forest Reserve (Perak). **Habitat:** Forests, including peat swamp. **Habits:** Frequents the canopy in small groups, flying vast distances between roosting and feeding grounds. Wingbeats continuous, producing a rather loud and sharp swishing sound, typical of the genus. Gathers to feed in fruiting fig trees.

338. Plain-pouched Hornbill, *Rhyticeros subruficollis* — Plate 42

Enggang Rimba

Description: 90 cm (36 in). Like respective sexes of Wreathed Hornbill (339) in terms of casque shape and body colouration, but distinguished by the lack of both the dark bar on the gular pouch and the corrugations on the sides of bill. Differences in size between both species apparent only when seen together, **Voice:** A trisyllabic *keh-kek-kehk*, similar in quality to the bark of the Wreathed Hornbill but higher pitched, more quacky, and less booming and apparently always three-noted (L and R). **Range:** North-eastern India? Myanmar, south to western Thailand and the Malay Peninsula. **VU**. **Status and distribution:** Uncommon and localized resident principally at low elevations, recorded from Ulu Muda Forest Reserve (Kedah) south to the Temengor Forest Reserve (north-west Perak) where flights of up to 2067 commuting daily between roosting and foraging grounds have been recorded seasonally; higher counts of up to 2365 have been recorded at Kenering (Perak), 60 km south-west of Temengor. Forest corridors are more or less continuous between the two areas, but fast shrinking (W2). See 'Conservation' chapter. **Habitat:** Forests. **Habits:** Similar to Wreathed Hornbill, including loud and sharp swishing sound of wingbeats.

339. Wreathed Hornbill, *Rhyticeros undulatus* — Plate 42

Enggang Gunung

Description: 100 cm (42 in). Low and flat topped small casque with ridges (resembling wreath) distinctive. All black wings contrasting with wholly white tail together with black bar on gular pouch diagnostic. Bill whitish with corrugations on basal sides of maxilla and mandible. Orbital skin red. **Male:** Crown and drooping nuchal crest brownish; gular pouch yellow; neck white. **Female:** Head black, gular pouch pale blue. In underwing flight pattern, all black wings and belly together with all white tail distinctive. **Voice:** A roaring double bark, *wur-wooh*, the second syllable carrying further (W1). **Range:** North-East India and Myanmar through South-East Asia to the

Greater Sundas and Bali. **Status and distribution:** Locally common resident from low elevations up to 1800 m, south to Johor. Also on Pulau Langkawi. **Habitat:** Forests. **Habits:** Frequents the canopy. Usually flies in compact flocks of 5–8. Often gathers at fruiting trees together with other species. Nomadic, dispersing over vast areas to forage. In flight, the wingbeats are continuous, producing a distinct loud and sharp swishing sound. Glides less than other hornbills.

BARBETS AND ALLIES

ORDER *Piciformes* **FAMILY** *Ramphastidae*

World: *122 species* **Peninsular Malaysia and Singapore:** *11 species*

(This cosmopolitan family also includes the toucans of the Neotropics and a host of other Neotropical and Afro-tropical barbets which do not occur in the review area. For this purpose only the Asian Barbets (Megalaiminae) will be fully described).

Robust and heavy-bodied with large heads and strong stout bills with a strong tuft of rictal bristles. All except one have largely bright green plumage (paler below) with distinctively marked head patterns. Juveniles are duller. Primarily denizens of rainforests, some also inhabit wooded areas close to human habitation. Excavate tree holes to nest in. Largely arboreal, they are difficult to spot in the foliage; their distinctive calls which are characteristic sounds of the rainforest are often the best identification clues (see Appendix 2). The *Megalaima* barbets (341–349) utter monotonous repetitive hoots or trills, which in many species is preceded by a series of rapid rolling trills. Principally frugivorous, diet comprises largely fruit, seeds and flowers. Figs are particularly relished. Most species are usually solitary but will frequently gather in mixed species flocks to feed in fruiting trees. Sexes alike in most species.

340. Fire-tufted Barbet, *Psilopogon pyrrolophus* Plate 43
Takur Jumbai-Api/Takur Api

Description: 28 cm (11 in). From all *Megalaima* barbets by proportionately longer and more graduated tail, pale greenish bill with blackish vertical band half way along sides, and yellow and black breast band together with grey cheek patch. Throat and supercilium pale green; crown, nape and eyeline maroon; tuft of bristles at base of bill black, broadly tipped with red. **Voice:** A series of shrill cicada like buzzing trills, beginning slowly and accelerating on an ascending scale before tapering off abruptly, reminscent of a small tinny motorcycle accelerating away around a corner. Also a sharp *kik*. **Range:** The Malay Peninsula and Sumatra. **Status and distribution:** Fairly common resident above 900 m throughout the Larut Hills (Perak) and the Titiwangsa Range, south to Selangor. Also on Gunung Benom and Gunung Tahan (Pahang). Most common between 900–1500 m. Common and conspicuous at the hill-stations. **Habitat:** Forests, the forest edge. **Habits:** Frequents the canopy. Usually solitary or in pairs. Small groups may gather to feed in fruiting trees. Flight strong and direct with wingbeats producing a whirring sound. Enters hill-station gardens at the forest edge.

341. Lineated Barbet, *Megalaima lineata*

Plate 43

Takur Kukup

Description: 29 cm (11.5 in). From all green barbets by light brown head with buff streaks (bolder on the breast) and pale straw coloured bill and orbital skin. Streaking on head becomes indistinct at a distance or in worn plumage. **Voice:** A disyllabic *po-prrp* with the second syllable stressed. Also the rolling generic trill. **Range:** The Himalayas through Myanmar and the Malay Peninsula to Java and Bali. **Status and distribution:** Fairly common resident at low elevations and spreading along the coastal plains of Peninsular Malaysia, south to Tanjung Tuan (Melaka) on the west coast and Pekan (Pahang) on the east coast. Individuals also at 660 m at the Bukit Larut Tea Garden (Perak) and Kuala Tahan (Pahang). In Singapore, a small feral population originating from escaped captive individuals has spread from Bukit Batok Nature Park and the Bukit Timah Nature Reserve to various other sites. A fairly common garden bird throughout most of northern Peninsular Malaysia. **Habitat:** Coastal scrub, orchards, coconut groves and (in Singapore) tall secondary forest. **Habits:** Generally keeps to the tree tops in both rural and urban areas, feeding in fruiting trees, usually solitarily. Sometimes calls from an open perch. An individual was observed bathing in a tree hole (SINAV 23: 1).

342. Gold-whiskered Barbet, *Megalaima chrysopogon*

Plate 43

Takur Pipi Kuning

Description: 30 cm (12 in). The largest barbet in the review area. Longish thickset bill and large bright yellow oval cheek patch diagnostic. Forecrown greyish buff; forehead and hindcrown red, scaled with blue in the latter; broad eyeline greyish brown; throat pale greyish brown with purplish blue below; bill blackish with base grey. **Voice:** A low pitched and rapid *tehup*, uttered repeatedly. Also the generic trill which is slower and lower pitched than the other barbets, starting with a long *trrrrr* repeated on one note, usually for about 20 seconds or more, each time more slowly until it breaks up into three to four note phrases. **Range:** The Malay Peninsula, Sumatra and Borneo. **Status and distribution:** Fairly common resident in hilly terrain from low elevations up to 1100m, south to Johor. Very rare in the flat lowlands where it is replaced by the Red-crowned Barbet (343). **Habitat:** Forests, the forest edge, densely wooded areas. **Habits:** Frequents the canopy. Probably the most common forest barbet. Also frequents small patches of forest and wooded areas around towns throughout Peninsular Malaysia. Frequently gathers at fruiting fig trees where it often displaces other smaller barbets. **Note:** Calls ascribed to this species from the Bukit Timah Nature Reserve on 30 December 1989 and 13 January 1990 were rejected by Lim (1990) due to lack of substantial evidence.

343. Red-crowned Barbet, *Megalaima rafflesii*

Plate 43

Takur Ubun Merah

Description: 27 cm (10 in). From Gold-Whiskered Barbet (342) by all red crown, smaller yellow cheek patch, pale blue throat and supercilium. Small patch under eye and patch on side of throat red; broad eyeline black; thickset bill blackish. **Voice:** Two evenly spaced *tuk* notes with a pause, and then followed by a series of rapid *tuk* notes for about ten seconds, the whole process repeated. **Range:** Tenasserim (Myanmar) through the Malay Peninsula, Sumatra, and Borneo. Declining due to loss and fragmentation of lowland rainforest. **NT. Status and distribution:** Locally common resident at low elevations, south to Singapore, where it is uncommon and confined to the Bukit Timah and Central Catchment Nature Reserves. Particularly common in the flat lowlands of central and southern Johor. **Habitat:** Forests, the forest edge. **Habits:** Frequents the canopy but sometimes descends to feed lower. Often gathers in fruiting trees with other barbets. Has also been reported to excavate dead wood to feed on insect larvae.

344. Red-throated Barbet, *Megalaima mystacophanos* Plate 43

Takur Tekak Merah

Description: 22 cm (9 in). Bill rather thickset. **Male:** Red throat diagnostic. Forehead yellow; rest of crown red; eyeline black; breast band and patch below eye blue. **Female:** Forehead and throat pale greenish; upper breast, sides of face and midcrown pale blue; hindcrown red. **Voice:** A series of 4 *tok* notes (with a gap between the 3rd and 4th), uttered persistently. Also the rolling generic trill. **Range:** The Malay Peninsula to Sumatra and Borneo. Declining due to loss and fragmentation of lowland rainforest. **NT. Status and distribution:** Locally common resident up to 1000 m, south to Johor. Locally extinct in Singapore since 1895 (L and G). **Habitat:** Forests. **Habits:** Frequents the canopy. Usually solitary or in pairs. Often detected by call which is usually delivered from a high perch. Will gather in fruiting trees to feed with other frugivores. **Note:** Calls ascribed to this species from the Bukit Timah Nature Reserve in March 1989 were rejected by Lim (1990) in the absence of actual sightings.

345. Golden-throated Barbet, *Megalaima franklinii* Plate 43

Takur Tekak Kuning

Description: 22 cm (9 in). Overall bright green with a yellowish tinge beneath. Forehead and small spot on hindcrown red; throat and crown yellow; sides of face and breast silvery grey; bill short, stubby, and blackish with bluish grey base. **Voice:** A trisyllabic *pi-priu* uttered rapidly. Also the rolling generic trill. **Range:** The Himalayas and southern China discontinuously into Myanmar and the Malay Peninsula. **Status and distribution:** Locally common resident above 1300 m in the Larut Hills (Perak) and the Titiwangsa Range south to Selangor. Also on Gunung Benom, Gunung Tahan (Pahang) and Gunung Rabong (Kelantan). Common above Brinchang (Cameron Highlands) and Gunung Ulu Kali (Genting Highlands), both in Pahang. Also seen at 1200 m on the higher points of Fraser's Hill (Pahang). **Habitat:** Ericaceous forest. **Habits:** The most common barbet at upper montane elevations where it forages in the low canopy. Often delivers its call from an exposed low perch. Usually solitary or very small groups may gather in fruiting trees.

346. Black-browed Barbet, *Megalaima oorti* Plate 43

Takur Kening Hitam

Description: 20 cm (8 in). Broad black supercilium and malar stripe together with bright yellow throat diagnostic. Crown paler yellow; hindcrown and patch on sides of neck red; ear coverts and upper breast blue; orbital skin pale greenish; Blackish bill short and stubby. **Voice:** A trisyllabic *to-ka-krrr*, repeated endlessly, with the third syllable stressed. **Range:** Southern China and Indo-China discontinuously through the Malay Peninsula and Sumatra. **Status and distribution:** Fairly common resident primarily between 760–1500 m throughout the Larut Hills (Perak) and the Titiwangsa Range south to Gunung Telapa Buruk (Negeri Sembilan). Also on Gunung Benom and Gunung Tahan (Pahang). Most common between 760–1200 m. Frequently heard at The Gap (Selangor), the lower parts of Genting Highlands (Pahang) and Bukit Larut (Perak). **Habitat:** Forests, the forest edge. **Habits:** Frequents the canopy. Usually solitary but small groups often gather in fruiting trees where they are conspicuous especially along roads at the forest edge. Sometimes enters gardens in the hill stations.

347. Yellow-crowned Barbet, *Megalaima henricii* — Plate 43
Takur Ubun Kuning

Description: 21 cm (8.5 in). Yellow forecrown and supercilium diagnostic. Hindcrown and throat blue. **Voice:** A brief trill followed immediately by four to five moderately paced *tok* notes; *trrr-tok-tok-tok-tok-trrr*... continuously repeated. **Range:** The Malay Peninsula to Sumatra and Borneo. Declining due to loss and fragmentation of lowland rainforest. **NT. Status and distribution:** Locally common resident primarily at low elevations but recorded up to 1200 m, south to Johor. Locally extinct in Singapore since 1879 (L and G). **Habitat:** Forests. **Habits** Frequents the canopy. Most often unobtrusive in the foliage where its call betrays its presence. Will gather at fruiting trees with other barbets.

348. Blue-eared Barbet, *Megalaima australis* — Plate 43
Takur Pipi Hitam

Description: 17 cm (7 in). Smallish. *M. a. duvaucelii*: Black ear coverts with broad red stripes above and below diagnostic. Midcrown and throat blue; forehead and narrow band on breast black. Blackish bill short and stubby. **Voice:** A disyllabic high pitched rattling trill, *kri-krik* uttered persistently. Also a monosyllabic trill, *priu*, repeated for long periods **Range:** North-East India through South-East Asia through the Greater Sundas and Bali. **Status and distribution:** Fairly common resident up to 1000 m, south to Johor. In Singapore locally extinct due to forest shrinkage, formerly rare resident (W and H). **Habitat:** Forests, the forest edge, wooded suburban gardens. **Habits:** Frequenting the middle storey and canopy, it frequently gathers in numbers at trees bearing small fruit at the forest edge or in rural gardens where it is sometimes conspicuous.

349. Coppersmith Barbet, *Megalaima haemacephala* — Plate 43
Takur Akar/Burung Tukang Besi

Description: 15 cm (6 in). Small size. Red legs and greenish white underparts with bold green streaks together with red forehead and upper breast band diagnostic. Sides of face black; throat, patch under eye and lower breast band yellow. **Voice:** An evenly spaced metallic *wout* with a resonant quality, uttered persistently. **Range:** India through southern China and South-East Asia through Sumatra, Java, Bali, and the Philippines. **Status and distribution:** Common and widespread resident at low elevations, south to Singapore. A commonly heard garden bird. **Habitat:** The mangrove edge, open country, wooded gardens in rural settlements, towns and cities. **Habits:** Perches on bare branches or high tension wires, sometimes huddled together in groups of up to 6 where it is conspicuous. Unlike its larger congeners, it does not always frequent the canopy but will often feed low on small fruit in trees and shrubs.

350. Brown Barbet, *Calorhamphus fuliginosus* Plate 43
Takur Coklat

Description: 18 cm (7.5 in). Smallish. From all other barbets by brown upperparts, whitish underparts and the lack of rictal bristles. Legs orange; bill thickset and blackish with maxilla extending over mandible in a hook at the tip. **Voice:** High pitched wheezing whistles and squeaks, *pseuoo*. Also a repeated *see..se..see...* (W1). **Range:** The Malay Peninsula to Sumatra and Borneo. **Status and distribution:** Fairly common resident up to 1000 m, south to Johor. In Singapore, locally extinct; formerly common resident on Pulau Ubin (W and H). **Habitat:** Forests, the forest edge. **Habits:** Gregarious, gathering in parties of up to six or more. Forages mainly in the middle and lower storeys. Moves around the ends of branches, sometimes upside down as parties feed on ripe fruit.

HONEYGUIDES

ORDER *Piciformes* **FAMILY** *Indicatoridae*

World: *17 species* **Peninsular Malaysia:** *1 species*

A distinct group of small nondescript birds superficially resembling bulbuls (453–477) with their centre of distribution in Africa. Have long pointed wings and short robust finch-like bills. Like barbets (340–350) and woodpeckers (352–377), they have zygodactylous feet. Largely arboreal. Most species are parasitic, laying their eggs in the nests of other birds. Solitary or in pairs. Sexes alike or nearly so.

351. Malaysian Honeyguide, *Indicator archipelagicus* Plate 44
Gembala Lebah/Musuh Lebah

Description: 18 cm (7 in). Bears superficial resemblance to *Pycnonotus* bulbuls (453–468) but bill robust, finch-like and much heavier. Upperparts dark olive brown; underparts whitish with greyish tinge on breast with flanks marked with indistinct dark streaks. Iris red. **Male:** Narrow lemon yellow shoulder patch diagnostic when visible. **Female (not illustrated):** Lacks yellow shoulder patch. **Voice:** A short nasal note, followed by a high-pitched ascending rattle, *miaw-kruuuu* or *miaw-miaw-kruuuuu*. **Range:** The Malay Peninsula to Sumatra and Borneo. Declining owing to loss and degradation of lowland rainforest. **NT. Status and distribution:** Rare and localized resident from low elevations up to 1050 m, south to the Pasoh Forest Reserve (Negeri

Sembilan). **Habitat:** Forests. **Habits:** Usually solitary, frequenting the canopy, from which it often calls repeatedly. Generally unobtrusive but will sometimes call in the presence of humans. Feeding and breeding behaviour have been seldom observed so little is known.

WOODPECKERS

ORDER *Piciformes* **FAMILY** *Picidae*

World: *220 species* **Peninsular Malaysia and Singapore:** *26 species*

Small to large sized birds characterized by long, straight chisel bills, specially adapted for boring and probing under bark of both dead and living wood for insects and grub on which they feed. Most have strong zygodactylous feet for clinging onto vertical tree trunks and climbing up trees; the only exceptions are the genera *Dinopium* (367, 368) and *Gecinulus* (370) which have three toes, two pointing forward and one backward. The tail feathers are stiff and act as props when climbing or clinging. Many species are colourful with some sporting prominent nuchal crests. Most species are territorial and many display a distinctive 'drumming' behaviour on dead wood. Nest in holes usually excavated in trees. Largely arboreal, foraging in trees though some species will take insects on the ground. Flight straight and undulating with wings briefly closed during short glides. Calls vary from high-pitched squeaks in small species to harsh and discordant shrieks in the larger ones. Usually solitary or in pairs; sometimes in small parties. Sexes alike or nearly so.

352. Eurasian Wryneck, *Jynx torquilla* Plate 43

Belatuk Tanah

Description: 18 cm (7.5 in). Small size with short pointed bill and cryptically coloured plumage distinctive. Has a brown eyeline and an indistinct whitish-grey supercilium. Has broad dark stripe running down from nape to mantle. Upperparts typically greyish-brown marked with dark vermiculations. Underparts whitish, often washed buff on the throat and breast, and marked with dark vermiculations. Wings and tail heavily vermiculated. Bill pale horn with culmen and tip of mandible dark. Legs greyish. **Juvenile (not illustrated):** Like adult but with duller and darker upperparts which are more heavily barred, with less distinct vermiculations on underparts. **Voice:** Unknown from review area. **Range:** Breeding throughout the Palaearctic, wintering south to Africa, the Indian subcontinent, southern China, and southern Japan, and continental South-East Asia. **Status and distribution:** Vagrant, with two records at low elevations, one each from Malim Nawar (Perak) on 17 October, 1990 (Lim, 2007), and Sungai Dua, Seberang Prai (Pulau Pinang) on 15 January, 2007 (Bakewell, 2007). **Habitat:** In review area, scrubland in a disused tin mine and on the fringe of a rice-field. **Habits:** Forages mostly on the ground. When alarmed, assumes a typical upright posture with bill pointed upwards and feathers sleeked down (Bakewell, 2007). Keeps to thick cover. Shy.

353. Speckled Piculet, *Picumnus innominatus* Plate 44
Belatik Bintik

Description: 10 cm (4 in). Tiny size with short bill and two white facial stripes with olive green upperparts and creamy white underparts marked with bold black spots diagnostic. **Male:** Has small orange patch on forecrown. **Female:** Entire crown green. **Voice:** A sharp *tsik*. **Range:** The Indian subcontinent and southern China discontinuously through South-East Asia to Sumatra and Borneo. **Status and distribution:** Locally common resident from 1000 m to at least 1400 m, recorded from Larut Hills (Perak) and the Titiwangsa Range from Cameron Highlands, south to Fraser's Hill (Pahang). May be more widespread along the Titiwangsa Range. Also on Gunung Tahan (Pahang). Sometimes seen along hill-station forest paths. **Habitat:** Forests, the forest edge. **Habits:** Usually solitary or in pairs, frequenting the middle storey, often within the vicinity of bamboo. Forages mostly by tapping on bamboo, vines, and thin branches. 'Drums' softly. Often participates in bird-waves.

354. Rufous Piculet, *Sasia abnormis* Plate 44
Belatik Jingga

Description: 9 cm (3.5 in). Tiny size with short bill and tail distinctive. Bright olive green upperparts contrasting with orange rufous underparts diagnostic. Orbital skin pinkish. **Male:** Yellow patch on forehead. **Female:** Rufous forehead. **Voice:** A high-pitched *kik-kik-kik-kik-kik*.. uttered persistently. **Range:** Tenasserim (Myanmar) and the Malay Peninsula to the Greater Sundas. **Status and distribution:** Fairly common resident from low elevations up to 1300 m, south to Johor. Also on Gunung Tapis (Pahang) and Gunung Rabong (Kelantan). **Habitat:** Forests, the forest edge. **Habits:** Usually solitary and unobtrusive, frequenting the middle and lower storeys. Taps softly on wood matter while foraging by gleaning insects from thin stems, creepers, and tall gingers.

355. Sunda Woodpecker, *Dendrocopus moluccensis* Plate 44
Belatuk Kecil Sunda

Description: 14 cm (5.5 in). From Grey-capped Woodpecker (356) by brown crown, dark brown upperparts, more prominent dark malar stripe and white underparts marked with dark streaks. Male has narrow red streak behind eye. **Voice:** A short wheezy trill, *ki-ki-ki-ki-ki-ki*.. **Range:** The Malay Peninsula to the Greater and Lesser Sundas to the Philippines. **Status and distribution:** Fairly common resident at low elevations along both coastal strips, south to Singapore, where it is the most common woodpecker. Sometimes extends inland in coastal districts. More common than Grey-capped Woodpecker in mangroves. **Habitat:** Mangroves, coastal scrub, plantations, gardens. **Habits:** Similar to Grey-capped Woodpecker.

356. Grey-capped Woodpecker, *Dendrocopos canicapillus* Plate 44
Belatuk Kecil Ubun Kelabu

Description: 15 cm (6 in). Small size with grey crown, black upperparts with white barring together with brownish buff underparts with dark streaks diagnostic. Dark malar stripe usually indistinct. Central tail feathers usually unbarred. Male has narrow red streak behind eye. **Voice:** A shrill trilling cadence, *pic-chiru-chiru-chihihih:* (M and W). **Range:** The Indian subcontinent to eastern Russia, China and Korea, south to South-East Asia, Sumatra and Borneo. **Status and distribution:** Locally common resident principally at low elevations, sometimes above 900 m, south to Johor. Locally extinct in Singapore (W and H). **Habitat:** Mangroves, the forest edge, beach scrub, plantations, wooded gardens. **Habits:** Usually frequents the canopy and middle storeys solitarily or in pairs. Probes for insects and grubs in wood bark and dead rotting wood. Usually unobtrusive. 'Drums softly'.

357. Rufous Woodpecker, *Celeus brachyurus* — Plate 45
Belatuk Biji Nangka

Description: 25 cm (10 in). Medium size with short nuchal crest and short black bill together with rufous plumage marked with black bars on upperparts diagnostic. Male has red cheek patch. **Voice:** A sharp and rapid *kwee-kwee-kwee-kwee*, on a descending scale. **Range:** The Indian subcontinent and southern China through South-East Asia to the Greater Sundas. **Status and distribution:** Fairly common resident from low elevations up to 1050 m, south to Singapore, where it is uncommon. **Habitat:** Forests, the forest edge, scrub, plantations, gardens. **Habits:** Frequents the canopy and middle storey. Generally unobtrusive. Sometimes very vocal. Excavates nest cavities in the nests of tree ants. 'Drums' loudly.

358. White-bellied Woodpecker, *Dryocopus javensis* — Plate 45
Belatuk Hitam-Putih

Description: 43 cm (17 in). Large size with black plumage and white belly together with red nuchal crest diagnostic. **Male:** Crown red and concolorous with crest. Has red malar stripe. **Female:** Crown black. Lacks red malar stripe. **Voice:** A loud *kiau*, on a falling tone. Also a sharp *kiau-kiau-kiau*.. in flight. **Range:** The Indian subcontinent, China and Korea, south to South-East Asia, the Greater Sundas, Bali and the Philippines. **Status and distribution:** Locally common resident at low elevations up to 300 m, south to Singapore where it is very rare, mainly in the Bukit Timah Nature Reserve, Nee Soon, and the MacRitchie Reservoir. Also on Pulau Pinang. **Habitat:** Forests, the forest edge, mangroves plantations. **Habits:** Frequents the canopy and middle storey solitarily or in pairs. Noisy and conspicuous. 'Drums' loudly.

359. Banded Yellownape, *Picus miniaceus* — Plate 44
Belatuk Merah Berjalur

Description: 25 cm (10 in). Red crest with yellow tip, red wings, dull greenish mantle with pale wavy barrings together with chestnut underparts marked with whitish barrings diagnostic. **Male:** Red sides of head. **Female:** Sides of head brownish with white spots. **Voice:** A short strident *pi-eew*, on a falling tone. **Range:** Tenasserim (Myanmar) and the Malay Peninsula to the Greater Sundas. **Status and distribution:** Fairly common resident principally at low elevations, south to Singapore. Has been recorded at 1200 m in secondary growth at Cameron Highlands (Pahang). **Habitat:** Forests, the forest edge, mangroves, coastal scrub, plantations. **Habits:** Usually frequents the middle and lower storeys in pairs or small groups. Forages largely among vines and rotting wood.

360. Crimson-winged Yellownape, *Picus puniceus* Plate 44

Belatuk Sayap Merah

Description: 25 cm (10 in). Bright olive green upperparts and underparts contrasting with red wings together with red crown and yellow nuchal crest diagnostic. Has white streaks on flanks. Male has red malar stripe. **Voice:** A descending cadence of up to 7 notes, *keew-keew-keew-keew*... Also a short distinct strident *pi-eew*. **Range:** Tenasserim (Myanmar) and the Malay Peninsula to the Greater Sundas. **Status and distribution:** Fairly common resident from low elevations up to 820 m, south to Johor. Locally extinct in Singapore since 1950 (L and G). Has been recorded above 1100 m at Fraser's Hill (Pahang). **Habitat:** Forests, the forest edge, coastal scrub, plantations. **Habits:** Usually frequents the canopy and middle storey, 'Drums' weakly.

361. Lesser Yellownape, *Picus chlorolophus* Plate 44

Belatuk Tengkuk Kuning Kecil

Description: 27 cm (10.5 in). From Greater Yellownape (363) by smaller size, red on head, long narrow white facial stripe, narrow whitish bars on underparts and unmarked chestnut flight feathers. **Male:** Red supercilium and malar stripe. **Female:** Lacks red malar stripe. Red on head less extensive, Narrow white facial stripe shorter and less distinct. **Voice:** A high pitched disyllabic *kweee-ee*, repeated at intervals. **Range:** The Indian subcontinent and southern China discontinuously through South-East Asia and the Malay Peninsula to Sumatra. **Status and distribution:** Locally common resident, recorded at 1060 m in the Larut Hills (Perak) and the Titiwangsa Range from Cameron Highlands south to Genting Highlands (Pahang). Common at the hill-stations and sometimes enters bungalow gardens. **Habitat:** Forests, the forest edge. **Habits:** Usually frequents the canopy and middle storey solitarily but sometimes in small loose groups. Participates in bird-waves.

362. Chequer-throated Yellownape, *Picus mentalis* Plate 44

Belatuk Leher Berpetak

Description: 28 cm (11 in). From Crimson-winged Yellownape (360) by the lack of red on head. From Banded Yellownape (359) by brighter olive green plumage without barrings on upperparts and underparts. Yellow nuchal crest, chestnut sides of head and collar together with streaked black and white throat diagnostic. **Male:** Malar area streaked as throat. **Female:** Malar area concolorous with sides of head and collar. **Voice:** A series of shrill disyllabic notes *ki-yee,* uttered on an ascending scale with the stress on the first syllable. **Range:** Tenasserim (Myanmar) and the Malay Peninsula to the Greater Sundas. **Status and distribution:** Fairly common resident principally at low elevations up to 1200 m, south to Johor. Locally extinct in Singapore since 1950 (L and G). **Habitat:** Forests, the forest edge, landward fringe of mangroves. **Habits:** Frequents the middle and lower storeys. 'Drums' in short bursts.

363. Greater Yellownape, *Picus flavinucha.* Plate 44

Belatuk Tengkuk Kuning Besar

Description: 34 cm (13.5 in). Large size with golden yellow nuchal crest and dark brown flight feathers marked with black bars diagnostic. Upperparts olive green; tail black; underparts greyish; crown dark brown; sides of head greyish brown. **Male:** Creamy yellow throat and malar area with small area of white with dark spots below throat. **Female:** Throat and malar area brown; white area with dark spots below throat more extensive. **Voice:** A loud *keeou*. **Range:** The Himalayas and southern China through South-East Asia to Sumatra. **Status and distribution:** Locally common resident above 900 m in the Larut Hills (Perak) and the Titiwangsa Range from Cameron Highlands, south to Gunung Ulu Kali, in the Genting Highlands (Pahang). Common at the hill stations. **Habitat:** Forests, the forest edge. **Habits:** Usually frequents the canopy and middle storey solitarily, sometimes in small loose groups. Very noisy and conspicuous.

364. Laced Woodpecker, *Picus vittatus* Plate 44

Belatuk Bakau

Description: 30 cm (12 in). Greenish upperparts, greyish sides of head, blackish malar stripe marked with fine whitish streaks, yellowish buff throat and upper breast together with buff lower breast and belly marked with dark lacey streaks diagnostic. Rump pale greenish yellow, conspicuous in flight Uppertail coverts dark green. **Male:** Red crown. **Female:** Black crown. **Voice:** A loud and shrill monosyllabic *kyauk*, or a disyllabic *chaakauk*, given singly or in short series (W1). **Range:** Myanmar through Indo-China and the Malay Peninsula to the Greater Sundas and Bali. **Status and distribution:** Locally common resident principally at low elevations south to Singapore. Confined mainly to coastal districts up to 15–20 km inland. On Pulau Langkawi (Kedah) occurs far inland and on the hills. **Habitat:** Mangroves, scrub, plantations, coconut groves, rural gardens. **Habits:** Usually solitary or in pairs. Forages mainly by tapping and probing into dead and living trunks, stems, and branches. Also feeds on the ground.

365. Streak-breasted Woodpecker, *Picus viridanus* Plate 44

Belatuk Hijau Berjalur

Description: 30 cm (12 in). Like respective sexes of Laced Woodpecker (364) but throat greener with dark streaks, breast greener with whitish scalings and black malar stripe less distinct. Uppertail coverts olive green. **Voice:** The most frequently heard call is a loud explosive down-inflected *kwiep* or *kiek* (W1). **Range:** Myanmar (except northern), Thailand and the Malay Peninsula. **Status and distribution:** Uncommon and localized resident in the extreme north of Peninsular Malaysia; recorded around secondary growth and plantations around Timah Tasoh reservoir (Perlis) and in forest at Sintok (northern Kedah), south to Air Itam Dalam, Sungai Dua, Seberang Prai (Pulau Pinang) (BiW). **Habitat:** Forests, the forest edge, overgrown plantations, coastal scrub, mangroves. **Habits:** Like Laced Woodpecker.

366. Grey-faced Woodpecker, *Picus canus* Plate 44
Belatuk Gunung

Description: 33 cm (13 in). Olive green woodpecker with greyish sides of head, blackish midcrown and nape together with dark malar stripe distinctive. Flight feathers blackish, marked with whitish spots. **Male:** Red forecrown. **Female:** Entire crown black. **Voice:** A clear descending whistling cadence of 5–6 notes, *piu-piu-piu-piu-piu*. **Range:** Europe, temperate Asia and the Himalayas discontinuously through South-East Asia to Sumatra. **Status and distribution:** Very rare resident, known from Gunung Tahan (Pahang) from 900–1800 m and in the Cameron Highlands (Pahang) at 1300 m. **Habitat:** Forests. **Habits:** Little known.

367. Olive-backed Woodpecker, *Dinopium rafflesii* Plate 45
Belatuk Pinang Rimba

Description: 28 cm (11 in). Olive green upperparts and underparts together with bold black and white facial pattern diagnostic. Has three toes. **Male:** Red crown and pointed crest. **Female:** Black crown and crest. **Voice:** Loud *chak* notes, uttered singly. Also a rapid, descending stacato laugh, *chak-chak-chak-chak*... which sometimes ends with one or two spaced notes. **Range:** Tenasserim (Myanmar) through the Malay Peninsula to Sumatra and Borneo. Being forest dependent, it is declining due to loss and degradation of primary evergreen forest. **NT. Status and distribution:** Uncommon and localized resident from low elevations up to 1200 m in the Titiwangsa Range, south to Johor. Also on Gunung Tahan and Gunung Tapis (Pahang). Locally extinct in Singapore since 1950 (L and G). **Habitat:** Forests, the forest edge. **Habits:** Frequents the middle and lower storeys, usually in pairs. Forages largely on rotten wood.

368. Common Flameback, *Dinopium javanense* Plate 45
Belatuk Pinang Kecil/Belatuk Mas

Description: 30 cm (12 in). Golden mantle and wings, red lower back and rump, white face with two black malar stripes and largely black hindneck with white sides of neck together with whitish underparts marked with blackish scales distinctive. Throat and face tinged rufous or buff. Has three toes. Iris brown. **Male:** Crown and pointed crest red. **Female:** Crown and crest black with whitish spots from forehead to tip of crest. **Voice:** A shrill trilling cadence of brief to rapid notes, *ki-ki-ki-ki-ki-* to *kikikikikikiki*. Also a rapid *di-i-i-i-dit*, in flight. **Range:** The Indian subcontinent and South-East Asia to the Greater Sundas, Bali and the Philippines. **Status and distribution:** Fairly common resident at low elevations, south to Singapore. The most common woodpecker in wooded suburban areas. **Habitat:** Mangroves, the forest edge, peat swamp forest, coastal scrub, plantations, gardens. **Habits:** Usually in pairs. Forages largely on tree trunks and stems, probing soft bark for insects and grubs. Very vocal, a pair will regularly call out to each other. **Note:** Generally difficult to separate from Greater Flameback (369) from lateral view, especially at a distance. Best separated by colour of iris and hindneck pattern (see Noske, 1991).

Common Flameback
Dinopium javanese

Greater Flameback
Chrysocolaptes lucidus

369. Greater Flameback, *Chrysocolaptes lucidus* — Plate 45

Belatuk Pinang Besar

Description: 33 cm (13 in). From Common Flameback (368) by larger size, largely white hindneck with black sides of neck, yellow iris, longer and more robust bill and thin double black malar stripes separated by small white patch. Crest more shaggy and less pointed. **Male:** Red crest. **Female:** Black crest with white spots. **Voice:** A distinct series of sharp notes, *di-di-di-di-di-di*. . . reminiscent of a large cicada. Flight call similar to Common Flameback. **Range:** The Indian subcontinent through South-East Asia to Sumatra, Java, Bali, and the Philippines. **Status and distribution:** Uncommon and localized resident at low elevations, principally in coastal districts, south to Johor. Less common than Common Flameback. Locally extinct in Singapore since 1904 (L and G). **Habitat:** Mangroves, scrub, plantations, forests (on Pulau Langkawi only). **Habits:** Frequents the tree-tops, usually in pairs, 'Drums' loudly.

370. Bamboo Woodpecker, *Gecinulus viridis* — Plate 45

Belatuk Buluh

Description: 28 cm (11 in). Crestless woodpecker with olive green upperparts, brownish buff head tinged yellow, red rump, dark brownish underparts and whitish bill diagnostic. Has three toes. **Male:** Entire crown red. **Female:** Crown brownish buff and concolorous with head. **Voice:** A shrill and rapid cadence, *kyeek-kyeek-kyeek-kyeek*. **Range:** Myanmar, Thailand and parts of Indo-China to the Malay Peninsula. **Status and distribution:** Uncommon and localized resident from low elevations up to 1370 m, south to Selangor. **Habitat:** Forests, the forest edge. **Habits:** Usually solitary or in pairs, foraging largely among bamboo thickets. Excavates nesting hollows in bamboo.

371. Maroon Woodpecker, *Blythipicus rubiginosus* Plate 45
Belatuk Punggur.

Description: 23 cm (9 in). Chestnut maroon upperparts, dark brown underparts and pale yellowish bill diagnostic. Flight feathers and tail marked with indistinct pale bars. Male has red sides of neck and nape. **Voice:** A high-pitched *kik*, uttered frequently. Also a disyllabic *kik-ik*, with an upward inflection on the second syllable. Also a shrill descending cadence. **Range:** Tenasserim (Myanmar) and the Malay Peninsula to Sumatra and Borneo. **Status and distribution:** Fairly common resident from low elevations up to 1500 m, south to Johor. Locally extinct in Singapore since 1950 (L and G). **Habitat:** Forests, the forest edge. **Habits:** Usually solitary or in pairs, frequenting the middle and lower storeys. Forages mainly on dead wood. Not known to 'drum'.

372. Bay Woodpecker, *Blythipicus pyrrhotis* Plate 45
Belatuk Merah Gunung

Description: 30 cm (12 in). Large size with dark rufous upperparts marked with bold blackish bars together with pale yellowish bill diagnostic. Male has red sides of neck. **Voice:** A loud *keek-keek-keek-keek-keek*, on a descending scale. **Range:** The Himalayas and southern China discontinuously through South-East Asia to the Malay Peninsula. **Status and distribution:** Uncommon and localized resident above 1066 m in the Titiwangsa Range, from Cameron Highlands, south to Genting Highlands (Pahang). May sometimes be observed at the hill stations, particularly, along forest paths overlooking valleys. Also on Gunung Benom (Pahang). **Habitat:** Forests. **Habits:** Usually solitary or in pairs, frequenting mainly the middle storey. Not known to 'drum'.

373. Orange-backed Woodpecker, *Reinwardtipicus validus* Plate 45
Belatuk Dada Merah

Description: 30 cm (12 in). Tall crested woodpecker with blackish upperparts marked with chestnut wingbars and pale line running down upperparts from nape to rump. **Male:** Red crown and crest, red underparts and orange line running down back. **Female:** Dark brown crown and crest, greyish brown underparts and whitish line running down back. Bill yellowish in both sexes. **Voice:** A cadence of high-pitched notes, *kit-kit-kit-kit-kit-it* with an upward inflection on the last note. Also a rapid trill, *ki-i-i-i-i-i-ik*. **Range:** The Malay Peninsula to the Greater Sundas. **Status and distribution:** Locally common resident from low elevations up to 1220 m, south to Johor. Locally extinct in Singapore since 1950 (L and G). **Habitat:** Forests, the forest edge, coastal scrub, plantations. **Habits:** Usually inhabits the canopy and middle storey in pairs, sometimes in small parties. Very noisy and conspicuous. 'Drums' softly.

374. Buff-rumped Woodpecker, *Meiglyptes tristis* Plate 45
Belatuk Pinggul Kuning

Description: 19 cm (7.5 in). Small woodpecker with short black bill and slight nuchal crest. Black plumage with white bars (finer on underparts), pale buff rump and fulvous head diagnostic. Male has red malar stripe. **Voice:** A rapid trill, *ki-i-i-i-i-i-i*. Also a single *chit* in flight. **Range:** Tenasserim (Myanmar) and the Malay Peninsula to the Greater Sundas. **Status and distribution:** Fairly common resident at low elevations up to 1300 m, south to Johor. Locally extinct in Singapore since 1950 (L and G). **Habitat:** Forests, the forest edge. **Habits:** Usually frequents the canopy solitarily or in pairs. Generally unobtrusive. Sometimes forages with mixed species flocks.

375. Buff-necked Woodpecker, *Meiglyptes tukki* Plate 45

Belatuk Leher Kuning

Description: 22 cm (8.5 in). Small woodpecker with short blackish bill. Uniform dark brown plumage marked with narrow buff bars on mantle, wings, tail and entire underparts together with buff neck patch diagnostic. Male has red malar stripe. **Voice:** A high-pitched trill, *ki-ti-ki-ti-ki-ti-ki-ti-ki*. **Range:** Tenasserim (Myanmar) and the Malay Peninsula to Sumatra and Borneo. This forest dependent species is declining owing to habitat loss and degradation. **NT. Status and distribution:** Locally common resident from low elevations up to 1200 m, south to Johor. Also on Gunung Benom (Pahang) and on Pulau Pinang. Locally extinct in Singapore since 1950 (L and G). **Habitat:** Forests, the forest edge. **Habits:** Frequents the middle and lower storeys, solitarily or in pairs. Often forages with mixed species flocks.

376. Grey-and-buff Woodpecker, *Hemicircus concretus* Plate 44

Belatuk Kecil Berjambul.

Description: 13 cm (5 in). Small crested and short-tailed woodpecker. Black upperparts with pale yellowish scales and white rump together with greyish head and underparts diagnostic. Bill dark grey. Male has red crown. **Voice:** High pitched rattles *ki-ki-ki-ki-ki*... Also a sharp disyllabic *ki-yow* **Range:** Tenasserim (Myanmar) to the Malay Peninsula and the Greater Sundas. **Status and distribution:** Locally common resident from low elevations up to 1100 m, south to Johor. Locally extinct in Singapore since 1950 (L and G). **Habitat:** Forests, the forest edge. **Habits:** Frequents the canopy and middle storey in loose flocks of up to 10. Mostly gleans for insects among foliage, sometimes hanging upside down. Seldom probes into wood matter. Often forages with mixed species flocks.

377. Great Slaty Woodpecker, *Mulleripicus pulverulentus* Plate 45

Belatuk Kelabu

Description: 50 cm (20 in). Very large size with uniform slaty grey plumage marked with whitish spots on nape and neck, yellowish white bill and buff throat diagnostic. Male has red malar stripe. **Voice:** A series of loud cackling brays. Also a short *wu-ik* uttered both in flight and at rest. **Range:** North India through South-East Asia to the Greater Sundas and the Philippines. Has declined rapidly due to loss of primary forest. **VU. Status and distribution:** Locally common resident from low elevations up to 200 m, south to Johor. Locally extinct in Singapore since 1950 (L and G); an individual in Changi in the 1970s was probably a non-breeding dispersant from Johor.

Habitat: Forests, the forest edge mangroves. **Habits:** Usually frequents the canopy in small parties. Very noisy and conspicuous.

BROADBILLS

ORDER *Passeriformes* **FAMILY** *Eurylaimidae*

World: *15 species* **Peninsular Malaysia:** *7 species*

Rather thickset with broad bills, large heads and relatively short tails. Sluggish, arboreal forest inhabitants. Feed mainly on insects and other small animals. Some species also take fruit. Flight is swift and direct over short distances. Calls loud and diagnostic. Nests are rather large distinctive hanging structures of dead vegetation (egg chamber regularly lined with fresh green leaves), very often over forest streams. Solitary or in pairs but may sometimes gather in larger groups. Sexes alike in most species. Most species in the review area are relatively secure but some have declined greatly over the years due to loss and fragmentation of lowland rainforest, and in global terms rank as **NT.** This is stated appropriately in each species account below. Represented by the following groups and genera: **Green-broadbills,** genus *Calyptomena* (378). A distinct group of small to large broadbills, characterized by chunky bodies, predominantly bright green plumage, short stumpy tails, and a tuft of feathers covering the maxilla and nostrils. Predominantly fruit eaters, but will also take insects. This genus is endemic to the Sunda subregion and also includes the Black-breasted Green-broadbill *C. whiteheadi* and Blue-bellied Green-broadbill *C. hosii*, which occur only in Borneo. Only one species occurs in the review area. **Typical broadbills,** genera *Cymbirhynchus* (379) and *Eurylaimus* (382, 383). Medium-sized with rather thickset bills and clad mostly in combinations of red, black, or yellow. Genus *Psarisomus* (380). Long-tailed with predominantly green plumage. Inhabits montane rainforests. Genus *Serilophus* (381). Largely grey and chestnut clad broadbill, inhabiting largely submontane and montane elevations. Genus *Corydon* (384). Large and sluggish with broad pinkish bill. Highly gregarious forest inhabitants.

378. Little Green-broadbill, *Calyptomena viridis* Plate 46

Takau Hijau/Burung Seluwit

Description: 19 cm (7.5 in). Chunky appearance with short stumpy tail and bright grass green plumage diagnostic. **Male:** Maxilla covered with feathers. Spot on neck and wingbars black. **Female:** Paler green and lacking the black neck spot and wingbars. **Voice:** A bubbling rattle, *ooo-turrr*, with an upward inflection on the second note. **Range:** Tenasserim (Myanmar) through the Malay Peninsula to Sumatra and Borneo. **NT. Status and distribution:** Locally common resident principally at low elevations and hills up to 760 m, south to Johor. Locally extinct in Singapore since 1941 (L and G). **Habitat:** Forests. **Habits:** Frequents the middle and lower storeys. Feeds mainly on fruit but will also take insects. Usually solitary or in pairs but small parties may frequently gather to feed in fruiting trees. Often builds nest very low, suspended from twigs above the forest floor, well away from water.

379. Black-and-red Broadbill, *Cymbirhynchus macrorhynchus* Plate 46
Takau Hitam-Merah/Burung Rakit

Description: 25 cm (10 in). Blackish upperparts, narrow white streak on wing, reddish maroon rump, black breast band separating reddish maroon throat patch and underparts together with blue maxilla and yellow mandible diagnostic. **Juvenile (not illustrated):** Throat and breast grey-brown. From juvenile Banded Broadbill (382) by lack of yellow on upperparts. **Voice:** Short and coarse grating notes, similar to cicadas. Also a brief rising trill like Black-and-Yellow Broadbill (383) but shorter and softer. **Range:** Myanmar and southern Indo-China through the Malay Peninsula to Sumatra, Bangka, Belitung and Borneo. **Status and distribution:** Locally common resident at low elevations and hills up to 300 m, south to Johor. Probably the most commonly seen broadbill in rural riverine habitats throughout the lowlands in Peninsular Malaysia. Locally extinct in Singapore since 1950 (L and G). **Habitat:** Forests, the forest edge, plantations, the mangrove fringe. **Habits:** Frequents the middle and lower storeys, usually near water courses. Generally unobtrusive, perching quietly in dense foliage. Will also perch on low branches overlooking water. Nests are constructed mostly on branches overhanging water but telephone wires are also utilized. Shy.

380. Long-tailed Broadbill, *Psarisomus dalhousiae* Plate 46
Takau Ekor Panjang

Description: 28 cm (11 in). Long blue tail, bright green plumage and black and yellow head pattern diagnostic. Bill and spot on side of head pale green; patch at base of primaries blue; underside of tail black. **Juvenile (not illustrated):** Tail shorter, plumage duller, head green. **Voice:** A shrill high pitched cadence of 5–8 notes, *pseeu-pseeu-pseeu*...on a descending scale. Also a short rasping *pseeup*. **Range:** The Himalayas and Myanmar, discontinuously through South-East Asia and the Malay Peninsula to Sumatra and Borneo. **Status and distribution:** Locally common resident from 820–1500 m in the Larut Hills (Perak) and the Titiwangsa Range south to Ulu Langat (Selangor); also as low as 250 m in the Royal Belum State Park (Perak). Also on Gunung Tahan (Pahang). Common and conspicuous at the hill-stations. **Habitat:** Forests. **Habits:** Frequents the canopy and middle storey solitarily or in pairs but may gather in larger groups during breeding. Flicks tail upwards while perched. Generally unobtrusive, feeding in the tree tops. Noisy during breeding. Shy.

381. Silver-breasted Broadbill, *Serilophus lunatus* Plate 46
Takau Dada Perak/Burung Tadah Hujan

Description: 18 cm (7 in). Broad black supercilium, greyish white crown, light blue bill and narrow lemon yellow orbital skin diagnostic. Upperparts greyish brown with chestnut rump; underparts greyish white; wings black with blue patch; tail black with white outer tail feathers. Female has narrow silvery white breast band. **Voice:** A clear whistle, *pee-yu*, while perched. **Range:** The Himalayas, southern China and Myanmar through South-East Asia and the Malay Peninsula to Sumatra. **Status and distribution:** Locally common resident from 250–1200 m, in the Larut Hills (Perak) and the Titiwangsa Range, including the Gombak Valley (Selangor) south to Gunung Telapa

Buruk (Negeri Sembilan). Common at Bukit Larut (Perak) and the lower elevations of Genting Highlands and Fraser's Hill (Pahang) and The Gap (Selangor). Also at 230 m in the Royal Belum State Park (Perak). **Habitat:** Forests, the forest edge. **Habits:** Frequents the canopy and middle storey. Usually solitary or in pairs but may sometimes gather in small parties. Generally unobtrusive, sitting quietly on a high perch. Will often perch on telephone or electricity cables at the lower elevations of the hill-stations. Shy.

382. Banded Broadbill, *Eurylaimus javanicus* Plate 46

Takau Berjalur/Takau Rimba

Description: 23 cm (9 in). From Black-and-Red Broadbill (379) by entirely red head, all blue bill, yellow streakings on upperparts and rump and paler vinous red underparts. From Black-and-Yellow Broadbill (383) by larger size and lack of white collar. Male has narrow black band on upper breast. **Juvenile (not illustrated):** Dull brownish head, speckled greyish white underparts. **Voice:** An explosive *wheeow*, followed by a rapid rising high pitched bubbling trill which slows down at its peak before tapering off, the whole process lasting for about six seconds. Often performed antiphonally. Also a loud *wiuk*. **Range:** Tenasserim (Myanmar) and Indo-China through the Malay Peninsula and the Greater Sundas. **Status and distribution:** Locally common resident in the lowlands and hills up to 1100 m, south to Johor. Locally extinct in Singapore since 1950 (L and G). **Habitat:** Forests. **Habits:** Frequents the canopy and middle storey, usually solitarily or in pairs. Often remains unobtrusive in dense foliage. Nests very high up in forest trees, often away from water.

383. Black-and-yellow Broadbill, *Eurylaimus ochromalus* Plate 46

Takau Hitam-Kuning

Description: 17 cm (6.5 in). Smallish with stumpy tail. Black head and narrow breast band separated by broader white collar diagnostic. Mantle, wings and uppertail coverts black with rump, lower back and streaks on wings yellow. Breast pale vinaceous pink grading into lemon yellow lower belly and undertail coverts. Iris yellow. **Male:** Complete breast band. **Female:** Breast band usually broken. **Juvenile (not illustrated):** Pale yellow supercilium, white throat and grey breast. Lacks breast band. **Voice:** A high pitched bubbling trill on an ascending scale, uttered for about ten seconds. Similar to that of Banded Broadbill (382) but lacks the initial explosive note with the acceleration rate more gradual. **Range:** Tenasserim (Myanmar) and the Malay Peninsula to Sumatra and Borneo. **NT. Status and distribution:** Fairly common resident at low elevations and hills up to 970 m, south to Johor. Locally extinct in Singapore since 1879 (L and G). **Habitat:** Forests. **Habits:** Frequents the canopy and middle storey. Usually solitary or in pairs, sometimes in small parties of 3–4. Generally unobtrusive, perched quietly on a high branch. Often nests far from water. A known brood host of the Indian Cuckoo (247).

384. Dusky Broadbill, *Corydon sumatranus* Plate 46

Takau Kelabu

Description: 28 cm (11 in). Large and sluggish with blackish plumage. Broad pinkish bill and orbital skin together with buff throat and subterminal tail band of white spots diagnostic. Broad white bar at base of primaries conspicuous in flight. Yellow-orange streak on back usually inconspicuous at rest but shows well in flight. **Juvenile (not illustrated):** Bill paler, throat blackish, lacks streak on back. **Voice:** A cadence of shrill, thin notes, *ki-ii, ki-ii, ki-ii*... with a rising inflection. Also a shrill *pseeeo*, on a descending tone. **Range:** Tenasserim (Myanmar) and Indo-China through the Malay Peninsula to Sumatra and Borneo. **Status and distribution:** Uncommon and localized resident principally at low elevations and hills up to 1050 m, south to Johor. Locally extinct in Singapore since 1879 (L and G). **Habitat:** Forests. **Habits:** Frequents the canopy and middle storey. Usually in small noisy groups of up to ten, even at nest. Sometimes solitary.

PITTAS

ORDER *Passeriformes* **FAMILY** *Pittidae*

World: *33 species* **Peninsular Malaysia and Singapore:** *7 species*

Medium to small sized birds of the forest understorey, characterized by rather plump bodies, short tails and long legs. Wings short and rounded but flight is strong. Move on the ground in a series of short bounds. Most are brilliantly coloured with patches of blue, red and green which are often lost in the poor light of the forest floor. Some have white wing patches which are conspicuous in flight. Many are secretive and difficult to observe. Their distinctive calls, some of which have explosive qualities, are often the indication of their presence. Roost off the ground in low branches vines, and liane stems. Sexes alike or nearly so in many species. Solitary or in pairs.

385. Rusty-naped Pitta, *Pitta oatesi* — Plate 47

Burung Pita Gunung

Description: 25 cm (10 in). **Male:** Bright rufescent head and underparts, black eyeline, whitish throat together with saturated dark green upperparts diagnostic. **Female:** The female of *P. o. deborah* of the Malay Peninsula has not been described and is probably similar to the nominate *P. o. oatesi* which resembles male but with head and underparts duller. Upperparts duller with a rusty tinge. This species shows a conspicuous blue rump in flight. **Voice:** A sharp disyllabic *cho-wit*, with an upward inflection on the second note. **Range:** South-West China through South-East Asia discontinuously to the Malay Peninsula. **Status and distribution:** Uncommon and localized resident, generally from 1000–1200 m recorded in the Larut Hills (Perak) and the Titiwangsa Range from the Cameron Highlands, Fraser's Hill and Genting Highlands (Pahang) south to Gunung Telapa Buruk (Negeri Sembilan). **Habitat:** Forests **Habits:** Frequents gullies where it forages on the forest floor, usually picking through leaf litter. Sometimes appears along hill-station forest paths but quickly disappears on approach. Very shy and secretive.

386. Giant Pitta, *Pitta caerulea* — Plate 47

Burung Pita Besar

Description: 29 cm (11.5 in). Large size with thickset bill. Blackish brown band around neck in both sexes diagnostic. **Male:** Upperparts bright blue; hindcrown and nape blackish brown. **Female:** Upperparts rusty brown; rump and tail bright blue; hindcrown and nape scaled blackish brown. Underparts greyish buff in both sexes. **Voice:** A soft disyllabic mournful whistle, descending in pitch and with a downward inflection, lacking the loud explosive qualities of those of many of its smaller congeners. **Range:** Tenasserim (Myanmar) through the Malay Peninsula to Sumatra and Borneo. Declining due to loss and degradation of lowland rainforest. **NT. Status and distribution:** Rare and localized resident at low elevations, south to Johor. **Habitat:** Forests. **Habits:** Forages on the forest floor like other pittas. Shy and will quickly fly off on approach. Known to utilize stones as anvils for breaking the shells of snails on which it feeds. Very secretive and other habits little known.

387. Banded Pitta, *Pitta guajana* — Plate 47

Burung Pita Berjalur

Description: 23 cm (9 in). Both sexes have a conspicuous broad yellow supercilium which grades into deep orange on the nape. Central crown stripe and broad eyeline black; mantle golden-brown; tail deep blue; streaks on wing and throat white with yellowish tinge on sides of neck. **Male:** entire underparts dark blue with orange bands across breast. **Female:** Entire underparts marked with narrower buff brown and black bands. **Voice:** An explosive *pauk* on a falling tone. Also a short and strident whirring note. **Range:** The Malay Peninsula, Sumatra, Borneo, and Java. **Status and distribution:** Locally common resident at low elevations, south to Johor. **Habitat:** Forests. **Habits:** A typical forest bird, it forages in the ground storey either singly or in pairs. Frequents areas with thickets. Will sometimes appear along forest paths. Shy and retiring but very vocal and often responds to an imitation of its call. Roosts fairly low in palms and thickets.

388. Garnet Pitta, *Pitta granatina* — Plate 47

Burung Pita Delima

Description: 18 cm (7 in). Forehead, throat and cheek black; mantle and wings dark electric violet-blue; wing coverts, narrow line below hindcrown, rump, uppertail coverts and wash on upper breast bright iridescent pale blue; crown, nape and underparts bright red. Appears all dark in poor light. **Voice:** A long drawn out ventriloquial monotone whistle, with a gradual rise at the end. Very similar to Rail-babbler (393) which is longer and level throughout. **Range:** Tenasserim (Myanmar) through the Malay Peninsula to Sumatra and Borneo. Declining due to loss and degradation of lowland rainforest. **NT. Status and distribution:** Locally common resident at low elevations, south to Johor. Locally extinct in Singapore since 1949 (L and G). **Habitat:** Forests. **Habits:** Frequents the wetter areas of forests. Will often perch on branches and vines. Frequently responds to an imitation of its call and will come closer to investigate.

389. Hooded Pitta, *Pitta sordida* — Plate 47

Burung Pita Hijau/Burung Gembala Pelanduk

Description: 19 cm (7.5 in). Black hood with brown crown contrasting with bright green mantle, breast, and upper belly diagnostic. Rump and patch on shoulder pale bright blue, primaries black, lower belly and undertail coverts red. White wing patch conspicuous only in flight. **Voice:** A fluty *raew-raew* (L and R) almost similar in quality to that of the Blue-winged Pitta (390). Also a short and squeaky woodpecker like *keew* when alarmed. **Range:** Breeding in northern India through Myanmar and southern China through South-East Asia and the Malay Peninsula. Migrants move on south to the Philippines, the Greater Sundas and New Guinea. **Status and distribution:** Uncommon and localized resident only in Kedah in the lowlands and hills up to 140 m. Elsewhere uncommon passage migrant and winter visitor, south to Singapore. Occurs above 900 m only during spring and autumn passage. **Habitat:** Forests **Habits:** Like other pittas, forages on the forest floor. Keeps more to forest cover especially in gullies and near bamboo thickets. Roosts in trees.

390. Blue-winged Pitta, *Pitta moluccensis* Plate 47

Burung Pita Sayap Biru

Description: 20 cm (8 in). Buff lateral coronal stripe distinctive. Broad eyeline and primaries black; mantle dark green; wings, rump and uppertail coverts bright blue; breast and belly bright rufous orange; throat white; undertail coverts red; white patch on wing conspicuous in flight. **Voice:** A high pitched disyllabic double whistle *wer-wow-wer-wou*. Also a sharp woodpecker like *keew* when alarmed. **Range:** Southern China through Myanmar and Thailand to the Malay Peninsula. Migrants move on south to Sumatra and Borneo with stragglers reaching Australia. **Status and distribution:** Uncommon and localized breeding summer visitor at low elevations only on Pulau Langkawi (Kedah), Perlis, and mainland Kedah, south to Kuala Tahan (Pahang). Elsewhere, locally common passage migrant and winter visitor at low elevations, south to Singapore. Occurs above 900 m only during autumn and spring migration. **Habitat:** Forests, mangroves, wooded suburbs, wooded areas in scrub and grassland. **Habits:** Frequently calls on breeding grounds, less frequently in wintering quarters. Like other pittas, forages on the ground and may sometimes be observed at the forest edge. Roosts in trees. **Note:** The similar looking Fairy Pitta *Pitta nympha* of China, Korea and Japan winters south to Borneo and could turn up in Peninsular Malaysia or Singapore. Has buff lateral coronal stripe but underparts are paler and greyer with wing and rump sky blue.

391. Mangrove Pitta, *Pitta megarhyncha* Plate 47

Burung Pita Bakau

Description: 20 cm (8 in). Like Blue-winged Pitta (390) but lacks the buff lateral coronal stripe, bill proportionately longer and heavier and orange-rufous breast and belly duller. **Voice:** A double whistle, *woo-e-oo weow*, with a rise on the initial note, slower and lower pitched than Blue-winged Pitta and most often delivered from a perch. Also a sharp woodpecker like *keew* when alarmed. **Range:** Tenasserim (Myanmar) through the Malay Peninsula to Sumatra and the Riau Archipelago. Being a mangrove specialist, populations are declining due to loss and degradation of mangroves. **NT. Status and distribution:** Uncommon and localized resident, confined mainly to the existing pockets of mangroves along the west coast, from the Merbok estuary (Kedah) south to the Kukup Mangrove Forest Reserve (Johor); on the east coast rarer with sightings from the Kertih estuary (Terengganu) south to the estuaries of the Sungai

Pahang and Sungai Kuantan (Pahang) and Mersing Bay (Johor) BiW). In Singapore, rare resident on Pulau Tekong and Pulau Ubin. **Habitat:** Mangroves. **Habits:** Hops along the muddy mangrove floor during low tide to feed largely on crabs and other crustacea. Frequently nests on the mounds of mud lobsters. Often calls from a high perch. Not shy and may sometimes allow close approach. Very vocal when breeding.

AUSTRALO-PAPUAN WARBLERS

ORDER *Passeriformes* **FAMILY** *Acanthizidae*

World: *62 species* **Peninsular Malaysia and Singapore:** *1 species*

(This cosmopolitan family has its centre of distribution in Australasia and also includes bristlebirds, mouse-warblers, scrubwrens, thornbills, and whitefaces, all of which do not occur in the review area. For this purpose only the gerygones are described below.)

Small sized active birds with relatively short tails and bills. Plumage is rather drab with shades of olive, brownish, or greyish upperparts with paler, sometimes yellowish underparts. Known for their calls, in many species, consist of pleasing repeated cadences. Construct globular hanging nests which consist largely of fibres and rootlets.

392. Golden-bellied Gerygone, *Gerygone sulphurea* Plate 61
Kelicap Perepat

Description: 9 cm (3.5 in). Tiny size. From *Phylloscopus* warblers (480–487) by smaller size and lack of pale supercilium and wingbars. Dark grey brown upperparts contrasting with yellow underparts diagnostic. Has subterminal band of white spots on underside of tail. **Juvenile (not illustrated):** Underparts whitish, tinged yellow. **Voice:** A soft, high-pitched wheezy song of up to 8 notes, uttered up and down the scale. **Range:** Southern Indo-China and southern Thailand to the Greater and Lesser Sundas. **Status and distribution:** Fairly common resident from low elevations up to 970 m, south to Singapore. **Habitat:** Mangroves, the mangrove edge, forests, the forest edge, plantations, wooded gardens. **Habits:** Frequents tree tops and forest canopies, usually solitarily or in pairs. Gleans insects from the foliage and branches. Very vocal and also sings during the heat of the afternoon. Very territorial and has been observed chasing away intruding migrant *Phylloscopus* warblers during the winter months. A known brood host of the Little Bronze-cuckoo (252).

RAIL-BABBLER AND ALLIES

ORDER *Passeriformes* **FAMILY** *Eupetidae*

World: *10 species* **Peninsular Malaysia:** *1 species*

(This cosmopolitan family has its centre of distribution in Australasia and also includes whipbirds and jewel-babblers, all of which do not occur in the review area.

For this purpose only the Rail-babbler will be fully described in the species account below.)

393. Rail-babbler, *Eupetes macrocerus* Plate 59

Kekicau-Sintar

Description: 29 cm (11.5 in). Long neck, legs and tail distinctive. Bill long with hooked tip. Plumage bright rufescent with broad white supercilium, black mask and band to the side of head with narrow blue streak below it. Forehead rufous buff, crown, nape and throat chestnut. **Voice:** A long drawn out ventriloquial whistle on a level note, similar to Garnet Pitta (388) but is longer, flatter and higher pitched. Also soft clicking sounds while foraging on the ground (Lee, 2009). **Range:** The Malay Peninsula, Sumatra, Borneo and the Natunas Declining due to small range together with forest clearance and degradation. **NT. Status and distribution:** Uncommon and localized resident principally at low elevations but recorded up to 1067 m, south to Johor. **Habitat:** Forests. **Habits:** Frequents the ground storey, foraging on the forest floor among leaf litter and under fallen timber. Sometimes calls on a fallen log. When vocalizing, extends neck and inflates bare blue throat patch, balloon-like from the sides of neck. Moves about with a walking gait with neck extended. Shy.

INCERTAE SEDIS

ORDER *Passeriformes* **FAMILY** *Uncertain*

World: *6 species* **Peninsular Malaysia:** *3 species*

Wood-shrikes, genus *Tephrodornis* (394). Medium size with long graduated tails, strong bills, and broad blackish eyelines. Sexes alike or nearly so. Usually in small flocks. Formerly placed together with Campephagidae (Cuckooshrikes, Trillers, Minivets, and Flycatcher-shrikes). **Philentomas,** genus *Philentoma* (395, 396). Stocky and stout-billed forest insectivores, usually inhabiting the lower and ground storeys of forest. Sexually dimorphic, the males are dark bluish, the females lighter blue or brownish. Usually solitary or in pairs. Formerly placed together with Monarchidae (Monarchs and Paradise-flycatchers).

394. Large Wood-shrike, *Tephrodornis virgatus* Plate 48

Rembah Rimba Besar

Description: 23 cm (9 in). Thickset appearance, strong black bill, broad blackish eyeline and white rump diagnostic. **Male:** Greyish upperparts contrasting with whitish underparts. **Female** (not illustrated): Upperparts browner than male and eyeline narrower and browner. **Voice:** A loud and rapid musical *pi-pi-pi-pi-pi...* **Range:** The Indian subcontinent and Myanmar through South-East Asia to the Greater Sundas. **Status and distribution:** Fairly common resident from low elevations up to 970 m, south to Johor. Locally extinct in Singapore since 1950 (L and G). **Habitat:** Mangroves, forests, the forest edge, wooded gardens. **Habits:** Frequents the canopy and crowns of smaller trees in small flocks. Usually noisy and conspicuous. Feeds largely on insects obtained from aerial swoops. Also frequents the edge of water courses where it will pick off insects from the water surface.

395. Rufous-winged Philentoma, *Philentoma pyrhoptera* Plate 51
Filentoma Sayap Merah

Description: 18 cm (7 in). Rufous mantle and wings distinctive in common morph male and female. **Male common morph:** Blue head, breast and mantle. **Male blue morph:** All blue plumage. Distinguished from female Maroon-breasted Philentoma (396) by smaller size and brighter blue plumage and from Pale Blue (601) and Verditer Flycatchers (609) by red iris. **Female:** Greyish head, brown back, and buff underparts. **Voice:** A soft disyllabic whistle *tu-huuu*, the second syllable stressed. Also a rising whistle *teu-ii*. Also harsh metallic chatters. **Range:** Tenasserim (Myanmar), southern Vietnam and the Malay Peninsula to Sumatra and Borneo. **Status and distribution:** Fairly common resident from low elevations up to 1000 m, south to Johor. Locally extinct in Singapore since 1950 (L and G). The blue morph male is very rare. **Habitat:** Forests. **Habits:** Frequents the middle and lower storeys, the latter more frequently. Very active, obtaining insect prey by both aerial hawking and gleaning from leaves and branches. Will sometimes descend to the ground.

396. Maroon-breasted Philentoma, *Philentoma velata* Plate 51
Filentoma Dada Ungu

Description: 20 cm (8 in). **Male:** Dark indigo blue plumage, black throat and sides of head together with dark maroon breast diagnostic. Appears all dark in poor light. **Female:** Duller and lacking the maroon breast. Iris red in both sexes. **Voice:** Territorial song is a long cadence of bell-like whistles on a descending scale. Also harsh grating metallic chatters. **Range:** Tenasserim (Myanmar) through the Malay Peninsula to the Greater Sundas. Being forest dependent, it has assumed to have experienced moderately rapid declines due to extensive loss of lowland rainforest throughout South-East Asia but is still not seriously threatened because of its ability to utilize secondary type habitats and also because it occurs in lower montane forest. **NT. Status and distribution:** Locally common resident from low elevations up to 1000 m, south to Johor. Locally extinct in Singapore since 1874 (L and G). **Habitat:** Forests. **Habits:** Usually hawks for insects in the middle storey. Often found within the vicinity of water.

WOODSWALLOWS

ORDER *Passeriformes* **FAMILY** *Artamidae*

World: *11 species* **Peninsular Malaysia:** *2 species*

A predominantly Australo-Papuan group of insectivorous birds, characterized by stocky bodies, thick conical bills, broad triangular wings and short square-cut tails. Feed largely on flying insects caught on the wing. Highly gregarious, they frequently gather in parties, huddling closely on bare branches. Often utilize telephone wires as vantage points.

397. Ashy Woodwallow, *Artamus fuscus* Plate 49
Tirjup-Lelayang Kelabu

Description: 16 cm (6.3 in). Head and upperparts brownish grey, lighter on breast and lower belly with white undertail coverts. From White-breasted Woodswallow (398) by browner lower breast and belly and narrower band on uppertail coverts. **Voice:** Unknown in review area. **Range:** The Indian subcontinent, south and south west China to South-East Asia. **Status and distribution:** Vagrant, an individual associating with a White-breasted Woodswallow on a service wire at Ulu Dedap (Perak) on 17 February 2008 (Ooi and Chiu, 2008). **Habitat:** In review area, ricefields. **Habits:** Unknown in review area. **Note:** The appearance of this largely sedentary species way off its normal range is a point that needs further observation and investigations. It is known to undertake local movements within its known range.

398. White-breasted Woodswallow, *Artamus leucorhynchus* Plate 49
Tirjup-Lelayang Dada Putih

Description: 17 cm (6.5 in). Brownish grey head, breast and upperparts contrasting sharply with white breast and rump. In flight, white rump diagnostic. Bill light bluish with dark tip. **Juvenile (not illustrated):** Throat and breast greyish with no demarcation between breast and belly. Upperparts browner, scaled white. **Voice:** A coarse rasping, rather shrike-like *wek-wek-wek-wek* uttered from a perch (W2). **Range:** The Andamans, The Malay Peninsula, the Sunda region to the Phillipines, Wallacea, New Guinea, north and east Australia, to the South-West Pacific. **Status and distribution:** Locally common resident confined to the rice growing areas of Sekinchan and Tanjung Karang, south to Kuala Selangor (Selangor) and north to Lumut, Teluk Intan, Bagan Datoh, Ulu Dedap, Hutan Melintang, and Bidor (Perak). The northernmost coastal limit is Kuala Gula (Perak) in October 1997 (W2). First seen in Sekinchan in January 1977 (Wells, 1983), presumably dispersants from Sumatra. Populations are still stable but seem to have been reduced after initial spread, perhaps due to widespread use of pesticides in the ricefields. **Habitat:** Open country, principally ricefields. **Habits:** Typically seen gliding effortlessly in the sky or skimming low over fields in pursuit of flying insects. Will often utilize service wires as perches. Wags tail sideways when alarmed. Aggressive and territorial, it will frequently mob larger birds.

IORAS

ORDER *Passeriformes* **FAMILY** *Aegithinidae*

World: *4 species* **Peninsular Malaysia and Singapore:** *3 species*

Small and slender arboreal insect eating birds with largely green and yellow plumage, with two conspicuous wing bars in most species. Plumage brighter in males. Bills short and pointed with maxilla black and mandible bluish grey. Legs short and thin. Have long white flank feathers which often overlap rump, making it look white. Forage mostly in the canopy by gleaning insects and caterpillars from the foliage. Nests are shallow and compact cup-shaped structures, suspended in the fork of branches.

399. Common Iora, *Aegithina tiphia* Plate 52
Burung Kunyit Kecil

Description: 15 cm (6 in). From Green Iora (400) by bright yellow to yellowish underparts. Shows great individual variation. **Breeding male:** Colour of crown and upperparts vary from dark green to black, underparts bright yellow. **Non-breeding male (not illustrated):** Black upperparts replaced with duller green (darker than female), yellow underparts duller. **Female:** Upperparts dull olive green; underparts duller yellow; wing bars white. **Juvenile (not illustrated):** Like female but duller with less distinct wing bars. **Voice:** A diagnostic long drawn out whistle, *wheeeeeee-piu*, the last note dropping abruptly. Also a descending *piu-piu-piu*... Also chattering notes while foraging. **Range:** India, southern China and Myanmar through South-East Asia to the Greater Sundas, Bali and Palawan (the Philippines). **Status and distribution:** Common and widespread resident at low elevations and hills up to 450 m, south to Singapore. Also on Pulau Pinang. A conspicuous garden bird. **Habitat:** Mangroves, open country, scrub, forested river banks, plantations, gardens. **Habits:** Frequents the tree tops usually in pairs or in small parties. Gleans insects from the foliage and branches. A known brood host of the Banded Bay Cuckoo (249) (W and H).

400. Green Iora, *Aegithina viridissima* Plate 52
Burung Kunyit Hijau

Description: 14 cm (5.5 in). From respective sexes of Common Iora (399) by greener plumage and yellow patch above and below eye (more distinct in male). **Male:** Plumage dark olive green; wings and tail black; wingbars white. **Female:** Like female Common Iora but yellow parts tinged green and wingbars yellowish. **Juvenile (not illustrated):** Like female but duller. **Voice:** A thin high-pitched wheezy *tse-tseew,tse-tseew*. **Range:** Tenasserim (Myanmar) through the Malay Peninsula to Sumatra and Borneo. Suspected to declining due to loss of habitat. **NT. Status and distribution:** Fairly common resident from low elevations up to 800 m, south to Johor. Locally extinct in Singapore since 1950 (L and G). **Habitat:** Forests, the forest edge, mangroves (on Pulau Langkawi only). **Habits:** Frequents the canopy in small parties. Gleans insects from the foliage and branches. A known brood host of the Banded Bay Cuckoo (249) (Khoo, 2001).

401. Great Iora, *Aegithina lafresnayei* Plate 52
Burung Kunyit Rimba

Description: 17 cm (6.5 in). From Common Iora (399) by larger size, longer bill and the lack of wingbars. **Male:** Colour of upperparts varies from dark olive green to black with green rump. **Female:** Olive green upperparts and yellow underparts duller. **Juvenile (not illustrated):** Like female but duller with an olive wash on underparts, particularly the flanks. **Voice:** A rapid and high pitched *chiu-chiu-chiu...* **Range:** Myanmar through Indo-China to the Malay Peninsula. **Status and distribution:** Uncommon and localized resident in the lowlands and hills up to 900 m, south to Johor. **Habitat:** Forests, the forest edge. **Habits:** Frequents the canopy in pairs or in small parties. Mainly gleans insects from the foliage and branches.

CUCKOOSHRIKES, TRILLERS, MINIVETS, AND FLYCATCHER-SHRIKES

ORDER *Passeriformes* **FAMILY** *Campephagidae*

World: *89 species* **Peninsular Malaysia and Singapore:** *11 species*

Largely arboreal with long pointed wings, relatively long and graduated tails, short bills and short legs. Feed mainly on insects gleaned from branches or captured in the air. Build shallow cup-shaped nests in the forks of high branches. Gregarious, foraging mostly in the canopy. Represented by the following groups and genera. **Cuckooshrikes,** genus *Coracina* (402–404). Large and robust with strong bills and predominantly soft grey plumage. Sexes differ slightly. Usually in small loose flocks. **Trillers,** genus *Lalage* (405). Small size with largely black and white plumage with prominent white wingbars. Female duller than male. Usually solitary or in pairs. **Minivets,** genus *Pericrocotus* (406–410). Medium size and slender build with long graduated tails. Most species have conspicuous wing bars and are brightly coloured; the males in most species are predominantly red and black while the females yellow and dark grey. Subadult males in most species have the red parts replaced by bright orange-yellow. **Flycatcher-shrikes,** genus *Hemipus* (411, 412). Small size and slender appearance with black and white plumage. Sexes differ slightly, the females having the black parts replaced with brown. Usually gather in small noisy flocks.

402. Javan Cuckooshrike, *Coracina javensis* Plate 48

Tirjup-Sewah Gunung

Description: 30 cm (12 in). Large size with thickset black bill. Entire upperparts grey with diffuse broad blackish eyeline and brown iris diagnostic. Lores black. Greyish breast merges into white belly and undertail coverts. **Voice:** A loud and piercing *shweep*, uttered both in flight and while perched. Also utters low whinnying notes in flight. **Range:** The Malay Peninsula, Java, and Bali. **Status and distribution:** Locally common resident above 1000 m in the Larut Hills (Perak) and the Titiwangsa Range south to Ulu Langat (Selangor). Also on Gunung Tahan and Gunung Benom (Pahang). Also on the Eastern Range (W2). Fairly regular at Cameron Highlands, Fraser's Hill and Genting Highlands (Pahang) but rare on Bukit Larut (Perak). **Habitat:** Forests, the forest edge. **Habits:** Frequents the canopy and crowns of tall trees, usually in pairs but often in groups of up to five. Noisy and conspicuous, flying aimlessly with deep and regular wingbeats. Prefers to perch on bare branches where it has the habit of raising each wing alternately, a good identification feature. Often perches on service wires at the hill-stations. Will frequently enter hill-station gardens.

403. Bar-bellied Cuckooshrike, *Coracina striata* Plate 48
Tirjup-Sewah Mata Putih

Description: 30 cm (12 in). From Javan Cuckooshrike (402) by white iris, barred underparts and the lack of diffused blackish eyeline (altitudinal ranges of both species do not overlap). **Male:** Greyish plumage with diffuse whitish barrings on lower belly. **Female:** Shows distinct grey and white barrings from breast to vent and also on rump. **Voice:** A loud ringing *kliu-kliu-kliu-kliu*, usually uttered in flight. **Range:** The Malay Peninsula, Sumatra and Borneo east to the Philippines. **Status and distribution:** Uncommon and localized resident at low elevations, south to Johor. Also on Pulau Tioman (Pahang). Locally extinct in Singapore since 1950 (L and G). **Habitat:** Forests, the forest edge, mangroves, and plantations. **Habits:** Frequents the canopy in small parties. Noisy and conspicuous, flying high at tree-top level.

404. Lesser Cuckooshrike, *Coracina fimbriata* Plate 48
Tirjup-Sewah Sayap Hitam

Description: 23 cm (9 in). Small and slender build with proportionately slender bill. **Male:** Overall dark grey with dark charcoal grey wings. Vent white with undertail coverts barred black and white. **Female:** Upperparts dark grey; underparts whitish with narrow greyish bars. Shows incomplete white eyering and whitish streaks on ear coverts. **Voice:** A clear whistled sequence, *whip-whip-whip-wait-wait-wait-wait-wait* (M and W). **Range:** Tenasserim (Myanmar) through the Malay Peninsula to the Greater Sundas and Bali. **Status and distribution:** Locally common resident from low elevations up to 970 m, south to Singapore, where it is very rare, confined to the Bukit Timah Nature Reserve (W and H). **Habitat:** Forests (including peatswamp), the forest edge. **Habits:** Frequents the canopy usually solitarily or in pairs. Often forages for insects with other arboreal birds.

405. Pied Triller, *Lalage nigra* Plate 48
Rembah Dahi Putih

Description: 18 cm (7 in). From Ashy Minivet (407) by proportionately shorter tail and the lack of white wingbars. Shows distinct white supercilium, narrow black eyeline, broad staggered white wingbars, greyish rump and white tip to tail. **Male:** Black crown, mantle, wings and tail with all white underparts. **Female:** Like male but black parts largely replaced with brown. Underparts and rump finely marked with narrow brownish bars. **Voice:** A rolling nasal *kyek..kyek..kyek..kyek..kyek*... on a descending scale. **Range:** The Malay Peninsula south to the Greater and Lesser Sundas, east to Sulawesi and the Philippines. **Status and distribution:** Fairly common resident at low elevations, south to Singapore. A common garden bird. **Habitat:** Mangrove edge, scrub, plantations, gardens. **Habits:** Usually solitary or in pairs. Keeps largely to leaf cover where it feeds largely on insects and small caterpillars.

Usually unobtrusive, its call often gives away its presence. Will also descend to forage on the ground.

406. Rosy Minivet, *Pericrocotus roseus* Plate 48
Burung Matahari Jambu

Description: 20 cm (8 in). In shape, structure, and tail length, appears to resemble Ashy Minivet (407) closest. **Male:** Brownish grey upperparts (darker grey on crown), reddish rump, red uppertail coverts together with whitish throat and rosy pink underparts diagnostic. Has pale grey supercilium which appears to be suffused rather than distinct. Also has broad red wing patch and red lines extending along the fringes of some tertiaries as well as some primaries. **Female:** Like male but red and pinkish areas replaced with pale yellow. The females of the other resident minivets in the review area have much brighter yellow underparts. **Juvenile or first winter male (not illustrated):** Like female but underparts have pink suffusion. Wings have some orange markings and uppertail coverts together with tail fringes are red. **Voice:** No information from review area. In Thailand, a whirring trill has been recorded from wintering birds (L and R) which is hardly distinguishable from the vocalizations of both Swinhoe's and Ashy Minivets (P.D. Round in W2). **Range:** Breeding in the foothills of the Himalayas from the the north and north-eastern part of the Indian subcontinent east to China, Myanmar, and northern Vietnam, wintering south to the rest of the Indian subcontinent and parts of continental South-East Asia to the Malay Peninsula. **Status and distribution:** Vagrant, first known from a single female at 350 m at the Bintang Hijau Forest Reserve in northern Perak on 1 January 2007 and a probable first summer male was also seen in the same area on 11 February 2007 (Bakewell and Tan, 2007). **Habitat:** In review area, logged forest. **Habits:** Like other minivets, frequents the canopy where it probably gleans insects from the foliage and branches. Observed associating with Ashy Minivet. Other habits little known.

407. Ashy Minivet, *Pericrocotus divaricatus* Plate 48
Burung Matahari Kelabu

Description: 20 cm (8 in). The only minivet without red or yellow markings. Mantle, rump and uppertail coverts pale grey. Underparts white. Tail and wings black. Has whitish wingbars which are indistinct at rest but conspicuous in flight. **Male:** Black hindcrown and nape together with white forecrown distinctive. **Female:** Has narrow white band from forehead to lores. Crown and nape grey, concolorous with mantle. Flanks and sides of neck have grey suffusion. See White Wagtail (662). **Voice:** A jingling trill with a metallic quality, frequently uttered in flight. **Range:** Breeding in eastern Russia, China, Korea and Japan (including the Ryukyu islands); wintering south to the Indian subcontinent, South-East Asia, Sumatra, Borneo, Sulawesi, and the Philippines. **Status and distribution:** Locally common passage migrant and winter visitor at low elevations, south to Singapore. **Habitat:** Mangroves, forests (including peatswamp), beach scrub, plantations, open country, wooded gardens. **Habits:** Frequents the canopy in small loose parties, foraging in the foliage or catching insects in the air. Flies jerkily from tree-top to tree-top.

408. Fiery Minivet, *Pericrocotus igneus* Plate 48
Burung Matahari Kecil

Description: 15 cm (6 in). **Male:** From male Scarlet Minivet (410) by smaller size, less robust appearance and lacking separate red patch on secondaries. Wingbar orange-yellow and not red. **Female:** From other female minivets by flame-orange rump and uppertail coverts. Underparts and band from forehead to lores bright yellow. **Voice:** A disyllabic *twee-eeet* on an ascending scale. **Range:** The Indian subcontinent east to South-East Asia, the Greater Sundas, Bali and Palawan (the Philippines). Populations likely to be declining due to forest clearance. **NT. Status and distribution:** Locally common resident from low elevations up to 1220 m, south to Johor. Locally extinct in Singapore since 1950 (L and G). **Habitat:** Forests, the forest edge, beach scrub especially with casuarinas. **Habits:** Frequents the canopy in pairs or in small loose parties. Feeds largely on insects caught in the air or gleaned from branches and foliage.

409. Grey-chinned Minivet, *Pericrocotus solaris* Plate 48
Burung Matahari Gunung

Description: 18 cm (7 in). **Male:** From male Scarlet Minivet (410) by greyish suffusion on throat and sides of head (often difficult to discern in poor or diffused lighting) and the lack of separate red patch on secondaries. **Female:** From female Scarlet Minivet by greyish forecrown and sides of head and the lack of yellow patch on secondaries. **Voice:** Low twittering notes while foraging. A rather high pitched *sri..sisi...* in flight. **Range:** The Himalayas and southern China discontinuously through South-East Asia, Sumatra and Borneo. **Status and distribution:** Fairly common resident above 975 m in the Larut Hills (Perak) and the Titiwangsa Range south to Ulu Langat (Selangor). Also on Gunung Tahan and Gunung Benom (Pahang). Common and conspicuous at the hill-stations where it sometimes overlaps with the Scarlet Minivet at the lower elevations especially at The Gap (Selangor) below Fraser's Hill and the lower elevations of Genting Highlands (Pahang). **Habitat:** Forests, the forest edge. **Habits:** Frequents the canopy in pairs or in small parties, foraging mainly in the foliage. Flies jerkily from tree-top to tree-top, usually uttering its call. Will sometimes descend to feed on insects and small caterpillars from the fronds of tree-ferns. Sometimes sallies after flying insects. Enters hill-station gardens. Sometimes tame.

410. Scarlet Minivet, *Pericrocotus flammeus* Plate 48
Burung Matahari Besar

Description: 18 cm (7 in). More robust than Fiery (408) and Grey-chinned Minivets (409). Both sexes show a roundish patch on inner secondaries of closed wing, just behind wingbar. **Male:** Head, throat, breast, upperparts and wings black; underparts, rump and wingbars and roundish wing patch scarlet. **Female:** Bright yellow forecrown diagnostic. Throat, entire underparts, rump, wingbars and patch bright yellow. Head and upperparts grey; wings black. **Voice:** A series of loud and piercing *sweeep* notes, usually uttered by foraging parties. **Range:** The Indian subcontinent and southern China east through South-East Asia to the Greater Sundas, Bali, Lombok, and the Philippines. **Status and distribution:** Fairly common resident from low elevations up to 1000 m, south to Singapore where it is very rare and confined to the Bukit Timah Nature Reserve and the Central Catchment Forest (W and H). **Habitat:** Forests, the forest edge. **Habits:** Frequents the canopy. Usually in small parties. Forages mainly in the foliage but will often hawk for flying insects at canopy level.

411. Bar-winged Flycatcher-shrike, *Hemipus picatus* Plate 48
Sambar-Tirjup Sayap Putih

Description: 15 cm (6 in). From male Little Pied Flycatcher (592) by proportionately longer tail and lacking the broad white supercilium. Pied Triller (405) is proportionately shorter tailed, has black eyeline and is less gregarious in habits. **Male:** Black or sometimes dark brown upperparts with long narrow white wingbar, white rump and narrow edge of tail. Underparts dusky white with buff suffusion on breast. **Female:** Like male but duller with black parts replaced with sooty brown. **Voice:** A short and high pitched musical trill. **Range:** The Indian subcontinent through southern China and South-East Asia to the Malay Peninsula, Sumatra and Borneo. **Status and distribution:** Fairly common resident at low elevations up to at least 900 m in Perlis, Kedah, and northern Perak; at least 240–1600 m from central Perak south to Negeri Sembilan and at low elevations in Johor where it is scarce. Particularly common at the lower elevations of the hill-stations, especially at The Gap (Selangor), below Fraser's Hill and the lower parts of Genting Highlands (Pahang). **Habitat:** Forests including peatswamp), the forest edge. **Habits:** Frequents the canopy in small noisy parties, frequently perching on bare branches, or service wires at the hill-stations. Feeds largely on insects obtained from aerial sorties or gleaned from branches. Often participates in bird-waves.

412. Black-winged Flycatcher-shrike, *Hemipus hirundinaceus* Plate 48
Sambar-Tirjup Sayap Hitam

Description: 15 cm (6 in). Like respective sexes of Bar-winged Flycatcher-shrike (411) but lacks narrow white wingbar. Mantle of male is always black. **Voice:** Similar to Bar-winged Flycatcher-Shrike but phrases shorter and harsher. **Range:** Tenasserim (Myanmar) through the Malay Peninsula to the Greater Sundas and Bali. **Status and distribution:** Locally common resident from low elevations up to 800 m, south to Johor. Far more common in the lowlands than Bar-winged Flycatcher-shrike. **Habitat:** Mangroves, forests, the forest edge, plantations, wooded gardens. **Habits:** Frequents the canopy and the crowns of smaller trees, usually in pairs but sometimes less often in small parties. Other habits similar to Bar-winged Flycatcher-shrike.

WHISTLERS

ORDER *Passeriformes* **FAMILY** *Pachycephalidae*

World: *44 species* **Peninsular Malaysia and Singapore:** *1 species*

A group of robust and large round headed birds with strong, thick, shrike-like bills, distributed mainly in the Australasian region. Most species are territorial and utter loud melodious whistling songs, often performed antiphonally by pairs. Glean insects from the foliage. Sexes differ in many species. Solitary or in pairs. Nests are cup shaped, usually placed in the fork of a branch.

413. Mangrove Whistler, *Pachycephala cinerea* Plate 49
Penyiul Bakau

Description: 17 cm (6.5 in). Nondescript and stocky with large round head, dark greyish legs and thick black bill distinctive. From *Malacopteron* babblers (515–518) and *Rhinomyias* flycatchers (607, 608) by shape and colour of bill, leg colour as well as the lack of strong rufescent tinge on uppertail coverts. Upperparts dull brown; head greyish with brownish ear coverts. Pale ashy grey breast band shows little contrast with white throat, belly and vent. **Voice:** A series of loud and rapidly uttered brief whistling notes, terminating in one or two clear whip-crack whistles. **Range:** North-East India, Myanmar, Thailand, and southern Vietnam to the Malay Peninsula, the Greater Sundas, and the Philippines. **Status and distribution:** Locally common resident at low elevations (including on most offshore islands), principally along the coast but also up to some 40 km inland, south to Singapore where it is uncommon with the main concentrations on Pulau Ubin, Pulau Tekong Besar, and some of the smaller islands in the south. Also on mainland at Changi, Jurong River, Khatib Bongsu, Kranji, Sungei Buloh Wetland Reserve, Sg. Mandai, Sg. Simpang, and Seletar (W and H). **Habitat:** Mangroves, scrub along mangrove fringe, plantations, wooded gardens. **Habits:** Frequents the middle and lower storeys solitarily or in pairs. Usually unobtrusive, gleaning for insects amongst the foliage. Sometimes perches on open branches, making brief sorties after flying insects. Very vocal. Not shy.

SHRIKES

ORDER *Passeriformes* **FAMILY** *Laniidae*

World: *31 species* **Peninsular Malaysia and Singapore:** *3 species*

Medium sized with large heads, marked with distinctive black facial masks, strong hooked bills, strongly notched on the maxilla and relatively long tails. Prefer mainly open country and scrub, where they prey on insects, small reptiles and frogs, by pouncing on them from an open perch. Species which breed in northern temperate latitudes are migratory, wintering south to tropical latitudes. Calls usually harsh discordant scolding chatters. Solitary.

414. Tiger Shrike, *Lanius tigrinus* Plate 49
Tirjup Harimau

Description: 19 cm (7.5 in). From Brown Shrike (415) by shorter tail and thicker bill. **Male:** Crown and nape grey; mantle and uppertail coverts rufous chestnut barred with black; broad eyeline black; underparts white. **Female** (not illustrated): Like male but duller with whitish lores and thin supercilium. Flanks marked with short blackish bars. **Juvenile and First Winter:** From juvenile Brown Shrike by lack of eyeline. Upperparts rufous brown with bold black barrings on both upperparts and underparts. Ear coverts barred blackish. Sometimes has narrow white eyering. Most birds on autumn passage are juveniles in first winter plumage, which moult into adult plumage before spring passage. **Voice:** Harsh chatters, similar to Brown Shrike but more rapid and intense. **Range:** Breeding in eastern Russia, northern China and Japan; wintering through South-East Asia to the Greater Sundas, Bali, the Philippines and Sulawesi. **Status and distribution:** Fairly common passage migrant and winter visitor from low elevations up to 1200 m, south to Singapore. Occurs above 900 m only during autumn and spring passage. Less common than Brown Shrike. **Habitat:** The forest edge, clearings in forest, plantations, wooded gardens. **Habits:** Territorial in wintering quarters. Shuns open country. Frequently perches on exposed branches at the forest edge. Sometimes keeps to dense underbrush fringing forest edge. General habits similar to Brown Shrike, but more retiring.

415. Brown Shrike, *Lanius cristatus* Plate 49

Tirjup Belukar

Description: 20 cm (8 in). Broad black eyeline, whitish supercilium and forehead distinctive. Upperpart colour varies from rufous brown to ash brown, depending on subspecies. Underpart colour varies from white to buff. Four subspecies occur locally. All except *L. c. superciliosus* are indistinguishable in first winter plumage during autumn passage. Only during spring moult are the distinguishing adult features apparent. *L. c. cristatus*: Dull chestnut upperparts tinged ash-brown with crown slightly bright chestnut. Supercilium and forehead greyish white; forehead not sharply demarcated from crown. Rump and uppertail coverts pale chestnut, distinct from back. *L. c. superciliosus* **(not illustrated):** Rich chestnut upperparts distinctive. White supercilium and forehead sharply demarcated from crown. *L. c. confusus* **(not illustrated):** Pale ash brown upperparts, white supercilium. Greyish white forehead not sharply demarcated from crown. Rump and uppertail coverts pale chestnut, distinct from back. *L. c. lucionensis:* Pale ash brown upperparts with grey tinge, head paler ash brown merging into pale grey on crown. Supercilium and forehead greyish white. Pale chestnut rump and uppertail coverts distinct from back. **Juvenile and First Winter:** Upperparts scaled with thin blackish bars with dark scalings on sides of breast and flanks. Eyeline dull blackish brown; supercilium buff and usually does not extend to forehead. In *L. c. supercilosus*, rich chestnut crown apparent. **Voice:** A loud, harsh crackling, *kreech-kreech-kreech*... Also high pitched screeching notes. Also a soft chuckling subsong, usually from the end of February onwards. **Range:** Breeding in temperate Asia; wintering through South and South-East Asia to the Greater and Lesser Sundas, the Philippines and the Moluccas. **Status and distribution:** Common and widespread passage migrant and winter visitor principally at low elevations, south to Singapore. Occurs above 900 m only during autumn and spring passage. A good proportion of autumn passage migrants are first winter birds. *L. c. superciliosus* and *L. c. confusus* are the most common winter visitors. **Habitat:** Scrub, open country, reedbeds, the forest edge, plantations, gardens. **Habits:** Territorial in wintering quarters, often chasing away other resident species of birds from established territory. Utilizes fence posts and telephone or electricity wires as vantage points to look for prey, which it sometimes skewers on barb-wire. Very vocal.

416. Long-tailed Shrike, *Lanius schach* Plate 49
Tirjup Ekor Panjang

Description: 25 cm (10 in). From other shrikes by long black tail and white wing patch (conspicuous in flight). Cap and back pale grey; forehead, broad eyeline and wings black; throat and belly white; flanks tawny. **Juvenile:** Duller with faint blackish barrings on flanks. **Voice:** A loud and harsh crackling chatter. **Range:** Iran through the Indian subcontinent to South-East Asia discontinuously to New Guinea. **Status and distribution:** Locally common resident at low elevations, up to 1550 m in the Cameron Highlands and Genting Highlands (Pahang), south to Singapore. **Habitat:** Scrub, open country, reedbeds, **Habits:** Like all shrikes, typically perches on exposed branches in open country. Will also utilize service wires. Sometimes moves long tail up and down.

SHRIKE-BABBLERS, ERPORNIS, AND ALLIES

ORDER *Passeriformes* **FAMILY** *Vireonidae*

World: *59 species* **Peninsular Malaysia:** *3 species*

(This cosmopolitan family has its centre of distribution in the Nearctic and Neotropical regions and also includes the Peppershrikes, Shrike-vireos, Vireos, and Greenlets, all of which do not occur in the review area. For this purpose only the Shrike-babblers and Erpornis will be described below. Both groups were previously grouped together with the babblers (Timaliidae) but molecular evidence suggests that they be placed in Vireonidae (Collar and Robson, 2007).

417. White-browed Shrike-babbler, *Pteruthius flaviscapis* Plate 60
Kekicau-Tirjup Dahi Putih

Description: 17 cm (6.75 in). Large headed and plump with thickset bill distinctive. **Male:** Head black with broad white supercilium; mantle grey; wings black with large reddish orange patch on tertiaries; edge of primaries white; tail black. **Female:** Head grey with less distinct white supercilium; wings bright greenish olive; tail bright olive. **Voice:** Territorial song is a high pitched *chip-chip-chip-chip-chiap*, with many variations, uttered persistently from a high perch. **Range:** The Himalayas and southern China through South-East Asia and the Malay Peninsula to the Greater Sundas. **Status and distribution:** Locally common resident above 900 m in the Larut Hills (Perak) and the Titiwangsa Range south to Gunung Telapa Buruk (Negeri Sembilan). Also on Gunung Benom and Gunung Tahan (Pahang), and on Gunung Rabong (Kelantan). Common at the hill-stations. **Habitat:** Forests, the forest edge. **Habits:** Solitary or in pairs. Frequents the canopy and middle storey. Feeds on both insects and small berries, the former gleaned from foliage and branches and the latter obtained by clambering onto small clusters. Participates in bird-waves. Sometimes enters hill-station gardens.

418. Black-eared Shrike-babbler, *Pteruthius melanotis* Plate 60
Kekicau-Tirjup Telinga Hitam

Description: 11 cm (4.5 in). From *Seicercus* warblers (488–490) by stocky appearance and short thick bill. Bold black streak behind ear coverts diagnostic. **Male:** Green crown and upperparts contrasting with yellow underparts and rufous throat and breast. Two broad white wingbars. **Female:** Duller and lacks the rufous throat and breast. Underparts paler. Wingbars pale buff. Eyering white in both sexes. **Voice:** A slow trill, *chip-chip-chuchuchuchu*..., the first two notes audible only at close range. **Range:** The Himalayas and southern China discontinuously through South-East Asia to the Malay Peninsula. **Status and distribution:** Locally common resident above 1200 m in the Titiwangsa Range south to Genting Highlands (Pahang). Also on Gunung Benom and Gunung Tahan (Pahang), and Gunung Rabong (Kelantan). Also on Gunung Gelemat and Gunung Lawit (Terengganu), both in the Eastern Range (W2). **Habitat:** Forests, the forest edge. **Habits:** Solitary or in pairs. Frequents the middle storey, sometimes lower. Usually gleans for insects from foliage and branches. Participates in bird-waves.

419. White-bellied Erpornis, *Erpornis zantholeuca* Plate 60
Erpornis Perut Putih

Description: 13 cm (5 in). From *Phylloscopus* warblers (480–487) by short erect crest and lack of supercilium. Upperparts greenish yellow contrasting with whitish underparts. Undertail coverts yellow. Bill pinkish. **Voice:** A short, high pitched ascending trill. **Range:** The Himalayas and southern China through South-East Asia and the Malay Peninsula to Sumatra and Borneo. **Status and distribution:** Fairly common resident at low elevations up to 1200 m, south to Johor. **Habitat:** Forests, the forest edge. **Habits:** Frequents the middle storey. Usually in small parties, gleaning insects from foliage. Often hangs upside down like a tit while foraging. Participates in bird waves.

OLD WORLD ORIOLES AND ALLIES

ORDER *Passeriformes* **FAMILY** *Oriolidae*

World: *30 species* **Peninsular Malaysia and Singapore:** *5 species*

(This family also includes the figbirds, *Sphecotheres* spp., an Australo-Papuan group, which does not occur in the review area. Only the Orioles, *Oriolus* spp. are described below.)

Robust, with bright and colourful plumage; most species in the review area are clad in varying combinations of bright yellow and black, brighter in the males and duller in females and juveniles, which are more streaked. Arboreal, feeding largely on insects and fruit. Solitary or in pairs, but small parties may gather at fruiting trees. Bills moderately large and strong, and in most species are reddish, and are specially adapted to tackle an assortment of fruit as well as other small vertebrates and invertebrates. Flight, swift and undulating. Songs, loud, rich and melodious fluty notes with a repertoire of phrases. Call notes harsh and nasal. Nests are cup shaped, suspended in the forks of branches.

420. Dark-throated Oriole, *Oriolus xanthonotus* Plate 50
Dendang Leher Hitam

Description: 20 cm (7.5 in). **Male:** Black head, throat and upper breast together with white underparts with bold blackish streaks diagnostic. Mantle, rump and undertail coverts bright yellow; wings and tail black; iris red. **Female:** Upperparts olive brown; throat, breast and belly white with blackish streaks from breast below; undertail coverts yellow, iris brown. Bill pinkish in both sexes. **Juvenile (not illustrated):** Like female. **Voice:** A rich and melodious fluty *tu-u-liu,* with an upward inflection on the last note. Also a four note *peu-peu-peu-poh,* the first three level with a liquid and melodious quality, the last a descending nasal rasp. **Range:** Tenasserim (Myanmar) through the Malay Peninsula to the Greater Sundas and the Philippines. Although still widespread and common, populations are declining moderately rapidly due to rapid and widespread deforestation. **NT. Status and distribution:** Locally common resident, principally at low elevations and hills usually not more than 300 m, south to Johor. Has also been reported at 1200 m. Locally extinct in Singapore since 1941 (L and G). **Habitat:** Forests, the forest edge. **Habits:** Frequents the canopy. Flits about in the tree tops, gleaning small insects and caterpillars from the foliage and branches. Also descends to the middle storey to feed on fruit. Sometimes gathers in fruiting shrubs at the forest edge.

421. Eurasian Golden Oriole, *Oriolus oriolus* Plate 50
Dendang Emas

Description: 25 cm (10 in). **Male:** From male Black-naped Oriole (422) by all bright yellow head, neck and mantle and entirely black wings from the carpal joint to the tips, with a small yellow patch on primary coverts. Black on head restricted only to the lores, appearing as a prominent short eyeline. Iris, brownish, legs blackish. **Female:** From female Black-naped Oriole by the lack of the black nape band, blackish wings, and streaked underparts. Duller than male with eyeline more diffused and less distinct. Yellow upperparts have a distinct greenish tinge. **Voice:** Unknown in review area. **Range:** Breeding throughout much of Europe, through temperate Asia and the northern part of the Indian subcontinent to southern China, migrating to Africa and the southern part of the Indian subcontinent, with stragglers reaching the Andamans. **Status and distribution:** Vagrant, an adult male at Burau Bay, Pulau Langkawi (Kedah) on 14 January, 2001 (Koji, 2001; RCS). **Habitat:** In review area, degraded mangroves. **Habits:** Little known from review area. **Note:** Rasmussen and Anderton (2005) split Indian subcontinent subspecies *kundoo* from nominate *oriolus* calling them Indian and European Golden Orioles respectively citing differences in morphology and vocalizations.

422. Black-naped Oriole, *Oriolus chinensis* Plate 50
Dendang Tengkuk Hitam

Description: 27 cm (10.5 in). **Male:** Bright yellow plumage with broad black eyeline and nape diagnostic. Wings (excluding carpal joints and secondaries) and tail black with lower half of outer tail feathers yellow. Extent of black on wings greater in resident *O. c. maculatus* and lesser in migrant *O. c. diffusus*. Iris red, legs blackish. **Female (not illustrated):** Like male but duller. Bill pinkish in both sexes. **Juvenile:** Upperparts dull olive yellow; eyeline and nape duller and less distinct; throat and breast whitish with streaks on breast; undertail coverts yellow; bill blackish. **Voice:** A rich and fluty *tu-u-u-liu,* the first note shortest and the last highest. Also a fluty *tu-u-liu* on a falling tone, the last note longest. Also a harsh and rasping *kyerrr*. **Range:** The Indian subcontinent through China, Korea and eastern Russia discontinuously through South-East Asia to the Greater and Lesser Sundas, Sulawesi, and the Philippines. Northern populations winter south in tropical zones. **Status and distribution:** Abundant and widespread resident, passage migrant and winter visitor at low elevations, south to Singapore. A common and conspicuous garden bird throughout much of its distribution. **Habitat:** Resident population frequents gardens, wooded suburbs, the mangrove edge and open country while migratory populations frequent mainly mangroves, forest canopies and the forest edge. **Habits:** Frequents tree tops, usually in pairs but sometimes in small parties. In towns and cities, often gathers to feed in low fruiting ornamental palms. Also takes morsels of ripe papaya and mango from fruiting trees. Aggressive and often mobs house crows. Vocal.

423. Black-hooded Oriole, *Oriolus xanthornus* Plate 50
Dendang Topeng Hitam

Description: 25 cm (10 in). Bright yellow body plumage with black head and breast together with black and yellow wings and tail diagnostic. Bill pinkish. **Juvenile:** Body plumage duller, forehead and narrow eyering yellow; throat and breast whitish with fine blackish streaks; head and nape dull blackish. **Voice:** A fluty four or three note sequence, *tu-u-u-liu*. Call note similar to that of Black-naped Oriole (422) but less harsh. **Range:** India through southern China through South-East Asia discontinuously to North-East Sumatra and North-East Borneo. Northern populations may be migratory. **Status and distribution:** Uncommon and localized resident at low elevations only on Pulau Langkawi (Kedah), confined to Kampung Kisap and Kampung Temoyang (Yeap, 2005). Rare and localized passage migrant or winter visitor, recorded at Sintok (Kedah), along the east-west Highway (Perak) and up to 1250 m on Fraser's Hill (Pahang) (W2). **Habitat:** Mangroves, the mangrove edge, rubber plantations, and forests. **Habits:** Keeps to the canopy and middle storey. Gleans insects from the foliage and branches. Also descends to the middle storey to feed on small fruit.

424. Black-and-crimson Oriole, *Oriolus cruentus* Plate 50
Dendang Hitam-Merah

Description: 23 cm (9 in). **Male:** Broad dark crimson breast patch separating black head, belly and undertail coverts diagnostic. Upperparts and wings black with dark crimson wing patches. **Female:** From Malaysian Whistling-thrush (568) by shorter tail, colour of bill, lack of iridescence in plumage and strictly arboreal habits Plumage dull black tinged greyish on lower breast and belly. Bill bluish grey in both sexes. **Voice:** A sharp and harsh, *keeeeu,* on a falling tone. **Range:** The Malay Peninsula and the Greater Sundas. **Status and distribution:** Locally common resident from 900 m up to 1500 m in the Larut Hills (Perak) and the Titiwangsa Range, south to Ulu Langat (Selangor). May sometimes wander below the montane ecotone to lower altitudes (W2). Also on Gunung Benom (Pahang). Common at the hill-stations. **Habitat:** Forests, the forest edge. **Habits:** Frequents the canopy and middle storey, flitting through the foliage after small insects. Generally silent and unobtrusive. Regularly participates in bird waves.

DRONGOS

ORDER *Passeriformes* **FAMILY** *Dicruridae*

World: *22 species* **Peninsular Malaysia and Singapore:** *6 species*

Medium sized birds characterized by an upright posture when perched, black plumage with a shiny iridescence, dark and stout, arched bills with a hooked tip. Legs and feet strong.. Tails rather long, broad-ended and forked in most species; the shapes of which are diagnostic in some. Arboreal, sallying forth after insects from exposed perches. Iris red in adults and brown in juveniles. Distinguished from similar looking Drongo Cuckoo (256) by more robust bills and proportionately longer tails without white barrings on undertail coverts. Highly territorial during nesting and usually in pairs. During migration and winter, some species may gather in loose flocks. Noisy, uttering harsh and metallic scolding notes; several species also utter melodious bell-like liquid notes. Often mimic the calls of other birds. Nests are cup shaped, suspended in the forks of branches.

425. Black Drongo, *Dicrurus macrocercus* Plate 50
Cecawi Hitam

Description: 28 cm (11 in). Deeply forked long tail with slightly upturned tips distinctive. Plumage black with slight gloss. **Voice:** A distinctive rather dry sounding *jeeez* (W2) lacking the characteristic bell-like notes of many species. **Range:** The Indian subcontinent and China through South-East Asia discontinuously to Java and Bali. **Status and distribution:** Fairly common passage migrant and winter visitor at low elevations, south to Singapore where it is uncommon. **Habitat:** Open country, ricefields. **Habits:** Gregarious, gathering in small groups. Hawks for insects from open perches, including small trees or service wires. Uses cattle as mobile perches, catching insects disturbed by the animals. Forms communal roosts.

426. Ashy Drongo, *Dicrurus leucophaeus* Plate 50
Cecawi Kelabu

Description: 29 cm (11.5 in). In *D.l. nigriescens*, plumage is blackish with slight gloss above and darker below. Resembles Black Drongo (425) in shape and proportions but is never as black. *D.l. leucogenis* is pale ashy grey with white cheek patch, including area surrounding eye. *D.l. salangensis* resembles *leucogenis* but white is confined to area surrounding eye. **Voice:** A distinct series of mewing calls. **Range:** Breeding in the Indian subcontinent and China through South-East Asia to the Greater Sundas, Bali, Lombok and the Philippines. Northern populations winter south to tropical latitudes. **Status and distribution:** *D.l. nigriescens* is a fairly common resident along the western coastal plain, south to Singapore where it is only a post-breeding dispersant. The migratory *D. l. leucogenis* is uncommon, occurring along both the eastern and western coastal plains, south to Singapore. *D. l. salangensis* which winters as far south as Peninsular Thailand is rarer and has been recorded at Nee Soon (Singapore) on 16 March 1997 and at Poyan on 22 March 1998 (W and H). **Habitat:** Mangroves, beach scrub, plantations, the forest edge, wooded gardens. **Habits:** Usually sits on an exposed perch from which it hawks for flying insects. Also engages in dives and climbs while pursuing insects in mid air.

427. Crow-billed Drongo, *Dicrurus annectans* — Plate 50

Cecawi Paruh Tebal

Description: 28 cm (11 in). From Bronzed (428) and Ashy Drongos (426) by more robust build, heavier bill and broader tail with widely splayed tips. **Juvenile (not illustrated):** Underparts marked with minute whitish scales. **Voice:** Loud bell-like notes and harsh churrs, typical of the family. Also a characteristic series of descending harp-like notes. **Range:** Breeding in the Himalayas, northern Myanmar and northern Thailand; wintering south throughout South-East Asia to the Greater Sundas. **Status and distribution:** Uncommon and localized passage migrant and winter visitor up to 1200 m (on passage only), south to Singapore. Wintering birds generally do not occur above 800 m. **Habitat:** Mangroves, forests, the forest edge, plantations, wooded gardens. **Habits:** Hawks for insects from open perches in typical drongo fashion.

428. Bronzed Drongo, *Dicrurus aeneus* — Plate 50

Cecawi Tembaga

Description: 23 cm (9 in). Small size with black plumage, heavily glossed with blue especially on breast. Proportionately slimmer than other drongos. From Lesser Racquet-tailed Drongo (429) with abraded tail feather shafts by deeply forked tail. **Voice:** Loud liquid bell-like notes. Also harsh calls, typical of the family. **Range:** India and southern China through South-East Asia to Sumatra and Borneo. **Status and distribution:** Locally common resident principally at low elevations up to 1500 m, south to Johor. At high altitudes, occurs only in disturbed forest leading to the hill-stations. Locally extinct in Singapore since 1950 (L and G). **Habitat:** Forests (including peatswamp), the forest edge. **Habits:** Frequents the canopy of forests, particularly where there are gaps. At the forest edge, perches on small trees and bamboo groves from which it hawks after prey, usually at mid-storey level. Mimics other bird calls.

429. Lesser Racquet-tailed Drongo, *Dicrurus remifer* Plate 50

Cecawi Anting-Anting Kecil/Cecawi Gunung

Description: 27 cm (10.5 in). Outer tail feather shafts and racquets 50 cm (20 in) more. From Greater Racquet-tailed Drongo (430) by a characteristic flat-headed appearance with crestless forehead and square-cut tail ending in elongate tail feather shafts with long ribbon-shaped racquets. See Bronzed Drongo (428). Plumage glossy black with metallic bluish gloss. Individuals with abraded outer tail feather shafts and racquets may resemble Large Niltava (597) in poor light but may be best distinguished by the fact it perches higher than the latter, usually at canopy and middle storey levels. **Voice:** A repertoire of melodious whistles. Also harsh churrs typical of the family. **Range:** The Himalayas and southern China through South-East Asia to Sumatra and Java. **Status and distribution:** Locally common resident above 760 m in the Larut Hills (Perak) and the Titiwangsa Range south to Ulu Langat (Selangor). Also on Gunung Lawit (Terengganu) in the Eastern Range, and on Gunung Tahan, Gunung Benom (Pahang) (down to 445 m) and Gunung Rabong (Kelantan). Common at the hill-stations, where it may sometimes overlap with the Greater Racquet-tailed Drongo at the lower elevations, especially at The Gap (Selangor), below Fraser's Hill. **Habitat:** Forests, the forest edge. **Habits:** Frequents the canopy and middle-storey. Keeps more to forest cover than the Greater Racquet-tailed. Mimics other bird calls. Frequently participates in bird waves. Has been observed taking an insect on the grassy verge of a road at Fraser's Hill (Pahang).

430. Greater Racquet-tailed Drongo, *Dicrurus paradiseus* Plate 50

Cecawi Anting-Anting Besar/Hamba Kera

Description: 32 cm (13 in). Outer tail feather shafts and racquets 30 cm (12 in) more. Black plumage with blue-black gloss. Tail slightly forked with outer tail feather shafts ending in spoon-shaped raquets (sometimes abraded or missing). Short crest on forehead. **Voice:** A loud metallic bell-like *kliyop*. Also harsh chatters, *chi-chi-kua-kua*... **Range:** The Indian subcontinent and southern China through South-East Asia to the Greater Sundas. **Status and distribution:** Fairly common resident from low elevations up to 820 m, south to Singapore. The most frequently seen drongo in heavily wooded urban suburbs. Also on Pulau Langkawi (Kedah), Pulau Pinang and Pulau Tioman (Pahang). **Habitat:** Forests, the forest edge, degraded mangroves, plantations, wooded gardens. **Habits:** Frequents the middle storey. Like most other drongos, hawks after insects from open perches. An excellent mimic of the calls of other birds but always has a metallic sound to each call. When territorial, is aggressive and will mob intruders, including large raptors and hornbills. Like Lesser Racquet-tailed Drongo (429), has also been observed taking insect prey on the ground.

FANTAILS

ORDER *Passeriformes* **FAMILY** *Rhipiduridae*

World: *44 species* **Peninsular Malaysia and Singapore:** *3 species*

A distinct group of small birds which are widely distributed throughout much of the Oriental, Australo-Papuan, and Pacific regions. They are characterized by long broad tails which are often distinctively cocked, fanned, or jerked, probably to flush insect prey, which the birds then capture using a variety of acrobatics. Wings short and rounded. Usually frequent the middle and lower storeys. Largely insectivorous, they have a strong tuft of rictal bristles. Hunt by moving rapidly through the undergrowth and foliage, picking up disturbed insects. Active, noisy, conspicuous, and tame. Sexes alike.

431. White-throated Fantail, *Rhipidura albicollis* Plate 51
Sambar-Kipas Leher Putih/Murai Gila Gunung

Description: 19 cm (7.5 in). Overall slaty grey with throat, supercilium and tip of tail white. **Voice:** Harsh rasping squeaks, *shrek..shrek*... while foraging. Also has a pleasing piping trill of six notes on a descending scale. **Range:** The Indian subcontinent and the Himalayas through southern China and South-East Asia to Sumatra and Borneo. **Status and distribution:** Locally common resident above 900 m in the Larut Hills (Perak) and the Titiwangsa Range, south to Ulu Langat (Selangor). Also on Gunung Tahan, Gunung Benom (both in Pahang) and Gunung Rabong (Kelantan). Also on Gunung Lawit (Terengganu) and Gunung Tapis (Pahang) in the Eastern Range. Very conspicuous at the hill-stations. **Habitat:** Forests, the forest edge. **Habits:** Frequents mostly the lower storey either singly or in pairs, attracting attention with its harsh squeaks. Will sometimes ascend to the middle storey. Often seen along roads or forest paths where it regularly participates in bird waves. Very tame and confiding.

432. Pied Fantail, *Rhipidura javanica* Plate 51
Sambar-Kipas Biasa/Murai Gila

Description: 18 cm (7 in). From White-throated Fantail (431) by dark greyish brown band across breast, separating white throat and belly (altitudinal ranges of the two do not overlap). Upperparts slaty grey. **Voice:** Harsh rasping squeaks like White-throated Fantail. Also a varied musical piping trill. **Range:** Tenasserim (Myanmar), southern Thailand, southern Indo-China, the Greater Sundas, and the Philippines. **Status and distribution:** Fairly common resident at low elevations and hills up to 450 m, south to Singapore. **Habitat:** Mangroves, the mangrove edge, peatswamp forest, scrub, open country and cultivated land. **Habits:** Found singly or in pairs, it is very active, moving restlessly in pursuit of insects, which it catches in mid air or gleans from the foliage. Will often enter rural gardens where it is particularly conspicuous. Tame and confiding. A known brood host of the Rusty-breasted Cuckoo (251).

433. Spotted Fantail, *Rhipidura perlata* Plate 51
Sambar-Kipas Berbintik/Murai Gila Berbintik

Description: 18 cm (7 in). White spots on dark grey throat and breast diagnostic. Upperparts slaty grey, belly, narrow supercilium and tip of tail white. **Voice:** Sparrow like chirps while foraging (Lewis *et al.* in W2). Also a variety of metallic jingles. **Range:** The Malay Peninsula to Sumatra, Borneo and Java. **Status and distribution:** Uncommon and localized resident principally at low elevations up to 760 m, south to Johor. Has been recorded up to 1100 m on Bukit Larut (Perak). **Habitat:** Forests. **Habits:** Frequents canopy and middle storey. Moves about exclusively on its own or in mixed species flocks.

MONARCHS

ORDER *Passeriformes* **FAMILY** *Monarchidae*

World: *94 species* **Peninsular Malaysia and Singapore:** *3 species*

Medium sized robust insectivores characterized by relatively long bills, ridged, with a broad flat base. Strong tuft of bristles at base of bill. Insect prey obtained both by gleaning from branches and foliage as well as hawking from perches. Plumage bright. Sexually dimorphic; males have brighter plumage or longer tails. Solitary or in pairs.

434. Black-naped Monarch, *Hypothymis azurea* Plate 51

Sambar Tengkuk Hitam

Description: 17 cm (6.5 in). **Male:** Bright blue plumage with whitish belly together with black nape and gorget diagnostic. **Female:** Head duller blue, breast greyish, wings and tail greyish brown. Bill bluish in both sexes. **Voice:** Song is a ringing *pwee-pwee-pwee-pwee* (L and R). Also harsh rasps, not unlike those of the Asian Paradise flycatcher (435). **Range:** India and southern China through South-East Asia to the Greater and Lesser Sundas, Sulawesi and the Philippines. **Status and distribution:** Locally common resident at low elevations up to 1000 m, south to Singapore where it is rare and confined to Pulau Ubin and Pulau Tekong (W and H). **Habitat:** Forests, the forest edge, plantations. **Habits:** Frequents the middle and lower storeys, solitarily or in pairs. Restless and inquisitive, flitting around and fanning out its tail in the manner of fantails (431–433). Forages by flushing insects from the undergrowth before capturing and feeding on them.

435. Asian Paradise-flycatcher, *Terpsiphone paradisi* Plate 51

Sambar-Ekor Gading Asia

Description: 21 cm (8.5 in). Elongate tail of male up to 23 cm (9 in) more. **Male dark morph:** Glossy blue-black crown contrasting with grey throat and sides of head (some subspecies have glossy black head and throat); upperparts rufous, brighter on tail. **Male pale morph:** All white plumage with glossy blue-black head and throat. Primaries and shafts of flight feathers black. **Female:** Like male but without elongate tail. Orbital skin and bill bright blue in both sexes. **Voice:** A ringing whistled song. Also harsh rasping notes. **Range:** The Indian subcontinent through China, eastern Russia and Korea through South-East Asia to the Greater and Lesser Sundas. Populations from northern latitudes migrate south to winter. **Status and distribution:** Locally common resident, passage migrant and winter visitor at low elevations up to 1500 m, south to Singapore, where it is a common passage migrant and rare winter visitor

(W and H). **Habitat:** Forests, the forest edge, wooded surburban gardens, mangroves, the mangrove edge. **Habits:** Frequents the middle and lower storeys where it hunts for insects in aerial sallies or grabs them while flushing them from the foliage. Often in mixed species flocks. **Note:** Some males lack elongate tail feathers.

436. Japanese Paradise-flycatcher, *Terpsiphone atrocaudata* Plate 51
Sambar-Ekor Gading Jepun

Description: 20 cm (8 in). Elongate tail of male up to 23 cm (9 in) more. **Male:** From male Asian Paradise-flycatcher (435) by all black tail. Plumage glossy black with dark maroon mantle and back. Belly white. **Female:** From female Asian Paradise-flycatcher by darker head and upperparts. Blackish head, dark rufous mantle and tail. Orbital skin and bill blue in both sexes. **Voice:** Like Asian Paradise-flycatcher. **Range:** Breeding in Japan (including the Ryukyus) and Taiwan; wintering south to South-East Asia through the Malay Peninsula, Sumatra and the Philippines. Has declined due to habitat loss and degradation in wintering range. **NT. Status and distribution:** Uncommon and localized passage migrant and winter visitor at low elevations up to 1200 m, south to Singapore where it is rare (W and H). **Habitat:** Forests, the forest edge, mangroves. **Habits:** Like Asian Paradise-Flycatcher (435).

JAYS, MAGPIES, TREEPIES, AND CROWS

ORDER *Passeriformes* **FAMILY** *Corvidae*

World: *123 species* **Peninsular Malaysia and Singapore:** *7 species*

A diverse group of medium to large sized perching birds with comparatively long tails and strong bills. Noisy with harsh and discordant calls. Omnivorous with a varied diet. Many species also take the eggs and young of nesting birds. Gregarious. Sexes alike. Represented by the following groups and genera:–

Jays and Magpies. genera *Platylophus* (437) *Platysmurus* (438), and *Cissa* (439). Generally brightly coloured in many species. Some have long tails. Largely forest inhabitants. Arboreal. Generally shy. **Treepies,** genus *Crypsirina* (440). Very long tailed inhabitant of wooded country. Arboreal. **Crows,** genus *Corvus* (441–443). Largely black. Flight direct with slow wingbeats Some species commensals of man. Feed mainly on the ground. Form large communal roosts. Bold and aggressive.

437. Crested Jay, *Platylophus galericulatus* Plate 52
Gagak Berjambul/Burung Menjerit

Description: 33 cm (13 in). Black plumage, white neck patch and very long erect crest diagnostic. **Juvenile** (not illustrated): Browner with dark grey underparts and rusty buff spots on wing coverts. **Voice:** A harsh and rapid stacatto chatter, *tat-tat-tat-tat-tat*. . . **Range:** Tenasserim (Myanmar) through the Malay Peninsula to the Greater Sundas. Being forest dependent, it is presumed to be declining throughout because of deforestation. **NT. Status and distribution:** Uncommon and localized resident from low elevations up to 750 m, sparsely up to 1220 m (W2), south to Johor. **Habitat:** Forests. **Habits:** Frequents the middle and lower storeys in small noisy groups. Constantly on the move, flying from branch to branch, occasionally raising long crest. Feeds largely on insects disturbed in the foliage.

438. Black Magpie, *Platysmurus leucopterus* — Plate 52

Megpai Hitam/Burung Kambing/Burung Temenggang

Description: 41 cm (16 in). Robust with long and heavy tail distinctive. Glossy black plumage, long white wingbar, red iris and short crest over base of bill diagnostic. **Voice:** A loud and rapid discordant metallic chatter, *heh-heh-heh-heh-heh-heh-heh…*, the notes almost running together like a goat's bleat. Also a distinct bell-like *keliyop*, not unlike that of Greater Racquet-tailed Drongo (430) but lower pitched and slightly descending. **Range:** Tenasserim (Myanmar) through the Malay Peninsula to Sumatra and Borneo. Being forest dependent, it is presumed to be declining throughout because of deforestation. **NT. Status and distribution:** Locally common resident from low elevations up to 200 m, south to Johor. **Habitat:** Forests (including peatswamp) the forest edge, mangroves. **Habits:** Frequents the canopy and middle storey singly or in pairs. Feeds largely on insects caught in the foliage. Very vocal.

439. Common Green Magpie, *Cissa chinensis* — Plate 52

Megpai Hijau/Gagak Gunung

Description: 38 cm (15 in). Bright pale green plumage with bright yellow forehead and crown, broad black mask, long graduated tail, drooping crest, chestnut red wings, bright red bill, orbital skin and legs diagnostic. Broad tips of inner secondaries white with black subterminal band. Terminal tail band white; undertail coverts marked with black and white bars. **Voice:** A loud metallic *kleeep*. Also other raucous calls. **Range:** The Himalayas and Myanmar through South-East Asia discontinuously to Sumatra and Borneo. **Status and distribution:** Locally common resident above 900 m in the Larut Hills (Perak) and the Titiwangsa Range south to Ulu Langat (Selangor). Recorded as low as 800 m at The Gap (Selangor). Also on Gunung Tahan (Pahang). Conspicuous at the hill-stations. **Habitat:** Forests, the forest edge. **Habits:** Frequents the middle storey in small parties. Very active, constantly fluttering about in the foliage and gliding from one patch to another. Feeds largely on insects but will also take small vertebrates. Sometimes forages on the ground. Known to predate on the young of nesting birds. Generally shy but sometimes tame and confiding. Very vocal.

440. Racquet-tailed Treepie, *Crypsirina temia* — Plate 52

Gagak Anting-Anting

Description: 33 cm (13 in). Long spatulate tail, thickset bill, black plumage glossed with bronze green, velvet black face and blue iris diagnostic. Tails of drongos (425–430) are shorter, narrower and usually deeply forked. **Voice:** A series of harsh chatters, *kek kek kek*...(Yang Chong, BiW). Also short nasal mewing notes (L and R). **Range:** South-East Asia discontinuously to the Greater Sundas and Bali. **Status and distribution:** Rare and localized resident in Kedah and Perlis, with sightings from Bukit Kayu Hitam in April 2003 and Bukit Jernih, Chuping on 28 February and 2 March 2007 (BiW). Reported to be common in Kedah at least in the 1930s and 1940s. **Habitat:** The forest edge, plantations, wooded gardens. **Habits:** Where common, moves singly or in pairs at middle storey level. Feeds largely on insects. Keeps to foliage cover most of the time. Very vocal.

441. House Crow, *Corvus splendens* Plate 52
Gagak Rumah

Description: 43 cm (17 in). From Slender-billed (442) and Large-billed Crows (443) by smaller size with greyish nape, back and upper breast contrasting with black forehead, face and throat. Black plumage shows purplish sheen in good light. Bill much lighter than Large-billed Crow. Albinos **(not illustrated)** are all white or pale buff with pale pinkish bill. **Voice:** A high pitched *kreeeah,* lower in volume than Large-billed Crow. **Range:** The Indian subcontinent, Myanmar, and central Thailand. Widely introduced elsewhere, where it is considered a pest. Successfully eradicated from the island of Socotra (Yemen). **Status and distribution:** Introduced. Common and widespread resident only in certain regions. Large feral populations from Klang (Selangor) established since 1903 have spread inland to Kuala Lumpur, Petaling Jaya, and Bangi and along the entire Selangor coast north to Sungai Besar and south to Sepang. Has effectively displaced the indigenous Large-billed Crow within this range. Individuals have also been sighted as far south as Tanjung Tuan and Batu Berendam (Melaka). New human settlements in Selangor since the 1960s have facilitated its population growth and hence the expansion of its range. Feral populations are also well established in Georgetown (Pulau Pinang), Prai (Seberang Prai), Alor Star (Kedah), Johor Bahru (Johor), and Singapore, where it is common. **Habitat:** Cities, towns, open country. **Habits:** Always in small loose parties when foraging. Very opportunistic, feeding largely on food scraps obtained from rubbish tips, eating stalls and wet markets. Also takes offal in harbours. Very bold and aggressive and will mob other larger birds such as the Brahminy Kite (87). Forms large compact flocks when roosting communally in trees. Often predates on the eggs and young of nesting birds. A known brood host of the Asian Koel (257).

442. Slender-billed Crow, *Corvus enca* Plate 52
Gagak Rimba

Description: 51 cm (19 in). Large size. From Large-billed Crow (443) by much slender bill and in flight by much shallower wingbeats. Plumage glossy black. **Voice:** A long *kaaaaa*, similar to Large-billed Crow (443) but less gruff. Also a shorter high-pitched *kaa*. **Range:** The Malay Peninsula through the Greater Sundas, Bali, Sulawesi, and the Philippines. **Status and distribution:** Uncommon and localized resident from low elevations up to 600 m, south to Johor. **Habitat:** Forests, the forest edge, mangroves. **Habits:** Frequents the canopy in small loose parties. Feeds largely on insects and other small vertebrates obtained in the canopy. Sometimes descends to the ground to catch prey. Has also been observed feeding on figs (F.R. Lambert in W2). Not a commensal of man. Shy.

443. Large-billed Crow, *Corvus macrorhynchos* Plate 52
Gagak Paruh Besar

Description: 51 cm (19 in). Large size, massive bill with arched culmen (prominent in male, less so in female) and all glossy black plumage diagnostic. **Voice:** A loud, harsh and wavering *kaaaaah*, louder, lower pitched and gruffer than Slender-billed Crow (442) (W2). Also a low *kah-kah-kah* and a harsh *gawah,gawah*... Also low gurgling noises and rattles in flight. **Range:** The Indian subcontinent to China and eastern Russia and Japan, south to South-East Asia, the Greater and lesser Sundas and the Philippines. **Status and distribution:** Fairly common resident from low elevations up to 1500 m, south to Singapore. Also in the cleared areas of Bukit Larut (Perak), Cameron Highlands and Fraser's Hill (Pahang). **Habitat:** The forest edge, mangroves, beach scrub, open country, plantations, wooded gardens. **Habits:** Generally frequents the canopy or the crowns of smaller trees in pairs or in small loose parties. Like the House Crow (441) is opportunistic and will live around human habitations, feeding at rubbish tips or at abbatoirs. Not as bold and aggressive as House Crow. When vocalizing, throat hackles are erected. A known brood host of the Asian Koel (257).

TYPICAL TITS

ORDER *Passeriformes* **FAMILY** *Paridae*

World: *55 species* **Peninsular Malaysia:** *2 species*

Small to large plump bodied birds with short, conical bills and rounded wings. Gregarious, actively moving in small parties through trees, often in association with other birds. Feed largely on small insects but will also take seeds and small berries. Display a variety of acrobatic postures while foraging, often hanging upside down from twigs or leaves. Sexes alike in most species.

444. Great Tit, *Parus major* Plate 64
Serai Bakau

Description: 13 cm (5 in). Smallish. White cheek patch contrasting with black head diagnostic. Mantle grey with greenish tinge; underparts white with a bold black line running down centre of breast and belly; wingbar and outer tail feathers white. **Voice:** A loud trisyllabic metallic whistle, *pi-chi-chew*. Also churrs. **Range:** Europe and temperate Asia through South and South-East Asia to the Greater and Lesser Sundas. **Status and distribution:** Locally common resident along the mangrove fringed west coast from Pulau Pinang, Perak and Selangor, south to Tanah Merah (Negeri Sembilan) (BiW). Fairly common in the Matang Mangrove Forest Reserve (Perak). **Habitat:**

Mangroves. **Habits:** Moves rapidly through the mangrove canopy in small parties. Feeds largely on insects gleaned from branches and foliage. Often hangs head downwards while foraging in the foliage.

445. Sultan Tit, *Melanochlora sultanea*

Plate 64

Serai Sultan/Serai Raja

Description: 20 cm (8 in). Large size. Shaggy sulphur yellow crown and crest, yellow belly and undertail coverts diagnostic. **Male:** Glossy black upperparts, head and breast. **Female:** Black of upperparts tinged olive, throat and breast dark olive. **Voice:** Territorial song is a loud and rapid piercing whistle, *peeo peeo peeo peeo peeo*, persistently uttered. **Range:** The Himalayas and southern China through South-East Asia to Sumatra. **Status and distribution:** Locally common resident from low elevations and hills up to 1580 m, south to Johor. Particularly conspicuous at the lower elevations of the hill-stations. **Habitat:** Forests, the forest edge. **Habits:** Frequents the canopy, usually in small parties of 3–4. Active and noisy, constantly flitting through the foliage. Usually gleans insects from branches and foliage but will also engage in aerial sorties to capture flying insects. Will often descend low to feed on small berries at the forest edge near roads or logging tracks.

SWALLOWS AND MARTINS

ORDER *Passeriformes* **FAMILY** *Hirundinidae*

World: *86 species* **Peninsular Malaysia and Singapore** *7 species*

Small and slender aerial feeding insectivorous birds characterized by tiny bills with large gapes; an adaptation to capture flying insects which form the bulk of their diet. Superficially resemble swifts (293–304) but have shorter and broader (not sickle-shaped) wings with a slower and more fluttery flight. Tail shapes vary from slightly notched to deeply forked with elongate outer tail feathers in some species. Often perch on bare branches or wires. Descend only to gather mud to build nests, which are usually attached to the underside of cliff ledges or buildings. Some species excavate nest burrows. Drink by swooping low over pools, scooping water from the surface. Calls are mainly high-pitched twittering notes. Highly gregarious, roosting communally in large flocks outside breeding. Sexes alike.

446. Collared Sand Martin, *Riparia riparia*

Plate 53

Layang-layang Awan

Description: 13 cm (5 in). Small size with slightly forked tail. Brown upperparts and white underparts separated by distinct brown breast band diagnostic. In underwing flight pattern, white vent contrasts with broad brown sides of tail. **Voice:** Unknown in review area. **Range:** Breeding throughout temperate latitudes in North America, Europe and Asia; wintering south to South America, Africa, South and South-East

Asia to the Malay Peninsula, Borneo and the Philippines. **Status and distribution:** Uncommon and localized passage migrant and winter visitor, principally at low elevations, south to Singapore. **Habitat:** Open country, reedbeds. **Habits:** Usually in small numbers, hawking for insects over open country. Sometimes perches on bare branches. Roosts in small numbers on service wires or in reedbeds, often with Barn Swallow (448).

447. Dusky Crag-martin, *Ptynoprogne concolor* Plate 53
Layang-layang Batu Kapur

Description: 14 cm (5.5 in). Dark martin with distinctive square-cut tail (slightly notched) marked with white subterminal spots diagnostic. From swifts by stocky appearance, broader wings and distinctive flight and tail patterns. Upperparts blackish brown; underparts dark brown with very fine rufous streaks on throat and breast. Appears all black in poor light. **Voice:** Not described in review area. **Range:** The Indian subcontinent and South-West China through South-East Asia to the Malay Peninsula. **Status and distribution:** Uncommon and localized resident confined to large limestone outcrops; recorded from Bukit Air Jernih (Perlis), Gua Musang (Kelantan), south to Bukit Sagu (Pahang) in the east and Bukit Takun and Batu Caves (Selangor) in the west. **Habitat:** Limestone cliffs. **Habits:** Feeds in small parties, usually high up along cliff faces, flying back and forth. Does not associate with other swallows or martins.

448. Barn Swallow, *Hirundo rustica* Plate 53
Layang-layang Hijrah

Description: 15 cm (6 in) with elongate outer tail feathers up to 5 cm (2 in) more. Body and wings more slender than Pacific Swallow (449). Dark steel blue upperparts and breast band; chestnut throat and forehead together with deeply forked tail marked with white subterminal spots diagnostic. Underparts (including wing coverts) vary from pure white to pale pinkish in *H. r. gutturalis* and bright chestnut in the rarer *H. r. tyleri* (not illustrated).**Juvenile and first winter (not illustrated):** Lack elongate outer tail feathers; blue upperparts and breast band duller with brownish tinge; throat white in some individuals. **Voice:** High pitched twittering calls at roost. **Range:** Breeding throughout temperate latitudes in North America, Europe and Asia; wintering south to South America, Africa, South and South-East Asia with stragglers reaching northern Australia. **Status and distribution:** Abundant and widespread passage migrant and winter visitor at all elevations south to Singapore. The most numerous swallow. Many first winter birds stay throughout summer. **Habitat:** Open country, scrub, forests (feeding over only), reedbeds, towns, cities. **Habits:** Feeds in small loose parties by skimming low over open fields. Often perches on bare branches, fence wires or service wires, resting and preening in between feeding. Gathers in large flocks in communal roosting sites on service wires in towns and cities as well as in reedbeds fringing disused mining pools, often in association with Collared Sand Martin (446) and Pacific Swallow. **Note:** Elongate outer tail feathers of adult are sometimes abraded or lost during moult.

449. Pacific Swallow, *Hirundo tahitica* Plate 53
Layang-layang Pasifik

Description: 14 cm (5.5 in). From Barn Swallow (448) by the lack of elongate outer tail feathers and dark breast band. Has more extensive chestnut forehead. Underparts pale smoky grey, contrasting with entire greyish underwing. Tail slightly forked with white subterminal spots. Undertail coverts marked with black and white scales. **Voice:** Low twittering notes when foraging. Alarm call is a high-pitched *tweee*, uttered persistently. **Range:** The Indian subcontinent and South-East Asia through Tenasserim (Myanmar) and the Malay Peninsula south to the Indonesian Archipelago and Australia east to the western Pacific. **Status and distribution:** Abundant and widespread resident from low elevations up to 1800 m, south to Singapore. Less numerous than Barn Swallow. **Habitat:** Open country, forests (feeds over only), scrub, towns, cities. **Habits:** General habits similar to Barn Swallow with which it associates at roosting sites. Usually feeds in small loose parties over open country. A known commensal of man, it often nests in occupied buildings as well as under bridges and culverts. Very territorial, nesting pairs will frequently utter alarm call at the presence of intruders and have been seen to harry House Crows (441).

450. Red-rumped Swallow, *Cecropis daurica* Plate 53
Layang-layang Pinggul Jingga

Description: 18 cm (7 in). Large swallow with deeply forked tail (lacking subterminal white spots) and orange-rufous rump contrasting with dark steel blue upperparts at rest and in flight diagnostic. Orange-rufous on head confined to sides of neck only. Underparts dirty white (including wing coverts) with fine blackish vertical streaks (heavier on breast) showing slight contrast with greyish flight feathers. Undertail coverts, black, contrasting sharply with the underparts and the rest of the undertail, very distinct in flight. **Voice:** A tinny *chee-ank* (W2). **Range:** Breeding in Africa and Europe across temperate Asia to south eastern Siberia, Japan, and Korea, the Himalayas, the Indian subcontinent and China. Northern populations migrate to tropical latitudes to winter in sub Saharan Africa, throughout the Indian subcontinent and South-East Asia to the Malay Peninsula and Sumatra. **Status and distribution:** Locally common passage migrant and winter visitor, principally at low elevations, south to Singapore. **Habitat:** On passage, mangroves and forests from low to high altitudes, winter visitors utilize ricefields, mangroves, the mangrove edge, and open country. **Habits:** Feeds in small parties, often with other swallows over open country, especially ricefields. Will sometimes gather at communal roosts with other swallows. **Note:** A subspecies of Striated Swallow (451), *C. s. mayri* is similar in plumage pattern but is larger with broader streaks on underparts with reddish rufous on head confined to ear coverts only. Southernmost range in South-East Asia is northern Myanmar. Could turn up in review area.

451. Striated Swallow, *Cecropis striolata* Plate 53
Layang-layang Api

Description: 19 cm (7.5 in). In flight and at rest, large size, robust build and deeply forked tail distinctive with rufous-chestnut rump contrasting with dark steel blue upperparts. *C. s. badia* differs from Red-rumped Swallow (450) by diagnostic rich rufous-chestnut face, underparts (marked with fine black streaks) and lower wing coverts (contrasting slightly with greyish flight feathers) and black undertail coverts. The chestnut subspecies of Barn Swallow *Hirundo rustica tyleri* (448) has brighter chestnut underparts, blue breast band, white subterminal tail spots and a more graceful flight pattern. **Voice:** A sharp metallic *cheenk* in flight. Also a vibrating *schwirrrr* when perched. **Range:** The hilly regions of north-east India, Sri Lanka, Yunan, and Taiwan (including neighbouring islands) through continental South-East Asia to the Greater Sundas east to the Philippines and the Lesser Sundas. *C. s. badia* is endemic to the Malay Peninsula. **Status and distribution:** Locally common resident from low elevations up to 1250 m, south to Johor. Also on Pulau Langkawi (Kedah). Fairly common at the lower elevations of the hill-stations, especially at The Gap (Selangor), where it habitually perches on the window sills of the stone rest house. **Habitat:** Limestone outcrops, forests (feeding over only), open country. **Habits:** Feeds in small parties, often with other swallows over open country and forests. Also joins Pacific Swallow (449) at puddles to gather mud for nests. Not known to utilize service wires as roosting site and not known to roost communally with other swallows. **Note:** Dickinson and Dekker (2001) split *badia* from *striolata* on just the basis of the former's larger size, calling the new taxon Rufous-bellied Swallow. DNA analysis would be the deciding factor.

452. Asian House-martin, *Delichon dasypus* — Plate 53
Layang-layang Pinggul Putih

Description: 13 cm (5 in). From other swallows by slightly notched tail and distinctive white rump (conspicuous in flight). Fluttery flight pattern and broader wings distinguish it from swifts. Upperparts dark steel blue contrasting sharply with white underparts, tinged dusky on breast. In underwing flight pattern, greyish flight feathers contrast with blackish wing coverts. **Voice:** High pitched wickering notes. **Range:** Breeding in the Himalayas, eastern temperate Asia and Japan; wintering south to South and South-East Asia, the Greater Sundas and the Philippines. **Status and distribution:** Uncommon and localized passage migrant and winter visitor principally at hilly elevations above 180 m but also at low elevations, south to Singapore. Often seen at the hill-stations during the winter months. **Habitat:** Forests (feeding over only), scrub, open country. **Habits:** Usually in small parties, feeding high above the forest canopy, usually higher than other swallows. Often soars.

BULBULS

ORDER *Passeriformes* **FAMILY** *Pycnonotidae*

World: *132 species* **Peninsular Malaysia and Singapore:** *25 species*

Medium sized birds with fluffy plumage and broad rounded wings. Most are dressed in combinations of brown, olive, yellow or grey. Sexes virtually identical. Most juveniles like adults but duller. Gregarious, often gathering in small parties of either single or mixed species to feed in fruiting trees. Feed largely on small fruit but insects may also be taken. Flight undulating and jerky. Represented by the following groups and genera: **Typical Bulbuls,** genus *Pycnonotus* (453–468). Bill light in most species. Some have prominent crests. Mostly nondescript but some have bright plumage with patches of red or yellow on the head or undertail coverts. Calls vary from monotonous chinking notes to melodious bubbling songs. Some are garden birds but most inhabit the forest and may enter adjacent cultivated land. **Bearded Bulbuls,** genus *Alophoixus* (469–472). Bill powerful and stubby. Crest superficial or absent. Underparts mostly yellowish or buff. Calls mainly whistles and harsh chatters. Puff out throat feathers conspicuously when vocalizing. Territorial birds of the forest interior, usually frequenting the lower and middle storeys. **Jungle Bulbuls,** genera *Tricholestes* (473), *Iole* (474), *Ixos* (475), *Hypsipetes* (476), and *Hemixos* (477). Bills longer and more slender than in preceding genera. Crests superficial or absent. Mostly dull coloured. Calls mainly monotonous piping and harsh nasal notes. Forest inhabitants, usually foraging in the canopy.

453. Straw-headed Bulbul, *Pycnonotus zeylanicus* — Plate 54
Barau-Barau

Description: 29 cm (11.5 in). Large size with strong bill. Yellowish orange crown and ear coverts together with white throat diagnostic. Bold streaks on brownish breast and back whitish; wings and uppertail coverts olive-green; narrow eyeline and malar stripe black; belly whitish. **Juvenile (not illustrated):** Duller with brownish head. From other smaller brown bulbuls by larger size. **Voice:** A rich melodious bubbling cadence uttered up and down the scale, often in duets or in chorus. Also low gurgles when foraging and roosting. **Range:** Tenasserim (Myanmar), the Malay Peninsula to the Greater Sundas. Declining rapidly due to the caged bird trade and habitat loss. **VU. Status and distribution:** In Peninsular Malaysia uncommon and localized resident, in most unprotected areas to locally common in protected areas, from low elevations up to 240 m. In Singapore, locally common resident on Pulau Ubin, uncommon elsewhere. **Habitat:** Forests, the forest edge (particularly along streams and river courses), rural gardens, plantations, wooded scrubby areas, mangroves (in Johor only). **Habits:** Associates in loose parties of up to six, frequenting the canopy and middle storey. May sometimes forage on the ground. Very vocal especially at dawn and before dusk. Common near riverine villages where it often flies low across rivers. Often perches on low branches overlooking water. Frequently gathers in fruiting trees.

454. Black-and-white Bulbul, *Pycnonotus melanoleucos* — Plate 54
Merbah Hitam-Putih

Description: 18 cm (7 in). Black plumage and white wing coverts and lining diagnostic. **Juvenile (not illustrated):** Lacks wing patch, brownish black above, greyish brown with faint blackish streaks below. **Voice:** An unobtrusive tuneless disyllable, *per-it* (L and R). **Range:** The Malay Peninsula to Sumatra and Borneo. Declining moderately rapidly owing to continuing habitat loss. **NT. Status and distribution:** In Peninsular Malaysia, uncommon and localized resident principally at low elevations but recorded up to 1800 m on Gunung Benom (Pahang), south to Johor. Locally extinct in Singapore since 1895 (L and G). **Habitat:** Forests, the forest edge. **Habits:** Moves in small parties, keeping to the canopy and middle storey. Nomadic with flock movements dispersing over a wide area. Gathers in fruiting trees with other bulbuls. Sometimes enters rural gardens and plantations at the forest fringe. Other habits little known.

455. Black-headed Bulbul, *Pycnonotus atriceps* Plate 54
Merbah Kepala Hitam/Merbah Siam

Description: 18 cm (7 in). Wedge-shaped tail with dull black subterminal band diagnostic. **Common morph:** Mantle and breast olive yellow; uppertail coverts and rest of underparts bright yellow; head (including chin, throat and neck) glossy blue-black; primaries and primary coverts blackish brown; iris blue. **Grey morph:** Breast, upper belly, and collar pale grey. **Juvenile (not illustrated):** Lacks the black head. **Voice:** A rather tuneless metallic *cheep*. Also lively musical whistles. **Range:** North-East India through South-East Asia and the Greater Sundas to Palawan (The Philippines). **Status and distribution:** Locally common resident chiefly at low elevations but recorded up to 1300 m on Fraser's Hill (Pahang), south to Singapore where it is rare and confined to the Central Catchment forest and the Bukit Timah Nature Reserve. The grey morph is very rare. **Habitat:** Forests, the forest edge, the mangrove edge. **Habits:** Frequents the lower and middle storeys. Nomadic, with flocks dispersing over a wide area. Gathers in small noisy parties to feed in fruiting trees and shrubs.

456. Black-crested Bulbul, *Pycnonotus melanicterus* Plate 54
Merbah Jambul Hitam

Description: 19 cm (7.5 in). From Black-headed Bulbul (455) by long erectile black crest, yellow iris and square-cut tail. Upperparts dark olive-yellow; underparts bright yellow; head (including chin, throat and neck) black. **Voice:** A slurred whistle, *chu-chi-chu-chi-wit*, the last note inflected. **Range:** India and southern China through South-East Asia and the Greater Sundas. **Status and distribution:** Locally common resident in hilly country primarily at submontane elevations up to 1200 m south to Gunung Tampin (Negeri Sembilan). Particularly common at the lower elevations of Cameron Highlands and Genting Highlands (Pahang), The Gap (Selangor) and Bukit Larut (Perak). Also on the foothills of Gunung Raya, Pulau Langkawi (Kedah) (BiW). In Singapore feral populations originating from cage specimens are restricted to Bukit Brown, Bukit Timah Nature Reserve and the Central Catchment forest. **Habitat:** Forests, the forest edge. **Habits:** Frequents the lower and middle storeys. Often seen in small parties in fruiting trees and shrubs with other bulbuls. Usually delivers its slurred song from low branches or bamboo groves and at the hill-stations, from service wires. **Note:** W and H have included this species on the Singapore list.

457. Scaly-breasted Bulbul, *Pycnonotus squamatus* Plate 54
Merbah Sisik Dada

Description: 16 cm (5.5 in). White underparts with black scales on breast and flanks diagnostic. Black head contrasts sharply with white throat. Mantle and wings rich olive-yellow; undertail coverts and rump yellow; tail black. **Voice:** High pitched chinking notes, *tit-tit-tit*... **Range:** Tenasserim (Myanmar) through the Malay Peninsula to the Greater Sundas. Being a forest dependent species, deforestation has caused its decline. **NT. Status and distribution:** Uncommon and localized resident principally in hilly country from 120–1000 m, south to Panti Forest Reserve (Johor). Also on Pulau Langkawi (W2). **Habitat:** Forests, the forest edge. **Habits:** Frequenting the canopy, this bulbul is usually seen foraging in small groups. Sometimes descends to feed in the middle storey. Often calls as it feeds and moves about in the foliage.

458. Grey-bellied Bulbul, *Pycnonotus cyaniventris* Plate 54
Merbah Perut Kelabu

Description: 16 cm (5.5 in). Bluish grey head and underparts sharply contrasting with rich olive yellow upperparts diagnostic. Undertail coverts yellow; edge of primaries and lores black. **Voice:** A lively trill, *pi-pi-pi-pi-pi*... **Range:** Tenasserim (Myanmar), the Malay Peninsula to Sumatra and Borneo. Declining due to loss and fragmentation of forests. **NT. Status and distribution:** Locally common resident from low elevations up to 1000 m south to Johor. Locally extinct in Singapore since 1950 (L and G). **Habitat:** Forests, the forest edge. **Habits:** Frequents the canopy. Usually seen in small parties at the forest edge. Regularly gathers with other bulbuls in fruiting trees.

459. Red-whiskered Bulbul, *Pycnonotus jocosus* Plate 54
Merbah Telinga Merah

Description: 20 cm (8 in). Tall erect black crest and earthy brown upperparts contrasting with white underparts diagnostic. Ear coverts white; patch below eye and undertail coverts red. **Voice:** A distinctive and melodious 2–3 note whistle. Also chattering notes. **Range:** India and southern China through South-East Asia to the Malay Peninsula. Widely introduced to many parts of the world. **Status and distribution:** Uncommon and localized resident in the north only at low elevations in Kedah, upper Perak, Kelantan and northern Terengganu which constitute its natural range. Elsewhere south to Singapore, populations are feral, originating from escaped cage specimens. **Habitat:** Open country, lightly wooded scrubland, gardens. **Habits:** A conspicuous garden bird in the north, it keeps more to trees but frequently descends to low bushes. Delivers song from an exposed perch, usually the top of a bush or a branch. Moves around in small parties. **Note:** W and H have included this species on the Singapore list.

460. Sooty-headed Bulbul, *Pycnonotus aurigaster* Plate 54
Merbah Jawa

Description: 20 cm (8 in). Head, chin, and nape black, throat, breast, belly and rump white, together with earthy brown mantle, wings, and tail with white subterminal band diagnostic. Undertail coverts vary from yellow to orange-yellow. **Voice:** Varied, consisting largely of loud and clear melodious whistles. Also a rather unmusical *chook-chook*. **Range:** Myanmar and South China discontinuously through continental South-East Asia to Java and Bali. Widely introduced to Sumatra, Singapore, Sulawesi, and the Lesser Sundas. **Status and distribution:** Introduced from Java, probably in the early twentieth century. Very rare feral resident with rather sporadic occurrences from the 1920s up to the 1950s and again in the late 1980s up to 2003; latest records include singles at Pulau Sentosa on 29 January, 1986, Seaside Park on 21 November 1988, and at Marina South on 31 August 1989, with small feral populations established at Tampines since 2003, with breeding recorded. Two were also observed at Lorong Halus in 2004 and another two were seen at Punggol on 12 December 2006 (W and H). **Habitat:** In review area, cultivated lowlands and gardens **Habits:** Usually conspicuous, active and noisy, singly or in small groups. Forages from ground level to tree tops. **Note:** W and H have included this species on the Singapore list. It still remains to be established whether the original introduced stock from Java (*P. a. aurigaster*) actually survived and whether later sightings in the late 1980s through the 2000s were escaped cage individuals from the bird trade, whose sources may still be unknown.

461. Puff-backed Bulbul, *Pycnonotus eutilotus* Plate 54
Merbah Coklat Berjambul

Description: 23 cm (9 in). From other brown bulbuls by larger size and superficial crest. Crown and upperparts brown with a rufescent tinge on the wings and tail; underparts white with a greyish wash on breast. **Voice:** A cheerful musical whistle, *chu-li-li, chu-li*. **Range:** Tenasserim (Myanmar) through the Malay Peninsula to Sumatra and Borneo. Declining owing to loss and degradation of lowland rainforest. **NT. Status and distribution:** Uncommon and localized resident at low elevations up to 200 m south to Johor. Also recorded up to 1200 m on Gunung Jerai (Kedah). Locally extinct in Singapore since 1836 (L and G). **Habitat:** Forests, the forest edge. **Habits:** Frequenting the middle storey, it is particularly conspicuous at the forest fringe. Often gathers in fruiting trees to feed with other bulbuls. Not highly gregarious, it is most often seen singly or in pairs.

462. Yellow-vented Bulbul, *Pycnonotus goaivier* Plate 54
Merbah Kapur

Description: 20 cm (8 in). Whitish throat and broad white supercilium contrasting sharply with black lores and narrow dark brown coronal stripe diagnostic. Upperparts dark brown; underparts whitish; undertail coverts yellow. **Voice:** A bubbling *tud-liu,tud-liu,tud-liu*, frequently uttered at dawn. Also a rapid *chic-chic-chic* with an unmistakable bubbling quality. **Range:** Tenasserim (Myanmar), southern Vietnam to the Malay Peninsula, the Sunda region to the Philippines. **Status and distribution:** Abundant and widespread resident south to Singapore in the lowlands and hill-stations up to 1800 m at Cameron Highlands and Genting Highlands (Pahang). At Fraser's Hill (Pahang), restricted to Jeriau but gradually expanding to other cleared areas. **Habitat:** Scrub, plantations, gardens, open country. **Habits:** The most commonly seen bulbul. Small parties often gather in fruiting trees and shrubs. A regular visitor to gardens, it is opportunistic, feeding on food scraps and utilizing potted plants to nest in.

463. Stripe-throated Bulbul, *Pycnonotus finlaysoni* Plate 54
Merbah Leher Berjalur

Description: 19 cm (7.5 in). Yellow streaks on forehead, chin and throat diagnostic. Upperparts greyish olive with wings and tail brighter greenish olive; underparts greyish; undertail coverts yellow. **Voice:** A rapid and melodious double whistle, *whik-ku, whik-ku,* uttered persistently. **Range:** Myanmar through South-East Asia and the Malay Peninsula. **Status and distribution:** Common and widespread resident at low elevations in the north. From Perak southwards, inhabits principally hilly country from 200–800 m south to Johor. Also in secondary forest on Bukit Gasing, Petaling Jaya (Selangor) and Bukit Persekutuan (Kuala Lumpur). Seen as high as 1770 m on Gunung Ulu Kali (Selangor/Pahang). **Habitat:** Forests, the forest edge, coastal scrub. **Habits:** Frequents the middle and lower storeys. Moves around in small parties, gathering to feed in fruiting trees and shrubs with other bulbuls. Often seen in the roadside scrub at the lower elevations of the hill-stations.

464. Olive-winged Bulbul, *Pycnonotus plumosus* Plate 54
Merbah Belukar

Description: 20 cm (8 in). Upperparts dull brown with a strong olive wash on wings and tail. Underparts pale dull brown; undertail coverts buff brown; indistinct streaking on ear coverts whitish; iris reddish (brown in juvenile). **Voice:** A series of tuneless dry chattering notes. Also low bubbling notes. **Range:** Tenasserim (Myanmar) through the Malay Peninsula to the Greater Sundas and Palawan (The Philippines). **Status and distribution:** Fairly common resident at low elevations, south to Singapore. Has also been recorded up to 800 m at The Gap (Selangor) (W2). **Habitat:** Forests, the forest edge, coastal scrub, plantations. **Habits:** Usually seen singly or in pairs in wooded scrub or at the forest edge. Often joins other bulbuls in fruiting trees and shrubs. A known brood host of the Drongo Cuckoo (256) (Lim, in prep.).

465. Streak-eared Bulbul, *Pycnonotus blanfordi* Plate 54
Merbah Telinga Luris

Description: 20 cm (8 in). Iris pale grey in male and greyish brown in female and juvenile. Separated with difficulty from Olive-winged Bulbul (464) by paler colouration, lighter olive wash on upperparts. eye colour and more distinct white streaking on ear coverts. Undertail coverts buff or pale yellow. **Voice:** A treble piping *brink-brink* (Madoc, 1956). A harsh rasping *which-which-which* (L and R). **Range:** Myanmar through Indo-China and Thailand to the Malay Peninsula. **Status and distribution:** Locally common resident at low elevations in Perlis, north Kedah, and Kelantan. **Habitat:** Plantations, gardens, coastal scrub. **Habits:** A fairly conspicuous bird in the extreme north, it moves around in small groups, foraging in fruiting trees and low bushes. Visits gardens regularly.

466. Cream-vented Bulbul, *Pycnonotus simplex* Plate 55
Merbah Mata Putih

Description: 18 cm (7 in). Dark brown upperparts contrasting with white throat, whitish brown breast and creamy white centre of belly and undertail coverts. Iris white (red in juvenile). **Voice:** Unmelodious chattering notes. **Range:** The Malay Peninsula to the Greater Sundas. **Status and distribution:** Locally common resident principally at low elevations south to Singapore where it is uncommon and restricted to the Bukit Timah Nature Reserve and the Central Catchment forest. Also recorded above 1200 m on Gunung Jerai (Kedah), Gunung Rabong (Kelantan) and Gunung Ledang (Johor) (W2). **Habitat:** Forests, the forest edge. **Habits:** Very common, usually seen in small parties along forest trails especially in the more open areas. Noisy and conspicuous, it frequents the middle and lower storeys. Will gather in fruiting trees and shrubs with other bulbuls.

467. Red-eyed Bulbul, *Pycnonotus brunneus* Plate 55
Merbah Mata Merah

Description: 19 cm (7.5 in). From Cream-vented Bulbul (466) by red iris (brown in juvenile) and less contrast between upperparts and underparts. Upperparts dark brown; underparts lighter brown with deep buff on the abdomen; undertail coverts ochraceous buff. **Voice:** High pitched bubbling notes. Also trills and a short strident chirrup. **Range:** Tenasserim (Myanmar) through the Malay Peninsula to the Greater Sundas. **Status and distribution:** Fairly common resident from low elevations up to 1000 m south to Singapore where it is rare, restricted to the Bukit Timah Nature Reserve and the Central Catchment forest. Also on Pulau Tioman (Pahang). **Habitat:** Forests, the forest edge, plantations. **Habits:** Frequents the middle and lower storeys. Commonly observed in small parties along forest trails and in cultivated land at the forest fringe. Gathers in fruiting trees and shrubs with other bulbuls.

468. Spectacled Bulbul, *Pycnonotus erythropthalmos* Plate 55
Merbah Mata Merah Kecil

Description: 16 cm (5.5 in). From Red-eyed Bulbul (467) by smaller size and proportionately smaller and more rounded head. Upperparts dull brown with rufescent tinge on rump; ear coverts brown suffused with grey; throat whitish with a greyish tinge; underparts ashy grey with undertail coverts tinged pale yellow; iris red (brown in juvenile); narrow bright orange-yellow eyering perceivable only at close range. **Voice:** High pitched metallic chinking notes. **Range:** Tenasserim (Myanmar) through the Malay Peninsula to Sumatra, Borneo and intervening islands. **Status and distribution:** Fairly common resident at low elevations up to 900 m, south to Johor. Locally extinct in Singapore since 1950 (L and G). **Habitat:** Forests, the forest edge. **Habits:** Frequents the middle and lower storeys. Less common than Cream-vented (466) and Red-eyed Bulbuls. Typically seen in small parties at fruiting trees and shrubs along forest trails, usually with other bulbuls.

469. Sulphur-throated Bulbul, *Alophoixus finschii* Plate 55
Merbah Tekak Kuning

Description: 17 cm (6.5 in). Bright yellow throat patch, centre of belly and undertail coverts contrasting sharply with olive brown breast and flanks diagnostic. Mantle and wings dark brown; sides of head and tail dark olive. **Voice:** No information from review area. **Range:** The Malay Peninsula, Sumatra and Borneo. Declining owing to loss and fragmentation of lowland rainforest. **NT. Status and distribution:** Locally common resident from lowlands and hills up to 600 m south to Johor. **Habitat:** Forests. **Habits:** Keeps to the middle storey. Gregarious, gathering in small parties. Will congregate in fruiting trees with other bulbuls. Puffs out throat feathers. Other habits little known.

470. Ochraceous Bulbul, *Alophoixus ochraceus* Plate 55
Merbah Berjanggut Gunung

Description: 23 cm (9 in). Stocky with shaggy crest. Upperparts greyish brown with olive wash on mantle and rufescent wash on rump and uppertail coverts; sides of head (including ear coverts) greyish brown; chin and throat white; underparts dusky grey with greyish tinge on flanks; undertail coverts rufescent buff. **Juvenile (not illustrated):** Uppertail coverts, wing coverts and secondaries rusty rufescent, very distinct from mantle. **Voice:** Harsh grating cackles. Also a series of melodious trisyllabic whistles. **Range:** Central Myanmar through the Malay Peninsula to Sumatra and Borneo. **Status and distribution:** On Pulau Langkawi (Kedah) and Perlis, fairly common resident from low elevations to hilly country. Elsewhere common resident only at submontane elevations up to 1200 m from Perak south to Negeri Sembilan. Common at the lower elevations of Cameron Highlands and Genting Highlands (Pahang) and The Gap (Selangor). **Habitat:** Forests, the forest edge. **Habits:** Associates in loose parties of 5–6 individuals, feeding in fruiting trees and shrubs, often displacing other smaller bulbuls. Often puffs out throat when calling.

471. Grey-cheeked Bulbul, *Alophoixus bres* Plate 55
Merbah Berjanggut Rimba

Description: 22 cm (8.5 in). Stocky appearance. From Ochraceous Bulbul (470) by pale yellow underparts and superficial crest. Head brown; rest of upperparts olive-brown with greenish tinge on mantle and rufescent wash on uppertail coverts; throat white; sides of head, ear coverts, lores and narrow strip above eye grey; upper breast and flanks olive-brown suffused with yellow; undertail coverts buff ochreous. **Juvenile** (not illustrated): Sides of face browner grey with primaries and wing coverts more rufous. **Voice:** Harsh cackling notes. Also a series of melodious whistles, more varied and richer than Ochraceous Bulbul. **Range** Tenasserim (Myanmar) through the Malay Peninsula to the Greater Sundas and Palawan (The Philippines). **Status and distribution:** Locally common resident from low elevations to 1000 m south to Johor. **Habitat:** Forests, the forest edge. **Habits:** Frequents the lower and middle storeys. Often observed along forest trails, where like other forest bulbuls, it will congregate in small parties in fruiting trees and shrubs.

472. Yellow-bellied Bulbul, *Alophoixus phaeocephalus* Plate 55
Merbah Perut Kuning

Description: 20 cm (8 in). Bright yellow underparts, grey head, white throat and lores diagnostic. Upperparts olive brown. **Voice:** Harsh grating notes. **Range:** Tenasserim (Myanmar) through the Malay Peninsula to Sumatra and Borneo. **Status and distribution:** Locally common resident from low elevations to 760 m south to Johor. Locally extinct in Singapore since 1950 (L and G). **Habitat:** Forests. **Habits:** Frequents the lower storey where it is most often seen in parties of 2–3. Forages in low fruiting shrubs alone or with other birds. Typically seen perched on a low branch, puffing out its throat while calling.

473. Hairy-backed Bulbul, *Tricholestes criniger* Plate 55
Merbah Bulu Tengkuk

Description: 17 cm (6.5 in). Whitish yellow sides of face contrasts sharply with olive-brown crown. Upperparts olive brown; throat and breast greyish white; belly and undertail coverts yellow. Long hair-like feathers on back inconspicuous in the field. **Voice:** A high-pitched *chui-i*, ending with an upward inflection (M and W). Also a scratchy warbling (L and R). **Range:** Tenasserim (Myanmar) to the Malay Peninsula, Sumatra and Borneo. **Status and distribution:** Fairly common resident in the lowlands and hills up to 900 m, recorded south to Johor. Also on Pulau Pinang and Pulau Tioman (Pahang). **Habitat:** Forests, the forest edge. **Habits:** Frequents the lower storey solitarily or mostly in small noisy parties.Often gathers in low fruiting shrubs to feed on small berries but will also take tiny caterpillars and small insects.

474. Buff-vented Bulbul, *Iole olivacea* Plate 55
Merbah Riang

Description: 20 m (8 in). Greyish white iris distinctive. Upperparts brownish tinged olive; throat whitish buff; breast and belly greyish; undertail coverts creamy buff. **Voice:** A nasal *wheak* persistently uttered when perched or in flight. **Range:** The Malay Peninsula to Sumatra and Borneo. Declining due to widespread deforestation and conversion to agriculture. **NT. Status and distribution:** Fairly common resident from low elevations up to 760 m south to Singapore where it rare, with single sightings in 1990 and 1992 at the Bukit Timah Nature Reserve. Also on Pulau Pinang. **Habitat:** Forests, the forest edge. **Habits:** Frequents the canopy and middle storey but will sometimes descend lower to feed in low fruiting trees and shrubs. Calls frequently. Common at the forest edge, especially along old logging tracks.

475. Streaked Bulbul, *Ixos malaccensis* Plate 55
Merbah Lorek

Description: 23 cm (9 in). Superficially similar to Mountain Bulbul (476) but lacks shaggy crest (altitudinal ranges of the two do not overlap).Greyish breast and throat with fine white streaking readily distinguishes it from other lowland bulbuls.

Upperparts dark olive; belly and undertail coverts white. **Voice:** Harsh metallic trills, sometimes uttered in flight. Also a short, descending warbled song; *chiri-chiri-chu* or *chiru-chiru* (L and R). **Range:** Tennaserim (Myanmar) through the Malay Peninsula to Sumatra and Borneo. Likely to be declining due to loss and degradation of lowland rainforest. **NT. Status and distribution:** Locally common resident principally at low elevations but recorded up to 1100 m, south to Johor. In Singapore, occurs as a rare non-breeding dispersant with records of individuals and small parties at the Bukit Timah Nature Reserve and the Sungei Buloh Wetlands Reserve (W and H). **Habitat:** Forests, the forest edge. **Habits:** Frequents the canopy, moving around in small parties in the foliage. Sometimes seen singly or in pairs along forest trails or in fruiting trees. Noisy and conspicuous.

476. Mountain Bulbul, *Hypsipetes mcclellandii* Plate 55
Merbah Gunung/Bebarau Bukit

Description: 24 cm (9.5 in). Stocky appearance. Fine whitish streaks on brown crown and greyish brown throat and breast diagnostic. Mantle, wings and tail rich olive yellow; belly whitish; undertail coverts yellowish. Crown feathers sometimes raised in short shaggy crest. **Voice:** A high pitched monotonous piping *tsip-tsip-tsip*. **Range:** The Himalayas and southern China through South-East Asia and the Malay Peninsula. **Status and distribution:** Fairly common resident above 1000 m in the Larut Hills (Perak) and the Titiwangsa Range south to Selangor. Also on Gunung Benom and Gunung Tahan (Pahang) Also on the Eastern Range (W2). Conspicuous at the hill-stations. **Habitat:** Forests, the forest edge. **Habits:** Rather sluggish. Usually in loose parties of up to five. Feeds in fruiting trees and shrubs together with other birds. Often obtains berries by hovering clumsily below strands of them. Usually keeps to leafy cover but will sometimes perch in the open. Tame and confiding.

477. Ashy Bulbul, *Hemixos flavala* Plate 55
Merbah Kelabu

Description: 20 cm (8 in). Short blackish crest; black mask and white throat contrasting with pale greyish white breast diagnostic. Upperparts grey; belly and undertail coverts white. **Voice:** A series of loud ringing nasal notes, *tree-tree-tree*... **Range:** The Himalayas and southern China through South-East Asia and the Malay Peninsula to Sumatra and Borneo. **Status and distribution:** Locally common resident principally in hilly country from low elevations to an upper limit of 2013 m, south to Johor. Particularly common at Bukit Larut (Perak), The Gap (Selangor) and the lower elevations of Fraser's Hill and Genting Highlands (Pahang). In Singapore, occurs sporadically as a non-breeding dispersant, singles on St. John's Island and Pulau Sentosa respectively on 23 October 1985 and 23 October 1986. Also at the Bukit Timah Nature Reserve on 17 October 2004 and a flock of seven at the Sime Road forest on 23 October 2005 (W and H). **Habitat:** Forests, the forest edge. **Habits:** Frequents the canopy and middle storey. Gregarious, gathering in parties of up to six or more. Congregates in fruiting trees with other bulbuls. Noisy and conspicuous, raising crest when vocalizing. Also puffs out throat feathers like *Alophoixus* bulbuls.

OLD WORLD WARBLERS

ORDER *Passeriformes* **FAMILY** *Sylviidae*

World: *281 species* **Peninsular Malaysia and Singapore:** *17 species*

A large and varied family of small to medium-sized insectivorous birds characterized by thin bills. Most are dressed in drab colours while some have bright plumage. Voices vary from harsh chacking notes and wheezy calls to high-pitched trilling songs. Occupy a variety of habitats from forests and mangroves to scrub and open country. Many are long distance migrants. Sexes alike or almost so. Represented by the following groups and genera: **Grasshopper-warblers,** genus *Locustella* (478, 479). Smaller than Reed-warbles. Tails long, graduated and strongly rounded. Undertail coverts extend beyond outermost tip of tail feathers. Skulk in grassland. **Leaf-warblers,** genus *Phylloscopus* (480–487). Small size with very thin bills. Plumage vary from brownish to greenish yellow with very pale underparts. Have prominent pale supercilium and narrow dark eyeline. Some species have conspicuous pale wingbars. Most species forage in the canopy while some frequent the lower storey. Some lack prominent distinctive markings and may appear confusingly alike; best separated by voice, habitat and distribution. **Flycatcher-warblers,** genera *Seicercus* (488–490) and *Abroscopus* (491). Small size with largely yellow underparts with distinct head patterns. From Leaf-warblers by generally thicker bills and lacking the prominent pale supercilium, so characteristic of the latter. Forage in small parties at the lower and middle storeys. **Reed-warblers,** genus *Acrocephalus* (492–494). Nondescript with rather longish rounded tails. Bills thicker than other warblers. Skulk in reedbeds and marshy grassland where their sharp chacking notes indicate their presence.

478. Lanceolated Warbler, *Locustella lanceolata* — Plate 56
Cekup-Belalang Berjalur

Description: 13 cm (5 in). From Rusty-rumped Warbler (479) by smaller size, smaller bill and fine streaks on breast, belly and undertail coverts and the lack of white tips to outer tail feathers; rump not rufous. Upperparts olive brown, heavily streaked black. Shows whitish throat and distinct dark ear coverts. **Voice:** A diagnostic explosive *rink-tink-tink* (L and R). Also a thin and rapid high-pitched trilling sequence, *tit-tit-tit-tit-tit*... Also chacking notes. **Range:** Breeding in eastern Russia, Korea, northern China and Japan; wintering south to the Indian subcontinent, southern China and South-East Asia to the Greater Sundas. **Status and distribution:** Uncommon and localized passage migrant and winter visitor principally at low elevations south to Singapore. Occurs above 900 m only during autumn and spring passage. **Habitat:** Ricefields, wet grassland, reedbeds, waterside vegetation. **Habits:** Usually solitary. Skulks and forages in thick cover. Able to run mouse-like. More secretive than Rusty-rumped Warbler.

479. Rusty-rumped Warbler, *Locustella certhiola* — Plate 56
Cekup-Belalang Pinggul Perang

Description: 15 cm (6 in). Olive brown upperparts marked with blackish streaks on crown and mantle, whitish supercilium, rufescent lower back and rump together with dark tail with narrow white tips to outer tail feathers diagnostic. Throat whitish with underparts and undertail coverts whitish, tinged buff brown. **Juvenile (not illustrated):** Like adult but underparts pale yellowish buff with faint dark spots across upper breast. **Voice:** A harsh disyllabic *chir-chirr*. Also a thin *tik-tik-tik-tik* when alarmed (M and W). Also chacking notes. **Range:** Breeding in eastern Russia, northern China and Mongolia; wintering throughout South and South-east Asia to the Greater Sundas. **Status and distribution:** Uncommon and localized passage migrant and winter visitor principally at low elevations up to 1600 m, south to Singapore. **Habitat:** Reedbeds, grassland, tall scrub near water. **Habits:** Usually solitary, skulking in thick vegetation.

480. Dusky Warbler, *Phylloscopus fuscatus* Plate 56

Cekup Daun Bakau

Description: 12 cm (4.8 in). From other leaf-warblers by dark grey-brown upperparts and the lack of pale wingbars. Underparts whitish with pale brown suffusion on flanks and undertail coverts. **Voice:** A hard *chack*, frequently uttered. **Range** Breeding in northern Asia; wintering south to southern China, South and South-East Asia to the Malay Peninsula. **Status and distribution:** Rare and localized winter visitor at low elevations, south to Singapore. **Habitat:** Mangroves, margins of freshwater swamps, flooded grassland. **Habits:** Frequents low bushes and small trees, usually within the vicinity of water, solitarily or in small parties, uttering call while foraging.

481. Radde's Warbler, *Phylloscopus schwarzi* Plate 56

Cekup Daun Paruh Tebal

Description: 14 cm (5.30 in). From Dusky Warbler (480) by stockier build and thicker bill, with mandible pinkish or straw-coloured. Upperparts dull olive brown showing slight contrast with pale brownish-buff underparts, tinged yellow on the belly. Undertail coverts yellowish-buff. Supercilium prominent and is whitish, tinged yellow above the lores. Legs pinkish. **Voice:** A soft *tcheck*, sometimes repeated several times with a very short break between calls (Petter Olsson). **Range:** Breeding in southern Siberia, North Korea and northern China, wintering south to southern China and South-East Asia with stragglers reaching the Malay Peninsula, North-East India and the Philippines. **Status and distribution:** Vagrant, one in scrub between Kampung Tekek and Kampung Juara, Pulau Tioman (Pahang) on 24 November 2009 (Petter Olsson, 2009; RCS). **Habitat:** In review area, low roadside scrub. **Habits:** Forages at the lower storey. Flits around low bushes, sometimes on the ground. Calls intermittently. Other habits little known.

482. Yellow-browed Warbler, *Phylloscopus inornatus* Plate 56

Cekup Daun Paruh Pendek

Description: 11 cm (4.5 in). From Two-barred Warbler (484) by shorter bill with base of mandible pale and broader and more contrasting wingbars together with white edge of tertiaries. Entire crown and rump concolorous with olive green mantle and wings. **Voice:** A sharp disyllabic *wee-eest*, with an upward inflection on the second syllable. **Range:** Breeding in the Himalayas and northern temperate Asia; wintering south to the Indian subcontinent, southern China and South-East Asia to the Malay Peninsula. **Status and distribution:** Locally common winter visitor from low elevations up to 2100 m, south to Singapore, where it is rare. **Habitat:** Forests, the forest edge, plantations. **Habits:** Usually frequents the canopy and crowns of smaller trees, foraging amongst the foliage.

483. Arctic Warbler, *Phylloscopus borealis* Plate 56
Cekup Daun Artik

Description: 13 cm (5 in). Upperparts olive green with greyish tinge on crown. Has long and distinct yellowish white supercilium. Underparts whitish tinged pale yellowish. Legs pale horn. From Two-barred Warbler (484) by slightly larger size, larger bill which appears slightly upturned and a single (sometimes two) thin yellowish white wingbar. Both species have distinct calls and different habitat preference; the Two-barred Warbler is very rare. **Voice:** A hard *zik*. Also a brief tuneless buzzing trilling song before spring passage. **Range:** Breeding throughout northern Europe, northern temperate Asia and Alaska; wintering south to southern China and South-East Asia throughout the Indonesian Archipelago to the Moluccas. **Status and distribution:** Common and widespread passage migrant and winter visitor from low elevations up to 1000 m, south to Singapore. The most common *Phylloscopus* warbler. **Habitat:** Forests, the forest edge, mangroves, plantations, wooded gardens. **Habits:** Frequents the canopy and crowns of smaller trees, gleaning insects from the foliage.

484. Two-barred Leaf-warbler, *Phylloscopus plumbeitarsus* Plate 56
Cekup Daun Rimba

Description: 12 cm (4.8 in). Like Arctic Warbler (483) but bill narrower. Entire upperparts olive green. Has usually two narrow yellowish white wingbars. See Yellow-browed Warbler (482). **Voice:** A soft disyllabic *chu-i* (M and W). **Range:** Breeding in eastern Russia and northern China; wintering south to South-East Asia and the Malay Peninsula. **Status and distribution:** Vagrant, known from individuals mist-netted at 1200 m at Fraser's Hill (Pahang) on 25 September 1971 and October 1972 and 1973. **Habitat:** Forests. **Habits:** Forages in small parties at canopy and middle storey levels. Other habits little known.

485. Pale-legged Leaf-warbler, *Phylloscopus tenellipes* Plate 56
Cekup Daun Dahi Putih

Description: 13 cm (5 in). From Arctic Warbler (483) by pale flesh legs, broader blackish eyeline, white supercilium and the lack of yellowish tinge on both upperparts and underparts. Crown dark greyish showing little contrast with olive green upperparts. Shows rufescent olive tinge on rump. Feeds at much lower levels than Arctic Warbler. **Voice:** A distinct and high-pitched metallic *tink*. **Range:** Breeding in eastern Russia, northern China, Korea and Japan; wintering south to southern China and South-East Asia to the Malay Peninsula. **Status and distribution:** Rare and localized winter visitor at low elevations, with individuals recorded on Pulau Langkawi (Kedah), Kuala Tahan (Pahang) and Kukup (Johor). **Habitat:** Mangroves, the forest and the forest edge, usually near streams. **Habits:** Frequents the ground storey usually solitarily. Other habits little known.

486. Eastern-crowned Warbler, *Phylloscopus coronatus* Plate 56
Cekup Daun Tongkeng Kuning

Description: 13 cm (5 in). From Arctic Warbler (483) by olive green median coronal stripe (concolorous with mantle), clean white breast and belly contrasting with yellow undertail coverts. **Voice:** A hard and distinct *zweet*. **Range:** Breeding in eastern Russia, northern China, Korea and Japan; wintering throughout South-East Asia to Sumatra and Java. **Status and distribution:** Locally common passage migrant and winter visitor from low elevations up to 1600 m, south to Singapore where it is uncommon. **Habitat:** Forests, the forest edge. **Habits:** Frequents the canopy in small parties where it forages in the foliage.

487. Mountain Leaf-warbler, *Phylloscopus trivirgatus* Plate 56
Cekup Daun Gunung

Description: 11 cm (4.5 in). Small leaf-warbler with yellowish median coronal stripe and supercilium together with yellowish underparts diagnostic. Upperparts greenish with no wingbars. **Voice:** A high-pitched and variable song, *see-chee-weet*... **Range:** The Malay Peninsula and the Greater Sundas. **Status and distribution:** Fairly common resident above 1300 m in the Larut Hills (Perak) and the Titiwangsa Range south to Genting Highlands (Pahang). Also on Gunung Tahan and Gunung Benom (Pahang). Common at the hill-stations. **Habitat:** Forests, the forest edge. **Habits:** Frequents the canopy solitarily or in small groups, foraging among foliage and epiphytes. Frequently participates in bird-waves. Has been observed feeding unidentified young cuckoo, leaving Large Hawk-cuckoo (244) and Oriental Cuckoo (248) as probable brood parasites. Will frequently feed on small moths stranded by lights at peak top telecommunication towers.

488. ?Plain-tailed Warbler, *Seicercus soror* Plate 56
Cekup Mata Kuning

Description: 13 cm (5 in). Olive green mantle and uppertail coverts contrasting with bright yellow underparts together with grey crown, black supercilium and narrow yellow orbital skin distinctive. **Voice:** Contact call is a rather high-pitched *tsi-dit* or *tsi–rit* (Alström and Olsson, 1999). **Range:** Breeding in Yunnan, Sichuan, and Fujian (China) and wintering through South-East Asia to southern Vietnam and Cambodia, with stragglers reaching the Malay Peninsula. Originally part of the Golden-spectacled Warbler, which ranged from high altitudes in the Himalayas. *S. burkii* was subsequently divided into six distinct species by Alsröm and Olsson (1999, 2000) based on evidence mainly of song types and distribution. *These replace one another geographically or by altitude over a collective Sino-Himalayan breeding range that penetrates marginally into northern S.E. Asia* (W2) **Status and distribution:** Vagrant, one mist-netted at Sintok (Kedah) on 13 October 1991. Originally identified as Golden-spectacled Warbler *Seicercus burkii* before the publications of revisions, the Sintok individual has been narrowed down to either Grey-crowned Warbler *S. tephrocephalus* or Plain-tailed Warbler *S. soror*, both migrants, and known to reach Thailand. *S. soror* was subsequently identified as the dominant representative at mid to low elevations, at least in Thailand (P.D. Round in W2). The decision to recognize only *soror* in the Malay Peninsula is strictly provisional (W2). **Habitat:** Forests. **Habits:** Little known.

489. Chestnut-crowned Warbler, *Seicercus castaniceps* Plate 56
Cekup Ubun Perang

Description: 10 cm (4 in). Tiny size distinctive. From Mountain Tailorbird (546) by shorter bill and tail, yellow rump, yellowish wingbars and the lack of white supercilium and black eyeline. Chestnut crown, grey head and breast together with narrow white orbital skin diagnostic. Maxilla horn coloured, mandible yellowish horn. **Juvenile (not illustrated):** Like adult but middle of crown dull brownish grey bordered with broad ashy grey stripes, pale grey supercilium and faint greyish eyeline. **Voice:** A distinct high-pitched metallic tinkling song. Also a disyllabic *chi-chi*, usually uttered while foraging. **Range:** The Himalayas and southern China throughout South-East Asia to Sumatra. **Status and distribution:** Locally common resident from 820–1280 m, recorded in the Larut Hills (Perak) and the Titiwangsa Range south to Genting Highlands (Pahang). Also on Gunung Tahan and Gunung Benom (Pahang). Also on the Eastern Range (W2). Common at the hill-stations. **Habitat:** Forests, the forest edge. **Habits:** Frequents the middle storey and crowns of smaller trees solitarily or in small groups. Very active participant of bird-waves, constantly flitting in the foliage in pursuit of small insects. Nests under the overhang of steep banks. A known brood host of the Oriental Cuckoo (248).

490. Yellow-breasted Warbler, *Seicercus montis* Plate 56
Cekup Kepala Perang

Description: 10 cm (4 in). Chestnut head, black supercilium, white orbital skin and bright yellow rump and underparts diagnostic. Mantle and uppertail coverts olive green with two narrow yellow wingbars. **Voice:** Unknown in the review area **Range:** The Malay Peninsula to Sumatra, Borneo to the Lesser Sundas and Palawan (the Philippines). **Status and distribution:** Locally common resident above 1160 m in the Titiwangsa Range from Cameron Highlands, south to Genting Highlands (Pahang). Most common above 1500 m. Also on Gunung Tahan and Gunung Benom (Pahang). **Habitat:** Ericaceous forests, forests, the forest edge. **Habits:** Frequents the middle and lower storeys. Forages largely among the foliage, sometimes on the ground amongst fallen timber. Participates in bird-waves. A known brood host of the Oriental Cuckoo (248) at higher altitudes.

491. Yellow-bellied Warbler, *Abroscopus superciliaris* Plate 56
Cekup Dahi Putih

Description: 11 cm (4.5 in). Grey head with white supercilium, white throat and upper breast, olive green upperparts together with yellow lower breast and belly diagnostic.

Voice: Short *trrrt* notes while foraging. Song is a thin tinkling cadence of 4–6 notes running down and up the scale. **Range:** The eastern Himalayas and South-East Asia to the Greater Sundas. **Status and distribution:** Fairly common resident principally in hilly country from low elevations up to 1200 m, south to Negeri Sembilan. Fairly common at the lower elevations of the hill-stations especially at The Gap (Selangor) below Fraser's Hill and the lower levels of Genting Highlands (Pahang). **Habitat:** Forests and the forest edge with bamboo thickets. **Habits:** Frequents the middle and lower storeys in small parties. Forages largely in the foliage and undergrowth.

492. Black-browed Reed-warbler, *Acrocephalus bistrigiceps* Plate 56

Cekup Paya Kecil

Description: 14 cm (5.5 in). Long pale buff supercilium bordering black lateral coronal stripe which contrasts with olive brown centre of crown diagnostic. Upperparts rufescent olive brown, brighter on rump; underparts whitish buff. **Voice:** A soft and distinct *chuck*, uttered in series. Song is a series of complex warbling sequences, usually uttered before spring passage. **Range:** Breeding in eastern Russia, northern China, Korea and Japan; wintering south to North-East India through South-East Asia to the Malay Peninsula. **Status and distribution:** Locally common passage migrant and winter visitor at low elevations, south to Singapore, where it is uncommon. **Habitat:** Reedbeds, ricefields, wet grassland. **Habits:** Like other reed-warblers, skulks in reedbeds and tall dense waterside vegetation. Usually delivers song from open perches.

493. Oriental Reed-warbler, *Acrocephalus orientalis* Plate 56

Cekup Paya Besar

Description: 20 cm (8 in). Large size with whitish supercilium and dark eyeline distinctive. Upperparts olive brown; underparts whitish with tawny buff tinge on flanks and faint greyish streaks on upper breast. **Voice:** A loud grating *chack*, often repeated. Also soft *churrs* and croaks in between repeated warbling sequences, uttered before spring passage. **Range:** Breeding in eastern Russia, China, Korea and Japan; wintering south to North-East India and South-East Asia through the Indonesian Archipelago and the Philippines to New Guinea with stragglers reaching northern Australia. **Status and distribution:** Fairly common passage migrant and winter visitor at low elevations, south to Singapore. Occurs above 900 m only during autumn and spring passage. **Habitat:** Reedbeds, ricefields, scrub, the mangrove edge, wet grassland. **Habits:** Skulks in thick vegetation. Will sometimes feed in low bushes and trees. Often sings on tops of reed stalks prior to spring departure. Puffs out throat feathers while singing.

494. Thick-billed Warbler, *Acrocephalus aedon* Plate 56

Cekup Paya Paruh Tebal

Description: 20 cm (8 in). Large size and longish graduated tail distinctive. From Oriental Reed-warbler (493) by shorter and thicker bill, shorter wings, longer tail and lacking both dark eyeline and whitish supercilium. Lores whitish. Mandible pale flesh. **Voice:** A loud and harsh *chack*. Also harsh chattering notes. **Range:** Breeding in eastern Russia and northern China; wintering south to South and South-East Asia to the Malay Peninsula. **Status and distribution:** Vagrant, first known from an individual mist-netted at Rantau Panjang, Klang (Selangor) on 22 April 1973. Subsequent sightings include individuals at Timah Tasoh (Perlis) on 27 and 28 December 2009 (Yang Chong, BiW). **Habitat:** In review area, scrub bordering mangroves, thick waterside sedges. **Habits:** Little known but probably similar to other reed-warblers.

BABBLERS

ORDER *Passeriformes* **FAMILY** *Timaliidae*

World: *374 species* **Peninsular Malaysia and Singapore:** *47 species*

A diverse and arbitrary grouping of largely insectivorous birds with soft. loose plumage. Flight weak and direct on short rounded wings. All species in review area are sedentary. Most species are gregarious, others solitary or in pairs. Very vocal with distinctive whistling songs and a variety of chatters, cackles and hoots. Sexes similar in most species. May be divided into 8 distinct groups based on physical characteristics and behaviour. **White-eyes,** genus *Zosterops* (495, 496). Small warbler-like birds with short tails, short pointed bills and largely greenish and yellow plumage. Most have conspicuous white eyerings and are highly gregarious and active, moving in large flocks, sometimes in mixed species, persistently uttering soft plaintive piping notes. Formerly in its own family, current DNA work has demonstrated that white-eyes are babblers (Collar and Robson, 2007). **Fulvettas,** genera *Pseudominla* (497) and *Alcippe* (498, 499). Both genera are not closely related but have been grouped together here for convenience. Rather small birds, inhabiting the understorey of forests, brightly coloured wings with distinctive head pattern in the former, and brownish and nondescript in the latter. **Scimitar-babblers,** genus *Pomatorhinus* (500, 501). Long down-curved bill and relatively long tail distinctive. Usually in small groups. Often frequent the ground. **Tree Babblers and Tit-babblers,** genera *Stachyris* (502–509) and *Macronous* (510, 511). Bills short and pointed. Move around in small groups in forest undergrowth and thickets. **Jungle Babblers,** genus *Pellorneum* (512–514). Frequents the ground and lower storeys solitarily or in pairs. Have bright plumage. Some have a characteristic walking gait. Genera *Malacopteron* (515–518) and *Ophrydornis* (519). Larger and longer tailed than the preceding and following genera. Have strong bills with a hooked tip. Frequent the middle and lower storeys in noisy flocks, often in mixed species groups. Forage mainly by gleaning invertebrates from the foliage. Genera *Trichastoma* (520–521), and *Malacocincla* (522 – 524). Skulkers in the undergrowth, most are clad in nonedescript plumage, some of which have strong rufescent tones. **Wren-babblers,** genera *Kenopia* (525), *Napothera* (526–529) and *Pnoepyga* (530). Small to large size with very short tails. Skulkers in the undergrowth on the forest floor. **Song Babblers:** The largest and most varied group. Contains some colourful species and notable songsters. May be further classified into the following groups. **Collared Babbler,** genus *Gampsorhynchus* (531). Slender with longish tail. Habits similar to Laughingthrushes. **Cutia,** genus *Cutia* (532). Chunky, short-tailed, and colourful. Clambers up and down tree trunks, probing bark for small invertebrates. **Laughingthrushes,** genus *Garrulax* (533–537). Large and robust, moving around in noisy flocks. Usually hop along branches or make short glides. **Mesia,** genus *Leiothrix* (538). Colourful and gregarious inhabitants of the middle and lower storeys. **Sibia,** genus *Heterophasia* (539). Slender with long graduated tail. Forages mainly at canopy and middle storey. **Minlas,** genus *Minla* (540, 541). Slender with long square-cut tails. Gregarious. Although placed ahead in the sequence, the fulvettas are considered part of the song babbler group.

495. Oriental White-eye, *Zosterops palpebrosus* Plate 61

Mata Putih Bakau

Description: 11 cm (4.5 in). Forehead bright yellow merging into yellow green on the crown and rest of upperparts (brighter on the rump). Lores black. Bold white eyering with narrow anterior black border. Chin, throat, upper breast and undertail coverts bright yellow. Belly pale grey with bright yellow mid ventral stripe from undertail coverts to upper breast. **Voice:** A disyllabic metallic *dza-dza*, repeated persistently. Also a shrill twittering *dzi-da-da*. **Range:** India through southern China and South-East Asia to the Malay Peninsula and the Greater and Lesser Sundas. **Status and distribution:** Fairly common resident at low elevations along the coastal plains, south to Johor. Locally extinct in Singapore since the 1970s due to progressive habitat loss, of both mangroves and forests (Lim, 1992) small feral populations probably derived from cage escapees occur throughout the mainland and south islands (L and G). **Habitat:** Mangroves, casuarinas, beach scrub, the forest edge, wooded areas. **Habits:** Moves around in large noisy flocks, feeding on nectar, small fruit and tiny insects, usually gleaned from branches. Will sometimes hang head downwards while foraging. **Note:** Feral populations of Japanese White-eye *Zosterops japonica*, derived from cage escapees, exist in certain parts of Singapore, particularly Mt. Faber, Telok Blangah Hill, Kent Ridge, Bukit Batok and Holland Woods. From Oriental White-eye by darker and less yellow-green upperparts and the lack of yellow mid ventral stripe.

496. Everett's White-eye, *Zosterops everetti* Plate 61

Mata Putih Rimba

Description: 11 cm (4.5 in). Like Oriental White-eye (541) but forehead same colour as crown, upperparts slightly darker with rump same shade as mantle and lower breast and belly darker grey. **Voice:** Soft twittering notes similar to Oriental White-eye. **Range:** The Malay Peninsula to Borneo and the Philippines. **Status and distribution:** Locally common resident principally in hilly country up to 2000 m, south to Negeri Sembilan. **Habitat:** Forests, the forest edge. **Habits:** Flocking habits precisely as Oriental White-eye. Will often probe the underside of tree-fern fronds for small insects. In the forest, usually keeps to the upper storey but will descend lower at the forest edge.

497. Rufous-winged Fulvetta, *Pseudominla castaneceps* Plate 60

Fulveta Sayap Merah

Description: 11 cm (4.5 in). Chestnut crown with narrow buff streaks; narrow white supercilium curving on rear margin of whitish scaled ear coverts; stripe behind eye and malar stripe black; wing patch black; primaries rufous. **Voice:** A wheezy descending sibilant trill, *si-si-si-si-si…* **Range:** The Himalayas and southern China discontinuously through South-East Asia to the Malay Peninsula. **Status and distribution:** Locally common resident above 1070 m in the Larut Hills (Perak) and the Titiwangsa Range south to Ulu Langat (Selangor). Also on Gunung Benom, Gunung Tahan (Pahang), and Gunung Rabong (Kelantan). More common in upper montane forest. **Habitat:** Forests, the forest edge. **Habits:** Moves in small noisy parties, keeping to undergrowth. Sometimes feeds vertically on tree trunks. Participates in bird waves. At the hill-stations, will enter the compounds of peak top telecommunication stations (Gunung Batu Brinchang, Cameron Highlands (Pahang) and Gunung Ulu Kali (Pahang/Selangor) in the early morning, feeding on stranded moths attracted by lights the previous night.

498. Brown Fulvetta, *Alcippe brunneicauda* Plate 60
Fulveta Coklat

Description: 15 cm (6 in). Rather nondescript. Upperparts rufescent brown; sides of head and neck greyish; underparts white with greyish tinge. **Voice:** Has a variety of loud and sweet notes. Also a 12 note cadence, *pi-pi-pi-pi…*, rising on the first eight notes before descending on the rest. **Range:** The Malay Peninsula, Sumatra, and Borneo. Declining moderately rapidly due to forest clearance in many areas but is still able to utilize secondary habitats and as such is not seriously endangered. **NT. Status and distribution:** Locally common resident at low elevations up to 900 m, south to Johor. Also on Pulau Pinang. **Habitat:** Forests, the forest edge. **Habits:** Frequents the middle and lower storeys in small parties, foraging for insects and small berries. Sometimes frequents small fruiting shrubs at the forest edge.

499. Mountain Fulvetta, *Alcippe peracensis* Plate 60
Fulveta Kening Hitam

Description: 16 cm (6.5 in). Smoky grey head and mantle with broad curving black supercilium diagnostic. Wings and upperparts dull brown. **Voice:** A pleasing and rapid whistle, *chu-chu-chu-chu*, the last note inflected. Also low chattering notes when foraging. **Range:** Laos and Vietnam through Thailand to the Malay Peninsula. **Status and distribution:** Fairly common resident above 800 m in the Larut Hills (Perak) and the Titiwangsa Range south to Gunung Telapa Buruk (Negeri Sembilan). Also occurs on Gunung Lawit (Terengganu) and Gunung Tapis (Pahang) on the Eastern Range. Also on Gunung Tahan, Gunung Benom, (Pahang), and Gunung Rabong (Kelantan). Common at the hill-stations. **Habitat:** Forests, the forest edge. **Habits:** Frequents both the middle and lower storeys. Gregarious, moving about in small parties, most often keeping to thick undergrowth. Participates in bird-waves. Enters hill-station gardens.

500. Large Scimitar-babbler, *Pomatorhinus hypoleucos* Plate 59
Kekicau Paruh Sabit Besar

Description: 28 cm (11 in). From Chestnut-backed Scimitar-babbler (501) by larger size, horn coloured bill and the lack of black on sides of head. Long streak behind eye white, ear coverts olive brown, neck patch rusty. Sides of neck and flanks heavily streaked whitish. Upperparts, flanks and undertail coverts rufous-brown with rufescent tinge on wings and tail. Throat, breast and belly white tinged buff. **Voice:** A deep fluty *hu-hoi-hu* with the middle syllable stressed. Often duets. **Range:** The Himalayas and southern China discontinuously through South-East Asia to the Malay Peninsula. **Status and distribution:** Uncommon and localized resident above 1000 m in the Larut Hills (Perak) and the Titiwangsa Range south to Ulu Langat (Selangor). Also on Gunung Benom and Gunung Tahan (Pahang). **Habitat:** Forests. **Habits:** Frequents the ground storey singly or in pairs, sometimes in small groups. Often feeds on the ground among leaf litter or fallen timber. Usually skulks in dense undergrowth. Very vocal.

501. Chestnut-backed Scimitar-babbler, *Pomatorhinus montanus* Plate 59
Kekicau Paruh Sabit Kecil

Description: 22 cm (8 in). Whitish bill, narrow white supercilium, black crown and broad eyeline together with bright chestnut mantle, rump and flanks diagnostic. Wings and tail blackish with a strong olive wash; throat, breast and belly white. **Voice:** A disyllabic, sometimes trisyllabic *hoo-hoo-hoo*. Also a disyllabic *hoo-hui*, with the second note inflected. **Range:** The Malay Peninsula to the Greater Sundas and Bali. **Status and distribution:** Locally common resident at low elevations up to 1300 m, south to Johor. **Habitat:** Forests. **Habits:** Usually frequents the middle storey solitarily or in pairs, sometimes in small groups. Probes bark crevices, rotten wood, and epiphytic root masses for invertebrates (W2). Will sometimes feed near or on the ground but less frequently than Large Scimitar-babbler (500).

502. Rufous-fronted Babbler, *Stachyris rufifrons* Plate 57
Kekicau Ubun Perang

Description: 13 cm (5 in). Rufous forehead and crown with grey sides of head, lores and underparts. Throat and centre of belly white; broad breast band buff brown. **Voice:** A cadence of soft and mellow hoots, *hu, hu-hu-hu-hu-hu*. . . . on an even pitch, with a pause after the first note. **Range:** The Himalayas through Myanmar and Thailand to the Malay Peninsula, Sumatra and Borneo. **Status and distribution:** Locally common resident principally in hilly country from low elevations up to 1000 m. Also occurs in flat lowlands, south to Johor. **Habitat:** Forest [illegible] he forest edge. **Habits:** Frequents the canopy and middle storeys in small parties. [illegible] the foliage by probing for insects, sometimes with other species. A known b[illegible] f the Drongo Cuckoo (256).

503. Golden Babbler, *Stachyris chrysaea* [illegible]e 57
Kekicau Emas

Description: 13 cm (5 in). Small size. From Mountain Leaf-warbler (487) by shorter tail, dark lores, brighter yellow plumage and the lack of both supercilium and wingbars. Bright golden yellow underparts and fine black streaked golden yellow crown diagnostic. Upperparts olive yellow, lores black, slightly extending below eye. **Voice:**

A cadence of soft and mellow hoots, similar to that of Rufous-fronted Babbler (502), but slower and on a lower pitch. **Range:** The Himalayas and southern China through South-East Asia and the Malay Peninsula to Sumatra. **Status and distribution:** Fairly common resident above 880 m in the Larut Hills (Perak) and the Titiwangsa Range south to Gunung Telapa Buruk (Negeri Sembilan). Also on Gunung Benom, Gunung Tahan, Gunung Tapis (Pahang), Gunung Padang (Terengganu) and Gunung Rabong (Kelantan). Very common at the hill-stations. **Habitat:** Forests, the forest edge. **Habits:** Frequents the middle and lower storeys in small parties. Very active, flitting in the tops of small trees or in the undergrowth in pursuit of small insects. Participates in bird waves. Sometimes tame and confiding.

504. Chestnut-winged Babbler, *Stachyris erythroptera* — Plate 57
Kekicau Sayap Perang

Description: 14 cm (5.5 in). Blue orbital skin and rictal area, greyish forehead, sides of head, throat and breast together with chestnut wings and tail diagnostic. Bare cobalt blue skin on sides of neck sometimes visible. **Voice:** A series of up to 12 hollow rapid *poop* notes with a drop in pitch at the end. Also harsh churrs. **Range:** Tenasserim (Myanmar) through the Malay Peninsula to Sumatra and Borneo. **Status and distribution:** Fairly common resident in the lowlands and hills up to 800 m, south to Singapore where it is uncommon and confined to the Bukit Timah Nature Reserve and the Central Catchment forest (W and H). **Habitat:** Forests, the forest edge. **Habits:** Frequents the lower storey in small parties. Skulks in the undergrowth. Feeds close to the ground. Brood host of Drongo Cuckoo (256).

505. Grey-throated Babbler, *Stachyris nigiceps* — Plate 57
Kekicau Tekak Kelabu

Description: 14 cm (5.5 in). Plumage olive brown, darker above; crown black with white streaks; supercilium greyish white; broad lateral coronal stripes black; throat grey with broad white malar stripe. **Voice:** A series of melodious high-pitched slurred whistles, *ti-hi-hi-hi-hi*... on a descending scale, the whole sequence lasting 3–4 seconds. Also a wheezy descending trill. **Range:** The Himalayas through South-East Asia and the Malay Peninsula to Sumatra and Borneo. **Status and distribution:** Fairly common resident in hilly country from 200–2000 m south, to Johor. Also on Pulau Pinang and on Pulau Tioman (Pahang). Fairly common at the hill-stations. **Habitat:** Forests, the forest edge. **Habits:** Frequents the lower storey. Moves about in small parties, skulking in the undergrowth. Feeds close to the ground. Sometimes briefly perches in the open.

506. Grey-headed Babbler, *Stachyris poliocephala* — Plate 57
Kekicau Kepala Kelabu

Description: 15 cm (6 in). Grey head with fine white streaks on forecrown and throat diagnostic. Plumage rufous brown with deep rusty tinge on breast; iris creamy yellow. **Voice:** A soft trisyllabic high-pitched whistle, *chew-weu-weu*, the second note highest. Also a series of descending churrs, *chu,chu,chu,chu*. **Range:** The Malay Peninsula,

Sumatra and Borneo. **Status and distribution:** Locally common resident at low elevations and hills up to 700 m, south to Johor. **Habitat:** Forests, the forest edge. **Habits:** Frequents the lower storey in small parties. Skulks in the undergrowth. Feeds close to the ground.

507. White-necked Babbler, *Stachyris leucotis* Plate 57

Kekicau Leher Putih

Description: 15 cm (6 in). Black throat, buff lores together with long narrow supercilium joining white spots behind grey ear coverts diagnostic. Upperparts rufescent olive brown; breast grey. **Voice:** A brisk *choi-oi-chu-u, choi-oi* (M and W). **Range:** The Malay Peninsula, Sumatra, and Borneo. Declining moderately rapidly due to forest clearance but its preference for slope forest implies that it is still relatively secure. **NT. Status and distribution:** Rare and localized resident at low elevations and hills up to about 1000 m, south to Johor. **Habitat:** Forests. **Habits:** Frequents the lower and ground storeys in small parties. Skulks in the undergrowth. Feeds near ground.

508. Black-throated Babbler, *Stachyris nigricollis* Plate 57

Kekicau Tekak Hitam

Description: 15 cm (6 in). Rufous chestnut upperparts, broad black throat with whitish streaked lower border and white malar stripe diagnostic. Narrow supercilium white. Sides of head, neck and belly grey. **Voice:** A series of 4–7 evenly spaced *pup* notes, at a rate of about two per second. Also a rapid version of up to 14 notes. Also descending chatters, *chi-chi-chiu-chiu*. **Range:** The Malay Peninsula, Sumatra and Borneo. Declining moderately rapidly due to forest clearance but its ability to colonize and utilize secondary habitats implies that it is still relatively secure. **NT. Status and distribution:** Uncommon and localized resident at low elevations and hills up to 500 m, south to Johor. Locally extinct in Singapore since 1895 (L and G). **Habitat:** Forests. **Habits:** Frequents the lower storey in small parties. Very active, moving about rapidly in the foliage and underbrush. Very vocal.

509. Chestnut-rumped Babbler, *Stachyris maculata* Plate 57

Kekicau Pinggul Perang

Description: 18 cm (7 in). Olive brown upperparts contrasting with rufous chestnut lower back and rump together with blackish throat and white breast, boldly spotted with black diagnostic. Flanks and belly dull grey brown. Iris pale yellow; orbital skin and rictal area blue. Bare blue skin around neck sometimes visible when calling. **Juvenile (not illustrated):** Throat and breast grey. **Voice:** A series of loud, mellow hoots, *hu-hup, hu-hup*... Also a rapid *pup-pup-pup-pup*... similar to that of Chestnut-winged Babbler (504) but richer and less hollow. **Range:** The Malay Peninsula, Sumatra and Borneo. Because of its preference for plains level rainforest, it is experiencing a moderately rapid decline due to deforestation. **NT. Status and distribution:** Fairly common resident principally at low elevations up to 500 m, south to Johor. Locally extinct in Singapore since 1895 (L and G). **Habitat:** Forests, the forest edge. **Habits:** Frequents the middle and lower storeys in small noisy parties, flitting through the foliage and clambering among lianes. Puffs out throat and neck feathers when calling.

510. Striped Tit-babbler *Macronous gularis* Plate 59
Kekicau Berjalur

Description: 14 cm (5.5 in). Small size. Crown and upperparts rufous brown; narrow supercilium and underparts pale yellow with fine blackish streaks on breast. **Voice:** A monotonous *chonk-chonk*, uttered repeatedly. Also scratchy notes. **Range:** India through South-East Asia to the Greater Sundas. **Status and distribution:** Common and widespread resident from low elevations up to 1220 m, south to Singapore. **Habitat:** Forests, the forest edge, mangroves, plantations. **Habits:** Gregarious, frequenting the middle and lower storeys at the forest edge. In primary forests, frequents the canopy. Generally keeps to thick undergrowth but will frequently come out to clearings. A known brood host of the Drongo Cuckoo (256).

511. Fluffy-backed Tit-babbler, *Macronous ptilosus* Plate 59
Kekicau Pong Pong/Burung Pong Pong

Description: 17 cm (6.5 in). Olive brown plumage with rufous chestnut crown, black throat and sides of head. White spot on sides of neck visible when calling. Orbital skin and rictal area blue. Long hair like feathers on lower back and flanks can often be seen protruding from below folded wings. Bare blue skin around neck perceivable only at close range. **Voice:** Scratchy notes followed by a hollow *pup, pup-pup-pup*. **Range:** The Malay Peninsula, Sumatra and Borneo. Experiencing moderately rapid population declines due to deforestation but is still able to utilize secondary forest. **NT. Status and distribution:** Fairly common resident at low elevations up to 300 m, south to Johor. Locally extinct in Singapore since 1854 (L and G). **Habitat:** Forests, the forest edge. **Habits:** Solitary or in pairs. Frequents the lower storey, keeping to dense undergrowth. More often heard. Seldom seen because of skulking habits.

512. Puff-throated Babbler, *Pellorneum ruficeps* Plate 58
Kekicau Tekak Kembang

Description: 17 cm (6.5 in). Crown rufous; supercilium pale buff; throat white; breast and flanks buff streaked with dark brown. Upperparts olive brown. **Voice:** A trisyllabic whistle, *tu-tu-teuw* with the second syllable stressed. Also a plaintive *hweet*. **Range:** India through southern China to South-East Asia and the Malay Peninsula. **Status and distribution:** Fairly common resident principally at low elevations but occurs at 1000 m on Gunung Jerai (Kedah). Southernmost distribution limits are up to southern Perak and northern Terengganu. Also on Pulau Langkawi. Fairly common in scrub fringing Timah Tasoh reservoir (Perlis). **Habitat:** Forests, beach scrub, bamboo scrub along river banks, the mangrove edge, mangroves (on Pulau Langkawi only). **Habits:** Frequents the ground storey. Usually solitary or in pairs. Walks or hops along the ground while foraging, most often among leaf litter. Often puffs out throat feathers.

513. Buff-breasted Babbler, *Pellorneum tickelli* Plate 58

Kekicau Dada Jingga

Description: 15 cm (6 in). Nondescript. Upperparts olive brown; breast and throat buff; sides of head buff brown. **Voice:** A rich disyllabic whistle repeated twice, *pi-chew, pi-chew*. **Range:** North-East India and southern China through South-East Asia and the Malay Peninsula to Sumatra. **Status and distribution:** Locally common resident from 700–1300 m in the Titiwangsa Range south to Gunung Telapa Buruk (Negeri Sembilan). Common at Fraser's Hill and Genting Highlands (Pahang). **Habitat:** Forests, the forest edge. **Habits:** Frequents the ground storey, usually in small groups of 3–4, gleaning insects from the forest floor. Generally keeps to thick cover, emerging only in clearings beside forest paths or roads. Very vocal.

514. Black-capped Babbler, *Pellorneum capistratum* Plate 58

Kekicau Ubun Hitam

Description: 18 cm (7 in). Black cap and malar stripe contrasting sharply with white throat, greyish white supercilium and bright rusty rufous underparts diagnostic. Upperparts dark rufous brown. **Voice:** A plaintive *puip* with an upward inflection. Also a powerful whistled cadence, *tri-ti-tu* on a descending scale followed by squeaky chatters. **Range:** Tenasserim (Myanmar) through the Malay Peninsula to the Greater Sundas. **Status and distribution:** Locally common resident at low elevations up to 700 m, south to Johor. Locally extinct in Singapore since 1855 (L and G). **Habitat:** Forests. **Habits:** Frequents the ground storey, usually skulking in the undergrowth. Walks on the forest floor, with head bobbing forward. Forages mostly among leaf litter. Shy.

515. Moustached Babbler, *Malacopteron magnirostre* Plate 58

Kekicau Bermisai

Description: 18 cm (7 in). Greyish sides of head with black malar stripe and rufous-brown tail diagnostic. Olive brown crown concolorous with mantle; underparts whitish with brownish wash on breast with fine greyish streaks on throat. Bill heavier than that of Sooty-capped Babbler (516). **Voice:** A cadence of 4–6 rich whistles on a descending scale, often terminating in a series of *swit* notes. Frequently delivered in duets. **Range:** Tenasserim (Myanmar) through the Malay Peninsula to Sumatra and Borneo. **Status and distribution:** Fairly common resident at low elevations and hills up to 900 m, south to Singapore where it is rare. **Habitat:** Forests. **Habits:** Frequents the lower and middle storeys. Moves in small parties. Gleans invertebrates from the foliage. Crown feathers sometimes raised in short crest. Sometimes puffs out throat feathers like *Alophoixus* bulbuls (469–472). Very vocal especially at dawn.

516. Sooty-capped Babbler, *Malacopteron affine* Plate 58

Kekicau Ubun Gelap

Description: 17 cm (6.5 in). From Moustached Babbler (515) by lack of moustachial streak, smaller size and thinner bill. Blackish crown darker than mantle. Shows pale supercilium. Underparts whitish suffused with pale grey on breast. **Voice:** A series of up to 8 whistled notes, rising and falling in pitch. **Range:** The Malay Peninsula to Sumatra and Borneo. Although relatively common and widespread, is experiencing a moderately rapid decline due to deforestation. **NT. Status and distribution:** Fairly common resident at low elevations up to 450 m, south to Johor. **Habitat:** Forests. **Habits:** Frequents the middle storey, moving about in small parties. Often responds to an imitation of its call. A brood host of the Drongo Cuckoo (256).

517. Scaly-crowned Babbler, *Malacopteron cinereum* Plate 58

Kekicau Ubun Sisik

Description: 17 cm (6.5 in). Rufous crown contrasting with black nape. Blackish scales on crown usually difficult to discern in field. Underparts whitish suffused with brownish grey on breast. Legs pinkish. **Voice:** A series of up to 6 wheezy high pitched whistles on an ascending scale. Also 3–4 clear whistles with the middle or second note lowest (L and R). Also high pitched chattering notes while foraging. **Range:** Indo-China through Thailand and the Malay Peninsula to the Greater Sundas. **Status and distribution:** Fairly common resident at low elevations and hills up to 790 m, south to Johor. Also on Pulau Pangkor (Perak). **Habitat:** Forests. **Habits:** Frequents the lower and middle storeys. Moves about in trees and shrubs in small noisy parties. Will sometimes raise crown feathers in short crest.

518. Rufous-crowned Babbler, *Malacopteron magnum* Plate 58

Kekicau Ubun Merah

Description: 18 cm (7 in). From Scaly-crowned Babbler (517) by larger size and greyish legs. Rufous crown contrasting with black nape. Mantle olive brown; tail rufous; underparts whitish with greyish streaks on breast. **Voice:** A series of up to 12 rich clear whistles, ascending in scale for most part before descending. **Range:** Tenasserim (Myanmar) through the Malay Peninsula to Sumatra and Borneo. Experiencing a moderately rapid decline due to habitat loss and degradation. **NT. Status and distribution:** Fairly common resident at low elevations and hills up to 500 m, south to Johor. **Habitat:** Forests. **Habits:** Frequents the middle and lower storeys. Travels in small parties through the foliage and undergrowth. Forages with other babblers.

519. Grey-breasted Babbler, *Ophrydornis albogulare* Plate 58

Kekicau Dada Kelabu

Description: 15 cm (6 in). Superficially resembles a flycatcher. Shorter-tailed than *Malacopteron* babblers. Dark grey crown and sides of head contrasting with thin white supercilium distinctive. White throat contrasting with diagnostic grey breast. Underparts whitish with tawny buff flanks and undertail coverts. Upperparts olive brown. **Voice:** A cadence of 4–10 clear, lightly held and slightly hesitant whistles ranging over the scale at a deliberate pace. Also a low pitched, purring rattle in alarm (W2). **Range:** The Malay Peninsula, Sumatra and Borneo. Being scarce and local throughout range, its habitat requirements are not fully understood. It is suspected to be undergoing moderately rapid population declines due to habitat loss. Currently ranked **NT** but if further research reveals that it prefers primary forest, its status may change to **VU. Status and distribution:** Rare and localized resident at low elevations up to 150 m. Distribution localized to Perak, Selangor, Pahang and Negeri Sembilan, south to Endau Rompin State Park (Johor). Locally extinct in Singapore since 1844 (L and G). **Habitat:** Forests, including peatswamp. **Habits:** Frequents the lower storey in small groups, usually exclusively on its own. Gleans small invertebrates from the understorey foliage. Other habits little known.

520. White-chested Babbler, *Trichastoma rostratum* — Plate 58

Kekicau Dada Putih

Description: 15 cm (6 in). Dark brownish upperparts contrasting sharply with white underparts diagnostic. Ear coverts olive brown, sides of neck slightly tinged grey. Bill narrower than other *Trichastoma* babblers. **Voice:** A distinctive slurred whistle of 3–4 notes, *chiri-biri-bee or chiri-biri-bee-riu* (L and R), strongly modulated and rather musical, delivered in sometimes long bouts at a rate of about 15 phrases per minute (W2). Also harsh chattering notes. **Range:** Tenasserim (Myanmar) through the Malay Peninsula to Sumatra and Borneo. Experiencing a moderately rapid population decline due to loss and degradation of lowland rainforest but it's tolerance for secondary habitats implies that it is likely to be at a reduced risk. **NT. Status and distribution:** Locally common resident at low elevations up to 200 m, south to Singapore (including Pulau Ubin and Pulau Tekong) where it is rare. **Habitat:** Mangroves, scrub bordering mangroves, forests (in wetter areas and peatswamp), the forest edge. **Habits:** Frequents the ground storey. Usually in pairs but sometimes in small parties, foraging in the wetter areas, among leaf litter, near shaded areas of streams. Has been reported to enter water, in torrent streams (W2). Skulks in the undergrowth. Sometimes calls from an exposed perch.

521. Ferruginous Babbler, *Trichastoma bicolor* — Plate 58

Kekicau Perang

Description: 18 cm (7 in). From other *Trichastoma* babblers by robust appearance and longer tail. Bright rufescent orange upperparts contrasting with white underparts diagnostic. Sides of head brownish, flanks tinged pale tawny. **Voice:** A loud whistle, *hweet,* rising sharply in pitch. Also a falling *piou*. Also chattering notes. **Range:** Tenasserim (Myanmar) through the Malay Peninsula to Sumatra and Borneo. **Status and distribution:** Locally common resident at low elevations up to about 270 m (exceptionally up to 660 m), south to Johor. **Habitat:** Forests. **Habits:** Keeps to the ground storey, skulking in the undergrowth. Usually in pairs or in small parties.

522. Abbott's Babbler, *Malacocincla abbotti* — Plate 58

Kekicau Riang

Description: 17 cm (6.5 in). Plump appearance with robust bill and moderately short tail. Upperparts uniformly olive brown with faint buff streaks on crown; sides of head and narrow supercilium grey; throat whitish; belly and centre of breast buff, merging into bright rufescent flanks and undertail coverts. **Voice:** A slurred trisyllabic whistle, *whee-woo-wheew*, the last note longer. **Range:** North-East India through Myanmar and South-East Asia to the Malay Peninsula, Sumatra and Borneo. **Status and distribution:** Fairly common resident at low elevations, south to Singapore. Has also been recorded at 300 m on Gunung Jerai (Kedah) and at 820 m on Gunung Raya, Pulau Langkawi (Kedah). **Habitat:** Forests, including peatswamp, the forest edge, scrub, plantations, the landward edge of mangroves, and mangroves (on Pulau Langkawi only). **Habits:** Frequents the ground storey in small parties. Skulks in thick cover, appearing only in clearings. Will sometimes ascend to the middle storey and sing. Otherwise sings from the underbrush.

523. Horsfield's Babbler, *Malacocincla sepiarium* — Plate 58
Kekicau Paruh Tebal

Description: 15 cm (6 in). From Abbott's Babbler (522) by much shorter tail and more robust bill with prominent hooked bill nail on tip of maxilla. Crown dark greyish olive brown grading into lighter olive brown upperparts; throat greyish white; breast, flanks and underparts greyish buff. **Voice:** A trisyllabic whistle, *chip-chop-chui*, the second syllable lowest, the last syllable stressed. **Range:** The Malay Peninsula to the Greater Sundas and Bali. **Status and distribution:** Locally common at low elevations up to 850 m, south to Johor. **Habitat:** Forests, usually in the vicinity of flowing streams. **Habits:** Frequents the ground storey. Moves in small parties in the undergrowth. Feeds largely invertebrates obtained from the forest floor. Sometimes calls from open perch. A known brood host of the Drongo Cuckoo (256).

524. Short-tailed Babbler, *Malacocincla malaccense* — Plate 58
Kekicau Ekor Pendek

Description: 15 cm (6 in). Short tailed appearance distinctive. From Horsfield's Babbler (523) by brighter plumage and less robust bill. Dark greyish brown crown, grey sides of head and narrow black malar stripe diagnostic. Upperparts dark olive- brown, throat, breast and belly white, flanks and undertail coverts tawny. **Voice:** A series of rich, descending whistles. Also a repeated *teu-teu...* **Range:** The Malay Peninsula to Sumatra and Borneo. Populations are suspected to be experiencing moderately rapid declines due to rapid clearance and degradation of lowland forest but its ability to still survive in logged forest and secondary growth implies that it is not in immediate danger. **NT. Status and distribution:** Locally common resident principally at low elevations south to Singapore, where it is uncommon and confined to the Bukit Timah Nature Reserve and the Central Catchment forests (W and H). **Habitat:** Forests, the forest edge. **Habits:** Frequents the ground storey, usually solitarily or in pairs or small groups. Keeps to thick undergrowth. Sometimes seen in clearings at the forest edge.

525. Striped Wren-babbler, *Kenopia striata* — Plate 57
Kekicau-Rimba Berjalur

Description: 14 cm (5.5 in). Orange buff lores and white streaks on blackish crown and rufous brown upperparts diagnostic. Sides of face pale greyish white; underparts white with rufous flanks. **Voice:** A high pitched whistle, *hweeeeee*, constantly repeated.

Also a low *chuk chuk chuk* (W2). **Range:** The Malay Peninsula, Sumatra and Borneo. Declining moderately rapidly due to deforestation. **NT. Status and distribution:** Uncommon and localized resident at low elevations, south to Johor. Also recorded up to 200 m on Gunung Benom (Pahang). Locally extinct in Singapore since 1842 (L and G). **Habitat:** Forests. **Habits:** Frequents the lower storey usually in pairs. Particularly favours swampy forest and palm thickets.

526. Large Wren-babbler, *Napothera macrodactyla* Plate 57
Kekicau-Rimba Dada Putih

Description: 19 cm (7.5 in). Large size and short tailed. Black ear coverts, white lores, throat and breast diagnostic. Brownish upperparts streaked buff and marked with dark scales. Greyish underparts streaked white. **Voice:** A variety of loud whistling notes, persistently uttered. Often heard is a disyllabic *pri-preeeeew*, the second note descends and then ascends slightly. **Range:** The Malay Peninsula, Sumatra and Java. Some populations are experiencing moderately rapid declines due to loss and degradation of lowland forest. **NT. Status and distribution:** Uncommon and localized resident principally at low elevations and hills up to 700 m, south to Johor. Locally extinct in Singapore since 1950 (L and G). **Habitat:** Forests. **Habits:** Frequents the ground storey solitarily or in small parties. Skulks in the forest undergrowth. Feeds among leaf litter on the forest floor. Vocal.

527. Marbled Wren-babbler, *Napothera marmorata* Plate 57
Kekicau-Rimba Telinga Perang

Description: 20 cm (8 in). Large size and short tailed. From Large Wren-babbler (526) by rufous ear coverts and black underparts heavily scaled with white. Upperparts dark brown with black scales. Lores white. **Voice:** A harsh disyllabic *preeeee-preeeee*, the first note ascending, the second descending and audible only at close range. **Range:** The Malay Peninsula and Sumatra. **Status and distribution:** Uncommon and localized resident from 700–1200 m in the Larut Hills (Perak) and the Titiwangsa Range south to Gunung Telapa Buruk (Negeri Sembilan) (BiW). Also on Gunung Benom (Pahang). **Habitat:** Forests. **Habits:** Frequents the ground storey, favouring steep-sided gullies with fallen timber and dense undergrowth. Forages on the forest floor.

528. Streaked Wren-babbler, *Napothera brevicaudata* Plate 57
Kekicau-Rimba Berlorek

Description: 13 cm (5 in). Upperparts olive brown with black scales and buff streaks; sides of head greyish brown with fine pale streaks; throat white with broad dark brown streaks; underparts rufescent brown. **Voice:** A long shrill whistle, *peu-ii or peu-ii-uu*. Also a falling *piu*. When alarmed, utters churring or rattling notes. **Range:** North-East India through South-East Asia to the Malay Peninsula. **Status and distribution:** Locally common resident up to about 50 m in Perlis. Elsewhere in Peninsular Malaysia not recorded below 760 m. Recorded on Gunung Jerai (Kedah), the Larut Hills and the Titiwangsa Range south to Gunung Tampin (Negeri Sembilan). Also on Gunung Benom, Gunung Tahan, Gunung Tapis (Pahang), and Gunung Rabong (Kelantan). Also on Pulau Tioman (Pahang). **Habitat:** Forests. **Habits:** Frequents the ground storey, usually in damp gullies. Skulks in thick vegetation. Feeds mostly among leaf litter on the ground. Usually solitary or in pairs.

529. Eyebrowed Wren-babbler, *Napothera epilepidota* Plate 57
Kekicau-Rimba Kening Putih

Description: 11 cm (4.5 in). From Streaked Wren-babbler (528) by smaller size and much shorter tail. Long and narrow buff supercilium diagnostic. Crown dark brown; eyeline blackish; streaks on back whitish buff; streaks on breast blackish; throat whitish; tips of wing coverts whitish, forming two narrow wingbars. **Voice:** A plaintive *peeeow* on a falling tone. Also a series of undulating squeaky chatters. Loud churrs when alarmed. **Range:** North-East India through southern China and South-East Asia to the Malay Peninsula and the Greater Sundas. **Status and distribution:** Uncommon and localized resident in hilly country at low elevations, usually below the montane zone, south to Johor. **Habitat:** Forests. **Habits:** Frequents the ground storey solitarily or in pairs. Skulks in the undergrowth. Feeds on or near ground. Shy.

530. Pygmy Wren-babbler, *Pnoepyga pusilla* Plate 57
Kekicau-Rimba Kerdil

Description: 9 cm (3.5 in). Tiny size with tailless appearance distinctive. Upperparts dark brown with pale spottings to wings and lower back. Underparts either whitish or buff, marked with blackish scales. **Voice:** A powerful, far carrying high-pitched thin disyllabic whistle, *peeeeeee-peeeeeee*; the second syllable (sometimes omitted) on a descending scale after about a two second interval. **Range:** The Himalayas and southern China through South-East Asia and the Malay Peninsula to Sumatra, Java and the Lesser Sundas. **Status and distribution:** Locally common resident above 1000 m in the Larut Hills (Perak) and the Titiwangsa Range south to Ulu Langat (Selangor). Also on Gunung Tahan and Gunung Benom (Pahang). Common at the hill-stations. **Habitat:** Forests. **Habits:** Frequents the ground storey solitarily or in pairs. Inhabits undergrowth in dark, damp gullies. Will also frequent forest paths. Frequently flicks wings. Feeds near or on ground. Very vocal.

531. Collared Babbler, *Gampsorhynchus torquatus* Plate 59
Kekicau Kepala Putih

Description: 24 cm (9.5 in). Long tail. White head contrasting with rufous brown plumage diagnostic. Band around neck blackish; tail feathers tipped white; bill pale pinkish. **Voice:** A loud grating chatter, *gunk-grrrrunk*, repeated persistently. **Range:** Yunnan (southern China) and western Myanmar through continental South-East Asia to the Malay Peninsula. **Status and distribution:** Uncommon and localized resident from 730–1100 m in the Titiwangsa Range (exceptionally at 1500 m) south to Genting Highlands (Pahang). Fairly common at the lower elevations of Fraser's Hill and Genting Highlands (Pahang). **Habitat:** Forests, the forest edge with tall bamboo stands. **Habits:** Gregarious, moving around in small parties, foraging among bamboo.

Vocal and conspicuous where common. Will also perch on electricity or telephone wires near bamboo stands.

532. Cutia, *Cutia nipalensis* Plate 60
Kiutia

Description: 19 cm (7.5 in). Chunky and short-tailed appearance together with slaty blue crown and wings distinctive. Underparts white with bold black barrings on flanks and sides of breast. **Male:** Black mask and bright unmarked rufous mantle and uppertail coverts. **Female:** Mask dull brown, mantle olive brown with dark spots. **Voice:** The male utters a loud, high pitched trill, *yeet-yeet-yit-yit*, the female an emphatic, woodpecker-like *pik* (W2). **Range:** The Himalayas and Myanmar discontinuously through South-East Asia to the Malay Peninsula. **Status and distribution:** Uncommon and localized resident above 1200 m in the Larut Hills (Perak) and the Titiwangsa Range south to Genting Highlands (Pahang). Also on Gunung Benom (Pahang). **Habitat:** Forests, the forest edge. **Habits:** Solitary or in pairs. Frequents the canopy and middle storeys. Probes for insects in creeper and moss covered branches. Climbs up and down vertical tree trunks in nuthatch fashion. Regularly participates in bird-waves. Pair members engage in allopreening when perched together.

533. White-crested Laughingthrush, *Garrulax leucolophus* Plate 59
Kekicau-Riang Jambul Putih

Description: 30 cm (12 in). White head and erect crest, throat and breast contrasting with broad black eyeline together with bright rufous upperparts diagnostic. Lower belly and undertail coverts rufous. **Voice:** A series of maniacal laughing hoots, usually uttered in chorus. Also a series of scolding cackles. **Range:** The Himalayas through South-East Asia to Sumatra. A popular cage bird, large numbers are trapped throughout its indigenous range for the caged bird trade. **Status and distribution:** Introduced. Feral populations, derived from caged escapees are well established throughout Singapore, where it is now fairly common and more numerous and widespread than the Hwamei (536). Feral populations in Kuala Lumpur and Pulau Pinang established in the 1970s have been greatly reduced by trapping and may exist in very small numbers. **Habitat:** Wooded areas. **Habits:** Moves about in small parties, generally keeping to tree cover but sometimes descending to feed on the ground. Like other laughingthrushes, makes short glides from place to place. Very vocal.

534. Black Laughingthrush, *Garrulax lugubris* Plate 59
Kekicau-Riang Hitam

Description: 27 cm (10.5 in). Dull blackish slate plumage with patch of bare blue skin around eye together with bright orange bill diagnostic. **Voice:** A series of rapid *poop* notes followed by loud discordant chuckles. **Range:** The Malay Peninsula, Sumatra and Borneo. **Status and distribution:** Locally common to uncommon resident from 800–1500 m in the Larut Hills and the Titiwangsa Range, south to Genting Highlands (Pahang). Also on Gunung Benom (Pahang). Most common between 800–1200 m. Conspicuous at the lower elevations of Fraser's Hill and Genting Highlands

(Pahang). **Habitat:** Forests, the forest edge. **Habits:** Moves around in small noisy groups. Keeps mostly to thick cover. Does not engage in long laborious flights, preferring to glide from place to place.

535. Spectacled Laughingthrush, *Garrulax mitratus* Plate 59
Kekicau-Riang Mata Putih

Description: 23 cm (9 in). Plumage bluish grey. Dark chestnut brown cap, orange bill and legs together with white eyering and wingbar diagnostic. Forehead blackish with fine white streaks; undertail coverts chestnut. **Voice:** A pleasing slurred whistle, *kwo- kwee-kwo-kwee*... Also harsh rasping notes. **Range:** The Malay Peninsula, Sumatra and Borneo. **Status and distribution:** Fairly common resident above 900 m in the Larut Hills (Perak) and the Titiwangsa Range south to Ulu Langat (Selangor). Also on Gunung Benom and Gunung Tahan (Pahang). Most common between approximately 1200–1400 m. Very common and conspicuous at the hill-stations. **Habitat:** Forests, the forest edge. **Habits:** Moves about in small parties, preferring to hop from branch to branch. Like all laughingthrushes, glides from one patch of foliage to another. Forages among branches and foliage as well as on the ground. Frequently enters hill-station gardens. Known to allopreen. Not shy.

536. Hwamei, *Garrulax canorus* Plate 59
Kekicau-Riang Halia

Description: 25 cm (10 in). Gingery brown plumage with conspicuous white outer eyering and streak behind eye diagnostic. Crown and nape streaked blackish, throat and breast narrowly streaked blackish with faint blackish barrings on tail. Inner orbital skin blue; bill yellow. **Voice:** A loud and melodious series of flute like notes. **Range:** Northern Indo-China through central and southern China to Hainan and Taiwan. A good songster, large numbers are trapped throughout its indigenous range for the caged bird trade. **Status and distribution:** Introduced. Very small feral populations of caged escapees occur throughout Singapore with breeding recorded on Sentosa Island. In Peninsular Malaysia, small non-breeding populations found in the suburbs of Kuala Lumpur and Petaling Jaya (Selangor) might have already been trapped or died out. **Habitat:** Gardens, scrub. **Habits:** Usually keeps to thick cover, feeding largely on insects and small fruit. Very vocal. Not shy.

537. Chestnut-crowned Laughingthrush, *Garrulax erythrocephalus*
Kekicau-Riang Ubun Merah Plate 59

Description: 27 cm (10.5 in). Overall dark olive grey with a chestnut tinge and bright olive wash on wings and tail. Has small black patch on wings. Crown dark chestnut; sides of face greyish; primary coverts black; narrow eyering white. **Voice:** A slurred but powerful and rapid *churr-wee-witt,* the last two closer. Also a series of low rasping nasal notes. **Range:** The Himalayas, Myanmar and southern China through South-East Asia to the Malay Peninsula. **Status and distribution:** Locally common resident above 1000 m in the Larut Hills (Perak) and the Titiwangsa Range south to Selangor. Also on Gunung Benom and Gunung Tahan (Pahang). Also on the Eastern Range

(Terengganu) where it is the only *Garrulax* species (W2). The most common laughingthrush above 1400 m and is especially common at the higher elevations of Bukit Larut (Perak), Cameron Highlands, Fraser's Hill, and Genting Highlands (Pahang). **Habitat:** Forests, the forest edge. **Habits:** Keeps to thick cover. Moves along branches and vines in a series of rapid hops like squirrels. Often enters gardens of hill-station bungalows where it is often tame.

538. Silver-eared Mesia, *Leiothrix argentauris* Plate 60

Mesia Telinga Perak

Description: 18 cm (7 in). Black cap contrasting sharply with bright yellow bill and silvery white ear coverts diagnostic. Upperparts olive-grey with red wing patch; underparts bright yellow. **Male:** Upper and undertail coverts red. **Female** (not illustrated): Upper and undertail coverts dull orange-yellow. **Voice:** A loud cheerful whistle uttered up and down the scale. Also a sharp rattle. Also low chuckling notes. **Range:** The Himalayas through South-East Asia and the Malay Peninsula to Sumatra. **Status and distribution:** Fairly common resident above 900 m in the Larut Hills (Perak) and the Titiwangsa Range south to Ulu Langat (Selangor). Also on Gunung Tahan (Pahang). Also on the Eastern Range (W2). Rare on Bukit Larut (Perak) but common on Cameron Highlands, Fraser's Hill and Genting Highlands (Pahang). Birds seen in Singapore are escaped caged individuals. **Habitat:** Forests, the forest edge. **Habits:** Frequents the lower and middle storeys. Moves through the underbrush in small groups, feeding largely on insects and small fruit. Often enters hill-station gardens.

539. Long-tailed Sibia, *Heterophasia picaoides* Plate 59

Sibia Ekor Panjang

Description: 30 cm (12 in). Slender build with long graduated tail, greyish plumage with brownish tinge on the upperparts together with white wing patch and whitish barrings underneath tail diagnostic. Iris red. **Voice:** A dry sibilant rattle, *churrr*. Also high pitched twittering notes uttered both while perched and in flight. **Range:** The Himalayas and Myanmar through South-East Asia to the Malay Peninsula and Sumatra. **Status and distribution:** Fairly common resident above 1200 m in the Larut Hills (Perak) and the Titiwangsa Range south to Ulu Langat (Selangor). Also on Gunung Benom and Gunung Tahan (Pahang). Very common at the hill-stations. **Habitat:** Forests, the forest edge. **Habits:** Gregarious, gathering in parties of up to 12. Generally forages in the canopy but often descends lower to feed in fruiting shrubs. Will also feed on food scraps in hill station gardens. Sometimes clumsily hovers below strands of small berries, picking them off. Flight rather wobbly, as noisy parties fly from tree to tree. Frequently enters hill-station gardens. Not shy.

540. Blue-winged Minla, *Minla cyanouroptera* Plate 60

Minla Sayap Biru

Description: 16 cm (6.5 in). Long square-cut tail distinctive. Upperparts dull brownish grey; underparts whitish; lining on wing and edge of tail blue; iris white. **Voice:**

A series of twittering notes, while foraging. Also a short and soft rattle. A rather sharp *wuick* in flight. **Range:** The Himalayas and southern China through South-East Asia to the Malay Peninsula. **Status and distribution:** Fairly common resident from 1000–1600 m in the Larut Hills (Perak) and the Titiwangsa Range south to Ulu Langat (Selangor). Very common at the hill-stations. **Habitat:** Forests, the forest edge. **Habits:** Frequents the canopy and middle storey in small parties, flitting through the foliage in pursuit of insects. Will also feed on berries in fruiting shrubs. Flight jerky. Often enters hill-station gardens. Not shy.

541. Chestnut-tailed Minla, *Minla strigula* Plate 60
Minla Ekor Perang

Description: 17 cm (6.75 in). Long square-cut tail distinctive. Forehead and fore-crown golden olive merging into dark smoky grey on mid and hind crowns and upper neck. Mantle and to uppertail coverts, slaty grey tinged olive, Chin pale yellow, throat white with black bars, Breast and flanks smoky yellowish grey. Belly and undertail coverts whitish. Uppertail coverts and patch on wings chestnut. **Voice:** A series of thin chattering and ripling notes, uttered by foraging parties. Also a reedy little song (Madoc in W2). **Range:** The Himalayas and southern China discontinuously through South-East Asia to the Malay Peninsula. **Status and distribution:** Locally common resident above 1585 m in the Larut Hills (Perak) and the Titiwangsa Range south to Genting Highlands (Pahang). Also on Gunung Tahan and Gunung Benom (Pahang). Particularly common at Gunung Batu Brinchang in the Cameron Highlands (Pahang) and Gunung Ulu Kali (Pahang/Selangor). **Habitat:** Ericaceous forests. **Habits:** Forages in small parties in low trees and branches. Gleans small invertebrates from the foliage and branches. Will also feed on the ground, where it is capable of making short bounds. Frequently gathers at peak top telecommunication towers in the early morning to feed on stranded moths attracted by lights the previous night. Not shy.

CISTICOLAS, PRINIAS, AND TAILORBIRDS

ORDER *Passeriformes* **FAMILY** *Cisticolidae*

World: *138 species* **Peninsular Malaysia and Singapore:** *9 species*

A distinct family of small-sized insectivorous birds characterized by thin bills and short to rather longish tails. Most are dressed in drab colours while some have bright plumage. Have loud distinct calls. Occupy a variety of habitats from forests and mangroves to associated secondary growth and open country. Some are conspicuous garden birds. All are sedentary. Sexes alike or almost so. Previously grouped together with Old World Warblers. Represented by the following groups and genera: **Cisticolas,** genus *Cisticola* (542). Small and nondescript warblers with relatively long tails and streaked plumage with the centre of distribution in Africa. One species is found here. Inhabits grassland and scrub. Often delivers song in display flight. **Prinias,** genus *Prinia* (543–545). Slim and longish appearance with distinctive long graduated tails which are often cocked. Inhabit grasslands and scrub. Generally skulk in thick vegetation but usually deliver songs from open perches. **Tailorbirds**, genera *Phyllergates* (546) and *Orthotomus* (547–550). Small size with long bills and relatively long tails which are frequently cocked in the latter. Have bright rufous crowns. All *Orthotomus* species stitch their purse-shaped nests on the underside of leaves while *Phyllergates* builds a pouched shaped nest, placing it thickets. Generally inhabit forests, the forest edge, open country, and gardens. Glean insects from foliage. **Note:** The monotypic genus *Phyllergates* superficially resembles *Orthotomus* in structure but differs in nest-building habits, plumage colouration and vocalizations, and might not be a true tailorbird or a cisticolid. Some authors place it in the family Cettidae together with *Abroscopus* warblers, *Cettia* bush-warblers, and tesias. Tentatively retained in Cisticolidae pending further research and evidence.

542. Zitting Cisticola, *Cisticola juncidis* Plate 61
Cekup Ekor Kipas/Burung Main Angin

Description: 11 cm (4.5 in). From Lanceolated (478) and Rusty-rumped Warblers (479) by much smaller size and shorter tail with more extensive white tips. Crown, mantle and wings brown with blackish streaks; rump rufous; throat white; underparts whitish tinged rufous buff on flanks and undertail coverts. Shows pale buff supercilium. **Voice:** A sharp *tik-tik-tik*... repeatedly uttered in flight. **Range:** Africa, southern Europe, temperate Asia, South and South-East Asia to northern Australia. **Status and distribution:** Locally common resident principally at low elevations south to Singapore but recorded up to 760 m on Bukit Larut (Perak). **Habitat:** Ricefields, open grassland, grassy riverbanks. **Habits:** Seldom perches in the open, preferring to skulk in long grass. During breeding, the male hovers and circles in the air, bobbing up and down with tail fanned, uttering high pitched call, before dropping back to cover.

543. Hill Prinia, *Prinia atrogularis* Plate 61
Prinia Gunung Ekor Panjang

Description: 18 cm (7 in). Very long tail with dark brown upperparts and whitish underparts with tawny buff tinge on flanks distinctive. Long whitish supercilium, grey sides of head, whitish throat together with blackish streaks on breast diagnostic. **Voice:** A loud and repeated disyllabic *chu-ip, chu-ip*,... Also loud *chup* notes. **Range:** The Himalayas and southern China discontinuously through South- East Asia to Sumatra. **Status and distribution:** Rare and localized resident from 1500–2100 m only on Gunung Tahan (Pahang). **Habitat:** Ericaceous grassland and scrub. **Habits:** Usually solitary or in pairs, generally keeping to thick cover but will also forage in the open. Other habits little known.

544. Rufescent Prinia, *Prinia rufescens* Plate 61
Prinia Belukar

Description: 12 cm (4.75 in). Dark rufescent brown upperparts and long graduated tail separate it from tailorbirds (546–550). Underparts whitish. Shows narrow white supercilium. **Breeding:** Grey head. **Non-breeding:** Rufescent brown head. **Voice:** A sharp and monotonous *chuip-chuip-chuip....*, persistently uttered. **Range:** The Himalayas through South-East Asia to the Malay Peninsula. **Status and distribution:** Locally common resident from low elevations up to 1500 m, south to Johor. More common than Yellow-bellied Prinia (545) in forest clearings and forest edge scrub. **Habitat:** Forest edge and plantations with long grass and undergrowth. **Habits:** Usually solitary or in pairs, skulking in the undergrowth. Less gregarious and more retiring than Yellow-bellied Prinia. Frequently calls from an open perch.

545. Yellow-bellied Prinia, *Prinia flaviventris* Plate 61
Prinia Perut Kuning

Description: 13 cm (5.5 in). From Rufescent Prinia (544) by longer tail, dark greenish olive upperparts together with yellow belly and undertail coverts which contrast with whitish throat and breast. Head slate grey with short narrow white supercilium in front of eye. **Voice:** A series of rapid bubbling trills, repeatedly uttered. Also a sharp nasal descending mew, *eeeah*. **Range:** The Indian subcontinent and southern China through South-East Asia to the Greater Sundas. **Status and distribution:** Fairly common resident from low elevations up to 1800 m at Cameron Highlands (Pahang), south to Singapore. More common than Rufescent Prinia in open country habitats. **Habitat:** Reedbeds, swampy grassland, open country with tall grass, the forest edge. **Habits:** Keeps to thick cover. Usually sings from the top of a reed stalk or grass stem. Brood host of Plaintive Cuckoo (250).

546. Mountain Tailorbird, *Phyllergates cuculatus* Plate 61
Perenjak Gunung

Description: 12 cm (4.75 in). Grey head, whitish throat and upper breast together with bright yellow lower breast, belly and undertail coverts separate it from other tailorbirds. Shows narrow black eyeline and narrow white supercilium (see Chestnut-crowned Warbler (489)). **Juvenile (not illustrated):** Dull olive upperparts and sides of head, pale yellowish supercilium and underparts. **Voice:** The song is a distinctive and melodious high-pitched tinkle of 4–6 ascending notes, uttered in a series of four sequences, starting on a low scale and ascending each time thereafter; each sequence is spaced by a two second interval. **Range:** The Himalayas and southern China throughout South-East Asia to the Greater and Lesser Sundas to Sulawesi and the Philippines. **Status and distribution:** Locally common resident above 1000 m, recorded in the Larut Hills (Perak) and the Titiwangsa Range south to Genting Highlands (Pahang). Also on Gunung Tahan, Gunung Benom and Gunung Rabong (Kelantan). On the Eastern Range, recorded on Gunung Lawit (Terengganu) and Gunung Tapis (Pahang). Fairly common at the hill-stations. **Habitat:** Forests, the forest edge. **Habits:** Frequents the understorey and bamboo thickets solitarily or in small parties. Neither cocks tail nor stitches nest on underside of leaves like other tailorbirds. Builds a pouch-shaped nest, usually in bamboo clumps. Mostly gleans insects from foliage. Will join bird waves. Skulks largely in the undergrowth from where its song is delivered.

547. Common Tailorbird, *Orthotomus sutorius* Plate 61
Perenjak Ekor Panjang

Description: 12 cm (4.75 in). Tail length of breeding male up to 3.75 cm (1.5 in) longer. Rufous forecrown, narrow whitish supercilium and olive green upperparts contrasting with pale buff underparts diagnostic. Dark feather bases on throat often appear as greyish speckles. **Voice:** Territorial song is a brief and rapid disyllabic trill, *ti-chi*, persistently uttered. **Range:** The Indian subcontinent and southern China throughout South-East Asia to the Malay Peninsula and Java. **Status and distribution:** Common and widespread resident from low elevations up to 1580 m, south to Singapore. A common and conspicuous garden bird. **Habitat:** Open country, scrub, the forest edge, mangroves, plantations, gardens. **Habits:** Frequents the understorey solitarily or in pairs. Sometimes feeds in more open areas. Habitually cocks tail. Often nests in potted plants in gardens.

548. Dark-necked Tailorbird, *Orthotomus atrogularis* Plate 61
Perenjak Leher Hitam

Description: 11 cm (4.5 in). From Common Tailorbird (547) by yellowish lower flanks and undertail coverts. **Male:** Rufous crown and nape together with blackish throat and sides of neck diagnostic. **Female:** Lacks blackish throat with rufous on head reduced to forecrown (more extenstive than both sexes of Common Tailorbird). **Voice:** A high-pitched nasal *kri-ii-ii*. Also a short *tiu*. **Range:** North-East India and South-East Asia to Sumatra and Borneo. **Status and distribution:** Fairly common resident from low elevations up to 1500 m, south to Singapore. Probably the most common tailorbird in disturbed forest habitats. **Habitat:** Forests, the forest edge, mangroves, scrub, gardens. **Habits:** Usually solitary or in pairs. Skulks in the undergrowth. Brood host of Plaintive Cuckoo (250).

549. Rufous-tailed Tailorbird, *Orthotomus sericeus* Plate 61
Perenjak Ekor Merah

Description: 12 cm (4.75 in). From female and juvenile Ashy Tailorbird (550) by whitish cheeks and underparts together with diagnostic bright rufous tail. **Voice:** A loud and distinct *tweet-to* or *to-tweet*, uttered repeatedly. Also a high-pitched wheezy *tzee-tzee-tzee-tzee*... **Range:** Tenasserim (Myanmar) and the Malay Peninsula to Sumatra and Borneo east to Palawan (the Philippines). **Status and distribution:** Locally common resident from low elevations up to 400 m, south to Singapore, where it is uncommon. Less common than Dark-necked (548) and Ashy Tailorbirds (550). **Habitat:** Mangroves, the forest edge, scrub. **Habits:** Usually solitary or in pairs. Like other tailorbirds, skulks in the undergrowth.

550. Ashy Tailorbird, *Orthotomus ruficeps* Plate 61
Perenjak Kelabu

Description: 12 cm (4.75 in). From other tailorbirds by rufous head, sides of head and chin. Tail grey-brown with white tips and very small and narrow dark subterminal markings. **Male:** Dark grey mantle, breast and belly. **Female (not illustrated):** Breast pale grey, belly whitish. **Juvenile:** Entire underparts whitish. **Voice:** A diagnostic high-pitched trill, *preee-yip*. repeatedly uttered. **Range:** Tenasserim (Myanmar) to the Malay Peninsula and the Greater Sundas. **Status and distribution:** Fairly common resident at low elevations south to Singapore. More common in mangroves than other habitats. **Habitat:** Mangroves, the forest edge, plantations, scrub, gardens. **Habits:** Usually solitary or in pairs frequenting the understorey and the low branches of trees. Very active, constantly flitting about in search of small insects. Very vocal.

NUTHATCHES

ORDER *Passeriformes* **FAMILY** *Sittidae*

World: *25 species* **Peninsular Malaysia:** *2 species*

Small and compact bodied birds with short tails and sharp, tapered bills. Arboreal, moving in small parties up and down vertical tree trunks with ease. Feed by probing into crevices in bark for insects and grub. Will also glean insects and tiny invertebrates from branches. Flight swift and undulating over short distances. Sexes alike or almost so.

551. Velvet-fronted Nuthatch, *Sitta frontalis* Plate 64
Pepatuk Dahi Hitam

Description: 12 cm (4.5 in). Deep violet-blue upperparts, whitish throat and breast, vinous underparts together with red bill and velvety black patch on forehead diagnostic. **Male:** Has narrow black supercilium. **Female:** Lacks black supercilium. **Juvenile (not illustrated):** Like adult but bill black. **Voice:** Thin sibilant notes, *chwit-chwit-chwit*, usually preceded by a high pitched trill, *chiri, chwit-chit-chit-chit-chit*... **Range:** India through South-East Asia to the Malay Peninsula and the Greater Sundas. **Status and distribution:** Locally common resident at low elevations and hills up to 1150 m, south to Johor. **Habitat:** Forests, the forest edge. **Habits:** Frequents the canopy and middle storey in small noisy parties or sometimes solitarily, rapidly moving up and down tree trunks while foraging. Often in mixed species flocks.

552. Blue Nuthatch, *Sitta azurea* Plate 64
Pepatuk Biru Gunung

Description: 12 cm (4.5 in). From Velvet-Fronted Nuthatch (551) by whitish blue bill and orbital skin. Upperparts black with bluish grey barrings on back; throat and breast white; belly black with blue sheen; undertail coverts blue. **Voice:** High pitched grating squeaks. **Range:** The Malay Peninsula, Sumatra and Java. **Status and distribution:** Locally common resident above 800 m in the Larut Hills (Perak) and the Titiwangsa Range, south to Ulu Langat (Selangor). Also on Gunung Tahan, Gunung Benom (both in Pahang), and Gunung Rabong (Kelantan). Common and conspicuous at the hill-stations. **Habitat:** Forests, the forest edge. **Habits:** Frequents the canopy and middle storeys in small parties. Forages in typical nuthatch fashion, probing into moss covered branches for insects and grub. Often walks upside down along the underside of branches. Regularly participates in bird-waves.

STARLINGS AND MYNAS

ORDER *Passeriformes* **FAMILY** *Sturnidae*

World: *116 species* **Peninsular Malaysia and Singapore:** *13 species*

Medium-sized stocky birds with strong, pointed bills, relatively long legs and short tails. Flight strong and direct. Highly gregarious, inhabiting a wide range of habitats from forests to open country. Feed mainly on fruit and an assortment of small invertebrates. Often roost communally in large flocks. Very noisy; calls vary from loud and raucous notes to pleasing whistles. Some are accomplished mimics. Nest in tree holes and cavities. Sexes alike in most species. Being popular cage birds, escaped non-indigenous species have formed feral populations in many areas. In this respect three species, the Asian Pied Starling *Sturnus contra*, Black-collared Myna *Gracupica nigricollis*, and the Vinous-breasted Myna *Acridotheres burmannicus* have occurred sporadically in the review area but being popular cage birds, their origins could not be verified and as such have not been included in the official checklists of both Peninsular Malaysia and Singapore. Represented by the following groups and genera: **Starlings,** genera *Aplonis* (553) and *Sturnus* (554–557). Largely arboreal, feeding in the canopy or crowns of trees. Plumage vary from glossy dark colours in the former to combinations of black, white or grey, or pink. Wings pointed and triangular. Most species are migratory. **Mynas,** genera *Acridotheres* (558–563) *Ampeliceps* (564) and *Gracula* (565). Generally dark coloured. Wings rounded with large white or pale patches, conspicuous in flight. *Acridotheres* species are terrestrial, inhabiting largely scrub and open country; some are commensals of man and live in towns and cities. Some species sport short or shaggy crests. *Ampeliceps* and *Gracula* are arboreal forest species, inhabiting the canopy. Legs yellow in most species. All species are sedentary.

553. Asian Glossy Starling, *Aplonis panayensis* Plate 49

Perling Mata Merah

Description: 20 cm (8 in). Glossy dark green plumage with red iris diagnostic. Appears all black in poor light. **Juvenile:** Upperparts blackish brown with slight greenish tinge; underparts whitish with bold blackish streaks. **Voice:** A sharp and clear metallic whistle, *teeeu*. Also a harsh *chreeek*, usually uttered in flight or intermittently among foraging flocks. **Range:** North-East India and South-East Asia to the Greater Sundas, Bali, Sulawesi and the Philippines. **Status and distribution:** Common and widespread resident at low elevations, south to Singapore, but recorded up to 1193 m on the summit of Gunung Telapa Buruk (Negeri Sembilan) (W2). Also on most islands off both coasts. **Habitat:** Mangroves, open country, the forest edge, plantations, gardens, cities. **Habits:** Gregrarious, moving and feeding in large noisy flocks, usually in trees but sometimes descending to the ground to feed on fallen fruit. Undertakes local dispersion, sometimes at night, and may also be encountered on far offshore islands and at high elevations. Flies fast in small compact flocks, turning in unison. Feeds largely on fruit and often gathers at fruiting trees and shrubs. Nests in tree holes, the eaves of buildings and cracks in concrete buildings.

554. Purple-backed Starling, *Sturnus sturninus* — Plate 49
Perling Belakang Ungu

Description: 19 cm (7.5 in). Small starling with metallic purple back and patch on nape. Has two narrow pale wingbars. Head and breast pale greyish, grading into whitish belly. Shows whitish or buff square patch on rump, conspicuous in flight. Bill and tail black. **Juvenile:** Like adult but metallic purple areas replaced with brown. **Voice:** A soft rasping *chweep*, uttered by foraging flocks. **Range:** Breeding in eastern Russia, northern China, and Korea; wintering south to southern China, North-East India through South-East Asia to Sumatra and Java, rarely Borneo. **Status and distribution:** Locally common passage migrant and winter visitor principally at low elevations, south to Singapore. Occurs above 900 m only during autumn and spring passage. **Habitat:** Coastal scrub, open country, reedbeds, wooded suburban parks and gardens, the forest edge. **Habits:** Usually feeds in tree-tops in small parties, sometimes in association with Asian Glossy Starling (553). Flocks in very large numbers at roosting sites. Sometimes descends to feed on the ground on fallen fruit. Often roosts communally in large numbers in reedbeds.

555. Chestnut-cheeked Starling, *Sturnus philippensis* — Plate 62
Perling Pipi Perang

Description: 18 cm (7 in). **Male:** Like Purple-backed Starling (554) but has chestnut ear coverts and lacks dark metallic purple patch on nape. **Female:** Mantle greyish brown; wings and tail blackish; underparts whitish. **Voice:** Unknown in review area. **Range:** Breeding in Japan; wintering south to Borneo and the Philippines. **Status and distribution:** Rare and localized passage migrant or winter visitor at low elevations south to Singapore. In review area first known from sightings of individuals at Loyang (Singapore) on 8 December 1987; an individual among a flock of Purple-backed Starlings and Petaling Jaya (Selangor) on 21 January 1988. **Habitat:** Wooded suburban parks and gardens. **Habits:** General habits similar to Purple-backed Starling with which it sometimes associates. Feeds largely in trees.

556. White-shouldered Starling, *Sturnus sinensis* — Plate 49
Perling Bahu Putih

Description: 20 cm (8 in). **Male:** Pale greyish plumage contrasting slightly with whitish crown but sharply demarcated from broad white upperwing coverts and scapulars together with black flight feathers diagnostic. Whitish parts of plumage often have strong rusty suffusion. **Female:** Like male but white on upperwing coverts and scapulars restricted to median coverts, appearing as a wingbar. Crown pale greyish. In both sexes, bill bluish white and iris whitish. Tail pattern distinctive with pale uppertail coverts, broad black subterminal tail band with narrow whitish tip. **Juvenile (not illustrated):** Like adult but pale greyish parts tinged brownish. Lacks white patch on shoulder (usually acquired in late summer). Best identified by distinctive tail pattern. **Voice:** Loud harsh squealing and squawking notes. **Range:** Breeding in southern China; wintering south to North-East India through South-East Asia and the Malay Peninsula to Borneo. **Status and distribution:** Uncommon and localized winter visitor in small numbers at low elevations, south to Singapore. **Habitat:** Open country, scrub. **Habits:** Frequents the canopy and crowns of trees in small loose flocks. Feeds largely on fruit and will congregate at fruiting trees and shrubs.

557. Rosy Starling, *Sturnus roseus* Plate 62

Perling Jambu

Description: 22 cm (8.5 in). **Adult:** Unmistakable with shaggy crest. Non-breeding bird shows dull buff-pink plumage contrasting with blackish hood and tail together with glossy greenish-black wings. Bill brownish-pink. During breeding males become brighter pink with hood turning glossy purplish-black. Bill turns pink with black base. Females are duller. **Juvenile:** Lacks crest. Upperparts light sandy buff, paler on rump, with darker tail and wings. Underparts, lighter buff with slightly darker buff-brown streaks on sides of breast. Bill yellow with dark tip. **Voice:** Unknown in review area. **Range:** Breeding throughout much of southern temperate Eurasia east across Iran, Afghanistan, southern Russia, to western China. Winters throughout much of the Persian Gulf region, the Indian subcontinent and South-East Asia with stragglers reaching the Malay Peninsula and Borneo. **Status and distribution:** Very rare winter visitor with sporadic records from Kuala Gula (Perak) and Sungai Balang, Muar (Johor), south to Singapore, where it has occurred at Tuas and Changi since 1994 (W and H);the latest being a small flock of six at the edge of ricefields at Sg. Balang (Johor) on 27 March 2010 (BiW). **Habitat:** Open country. **Habits:** In review area, solitary or in very small groups. Forages mostly on the ground. Other habits little known.

558. Black-winged Myna, *Acridotheres melanopterus* Plate 62

Tiong Sayap Hitam

Description: 23 cm (9 in). White head with short crest and entire white underparts together with yellow bill, broad tear-shaped orbital skin and legs diagnostic. Mantle pale grey; wings black with small white wing patch. Tail black with narrow white terminal band. **Juvenile (not illustrated):** White parts of plumage replaced with pale brown. Orbital skin whitish. **Voice:** A repertoire of loud and harsh whistling notes. **Range:** Java, Bali and Lombok. Once common, the scarcity of sightings in the field coupled with the increased rarity of individuals in the caged bird trade suggests that it has undergone a rapid decline since 1997 and has hence been up listed to **CR.** Urgent conservation action is therefore required to stop its exploitation for the caged bird trade. **Status and distribution:** Introduced. Small feral populations used to exist on St. John's Island (Singapore) but have disappeared due to trapping. Apart from occasional cage escapees, there seems to be no evidence of breeding populations in recent times. **Habitat:** Open country, scrub. **Habits:** Usually in pairs or small loose parties, feeding on open ground or in trees. More arboreal than Common (559) and Javan Mynas (561). Roosts largely in trees. Very vocal.

559. Common Myna, *Acridotheres tristis* Plate 62

Tiong Gembala Kerbau

Description: 25 cm (10 in). From all *Acridotheres* mynas by blackish head, throat and upper breast contrasting with dark brown body plumage and broad yellow orbital skin and all white underwing (conspicuous in flight). Lacks the crest, so typical of the genus. Flight feathers blackish; lower belly and undertail coverts whitish; tail blackish with very narrow white tip; bill yellow. During moult, head feathers are lost showing bare yellow skin of head and neck. At rest, white wing patch stands out as a white line along folded wing, a feature not often seen in the other *Acridotheres* mynas in the review area. **Juvenile (not illustrated):** Duller with blackish head, throat and upper breast replaced with dark brown. **Voice:** A repertoire of loud resonant musical whistling notes and a variety of harsh chuckles, louder and more varied than the other *Acridotheres* mynas in the review area. Contact calls include a rather creaky *krr..krr..krr..krr…* by perched or foraging individuals. Also a low and brief *shreeww* on take off when flushed. **Range:** Iran and southern Russia east to the Indian subcontinent and southern China through South-East Asia and the Malay Peninsula. Widely introduced on other continents and islands. **Status and distribution:** Common and widespread resident principally at low elevations, south to Singapore where it is greatly outnumbered by the Javan Myna (561). Has been recorded at 1500 m in the settled areas of Cameron Highlands (Pahang). Also on Pulau Langkawi (Kedah), Pulau Pangkor (Perak), Pulau Tioman (Pahang) and most of the smaller islands off Singapore. **Habitat:** Open country, scrub, grassy verges of roads, cleared river banks, lawns, plantations, gardens. **Habits:** Frequents the ground in pairs, sometimes in large flocks. Struts on lawns, feeding on small invertebrates or picking up food scraps. Also frequents rubbish tips. Nests in a variety of situations; from the tops of coconut palms to eaves of buildings and large cracks in concrete buildings. Very vocal. Roosts communally in trees and on buildings, but not in great numbers as Javan Myna.

560. Jungle Myna, *Acridotheres fuscus* Plate 62

Tiong Sawah

Description: 24 cm (9.5 in). From Common Myna (559) by short crest, darker grey-brown body plumage (can appear black on wet birds) greyish underwing and the lack of yellow orbital skin. Resembles Javan Myna (561) in shape, structure, and size and shape of crest but differs by paler grey body plumage (often tinged brown on mantle and wings) which shows more contrast with darker head and breast, and the presence of blue-black mark at base of mandible of yellow bill. This mark does not stand out clearly and can only be seen at close range. Greyish belly gradually merges into whitish vent and undertail coverts without showing sharp demarcation. Iris yellow. **Juvenile (not illustrated):** Duller with entirely brown head with white iris, pale throat and lacking crest. Genetic freaks **(not illustrated)** with wholly white throat patches sometimes occur. **Voice:** Not adequately described but consists of harsh whistles and creaky notes, typical of the genus. **Range:** The Indian subcontinent through Myanmar and central Thailand to the Malay Peninsula. **Status and distribution:** Locally common resident at low elevations, south to Melaka. Has been gradually displaced by the Javan Myna in the review area and is beginning to become much less common, especially in the southern half of Peninsular Malaysia. Fairly common and secure in areas not yet reached by Javan Myna, especially in central and northern Perak, Kedah, Perlis, central and northern Pahang, Terengganu, and Kelantan. **Habitat:** Mangroves, the mangrove edge, the forest edge, ricefields, pasture land. **Habits:** General habits similar to other mynas. Associates with grazing livestock in ricefields and cattle pastures.Not a commensal of man.

561. Javan Myna, *Acridotheres javanicus* Plate 62

Tiong Jawa

Description: 24 cm (9.5 in). From Jungle Myna (560) by dark grey (almost black) plumage with white vent sharply demarcated from dark underparts, and the lack of blue-black mark on base of mandible. Iris creamy white, contrasting with dark head. Bill and legs bright chrome yellow. Leucistic individuals (**not illustrated**) sometimes occur; overall plumage dirty white with yellow bill and legs. **Voice:** Varied and strident notes and whistles, typical of the genus, not as varied and loud as Common Myna (559). Also a low and brief *shreeww* on take off when flushed. **Range:** Java and Bali. Widely introduced to Sumatra, Sarawak, the Malay Peninsula, Sumbawa, Flores and Christmas Island. **Status and distribution:** Introduced. from Java to Singapore in the early 1900s and through successful breeding, has become the most common myna there since 1925. Started spreading across the Straits of Johor into Johor in the 1970s, presumably when suitable habitats became available and roosting sites in Singapore had exceeded capacity. Along the west coast plain, with the construction of the north-south highway and the subsequent creation of suitable habitats, it has gradually spread northwards into Negeri Sembilan and Melaka and large populations are now established in and around Kuala Lumpur and throughout much of Selangor, steadily expanding northwards, with individuals sighted in southern Perak by 2001. A roost at Genting Sempah (Pahang) at 634 m, on 6 June 2009 and subsequent sightings of individuals at Gohtong Jaya, just below Genting Highlands (Pahang) on 11 December 2009, is evidence that it has crossed the Titiwangsa Range from the west coast plain east into west Pahang, probably by following the Kuala Lumpur–Karak Highway. Along the east coast plain in Johor, has spread north from Kota Tinggi, into Mersing, probably augmented by populations moving east from Kluang, along the main trunk road. From Segamat, populations have presumably moved east along the highway into FELDA Keratong and Bandar Muazam Shah (Pahang), and have reached Rompin. These populations are likely to work their way up the coastal plain and will eventually reach Pekan and Kuantan, linking up with populations moving east from west Pahang, and setting the stage for a potential invasion of Terengganu and Kelantan. Populations in Bahau (Negeri Sembilan) too seem to have moved into south-west Pahang, and have spread north-east, with individuals sighted at Kampung Sangang, Temerloh (Yang Chong) and Sri Jaya, Kuantan District (D.N. Bakewell), both in Pahang (BiW). **Habitat:** Open country, parks, gardens, orchards, oil-palm plantations, the forest edge. **Habits:** Usually gathers from small to large flocks, catching insects and other smaller invertebrates on the ground or in the air. Often feeds among grazing livestock, catching insects disturbed by trampling hooves. Will also gather in flowering trees, notably *Erythrina* sp., presumably to feed on nectar. Also feeds on the fruit of the MacArthur Palm *Ptycosperma macarthuri*, *a* common ornamental plant throughout most of the review area. Also frequents rubbish dumps together with Common Myna and House Crow (441) to feed on the abundance of insects and maggots from decomposing organic matter. Opportunistic and will even forage in fading light before roosting for the evening. Very vocal, forming huge communal roosts in trees. **Note:** Probably conspecific with Jungle Myna and Sulawesian Pale-bellied Myna *A. cinereus*, More conclusive DNA work will eventually determine this. Many individuals within the former range of Jungle Myna have been seen to have a yellow iris and a strong brownish wash on the mantle (characteristics of Jungle) while lacking the bluish-black mark on the base of the mandible, suggesting hybridization or genetic swamping, depending on what molecular biologists and systemists decide in future.

562. White-vented Myna, *Acridotheres grandis* Plate 62

Tiong Berjambul

Description: 26 cm (10.5 in). Uniform slaty black plumage with diagnostic tall wispy crest and amber brown iris, showing very little contrast with black head. Bill and legs both bright chrome yellow. Crest is held vertically and is 30 mm over the culmen in males and 20 mm in females. **Voice:** Garrulous notes and whistles, typical of the genus, which are less noisy than both Common (559) and Javan Mynas (561). **Range:** North-East India and southern China throughout much of continental South-East Asia to the Thai provinces of the Malay Peninsula. **Status and distribution:** Introduced. Locally common feral populations, around Kuala Lumpur and its suburbs (particularly around Sungai Besi and Serdang); are the descendents of likely caged escapees in the 1980s Populations around Kuala Lumpur are fairly stable but have probably not expanded much due to the subsequent arrival of the Javan Myna. In Kelantan, a few seen around Kota Bharu. Captive individuals on sale at a market at Tumpat (Kelantan) in December 2005 were said to have been brought in from Thailand. **Habitat:** Playing fields, the grassy verges of road, the underside of concrete bridges, and concrete retaining walls of buildings. **Habits:** Usually in pairs or small groups, catching insects and other smaller invertebrates along roads, usually exclusively on its own but sometimes with other mynas. Like other mynas struts on the ground but will also move along in long bounds. Known to nest in seepage holes of concrete retaining walls.

563. Crested Myna, *Acridotheres cristatellus* Plate 62

Tiong Jambul Cina

Description: 26 cm (10.5 in). Fairly thickset myna with uniform slaty black plumage and prominent crest which is floppy and flat-topped. From White-vented Myna (562) by ivory-white bill with reddish or pinkish base, flesh-pink legs and black undertail coverts with narrow whitish scales. Iris amber-yellow, contrasting with black head.

Voice: Not adequately described from the area but consists of harsh garrulous notes and whistles, typical of the genus. **Range:** Southern China (including Hainan and Taiwan), eastern Myanmar and Vietnam. **Status and distribution:** Introduced. A small feral population is locally common on Pulau Pinang, resulting largely from caged escapees. Also a few pairs around Kuala Lumpur and in Singapore where it is now very rare, and declining. **Habitat:** Open country, gardens, parks. **Habits:** Frequents the lawns of parks and gardens in pairs or in small loose flocks. Other habits similar to White-vented Myna.

564. Golden-crested Myna, *Ampeliceps coronatus* Plate 62

Tiong Jambul Emas

Description: 22 cm (8.5 in). Small glossy black myna with yellow orbital skin, pale yellow wing patches (conspicuous in flight) and pinkish bill with dark base. **Male:** Entire crown from base of maxilla to nape bright yellow and concolorous with orbital skin. Broad patch on throat bright yellow. **Female:** Extent of yellow on crown much reduced and separated from orbital skin by narrow black line above eye. Yellow patch on throat very much smaller and less conspicuous. **Juvenile (not illustrated):** Head blackish; throat whitish; breast faintly streaked whitish. **Voice:** Not described from the review area. L and R describe it as a higher-pitched, more metallic whistle than Hill Myna and a bell-like note. **Range:** North-East India through South-East Asia to the Malay Peninsula (presently confined to the Thai provinces). **Status and distribution:** Vagrant, a small flock at Grik (Perak) in December 1939. No further sightings since and hence status should be reassessed. **Habitat:** In review area, semi open parkland. **Habits:** Frequents the canopy in small parties. Other habits little known.

565. Hill Myna, *Gracula religiosa* Plate 62

Tiong Emas

Description: 30 cm (12 in). Large and thickset with glossy black plumage (has purplish sheen in good light), yellow facial wattles and robust pinkish orange bill diagnostic. In flight, broad white wing patches very conspicuous. **Voice:** A diagnostic piercing disyllabic whistle, *ti-ong*. Also a repertoire of low and harsh whinnying notes. **Range:** The Indian subcontinent and southern China throughout South-East Asia to the Greater and Lesser Sundas and Palawan (the Philippines). A popular cage bird, the cage bird trade has virtually extirpated populations throughout its range but is still classified as **LC** (Least Concern). **Status and distribution:** Fairly common resident principally at low elevations, south to Singapore where it is uncommon with small indigenous populations surviving in the Bukit Timah and Bukit Batok Nature Reserves, the Central Catchment Forest, Pulau Ubin, and Pulau Tekong. Individuals seen in urban Singapore are likely to be cage escapees. Also on Pulau Langkawi (Kedah), Pulau Pinang, Pulau Pangkor (Perak), Pulau Tioman (Pahang), and Pulau Tinggi and Pulau Aur (Johor). Has a higher altitudinal range on the islands than on the mainland, occurring up to 900 m on Pulau Tioman. **Habitat:** Forests, the forest edge, mangroves, plantations, wooded areas. **Habits:** Frequents the canopy, usually in pairs but sometimes in small parties, often perching on tall dead trees. Has been observed aerial foraging for termites. Flight fast and direct with whirring wingbeats. An excellent mimic of other bird calls as well as human speech. Very vocal.

THRUSHES

ORDER *Passeriformes* **FAMILY** *Turdidae*

World: *348 species* **Peninsular Malaysia and Singapore:** *22 species*

A large and varied family of relatively long-legged birds, characterized by rounded heads, relatively large eyes and an upright stance. Show great variation in structure and size; from small to fairly large and plump. Tails vary from short to long and are usually cocked periodically. Bills usually slender and pointed in most species. Differ from flycatchers (588–613) and warblers (478–494) by longer and heavier tarsi. Many species have terrestrial habits, feeding upon insects, small invertebrates and small fruit on or near the ground. Very vocal, uttering a repertoire of loud and powerful melodious phrases. Some species are accomplished songsters. Nests are solid cups of fibrous material, usually reinforced with mud, moss and lichen. Juveniles have spotted plumage. Sexually dimorphic in most species; the males are brightly coloured, the females duller. Represented by the following groups and genera: **Rock-Thrushes,** genus *Monticola* (566, 567). Larger and plumper than chats. Males have parts of plumage blue; females largely scaly. Inhabit forests. **Whistling-thrushes,** genus *Myophonus* (568, 569). Large and plump with broad rounded tails and robust bills. Plumage dark purplish-blue with iridescent blue spangles, appearing all black in poor light. Inhabit forests, keeping to moist shady gullies with fast flowing streams. Have loud piercing whistling notes and habitually jerk and splay tails. Differentiating the two resident species in the mountains of the review area has always been problematic. See Bakewell *et al.* (2010). **Typical Thrushes,** genera *Zoothera* (570–573) and *Turdus* (574, 575). Medium to large size. Rather plump. Inhabit forests, usually skulking in the undergrowth. Often feed in fruiting trees. *Zoothera* has characteristic two broad white bands on underwing, conspicuous in flight. **Shortwings,** genus *Brachypteryx* (576). Small size with short rounded wings and short tails. Skulk in the forest understorey. **Chats,** genus *Saxicola* (577). Very small inhabitants of open country. Sit on low perches with an upright stance. Habitually flick tail and wings. **Robins,** genera *Luscinia* (578–580), *Copsychus* (581, 582), *Trichixos* (583) and *Myiomela* (587). Small to medium-sized skulkers in the forest understorey. Shy and secretive, except Oriental Magpie Robin (581). **Forktails:** Genus *Enicurus* (584–586). Medium-sized with largely black and white plumage with long scissor-like tails. Inhabit fast flowing rocky streams, feeding largely on aquatic invertebrates. Flight fluttery and usually low over water or ground. Utter high-pitched sibilant whistles.

566. White-throated Rock-thrush, *Monticola gularis* Plate 63
Murai Tarum Tekak Putih

Description: 19 cm (7.5 in). Small, slim thrush. **Male:** Bright blue crown and nape, blackish mantle with brown fringes, blackish brown wings with small white patch together with orange rufous lores, rump and underparts with small and narrow white throat stripe (sometimes inconspicuous) diagnostic. **Female:** From female Blue Rock-thrush (567) by white throat patch and slimmer build. Upperparts brown with blackish scales; underparts whitish with dense blackish scalings. Shows pale lores and malar stripe which contrast with darker ear coverts. **Voice:** Unknown in review area. **Range:** Breeding in northern China, Korea and eastern Russia; wintering south to southern China and South-East Asia to the Malay Peninsula. **Status and distribution:** Rare and localized winter visitor recorded above 760 m in the Larut Hills (Perak) and in the Titiwangsa Range at Fraser's Hill (Pahang), south to Singapore at low elevations. **Habitat:** Forests, the forest edge, wooded areas. **Habits:** Solitary, usually foraging in small trees or on the ground. Other habits little known.

567. Blue Rock-thrush, *Monticola solitarius* Plate 63
Murai Tarum Biru/Burung Tarum

Description: 23 cm (9 in). **Male:** *M. s. madoci* and *M. s. pandoo*: Bright blue-grey plumage marked with whitish and blackish scales all over distinctive. Both are indistinguishable in the field but *M. s. pandoo* has slightly darker throat; in the hand *M. s. madoci* has more rounded wing with shorter wing length. *M. s. philippensis* **(not illustrated):** Bright blue head, throat, mantle and rump contrasting with chestnut breast and belly diagnostic. Wings and tail blackish. **Female:** From female Siberian Thrush (572) by the lack of pale supercilium and white bands on underwing in flight. Plumage dark grey-brown scaled grey and white (sometimes has bluish cast). **Voice:** The male *M. s. madoci* has a loud, melodious, whistling song (M and W). **Range:** Breeding throughout southern Europe and temperate Asia discontinuously through South-East Asia to the Malay Peninsula. Northern populations are migratory, wintering south to Africa, South and South-East Asia to Sumatra, Borneo, Sulawesi and the Philippines. **Status and distribution:** *M. s. madoci* is a locally common resident at low elevations, south to Negeri Sembilan, where it has nested in Seremban town. *M. s. pandoo* and *M. s. philippensis* are uncommon passage migrants and winter visitor, principally at low elevations, south to Singapore. The latter has been mist-netted at 1300 m at Fraser's Hill (Pahang) and is the less common migrant subspecies. **Habitat:** Limestone outcrops, towns, coastal breakwater. **Habits:** Usually solitary or in pairs. Perches on rocks and buildings with a typical upright posture. Often bobs body and flicks tail. Hops along the ground or along cliff leges, foraging for small invertebrates.

568. Malaysian Whistling-thrush, *Myophonus robinsoni* Plate 63
Murai Bersiul Titiwangsa

Description: 25 cm (10 in). Plumage very similar to the larger Blue Whistling-thrush, subspecies *M. c. dicrorhynchus* (569) but size alone cannot be relied as the main criterion to separate the two where they overlap, especially when encountered in isolation. Bakewell *et al.* (2010) suggest shape of body, head, proportionate eye size, bill size and shape, together with behaviour as important features to distinguish the two in the field. Is squatter than Blue-Whistling-thrush with a shorter neck and a deeper and more rotund body. Head shape more rounded and proportionate, bill more slender and less hooked. Eye proportionately larger than Blue-Whistling-thrush with an obvious yellowish tinge to narrow orbital ring, which is absent in the Blue. These features while obvious in the hand may however not be clear in the field, especially if only fleeting glimpses of the bird are obtained. In addition they may not be obvious if the bird is at an awkward angle in poor light. Have different behaviour which is also important to make positive identification (see habits below).**Voice:** Contact call is a single pitched *chzzz-eee*, preceded by a rasping start, longer than that of Blue, and seems to hang in the air rather than taper off (Bakewell *et al.*, 2010). The territorial song has not been described. (W2). **Range:** Endemic to the Malay Peninsula. **Status and distribution:** Uncommon and localized resident from 760–1700 m in the Titiwangsa Range from Cameron Highlands to Fraser's Hill, south to Gunung Menuang Gasing (Selangor), where the type specimen was collected. **Habitat:** Forested gullies usually in the vicinity of small mountain streams. **Habits:** Frequents the ground storey, keeping close to forest streams. Forages largely on moist ground, moving about in short hops. Flight is swift and direct and usually close to the ground. While foraging, head is frequently cocked with bill tilted upwards at approximately 45 degrees, a characteristic not noted in Blue. Both species also lower and splay their tails, the Malaysian tends to only half splay its tail while the Blue splays tail fully. At Fraser's Hill, sometimes appears along roadsides at dawn and dusk. Shy. **Note:** Since there is an obvious overlap in the altitudinal ranges of both Blue and Malaysian Whistling-thrushes in the Titiwangsa Range, only very careful and meticulous observations of the physical, behavioural, and acoustic characteristics of both species are all essential in making a positive identification. Making on the spot identifications by inexperienced observers in unfavourable conditions are often difficult. Digital photographs and videos record these characteristics and positive identification can be made later by viewing relevant images and video clips.

569. Blue Whistling-thrush, *Myophonus caeruleus* Plate 63
Murai Bersiul Besar

Description: 33 cm (13 in). Large size with broad, rounded tail, dark purplish blue plumage marked with lighter blue spangles, relatively long black legs black and yellow bill, diagnostic in areas where Malaysian Whistling-thrush (568) does not occur. Two subspecies occur in the review area. *M. c. dicrorhynchus*: Blue-black to sooty black plumage washed violet, with white spotting on median coverts. Bill large, with tip of maxilla strongly hooked. Ridge of culmen black. *M. c. crassirostris:* Smaller than *M. c. dicrorhynchus* with a less massive bill, with tip of maxilla curved rather than hooked. Bill is wholly yellow including ridge of culmen. Plumage is bluer with larger and bluer spangles. **Voice:** Song is a cadence of rich mellow fluty notes and rather harsh scratchy notes. Contact call is a high-pitched whistle, *screee* when foraging. **Range:** Central Asia through the Himalayas to China, through South-East Asia to Sumatra and Java. **Status and distribution:** *M. c. dicrorhynchus* is a locally common and more widespread resident from low elevations up to 1200 m, south to Selangor and Pahang. *M. c. crassirostris* **(not illustrated)** is locally common in Kedah, including Pulau Langkawi. Fairly common around limestone outcrops. **Habitat:** Forests, the forest edge, limestone outcrops and associated caves, and mangroves (on Pulau Langkawi only). **Habits:** Feeds on the ground, singly, or in small parties. Fully splays tail. Often delivers song from a perch.

570. Chestnut-capped Thrush, *Zoothera interpres* Plate 63
Murai Tarum Ubun Perang

Description: 18 cm (7 in). Small and short-tailed thrush. Chestnut crown and nape, slaty grey mantle, black throat and upper breast together with whitish lores and patch on sides of head diagnostic. Wings blackish with broad white wingbars. Underparts white with blackish spots on lower breast and flanks. **Voice:** A series of short and rich fluty whistles, reminscent of White-rumped Shama (582) but with very high notes in between. **Range:** The Malay Peninsula to the Greater and Lesser Sundas to the Philippines. Loss and degradation of lowland forest together with trapping for the bird trade are the causes of its decline. **NT. Status and distribution:** Rare and localized resident from low elevations up to 760 m, south to Tampin (Negeri Sembilan). Also on Pulau Pinang. **Habitat:** Forests. **Habits:** Frequents the ground storey, skulking in thick vegetation. Has a hopping gait as it forages on the forest floor. Will sometimes gather to feed in fruiting trees. Shy.

571. Orange-headed Thrush, *Zoothera citrina* Plate 63
Murai Tarum Dada Jingga

Description: 22 cm (8.5 in). Chunky appearance distinctive. **Male:** Bright orange rufous head and underparts together with blue-grey upperparts diagnostic. Lower belly and undertail coverts whitish. Some show white wingbar. **Female:** Like male but upperparts olive brown. Shows two dark vertical bars on side of head. **Juvenile (not illustrated):** Dark orange-brown head and mantle with spots and scales with underparts pale orange buff marked with dark scales. **Voice:** A thin *dzeeet* in flight. Also a deep *chak* when alarmed. Has a rich melodious song, not heard in wintering range. **Range:** Breeding in the Himalayas and southern China, discontinuously through South-East Asia to the Greater Sundas. Northern populations are migratory. **Status and distribution:** Locally common winter visitor at low elevations up to 1500 m, south to Singapore, where it is rare. **Habitat:** Forests, wooded gardens. **Habits:** Frequents the ground storey. Generally skulks in the undergrowth but will sometimes feed in the open. Often feeds in fruiting trees with other migrant thrushes.

572. Siberian Thrush, *Zoothera sibirica* Plate 63
Murai Tarum Siberia

Description: 24 cm (9.5 in). **Male:** Slaty black plumage with white supercilium diagnostic. Centre of belly whitish; undertail coverts white with black bars. **Female:** Long buff supercilium curving around behind ear coverts diagnostic. Shows black eyeline which contrasts with streaked ear coverts. Upperparts rufous brown with underparts buff with dark brown scales on breast and upper flanks. Tips to outer tail feathers whitish. **Voice:** A low and subdued *tsit*, like that of Eye-browed Thrush (574) but thinner and shorter. **Range:** Breeding in eastern Russia, Mongolia, northern China and Japan; wintering south to eastern India, Myanmar and South-East Asia to the Greater Sundas. **Status and distribution:** Locally common passage migrant and winter visitor from low elevations up to 1800 m, south to Singapore, where it is uncommon. **Habitat:** Forests, the forest edge, wooded areas. **Habits:** Usually solitary or in pairs, frequenting the ground storey where it usually skulks in the undergrowth. Sometimes feeds in the open. Often feeds in fruiting trees with other migrant thrushes.

573. Eurasian Scaly Thrush, *Zoothera dauma* Plate 63
Murai Tarum Bersisik

Description: 28 cm (11 in). Large chunky thrush, with distinctive scaly appearance. Olive brown upperparts and whitish underparts marked with bold crescentic black and golden buff scales diagnostic. Shows prominent dark crescent on ear coverts. Lores buff white and outer tail feathers tipped white. **Voice:** A short and thin *zeeet*. **Range:** Breeding in the Himalayas from northern Pakistan, south-east Tibet east to southern China and Taiwan, and northern continental South-East Asia to Vietnam. Also in the Greater Sundas east to Lombok and Sumbawa in the Lesser Sundas. Northern populations are migratory, wintering south to the northern part of the Indian subcontinent, southern China, and South-East Asia to the Malay Peninsula and the northern Philippines, straggling to Sumatra and Borneo. **Status and distribution:** Vagrant, known from individuals on Pulau Rembia (the Sembilan group, Perak) in the Straits of Melaka in November 1918, and at higher altitudes at Bukit Larut (Perak) and Fraser's Hill (Pahang). **Habitat:** Forests, the forest edge. **Habits:** Usually solitary, keeping to the forest under storey, sometimes emerging briefly on the sides of roads. Forages among leaf litter. Other habits little known.

574. Eyebrowed Thrush, *Turdus obscurus* Plate 63
Murai Tarum Dahi Putih

Description: 24 cm (9.5 in). White supercilium, white patch below eye, whitish chin and pale rufous orange lower breast and flanks distinctive. Upperparts brown; centre of belly and vent white. **Male:** Grey head, throat and upper breast. **Female:** Olive brown head, whitish throat marked with fine dark streaks and duller rufous orange underparts. **Voice:** A rather thin drawn out *dzeeep*. **Range:** Breeding in eastern Russia, Mongolia and Japan; wintering south to North-East India and southern China through South-East Asia to the Greater Sundas and the Philippines. **Status and distribution:** Locally common passage migrant and winter visitor from low elevations up to at least 2000 m, south to Singapore, where it is uncommon. **Habitat:** Forests, the forest edge, mangroves, beach scrub, wooded gardens. **Habits:** Usually in small flocks, frequenting the canopy and crowns of smaller trees. More arboreal than other thrushes. Feeds principally on fruit and will frequently gather in fruiting trees together with other frugivorous birds. Will also feed on the ground.

575. Dark-throated Thrush, *Turdus ruficollis* Plate 64
Murai Tarum Tekak Merah

Description: 25 cm (10 in). *T. r. ruficollis.* From other thrushes by rufous outer tail feathers in both sexes. **Male (Breeding):** Rufous-chestnut supercilium, throat, and upper breast sharply demarcated from white underparts diagnostic. Upperparts greyish-brown, strongly tinged grey on crown and sides of neck. **Male (Non Breeding) (not illustrated):** Rufous-chestnut supercilium, throat and upper breast have narrow whitish feather fringes. **Female:** Plumage variable but has distinct pale buff supercilium. Rufous confined to upper breast and is much paler and marked with whitish scales. Throat and sides of neck pale, often marked with blackish longitudinal streaks. Bill yellowish in both sexes, often tipped black. **Voice:** Unknown from review area. **Range:** *T. r. ruficollis* breeds from south-central Siberia, and Mongolia and winters south to Turkmenistan, Afghanistan and the northern part of the Indian subcontinent, east to Tibet, China, and Myanmar, straggling into north-west Thailand and the Malay Peninsula. **Status and distribution:** Vagrant, a male at Byram, near Nibong Tebal, Seberang Prai (Pulau Pinang) on 15 February, 2010 (Lim, B.K. 2010; RCS). **Habitat:** In review area, scrubland on the fringe of degraded mangroves. **Habits:** Perched on a tree, the Byram individual descended into undergrowth. Other habits little known from review area. **Note:** Dickinson (2003) split the black throated *T. r. atrogularis* from *T. r. ruficollis*, calling them Black-throated and Red-throated Thrushes respectively. A conclusive DNA analysis will be the deciding factor.

576. Lesser Shortwing, *Brachypteryx leucophrys.* Plate 64

Murai Kerdil Gunung

Description: 13 cm (5 in). Small size with short tail and relatively long flesh-coloured legs distinctive with throat and centre of belly whitish. **Male:** Dark slate grey upperparts with prominent white supercilium distinctive (see Snowy-browed Flycatcher (591)). Underparts greyish with flanks and band across breast dark grey. **Female:** Rufescent olive brown upperparts and buff brown underparts with soft mottling on upper breast. Supercilium narrower and less distinct. Pygmy Wren-babbler (530) lacks white supercilium, is almost tailless and has more distinct mottles on underparts. **Voice:** A powerful and melodious warbling cadence of 10–12 notes, with a brief pause after the initial note, the first few rising up and down the scale and the remaining accelerating and terminating in a rapid jingle; a very characteristic sound of montane rainforest. Also a harsh *tack* when alarmed. **Range:** The Himalayas and China through South-East Asia to Sumatra and Java. **Status and distribution:** Locally common resident above 970 m in the Larut Hills (Perak) and the Titiwangsa Range, south to Ulu Langat (Selangor). Also on Gunung Tahan (Pahang) and Gunung Rabong (Kelantan). Commonly heard at the hill-stations. **Habitat:** Forests. **Habits:** Solitary, frequenting the ground storey. Skulks in thick undergrowth, from where it often delivers its melodious song. Very shy.

577. Common Stonechat, *Saxicola torquatus* Plate 64

Murai Sawah

Description: 14 cm (5.5 in). **Male:** Black head, throat and upper breast contrasting with white neck patch together with orange rufous breast and belly diagnostic. Upperparts blackish with brownish fringes. Has white rump. White patch on inner wing coverts shows as streak. **Female:** Lacks black head and upperparts. Throat whitish; underparts pale orange rufous; patch on uppertail coverts pale orange. Shows white wingbar and pale buff supercilium. **Voice:** A sharp *chak-chak-chak*, like two stones hitting together. **Range:** Breeding in Africa (including Madagascar), Europe and temperate Asia. Northern populations are migratory, wintering south to Africa,

South and South-East Asia to the Malay Peninsula and Borneo. **Status and distribution:** Uncommon and localized winter visitor at low elevations, south to Singapore, where it is rare. **Habitat:** Open country, ricefields, freshwater swamps. **Habits:** Usually solitary. Frequently sits conspicuously on low perches from which it drops to the ground to catch small invertebrate prey.

578. Siberian Rubythroat, *Luscinia calliope* Plate 64
Murai Tekak Merah

Description: 16 cm (6.5 in). Rather plump and heavy bodied. **Male:** Red throat, white supercilium and white malar stripe diagnostic. Upperparts olive brown; breast and flanks greyish brown; belly whitish. **Female:** Like male but throat whitish. Some individuals may show some pink on throat. **Voice:** Unknown in review area. **Range:** Breeding in China and eastern Russia; wintering south to India and South-East Asia to the Malay Peninsula and the Philippines. **Status and distribution:** Vagrant, an adult male (a probable spring migrant) mist netted in lower montane forest at 1300 m at Fraser's Hill (Pahang) on 2 May 1968. **Habitat:** Wintering habitat unknown. **Habits:** Skulks in dense vegetation. Other habits unknown.

579. Rufous-headed Robin, *Luscinia ruficeps* Plate 64
Murai Kepala Jingga

Description: 15 cm (6 in). **Male:** Bright orange rufous crown and nape contrasting with broad black stripe below eye which encircles white throat as a gorget diagnostic. Upper breast below gorget is slaty grey, contrasting with white lower breast and belly. Mantle and wings slaty grey. Tail orange rufous with central pair of tail feathers slaty black. **Female:** Sides of head and neck dark brown with scalings on throat and upper breast, olive brown flanks and dark buff undertail coverts. **Voice:** Unknown in review area. **Range:** Known to breed in China in the Qinling mountains (Shaanxi Province) and North-East Sichuan Province. Wintering range unknown. This species is poorly known and because of a restricted range and small population, is declining due to forest destruction and possibly dam construction. **VU. Status and distribution:** Vagrant, an adult male (a probable spring migrant) mist netted on Gunung Batu Brinchang, Cameron Highlands (Pahang) at 2031 m on 15 March 1963. **Habitat:** Wintering habitat unknown. **Habits:** Unknown. **Note:** In breeding range, favours steep sided valleys with streams.

580. Siberian Blue Robin, *Luscinia cyane* Plate 64
Murai Biru Siberia

Description: 15 cm (6 in). Relatively short-tailed Robin. **Male:** Dull blue upperparts and black from lores to sides of neck contrasting with white underparts diagnostic. **Female:** Olive brown crown, mantle and wings contrasting with dull bluish rump and uppertail coverts, Throat buff; breast mottled dark brown. **Juvenile male (not illustrated):** Like female but wing coverts and mantle mottled blue and brown with underparts rufous buff. **Voice:** A hard *tuk* note, repeatedly uttered. Also a loud *se-ic*. **Range:** Breeding in eastern Russia, northern China, Korea and Japan; wintering south to

North-East India and southern China through South-East Asia to Sumatra, Borneo and the Philippines. **Status and distribution:** Locally common passage migrant and winter visitor principally from low elevations up to 800 m south to Singapore. Occurs above 900 m only during autumn and spring passage. **Habitat:** Forests, scrub, reed-beds. **Habits:** Frequents the ground storey. Skulks in the thick undergrowth. Flicks tail when alarmed.

581. Oriental Magpie Robin, *Copsychus saularis* Plate 65

Murai Kampung

Description: 23 cm (9 in). Pied plumage unmistakable. **Male:** Glossy black head, breast and upperparts contrasting with white underparts. Shows long white bar on wing. Tail black with white outer tail feathers. **Female:** Like male but black of breast and upperparts replaced with dark grey. **Juvenile (not illustrated):** Plumage like female with brownish mottling on breast. **Voice:** A repertoire of melodious high-pitched whistled cadences, sometimes performed antiphonally by pairs. Also a sweet whistle, *peeeeew*, with an upward inflection; sometimes on a descending scale. Also a harsh, rasping *churrrrr* in alarm. **Range:** The Indian subcontinent and southern China through South-East Asia to the Greater Sundas, Bali and the Philippines. A popular cage bird, populations throughout its range have been reduced by the cage bird trade. **Status and distribution:** Common and widespread resident from low elevations up to 1500 m in the cleared areas of Cameron Highlands and Fraser's Hill (Pahang), south to Singapore, where it is uncommon. A very conspicuous garden bird. **Habitat:** Forest clearings, the forest edge, mangroves, open country, plantations, gardens. **Habits:** Usually solitary or in pairs. Forages mostly on the ground, hopping from place to place. Often flicks tail. Sings from open perches. Very territorial. Not shy.

582. White-rumped Shama, *Copsychus malabaricus* Plate 65

Murai Rimba/Murai Batu

Description: 28 cm (11 in), of which 6 cm (2.4 in) is tail length of male. Long black wedge-shaped tail with white outer feathers and white rump distinctive. **Male:** Glossy blue-black head, upperparts and upper breast contrasting with deep orange rufous lower breast and belly diagnostic. **Female:** Like male but tail shorter with black parts replaced with slaty grey. Orange rufous parts paler. **Voice:** A loud, distinct and rich, fluty melodious song with many variations. Also a harsh *chak* uttered in series. **Range:** The Indian subcontinent and southern China through South-East Asia to the Greater Sundas. A very popular cage bird, the bird trade has extirpated populations throughout its range, making it a threatened species but still classified as **LC** (Least Concern). **Status and distribution:** Fairly common resident from low elevations up to 600 m, south to Singapore, where it is very rare on the mainland but appears to have a viable population on Pulau Tekong (W and H). Also on islands of the Langkawi archipelago (Kedah), Pulau Pangkor (Perak), Pulau Pisang (Johor), Pulau Kapas (Terengganu), and Pulau Tioman (Pahang), where it has been recorded up to 900 m. Individuals seen in parts of urban Singapore are likely to be cage escapees. **Habitat:** Forests, the forest edge, mangroves, overgrown plantations. **Habits:** Frequents the lower storey solitarily or in pairs. Keeps to thick vegetation. Very vocal, usually singing from a low perch. Often hops about on the ground with tail held high. Mimics other birds. Shy.

583. Rufous-tailed Shama, *Trichixos pyrropygus* Plate 65
Murai Ekor Jingga

Description: 22 cm (8.8 in). From White-rumped Shama (582) by shorter and narrower diagnostic bright orange rufous tail with dark terminal band. **Male:** Blackish grey head, throat, upper breast, mantle and breast contrasting with bright orange rufous underparts and rump which is concolorous with tail. Shows short indistinct white supercilium in front of eye. **Female:** Like male but black parts browner. Entire underparts buff rufous. Lacks supercilium. **Voice:** A series of loud and slow mournful whistles, *pi-uuu*, uttered repeatedly. Also a harsh *chrrrr* when alarmed. **Range:** The Malay Peninsula, Sumatra and Borneo. Populations are undergoing a moderately rapid decline due to habitat loss and destruction. **NT. Status and distribution:** Uncommon and localized resident from low elevations up to 900 m, south to Johor. **Habitat:** Forests **Habits:** Frequents the lower storey, keeping to thickets. Often flicks tail.

584. Chestnut-naped Forktail, *Enicurus ruficapillus* Plate 65
Cegar Tengkuk Merah

Description: 20 cm (8 in). Chestnut crown and nape together with blackish scalings on breast diagnostic. Wings black with white wingbar. Underparts and rump white; tail black with narrow white bars and white tip. Lores white. Legs flesh pink. **Male:** Chestnut extends from crown and nape to upper back. **Female:** Entire back chestnut. **Voice:** A thin shrill metallic whistle, uttered in series. **Range:** Tenasserim (Myanmar) and the Malay Peninsula to Sumatra and Borneo. Populations are experiencing a moderately rapid decline owing to loss and destruction of lowland rainforest. However is not considered more threatened because of its ability to utilize secondary habitats and also occurs in lower montane forest. **NT. Status and distribution:** Locally common resident principally in hilly country from low elevations up to 900 m, south to Johor. Fairly common in foothill forests. **Habitat:** Rocky forested streams. **Habits:** Frequents the ground storey, keeping along stream edges. Hops along rocks and boulders, picking up small aquatic invertebrates. Often flicks tail when calling. Shy.

585. Slaty-backed Forktail, *Enicurus schistaceus* Plate 65
Cegar Tengkuk Kelabu

Description: 25 cm (10 in). From Chestnut-naped Forktail (584) by larger size, longer tail and slaty grey crown, nape and mantle. Lacks blackish scalings on breast. Lores white; legs flesh pink. **Juvenile (not illustrated):** Slaty grey parts replaced by dark brown with dark scalings on breast. **Voice:** A series of thin shrill metallic whistles, similar to Chestnut-naped Forktail. **Range:** The Himalayas and southern China through South-East Asia to the Malay Peninsula. **Status and distribution:** Locally common resident from 600–1700 m in the Larut Hills (Perak) and the Titiwangsa Range, south to Gunung Bunga Buah (Selangor). Also on Gunung Tahan (Pahang). More common than Chestnut-naped Forktail, generally above 800 m. Fairly conspicuous along rocky streams in the hill-stations, sometimes appearing along roads. **Habitat:** Rocky forested streams. **Habits:** Like Chestnut-naped Forktail.

586. White-crowned Forktail, *Enicurus leschenaulti* Plate 65
Cegar Ubun Putih

Description: 20 cm (8 in). From other forktails by white forecrown (feathers sometimes raised to a slight crest), black back and breast. Legs flesh pink. **Voice:** Alarm and contact calls, pure tone whistles (W2). **Range:** The Himalayas and China through South-East Asia to the Greater Sundas and Bali. **Status and distribution:** Uncommon and localized resident from low elevations up to 760 m, south to Johor. **Habitat:** Forested streams, flooded forests. **Habits:** Like other forktails, forages along the water's edge for small aquatic prey. Often sits on rocks, constantly flicking tail. Very vocal.

587. White-tailed Robin, *Myiomela leucura* Plate 65
Murai Ekor Putih

Description: 18 cm (7 in). Medium size with moderately long, broad tail with slightly rounded tip. Shows diagnostic white crescentic lateral patches when splayed. **Male:** Blue-black plumage with glossy blue forehead and carpal joint. Has white spot on neck which is usually concealed. **Female:** Dark olive brown plumage, paler underneath with whitish or buff throat. **Voice:** A sweet and powerful warbling sequence like Lesser Shortwing (576) but has a fluty quality and is deeper and faster with less phrases. **Range:** The Himalayas and southern China discontinuously through South-East Asia to the Malay Peninsula. **Status and distribution:** Locally common resident from 1200–1700 m in the Larut Hills (Perak) and the Titiwangsa Range at Cameron Highlands and Fraser's Hill (Pahang). Probably more extensively distributed along the higher parts of the Titiwangsa Range. Far more common at Cameron Highlands than Fraser's Hill (Pahang). Also on Gunung Gelemat and Gunung Lawit (Terengganu) on the Eastern Range, where it occurs at 760 m on the latter (W2). **Habitat:** Forests. **Habits:** Frequents the ground storey, usually in forested gullies and streams, solitarily or in pairs. Skulks in the undergrowth, but also appears along roadsides or at the edges of plantations adjoining forest. Often splays tail.

OLD WORLD FLYCATCHERS

ORDER *Passeriformes* **FAMILY** *Muscicapidae*

World: *122 species* **Peninsular Malaysia and Singapore:** *26 species*

A distinct and diverse group of largely insectivorous birds, characterized by flattened triangular broad based bills with long rictal bristles at the gape, specially adapted for catching insects on the wing. Legs short and slender with small feet. Wings long in some species and short in others. Typically sit in an upright position on exposed perches, sallying after flying insects, catching them with an audible snap of the bill before returning to devour them on the same perch. Will also descend from regular perches to the ground to catch insects. Most species are sexually dimorphic; the males usually brighter coloured with females usually drab and difficult to identify; their association with the unmistakable male is often the clue. Juveniles are usually dressed in combinations of mottled and spotted plumage. Calls vary from thin scratchy notes and harsh alarm calls to sweet melodious warbling. Solitary or in pairs. Represented by the following groups and genera: **Typical Flycatchers,** genera *Ficedula* (588–595), *Cyanoptila* (596), *Eumyias* (609), and *Muscicapa* (610–612). *Ficedula* flycatchers are small with short bills. Many species sexually dimorphic; the males identified by distinctive colouration, supercilium and wingbars. Wing length of species varies from short to long; the former usually frequent the lower storey while the latter hunt at middle storey. *Cyanoptila* is a monotypic genus and is large, robust and proportionately short billed. Sexually dimorphic; the male being blue and white while the female largely brown. *Euymais* is short-billed with longish tail and distinctive bluish plumage, brighter in the male. *Muscicapa* species are small and rather drab in most species. Bills thin and short. Typically erect posture while perched. Frequently sit on exposed perches and make sorties after flying insects. Sexes alike. These genera are grouped together because of very similar behavioural characteristics and are not strictly bound by species sequence, overlapping with other groups of flycatchers. **Niltavas,** genus *Niltava* (597–598). Large and robust with proportionately short bills. Inhabitants of montane rainforests. Males predominantly dark blue with brilliant iridescent blue crowns and neck patches. One species has rufous orange underparts. Females largely brown with iridescent blue neck or shoulder patches. Will also feed on small fruit. **Blue-flycatchers,** genera *Muscicapella* (599) and *Cyornis* (600–606). The former is a monotypic genus and is tiny with a thin bill, with the male brightly coloured and the female drab. The latter has proportionately long and heavy bill and short, rounded wings. Tails are longish and often cocked. Males predominantly blue above (shining on forecrown and carpal joint) with varying amounts of rufous-orange on underparts in many species. Females largely brownish with pale underparts. Many species have melodious songs. **Jungle Flycatchers,** genus *Rhinomyias* (607, 608). Bill proportionately long and heavy, wings short and tail of moderate length and often cocked. Plumage brownish with rufous tails. Shape and plumage patterns resemble some *Malacopteron* babblers (515–518) and best distinguished by flycatcher habits. Sexes alike. **Canary-flycatchers** genus *Culicicapa* (613). Long-tailed with predominantly yellow and olive plumage. Sibley and Monroe (1990) on the basis of DNA-hybridization studies placed it into the Australian Robin family Eopsaltriidae. However Sibley and Monroe (1993) reassigned it to Muscicapidae without providing any reason.

588. Taiga Flycatcher, *Ficedula albicilla* — Plate 66

Sambar Tekak Merah

Description: 13 cm (5.25 in). Tail in both sexes black with white base of outer tail feathers (conspicuous in flight and when flicked). Both sexes also show a thin whitish orbital skin. **Male breeding:** Greyish brown upperparts, slaty grey sides of head and breast together with brick red throat and upper breast diagnostic. Rest of underparts whitish. **Female and non-breeding male:** Upperparts greyish brown, paler on flanks; throat whitish tinged cream; entire underparts whitish. **Voice:** A grating *trrrt*. Also a harsh metallic *tic*. Alarm call is a harsh *ziik*. **Range:** Breeding in Russia from the Urals east to the Kamchatka peninsula and Ussuriland, wintering south to the Indian subcontinent, the Andamans, south China and South-East Asia to the Malay Peninsula and Borneo. **Status and distribution:** Rare and localized winter visitor at low elevations, recorded in Perlis, Kelantan, Seberang Prai (Pulau Pinang), and Pahang, south to Selangor. Also on Pulau Langkawi (Kedah) and Pulau Redang (Terengganu). **Habitat:** Mangroves, the forest edge, overgrown plantations. **Habits:** Solitary during winter. Usually frequents the lower storey. Often sallies after flying insects but will also feed on the ground. Flicks open tail when alarmed. **Note:** Birds seen in the review area have always been in non-breeding plumage. Northward bound spring migrants may show traces of breeding plumage.

589. Rufous-browed Flycatcher, *Ficedula solitaris* Plate 66

Sambar Kening Perang

Description: 13 cm (5.25 in). Olive brown plumage with white triangular throat patch together with buff rufous forehead and lores diagnostic. **Voice:** A thin and high pitched whistled sequence of up to 6–7 notes on an ascending scale which is highly variable. Also a harsh *churrr* when alarmed. **Range:** South-eastern Laos and southern Vietnam and Tenasserim (Myanmar) and the Malay Peninsula to Sumatra. **Status and distribution:** Fairly common resident above 900 m in the Larut Hills (Perak) and the Titiwangsa Range south to Gunung Telapa Buruk (Negeri Sembilan). Also on Gunung Benom, Gunung Tahan, Gunung Tapis (Pahang) and Gunung Rabong (Kelantan). Common along hill-station forest paths. **Habitat:** Forests, the forest edge. **Habits:** Frequents the lower storey, keeping close to the forest floor. Usually sits on an exposed perch from which it catches insect prey and also delivers its song. Sometimes descends to the ground. Tame and confiding.

590. Snowy-browed Flycatcher, *Ficedula hyperythra* Plate 66

Sambar Dahi Putih

Description: 12 cm (4.5 in). Tiny size and short winged. **Male:** From male *Cyornis* flycatchers with rufous orange underparts by conspicuous white supercilium and proportionately shorter bill. Upperparts dull slaty blue with brown flight feathers; throat, breast and flanks orange buff; belly and undertail coverts white. Male Lesser Shortwing (576) also has conspicuous white supercilium but is shorter tailed and dark slaty grey above and below. **Female:** Upperparts brown; underparts buff with pale buff supercilium. Both sexes show white at the base of tail (barely discernible in the field). **Voice:** No information from the review area. A soft and high pitched wheezy cadence, *si-si-si-si*. Also a short *sip*, persistently uttered (extralimital data). **Range:** The Himalayas, southern China and Myanmar discontinuously through South-East Asia to the Greater Sundas, Sulawesi, the Philippines and the Moluccas. **Status and distribution:** Locally common resident above 1400 m in the Titiwangsa Range, south to Genting Highlands (Pahang). Also on Gunung Benom and Gunung Tahan (Pahang). **Habitat:** Ericaceous forests. **Habits:** A bird of the high mountain ridges. Frequents the lower storey solitarily or in pairs. Generally unobtrusive and will sit motionless on a perch for long periods. Seldom sallies after flying insects, preferring to flit down and catch insects on the forest floor. Sometimes hops on the ground like a thrush.

591. Rufous-chested Flycatcher, *Ficedula dumetoria* Plate 66
Sambar Dada Jingga

Description: 12 cm (4.5 in). From both sexes of Mugimaki Flycatcher (595) by shorter appearance, shorter wings and tail and proportionately longer and thicker bill. **Male:** Upperparts blackish with long narrow white supercilium and long narrow wingbar in secondaries. Rufous orange breast sharply demarcated from white belly. Throat pale. **Female:** Brown upperparts with thin buff wingbar and prominent buff streak along edge of secondaries. Rufous orange underparts duller. **Voice:** A distinct and high pitched cadence of about 4 syllables, *chi-chi-chi-chi*, uttered up and down the scale. **Range:** The Malay Peninsula to the Greater and Lesser Sundas. Is assumed to have experienced moderately rapid declines owing to extensive clearance of lowland rainforest within this range but is not considered more threatened because of its ability to utilize secondary habitats. **NT. Status and distribution:** Uncommon and localized resident from low elevations up to 800 m, south to Johor. **Habitat:** Forests. **Habits:** Usually solitary, occasionally in pairs. Frequents the forest understorey, feeding close to the ground, especially within the vicinity of small streams.

592. Little Pied Flycatcher, *Ficedula westermanni* Plate 66
Sambar Kecil Hitam-Putih

Description: 12 cm (4.5 in). Tiny size and short tail distinctive. **Male:** Glossy black upperparts with broad white supercilium, white wingbar together with white edge of base of black tail diagnostic. Underparts white, sometimes with black suffusion on breast. **Female:** Upperparts pale bluish grey and dull brown; underparts whitish grey; rump and uppertail coverts tinged rufescent brown. Shows a thin pale orbital skin. **Voice:** A high pitched buzzing trill, *chi, chi-chi-chi-chi-chi-churrrr,* rapidly uttered up and down the scale. Also a high pitched *tik-tik-trrr*. **Range:** The Himalayas and southern China through South-East Asia to the Greater and Lesser Sundas and the Philippines. **Status and distribution:** Fairly common resident above 1000 m in the Larut Hills (Perak) and the Titiwangsa Range south to Genting Highlands (Pahang). Also on Gunung Benom, Gunung Tahan (Pahang), Gunung Padang (Terengganu) and Gunung Rabong (Kelantan). Common and conspicuous at the hill-stations. **Habitat:** Forests, the forest edge. **Habits:** Frequents the middle storey. Usually vocal but is often discreet. Frequently participates in bird-waves. Will sometimes feed on the ground. Not shy.

593. Yellow-rumped Flycatcher, *Ficedula zanthopygia* Plate 66
Sambar Pinggul Kuning

Description: 13 cm (5.25 in). Long wings distinctive. **Male:** Black upperparts with white supercilium, bold white wing patch with thin white streak on inner secondaries and bright yellow rump diagnostic (see Narcissus Flycatcher (594)). Underparts bright yellow with white undertail coverts. **Female:** Olive brown upperparts with yellow rump diagnostic. Shows thin wingbar and thin pale streak on inner secondaries. Underparts pale yellowish buff with fine olive brown scalings on throat and breast. Shows distinct white orbital skin and pale lores. **Juvenile male (not illustrated):** Resembles female but tail black. **Voice:** A thin *pirip-pirip-piip* **Range:** Breeding in eastern Russia, Mongolia, northern China and Korea; wintering south through South-East Asia to Sumatra and Java. **Status and distribution:** Fairly common passage migrant and winter visitor from low elevations up to 400 m, south to Singapore. Occurs above 900 m during autumn and spring passage. **Habitat:** Forests, the forest edge, mangroves, scrub, plantations, wooded gardens. **Habits:** Solitary, frequenting the canopy and crowns of trees but will also feed in the middle and lower storeys. Often quiet and unobtrusive.

594. Narcissus Flycatcher, *Ficedula narcissina* Plate 66
Sambar Tengkuk Hijau

Description: 13 cm (5.25 in). *F. n. elisae* is the subspecies recorded in the review area. **Male:** Upperparts olive green with yellow rump and lores, yellowish eyering, bold white patch on wing (without streaks on inner secondaries) and blackish tail. Underparts bright yellow. **Female:** Like male but much duller, lacking the yellow rump and white wing patch. Throat paler. Tail brownish. From Common Iora (399) by longer tail and wings and distinctive upright stance. **Voice:** Usually silent in wintering quarters but has been observed giving a twittering subsong (A. Helbig) Also a low *tchook tchook tchook*, with a purring quality (W2). **Range:** Breeding in eastern Russia, northern China and Japan (including the Ryukyus); wintering south through southern China and South-East Asia to Borneo and the Philippines. **Status and distribution:** Uncommon and localized winter visitor at low elevations, south to Singapore, where it is rare (W and H). **Habitat:** Forests, the forest edge, wooded areas. **Habits:** Usually solitary. Frequents the lower storey but also feeds high up in the canopy. Sallies after insects from exposed perches but will also glean small caterpillars from leaves. **Note:** This subspecies which has an isolated breeding population in the Hebei and Shaanxi provinces in China separated from the main cline of the nominate. *narcissina*, is distinct morphologically (Vaurie 1959) and differs in vocalization patterns. Further research particularly in DNA analysis would decide whether or not it would be moved up to species rank. *F. n. narcissina* (not illustrated) has been recorded from Borneo and could turn up here. Male is similar to male Yellow-rumped Flycatcher *F. zanthopygia* (593) but has conspicuous long yellow supercilium. Yellow underparts has orange suffusion on throat and upper breast.

595. Mugimaki Flycatcher, *Ficedula mugimaki* Plate 66
Sambar Mugimaki

Description: 13 cm (5.25 in). From both sexes of Rufous-chested Flycatcher (591) by longer and more streamlined appearance, shorter and more slender bill and longer wings. **Male:** Upperparts vary from blackish to dark slaty grey with short white supercilium behind eye, broad white wing patch with two or three thin streaks on inner secondaries and narrow white edge of base of tail. Underparts orange rufous from throat to upper belly with white lower belly and vent. **Female:** Upperparts olive brown with faint narrow whitish buff wingbar. Throat and breast duller orange rufous. **Juvenile male (not illustrated):** Like female but shows short white supercilium, thin white wing bar and white edge of base of tail. **Voice:** A soft rattling *trrrrr*. **Range:** Breeding in eastern Russia and northern China; wintering south to South-East Asia, the Greater Sundas and the Philippines. **Status and distribution:** Locally common passage migrant and winter visitor from low elevations up to 2000 m, south to Singapore, where it is uncommon (W and H). During winter months,very common at the lower elevations of Fraser's Hill, Genting Highlands (Pahang) and Bukit Larut (Perak). **Habitat:** Forests, the forest edge. **Habits:** Usually solitary on wintering grounds, frequenting the lower storey. Will also ascend to feed at middle storey and canopy levels. Generally unobtrusive. Frequently sallies after flying insects. Will also enter the compounds of hill-station telecommunication stations early in the morning to feed on stranded moths, attracted to lights the previous night.

596. Blue-and-white Flycatcher, *Cyanoptila cyanomelana* — Plate 67
Sambar Biru-Putih

Description: 18 cm (7 in). **Male:** Dull black face and breast, glossy blue upperparts with white base of tail together with white belly diagnostic. **Female:** Upperparts greyish brown with warmer brown on wings and uppertail coverts; breast, flanks, and undertail coverts pale brown with whitish throat and centre of belly. **Voice:** Unknown from review area. **Range:** Breeding in eastern Russia, northern China, Korea and Japan; wintering south to southern China, South-East Asia to the Greater Sundas and the Philippines. **Status and distribution:** Uncommon and localized passage migrant and winter visitor from low elevations up to 900 m, south to Singapore where it is rare (W and H). Sometimes occurs above 900 m during autumn and spring passage. **Habitat:** Forests, wooded areas. **Habits:** Frequents the canopy and middle storeys, usually keeping to the foliage. Generally feeds on insects but will also take small fruit. Other habits little known.

597. Large Niltava, *Niltava grandis* — Plate 67
Sambar Niltava Besar/Sambar Kubung Padi

Description: 22 cm (8.5 in). Robust with upright stance distinctive (see Lesser Racquet-tailed Drongo (429)). **Male:** Overall dark shining blue plumage with iridescent brilliant blue crown and neck patch contrasting with blackish face and sides of neck diagnostic. Appears all black in poor light. **Female:** Plumage rusty brown; crown and nape ashy blue; neck patch iridescent brilliant blue. **Juvenile (not illustrated):** Mottled brown plumage with buff spots above; male shows blue wings and tail. **Voice:** A musical whistle of 3–4 notes on an ascending scale. Also a harsh *trrrt* when alarmed. **Range:** The Himalayas and Myanmar through South-East Asia to the Malay Peninsula and Sumatra. **Status and distribution:** Locally common resident above 1200 m in the Larut Hills (Perak) and the Titiwangsa Range south to Ulu Langat (Selangor). Also on Gunung Benom and Gunung Tahan (Pahang).

Common at the hill-stations. **Habitat:** Forests, the forest edge. **Habits:** Frequents the lower and middle storeys, often along trails, streams and gullies. Usually perches on a low branch or tree-fern frond, from which it sallies after insects. Also utilizes service wires. Sometimes descends to the ground. to catch insects. Fans out tail when alarmed. Often tame and confiding.

598. Rufous-vented Niltava, *Niltava sumatrana* Plate 67
Sambar Niltava Perut Jingga

Description: 15 cm (6 in). **Male:** From male *Cyornis* flycatchers with orange rufous underparts by more robust build and proportionately shorter bill. Dark blue upperparts with brilliant iridescent blue crown, neck patch, shoulder patch, rump and uppertail coverts together with orange underparts diagnostic. **Female:** Brown with narrow white crescentic patch on upper breast and brilliant iridescent blue shoulder patch and orange-rufous undertail coverts diagnostic. Throat brownish grey; breast, belly, and flanks greyish. From female Large Niltava (597) by much smaller size and different colouration. **Juvenile (not illustrated):** Overall brown spotted rusty buff above and scaled blackish below. **Voice:** A hard *chik*. Other calls unknown. **Range:** The Malay Peninsula and Sumatra. **Status and distribution:** Uncommon and localized resident above 1500 m in the Titiwangsa Range, known from two locations; Gunung Korbu (Perak) and the Cameron Highlands (Pahang). May also occur on the other higher peaks in the Titiwangsa Range. **Habitat:** Ericaceous forests. **Habits:** A bird of the high mountain ridges, frequenting the lower storey, usually solitarily or in pairs. Will also frequent the canopy. Keeps to shaded areas. On Gunung Batu Brinchang at the Cameron Highlands, sometimes comes out to open spaces near roads, where it often utilizes the fence of the telecommunications station as a perch. Will also enter the facility in the early mornings to feed on insects stranded there by lights the previous night.

599. Pygmy Blue-flycatcher, *Muscicapella hodgsoni* Plate 67
Sambar Biru Kerdil

Description: 10 cm (4 in). Tiny size. **Male:** Upperparts blue with shining blue forecrown; underparts pale rufous. From male *Cyornis* flycatchers with orange rufous underparts by much smaller size, whitish centre of belly and thinner bill. From male Fire-breasted Flowerpecker (627) by longer, thinner bill and rufous underparts. **Female:** Upperparts olive brown with bright rufescent rump and uppertail coverts; underparts white, tinged rufous on breast, throat and belly. **Voice:** Unknown from the review area. **Range:** The Himalayas, Myanmar, and Thailand discontinuously to the Malay Peninsula, Sumatra, and Borneo. **Status and distribution:** Uncommon and localized resident from 900–1800 m from the Larut Hills (Perak) and the Titiwangsa Range south to The Gap (Selangor). May also occur further south along the Titiwangsa Range. Also on Gunung Benom and Gunung Tahan (Pahang). **Habitat:** Forests, the forest edge. **Habits:** Usually solitary or in pairs, frequenting the middle and lower storeys. When foraging and facing intruders, down flicks wings and cocks tail vertically.

600. White-tailed Flycatcher, *Cyornis concretus* Plate 67

Sambar Ekor Putih

Description: 19 cm (7.5 in). Largest of the *Cyornis* flycatchers with heavier bill. **Male:** Dark blue plumage tinged purplish with whitish belly. **Female:** Rich buff brown plumage with white gorget on throat and breast diagnostic. In both sexes, whole edge of tail white in a narrow straight line, conspicuous when spread. Male White-tailed Robin (587) is darker blue with white neck patch. White edge of tail is broader and almost crescent shaped, terminating near the tip. Altitudinal ranges of both species usually do not overlap. **Voice:** A series of variable loud and sharp sibilant whistles, which turn into warbling sequences, some of which have a rippling quality. Also a soft and clear *pweee* when alarmed, or when in contact. **Range:** North-East India and Myanmar through South-East Asia to Sumatra and Borneo. **Status and distribution:** Rare and localized resident principally in hilly country from low elevations up to 1000 m, south to Johor. **Habitat:** Forests. **Habits:** Frequents the lower and middle storeys solitarily or in pairs. Feeds close to the ground. Sometimes splays tail.

601. Pale Blue-flycatcher, *Cyornis unicolor* Plate 67

Sambar Biru Muda

Description: 18 cm (7 in). Relatively long tailed and streamlined *Cyornis* flycatcher. **Male:** Dull pale blue upperparts and breast contrasting with pale greyish belly diagnostic. Forecrown shining blue; lores dusky. From Verditer Flycatcher (609) by proprtionately longer bill and lacking greenish tinge to plumage. **Female:** Brownish upperparts, tinged greyish on head and nape contrasting with all grey underparts distinctive. Tail has strong rufescent wash. **Voice:** A rich and melodious cadence of descending phrases, often terminating in a harsh and abrupt *chizz*. **Range:** The Himalayas and Myanmar through South-East Asia to the Greater Sundas. **Status and distribution:** Locally common resident from low elevations up to 1350 m, south to Johor. **Habitat:** Forests, the forest edge **Habits:** Frequents the canopy and middle storey. Hawks after insects from open perches. Sometimes flicks wings and cocks tail vertically. Shy.

602. Blue-throated Flycatcher, *Cyornis rubeculoides* Plate 67

Sambar Tekak Biru.

Description: 15 cm (6 in). **Male:** From male Hill (603) and Tickell's Blue-flycatchers (604) by blue throat with a narrow triangular wedge of orange rufous extending upwards from breast. Demarcation between orange rufous breast and white underparts gradual; less distinct than Tickell's Blue but sharper than Hill Blue. **Female:** Whitish throat contrasting with pale orange rufous breast diagnostic. From female Hill Blue by stronger rufescent wash on tail and uppertail coverts. Greyish brown sides of head gradually merge into pale orange rufous breast. **Voice:** Like Tickell's Blue Flycatcher but phrases quicker, higher pitched and more varied with trills. Seldom calls in wintering quarters. **Range:** Breeding in the Himalayas through China, Myanmar, northern Thailand and Indo-China; northern populations winter south to South and South-East Asia to the Malay Peninsula. **Status and distribution:**

Uncommon and localized passage migrant and winter visitor from low elevations up to 1100 m, south to Singapore, where it is rare. Also occurs above 1100 m during autumn and spring passage. **Habitat:** Forests. **Habits:** Frequents the middle and lower storeys solitarily. General habits similar to Hill Blue and Tickell's Blue-flycatchers.

603. Hill Blue-flycatcher, *Cyornis banyumas* Plate 67

Sambar Biru Bukit

Description: 15 cm (6 in). From both sexes of Tickell's Blue-flycatcher (604) by gradual demarcation between orange rufous breast and white underparts. **Male:** From male Blue-throated Flycatcher (602) by orange rufous chin, throat and breast. **Female:** From female Blue-throated Flycatcher by sharper demarcation between greyish brown sides of head and denser orange-rufous breast. Tail and uppertail coverts rufescent brown. **Juvenile (not illustrated):** Brown upperparts with buff spots distinctive. Throat and breast brownish, merging into whitish belly. Tail blue in male and rufous in female. **Voice:** A cadence of rich and melodious warbling sequences, quicker and more varied than Tickell's Blue-flycatcher. An abbreviated version is also given by both sexes in alarm. Also utters a harsh *tac* and rattling *trrrt* notes. **Range:** Breeding in the Himalayas, southern China through South-East Asia discontinuously to Java, Borneo and the Philippines. Northern populations winter south to the Malay Peninsula. **Status and distribution:** Locally common resident, passage migrant and winter visitor, south to Selangor and Pahang. Resident populations occur from 450–1200 m while migrant populations have been recorded from low elevations up to 600 m. **Habitat:** Forests. **Habits:** Usually frequents the middle storey solitarily or in pairs. Favours shady areas, often sallying for insects from a low perch. Very vocal. Sometimes unobtrusive.

604. Tickell's Blue-flycatcher, *Cyornis tickelliae* Plate 67

Sambar Biru Hutan

Description: 15 cm (6 in). From both sexes of Hill Blue-flycatcher (603) by a sharp demarcation between orange rufous breast and white underparts. **Male:** Blue upperparts slightly paler than Hill-Blue-flycatcher. **Female:** From female Hill-Blue Flycatcher by greyer upperparts tinged bluish. Tail greyish-blue. **Juvenile (not illustrated):** Indistinguishable from juvenile Hill Blue-flycatcher in the field. **Voice:** Similar to Hill Blue-flycatcher but slower with shorter phrases. Both sexes also give an abbreviated and simpler version in alarm. Also utters a harsh *tac* and rattling *trrrt* notes. **Range:** The Indian subcontinent through South-East Asia to Sumatra. **Status and distribution:** Fairly common resident from low elevations up to 900 m, south to northern Johor. **Habitat:** Forests, the forest edge. **Habits:** Frequents the middle and lower storeys solitarily or in pairs. Hunts from a low perch. Other habits similar to Hill Blue-flycatcher. A known brood host of the Rusty-breasted Cuckoo (251).

605. Mangrove Blue-flycatcher, *Cyornis rufigastra* Plate 67

Sambar Biru Bakau

Description: 15 cm (6 in). Like Hill Blue-flycatcher (603) but blue upperparts duller and lacks the metallic sheen. Orange rufous underparts extends all the way down to belly. **Male:** Black lores and chin. **Female:** Buff rufous lores, chin and throat. **Voice:** High pitched melodious warbling notes, similar to Hill Blue-flycatcher. **Range:** The Malay Peninsula to the Greater Sundas, Sulawesi, and the Philippines. **Status and distribution:** On the west coast, locally common resident at low elevations from Kedah and Perak, south to Singapore, where it is rare and confined to Pulau Tekong Besar. On the east coast, recorded at Mersing and Teluk Mahkota (Johor), Kuala Rompin (Pahang) and at the confluence of the Sungai Keluang and Sungai Besar (Terengganu). **Habitat:** Mangroves and adjacent scrub. **Habits:** Frequents the middle and lower storeys. Usually solitary or in pairs. Often hunts close to the ground. Frequently sings from a low perch.

606. Malaysian Blue-flycatcher, *Cyornis turcosus* Plate 67
Sambar Biru Malaysia

Description: 13 cm (5.25 in). Smallest *Cyornis* flycatcher. **Male:** Dark blue upperparts with bright blue throat blackish lores and shining blue rump (very conspicuous in flight) diagnostic. Breast and flanks orange rufous gradually merging into whitish buff belly. **Female:** Like male but duller blue with whitish buff throat. Orange rufous on breast is greatly diffused, merging with whitish buff belly. **Voice:** The song is a weak whistled sequence of about 5–6 phrases, unlike the typical *Cyornis* type songs. Also utters a harsh grating *chrrk* when alarmed. **Range:** The Malay Peninsula, Sumatra and Borneo. Is assumed to have undergone moderately rapid declines due to extensive loss of lowland rainforest but is not considered more threatened because of its ability to utilize secondary habitats. **NT. Status and distribution:** Uncommon and localized resident from low elevations up to 800 m; known only from Perak and Pahang, south to Johor. **Habitat:** Riverine forests. **Habits:** Frequents the lower storey solitarily or in pairs, particularly in the vicinity of small streams. Hunts by sallying from low forest perches, usually overlooking streams. Other habits little known.

607. Brown-chested Jungle-flycatcher, *Rhinomyias brunneatus* Plate 66
Sambar Hutan Dada Coklat

Description: 15 cm (6 in). Upperparts olive brown with rufescent tail and uppertail coverts. Bill black with yellow mandible. Whitish throat with faint scalings is not sharply demarcated from brownish breast. **Voice:** Unknown from review area. **Range:** Breeding in eastern China; wintering south to the Malay Peninsula and probably Sumatra. Its small population is declining due to destruction of lowland forest in both its breeding and wintering grounds due to logging and conversion to agriculture. **VU. Status and distribution:** Rare and localized passage migrant and winter visitor from low elevations, south to Singapore. Occurs above 900 m only during autumn and spring passage. **Habitat:** Forests, the forest edge, coastal scrub, suburban parkland. **Habits:** Frequents the lower storey solitarily. Keeps to undergrowth but will sometimes sally after insects from exposed perches. Sometimes drops to forage among leaf litter on the forest floor. Other habits little known.

608. Grey-chested Jungle-flycatcher, *Rhinomyias umbratilis* Plate 66

Sambar Hutan Dada Kelabu

Description: 15 cm (6 in). From Brown-chested Jungle-flycatcher (607) by all white throat, sharply demarcated from grey breast band. Bill all black. **Voice:** A cadence of up to 4–6 tinkling notes on a descending scale, *ti, ti-ti-tu-ti-tu*. Also a low clicking *tchk-tchk,* while jerking tail (W2). **Range:** The Malay Peninsula to Sumatra and Borneo. Assumed to have experienced moderately rapid population declines due to extensive loss of lowland forests within range but is not considered more threatened because of its ability to utilize secondary habitats and also because it occurs in lower montane forest. **NT. Status and distribution:** Uncommon and localized resident from low elevations up to 1160 m, south to Johor. Also on Gunung Tahan (Pahang), Gunung Jerai (Kedah) and Gunung Rabong (Kelantan). Also on Pulau Pinang. **Habitat:** Forests. **Habits:** Frequents the lower storey solitarily. Known to aerial sally after flying insects. Other habits little known.

609. Verditer Flycatcher, *Eumyias thalassinus* Plate 66

Sambar Biru-Hijau

Description: 17 cm (6.25 in). Short bill and longish tail distinctive. **Male:** Uniformly greenish blue plumage with black lores diagnostic. **Female:** Like male but duller and greyer especially on underparts. Lores dusky. Both sexes have whitish scalings on undertail coverts, perceivable at close range. Iris brown in both sexes. Male Pale Blue-flycatcher (601) has longer bill, shows contrast between breast and belly and lacks whitish scalings on undertail coverts. Blue morph male Rufous-winged Philentoma (395) is more robust, has bright blue plumage, conspicuous red iris and longer bill. **Voice:** The song is a distinct high-pitched cadence of soft, thin and sweet warbling notes, rapidly uttered up and down the scale. **Range:** North India, the Himalayas and southern China through South-East Asia to Sumatra and Borneo. **Status and distribution:** Fairly common resident principally in hilly country from low elevations up to 1300 m, south to Johor. Common at the lower elevations of the hill-stations, especially at The Gap (Selangor) below Fraser's Hill and the foothills of Genting Highlands (Pahang). **Habitat:** Forests, the forest edge. **Habits:** Frequents the canopy and crowns of smaller trees. Utilizes exposed perches to sally after flying insects and also deliver its song.

610. Dark-sided Flycatcher, *Muscicapa sibirica* Plate 66

Sambar Siberia

Description: 13 cm (5.25 in). From Asian Brown Flycatcher (612) by longer wings, the tips reaching two thirds of the way down the tail, proportionately shorter and all dark bill as well as sooty grey-brown flanks and band across breast. Throat and half collar together with underparts whitish. Shows conspicuous white orbital skin and indistinct buff wingbar. **Voice:** A pleasing disyllabic *chi-rup-chi-rup*….with a metallic quality, less harsh and dry than Asian Brown Flycatcher. **Range:** Breeding in the Himalayas, eastern Russia, northern China and Japan; wintering south to the Indian subcontinent, southern China through South-East Asia to the Greater Sundas. **Status and distribution:** Locally common passage migrant and winter visitor principally in hilly elevations from 150–1000 m, south to Singapore, where it is uncommon (W and H). Less common than Asian Brown Flycatcher. **Habitat:** Forests, the forest edge. **Habits:** Usually solitary, frequenting the canopy and middle storey. Typically sits upright on an exposed perch, sallying after flying insects.

611. Ferruginous Flycatcher, *Muscicapa ferruginea* Plate 66
Sambar Perang

Description: 13 cm (5.25 in). From other *Muscicapa* flycatchers by rufous rump, uppertail coverts and buff rufous flanks. Head slaty grey; mantle brown; throat and upper breast whitish. Narrow orbital skin and narrow wingbars pale buff rufous. Shows indistinct rufous markings on lores. **Voice:** Generally silent in wintering quarters. **Range:** Breeding in the Himalayas and southern China; wintering south through North-East India and South-East Asia to the Greater Sundas. **Status and distribution:** Locally common passage migrant and winter visitor from low elevations up to 1000 m, south to Singapore, where it is uncommon (W and H). **Habitat:** Forests, the forest edge. **Habits:** Usually solitary. Frequents the middle and lower storeys especially near streams in shady gullies. Like other flycatchers, aerial sallies after flying insects. Perches vary from high branches to fallen timber on the forest floor.

612. Asian Brown Flycatcher, *Muscicapa dauurica* Plate 66
Sambar Asia

Description: 13 cm (5.25 in). Two subspecies have been recorded in the review area *Muscicapa dauurica dauurica:* Nondescript with dull grey-brown upperparts and whitish underparts suffused with pale grey-brown on upper breast. Whitish orbital skin and pale lores give it a wide eyed appearance. In fresh plumage during autumn passage, shows thin pale wingbar and edge of tertiaries which are worn out during the winter months. Bill proportionately longer than Dark-sided (610) and Ferruginous Flycatchers (611) with pale basal half of mandible with dark tip. *Muscicapa daaurica williamsoni*: From nominate *dauurica* by brown upperparts tinged rufescent on entire uppertail; underparts and throat whitish suffused with brownish buff on breast and flanks. In fresh plumage, shows broad brown streaks on breast and flanks (usually indistinct in the field) as well as thin buff wingbars and edges of tertiaries. These are all lost in worn plumage. **Voice:** A rather dry and rapid rattling *tit-tit-tit-tit*. Also a soft and thin song, usually audible at close range. **Range:** Nominate *dauurica* breeds in eastern Russia, northern China, Korea and Japan; wintering south through the Indian subcontinent, southern China and South-East Asia to the Greater Sundas and the Philippines. *M. d. williamsoni* ranges from Myanmar and southern Vietnam through the Malay Peninsula to Sumatra. **Status and distribution:** *M. d. dauurica* is a common and widespread passage migrant and winter visitor from low elevations up to 700 m, south to Singapore. Occurs above 900 m only during autumn and spring passage. *M.d. williamsoni* is an uncommon and localized resident as well as winter visitor at low elevations, south to Singapore, where it is a rare winter visitor (W and H) Breeding has been recorded in Kedah, and Perlis, south to the Gombak Valley in Selangor (W2). **Habitat:** Forests, the forest edge, mangroves, plantations, wooded gardens. **Habits:** Solitary, frequenting the canopy and middle storey and the crowns of small trees. Often sallies after flying insects from exposed perches. Sometimes unobtrusive.

613. Grey-headed Canary-flycatcher, *Culicicapa ceylonensis* Plate 67

Sambar-Kenari Kepala Kelabu

Description: 13 cm (5.25 in). Small size and slender build with pale grey head and throat together with olive yellow upperparts and brighter yellow underparts diagnostic. From *Seicercus* (488–490) and *Abroscopus* (491) warblers by upright stance when perched and the lack of wingbars and supercilium. **Voice:** A 5-note cadence, *chi-chi-chi-chu-chi,* the last note rising. **Range:** The Indian subcontinent and southern China through South-East Asia to the Greater and Lesser Sundas and the Philippines. **Status and distribution:** Fairly common resident from low elevations up to 1700 m, south to Johor. **Habitat:** Forests, the forest edge, tea plantations (Cameron Highlands only). **Habits:** Frequents the middle storey solitarily or in pairs. Flits around rapidly in the foliage, moving quickly from branch to branch. Will sometimes descend to a lower perch from which it will carry out sorties after flying insects.

FAIRY-BLUEBIRDS

ORDER *Passeriformes* **FAMILY** *Irenidae*

World: *2 species* **Peninsular Malaysia and Singapore:** *1 species*

Robust, with blue and black plumage. Arboreal forest birds, feeding largely on fruit and insects. Gregarious, moving in small parties. Flight direct and undulating.

614. Asian Fairy-bluebird, *Irena puella* Plate 68

Murai Gajah

Description: 25 cm (10 in). **Male:** Shining blue crown, mantle, rump and undertail coverts contrasting with velvety black head, throat, belly and tail diagnostic. **Female:** Verditer blue plumage with blackish primaries. Iris red in both sexes. **Voice:** Song consists of liquid and melodious notes. Also a sharp *wi-wick, wi-wick,* uttered both while perched and in flight. **Range:** India and Myanmar through South-East Asia to the Greater Sundas and the Philippines. **Status and distribution:** Fairly common resident in the lowlands and hills up to 900 m, south to Singapore where it is restricted to the Bukit Timah Nature Reserve and the Central Catchment Forest. Has also been recorded up to 1900 m in the Titiwangsa Range (W2). Also on Pulau Pinang and Pulau Langkawi (Kedah). **Habitat:** Forests, the forest edge. **Habits:** Frequents the canopy and middle storey. Frequently congregates in fruiting trees and shrubs together with other frugivores. Sometimes hawks after flying insects at the forest edge.

LEAFBIRDS

ORDER *Passeriformes* **FAMILY** *Chloropseidae*

World: *8 species* **Peninsular Malaysia and Singapore:** *4 species*

Small to medium sized birds with fairly slender bills and rather short tails. Have bright grass green plumage with patches of blue and yellow in some species. Sexually dimorphic, males are brighter green and have conspicuous black faces and throat patches. Both sexes show conspicuous iridescent blue malar stripes. Juveniles resemble females but are usually duller and lack the blue malar stripes. Largely arboreal, they, forage mostly in thick foliaged canopies primarily on insects gleaned from the leaves and branches but will also take small fruit and nectar. Vocalizations are loud and melodious liquid phrases and involve some mimicry. Nests are loose cups, suspended in the forks of branches

615. Greater Green Leafbird, *Chloropsis sonnerati* Plate 68
Burung Daun Besar

Description: 21 cm (8 in). **Male:** From male Lesser Green Leafbird (616) by larger size, heavier bill and lacking the narrow yellow border along black throat patch. **Female:** Yellow throat and eyering diagnostic. **Voice:** A series of liquid musical whistling notes, with brief chatters in between. Some notes like Asian Fairy Bluebird (615). **Range:** Tenasserim (Myanmar) and the Malay Peninsula to the Greater Sundas. **Status and distribution:** Fairly common resident at low elevations and hills up to 1100 m, south to Singapore, where it is rare and confined to the Bukit Timah Nature Reserve and the Central Catchment Forest. **Habitat:** Forests, the forest edge, mangroves. **Habits:** Frequents the canopy in small noisy parties. Gathers in fruiting trees and shrubs with other frugivores.

616. Lesser Green Leafbird, *Chloropsis cyanopogon* Plate 68
Burung Daun Kecil

Description: 18 cm (7 in). From Blue-winged Leafbird (617) by lack of blue in plumage. **Male:** From male Greater Green leafbird (615) by smaller size and narrow yellow border along black throat patch. **Female:** From female Greater Green Leafbird by the lack of yellow throat patch and eyering. **Voice:** A series of loud liquid musical whistles, typical of the genus, interspersed by more deep mellow notes. **Range:** Tenasserim (Myanmar) through the Malay Peninsula to Sumatra and Borneo. Assumed to have moderately rapidly declined due to loss and degradation of lowland rainforest but can still adapt to secondary forest. **NT. Status and distribution:** Locally common resident from low elevations up to 700 m, south to Singapore where it is uncommon and confined to the Bukit Timah Nature Reserve, the Central Catchment forest, and the Botanic Gardens. Also recorded up to 1100 m on Gunung Jerai (Kedah) (W2). Also on Pulau Pinang. **Habitat:** Forests, the forest edge. **Habits:** Frequents the canopy in small parties. Gathers in fruiting trees and shrubs with other frugivores.

617. Blue-winged Leafbird, *Chloropsis cochinchinensis* Plate 68

Burung Daun Sayap Biru

Description: 19 cm (7.5 in). Blue flight feathers and sides of tail diagnostic. **Male:** Crown and border around black throat patch bright yellow. **Female:** Lacks yellow parts, blue parts duller. **Voice:** A distinct 3 note, 4 syllable liquid call, *wee wee wee-eee*, with the fourth syllable dropping and rising again on the fifth. **Range:** India and Myanmar through South-East Asia to the Greater Sundas. **Status and distribution:** Fairly common resident from low elevations up to 1250 m. Frequently seen at the lower elevations of the hill-stations. Also on Pulau Pinang. In Singapore, common resident confined to the Bukit Timah Nature Reserve, the Central Catchment Forest, Kent Ridge Park, and the Botanic Gardens. Sightings outside these areas are probably escapees (W and H). **Habitat:** Forests, the forest edge. **Habits:** Frequents the canopy usually in small noisy parties. Sometimes comes down low to feed in fruiting shrubs at the forest edge.

618. Orange-bellied Leafbird, *Chloropsis hardwickii* Plate 68

Burung Daun Perut Jingga

Description: 20 cm (8 in). From other leafbirds by orange-yellow belly and undertail coverts. Black throat patch of male extends to upper breast. Flight feathers and uppertail coverts blackish in male, greenish in female. **Juvenile (not illustrated):** All green without orange underparts. **Voice:** A series of loud melodious rippling phrases with many variations. Also a sharp *chisikik*, very similar in quality to that of the Streaked Spiderhunter (646). **Range:** The Himalayas and southern China discontinuously through South-East Asia to the Malay Peninsula. **Status and distribution:** Locally common resident from 820–1900 m in the Larut Hills (Perak) and the Titiwangsa Range south to Ulu Langat (Selangor). Also on Gunung Benom (Pahang). Common and conspicuous at the hill-stations. Individuals seen in Singapore are escaped captive individuals. **Habitat:** Forests, the forest edge. **Habits:** Frequents the canopy solitarily or in pairs. Seldom in larger parties. Gathers in fruiting trees and shrubs. Also enters hill-station flower gardens to feed on nectar. Very vocal and will sing from an open perch.

FLOWERPECKERS

ORDER *Passeriformes* **FAMILY** *Dicaeidae*

World: *45 species* **Peninsular Malaysia and Singapore:** *10 species*

Tiny and compact with short tails. Bills thick and stubby in genus *Prionochilus* (619–621), longer, more slender and slightly down-curved in most species of genus *Dicaeum* (622–628). Some species in review area are sexually dimorphic: the males colourful, the females drab. Juveniles of most species have pale or bright orange bills and are often difficult to identify. Most species feed in the canopy on the fruits of epiphytic mistle-toes (Loranthaceae) and are effective seed dispersers. Many are also fond of fruits of the exotic *Muntingia calabura*, a garden tree and *Melastoma malabathricum*, a shrub which thrives in open country and the forest edge. Figs (*Ficus* sp.) are also tackled either by pecking holes and sucking out contents from ripe fruit or pecking off and consuming smaller morsels of the fruit which are too large to be swallowed whole. May also take small insects. Flight jerky and erratic, usually at treetop level. Calls harsh and sharp metallic clicking notes, distinct in each species, but sometimes very difficult to differentiate. Solitary or in pairs.

619. Yellow-breasted Flowerpecker, *Prionochilus maculatus* Plate 69
Sepah Bunga Dada Kuning

Description: 10 cm (4 in). Crown orange rufous; upperparts olive green; underparts yellow with olive green streaks on sides of throat and upper flanks. **Voice:** A rasping *chwit-chwit*. **Range:** Tenasserim (Myanmar) through the Malay Peninsula to Sumatra and Borneo. **Status and distribution:** Fairly common resident from low elevations up to 1500 m, south to Johor. Locally extinct in Singapore since 1950 (L and G). **Habitat:** Forests (including peatswamp), the forest edge, scrub. **Habits:** Frequents the middle and lower storeys. Usually keeps to dense vegetation but emerges to feed in more open areas. Very fond of *Melastoma* fruits.

620. Crimson-breasted Flowerpecker, *Prionochilus percussus* Plate 69
Sepah Bunga Pelangi

Description: 10 cm (4 in). **Male:** Slaty blue upperparts with scarlet patch on crown together with crimson breast patch on bright yellow underparts diagnostic. Thin malar stripe white. **Female:** Overall dull olive green, greyer below with yellowish streak running vertically along centre of breast and belly. Patch on crown dull orange; undertail coverts whitish. **Voice:** A monosyllabic *weg* (Madoc in W2). Other vocalizations unknown. **Range:** Tenasserim (Myanmar) and the Malay Peninsula to the Greater Sundas. **Status and distribution:** Locally common resident from low elevations up to 1200 m, south to Johor. Also on Pulau Pinang and Pulau Tioman (Pahang). **Habitat:** The forest edge, mangroves, scrub, plantations. **Habits:** Frequents the middle and lower storeys. Very fond of *Melastoma* fruits. Also known take figs.

621. Scarlet-breasted Flowerpecker, *Prionochilus thoracicus* Plate 69
Sepah Bunga Dada Merah

Description: 10 cm (4 in). **Male:** Black head, read breast and crown patch together with yellow underparts diagnostic. Mantle olive green; wings and tail black. In flight, yellow rump and uppertail coverts conspicuous. **Female:** From female Crimson-breasted Flowerpecker (620) by greyer head, white throat, olive green upperparts with yellowish rump and uppertail coverts and the lack of whitish malar stripe. Underparts yellowish with orange tinge on breast. **Voice:** Unknown in review area. **Range:** Southern Vietnam through the Malay Peninsula to Belitung and Borneo. Is assumed to have experienced moderately rapid population declines owing to extensive loss of lowland rainforest but is not considered more threatened because of its ability to utilize secondary habitats. **NT. Status and distribution:** Uncommon and localized resident from low elevations up to 1200 m, south to Johor. **Habitat:** Forest, the forest edge. **Habits:** Frequents the canopy and upper middle storeys. Sometimes feeds in fruiting shrubs at the forest edge.

622. Thick-billed Flowerpecker, *Dicaeum agile* Plate 69
Sepah Bunga Paruh Tebal

Description: 10 cm (4 in). Thick bill distinctive. Upperparts grey brown washed with olive green on wings; underparts whitish with heavy but faint greyish streaks. Tip of tail feathers whitish (difficult to discern in field). Iris orange to reddish. **Voice:** A thin *pseeou*, weaker than its other congeners. **Range:** India through South-East Asia to the Lesser Sundas. **Status and distribution:** Uncommon and localized resident from low elevations up to 1280 m, south to Singapore, where it is a rare non-breeding dispersant. **Habitat:** Forests. **Habits:** Frequents the canopy and middle storey, to feed on fruiting mistle-toes in the former. Flicks tail from side to side. Often calls in flight. **Note:** Voice and tail flicking are important characteristics to separate it from similar Brown-backed Flowerpecker (623).

623. Brown-backed Flowerpecker, *Dicaeum everetti* Plate 69
Sepah Bunga Paruh Serindit

Description: 10 cm (4 in). Like Thick-billed Flowerpecker (622) but bill thicker and parrot-like (usually not apparent in field) and iris pale yellowish to orange. Underparts darker and greyer with fainter streaking; upperparts dark rich earth brown. Tail feathers lack white tips. **Voice:** No information from review area. **Range:** The Malay Peninsula and Borneo. Is assumed to have experienced moderately rapid population declines owing to extensive loss of lowland rainforest but is not considered more threatened because of its ability to utilize secondary habitats. **NT. Status and distribution:** Very rare resident at low elevations, recorded only in central Selangor. **Habitat:** Forests. **Habits:** Little known.

624. Yellow-vented Flowerpecker, *Dicaeum chrysorrheum* Plate 69
Sepah BungaTonkeng Kuning

Description: 10 cm (4 in). Whitish underparts with bold dark streaks on breast and upper belly together with orange-yellow undertail coverts diagnostic. Upperparts olive green; narrow malar stripe black. **Voice:** A rasping *zet-zet-zet-zet*, while foraging. A longer *zeet-zeet-zeet* in flight (W2). **Range:** The Himalayas and southern China through South-East Asia to the Greater Sundas. **Status and distribution:** Locally common resident from low elevations up to 870 m, south to Singapore where it is a rare resident, confined to the Bukit Batok and Bukit Timah Nature Reserves. Also on Pulau Ubin (W and H). **Habitat:** The forest edge, scrub, plantations. **Habits:** Generally frequents the canopy, middle and lower storeys. Gathers at fruiting *Melastoma* shrubs. Also takes figs.

625. Orange-bellied Flowerpecker, *Dicaeum trigonostigma* Plate 69
Sepah Bunga Perut Jingga

Description: 9 cm (3.5 in). **Male:** Slaty blue head, mantle and wings contrasting with orange back, rump, lower breast and belly diagnostic. Throat and upper breast pale greyish. **Female:** From other female flowerpeckers by dull orange rump. Bill noticeably longer and more slender than female Scarlet-breasted (621) and Crimson-breasted Flowerpeckers (620). Upperparts dull olive green; breast greyish white; rest of underparts whitish yellow. **Voice:** A short *zit*. Song is a *si-si-si-si-see,* on a descending scale. **Range:** North-East India, Myanmar, Thailand, southern Vietnam through the Greater Sundas, Bali and the Philippines. **Status and distribution:** Fairly common resident from low elevations up to 1200 m, south to Singapore. Also on Pulau Pinang, Pulau Langkawi (Kedah) and Pulau Tioman (Pahang). **Habitat:** The forest edge, scrub, plantations, wooded gardens. **Habits:** At the forest edge, frequently gathers in fruiting *Melastoma* shrubs. Also enters gardens to feed in fruiting *Muntingia* trees. Will sing from an exposed perch.

626. Plain Flowerpecker, *Dicaeum concolor* Plate 69
Sepah Puteri Bongsu

Description: 9 cm (3.5 in). Nondescript with slightly down-curved bill. Upperparts dull olive; underparts pale greyish with creamy centre of belly. Bill greyish, darker on the tip. **Voice:** A high pitched *dik-dik-dik-dik* while foraging and also in flight. **Range:** India to southern China through South-East Asia to the Greater Sundas and Bali. **Status and distribution:** Locally common resident from low elevations up to 760 m, south to Johor. Also on Pulau Pinang. Locally extinct in Singapore since 1893 (L and G). **Habitat:** The forest edge, scrub, wooded gardens. **Habits:** Frequents the canopy and middle storey, usually among mistletoe clumps in the former. Also known to take figs.

627. Fire-breasted Flowerpecker, *Dicaeum ignipectus* Plate 69
Sepah Bunga Dada Api

Description: 9 cm (3.5 in). **Male:** Dark blue-green upperparts (with slight gloss) together with buff underparts and thin black line running down from crimson patch on upper breast through centre of lower breast and belly diagnostic. **Female:** Upperparts dull olive (except for black tail); underparts yellowish buff (Resembles Plain Flowerpecker (626) but voice diagnostic. Altitudinal ranges of both species do not overlap). **Voice:** A sharp metallic *tik* while foraging. In flight, the same note is repeated in series. Song is a repeated, *tissit, tissit,* … similar to Scarlet-backed Flowerpecker (628). **Range:** The Himalayas and southern China discontinuously through South-East Asia and the Malay Peninsula to Sumatra. **Status and distribution:** Fairly common resident principally from 820 m to 1700 m in the Larut Hills (Perak) and the Titiwangsa Range south to Gunung Telapa Buruk (Negeri Sembilan). May sometimes occur at much lower elevations. Also on Gunung Benom, Gunung Tahan (Pahang) and Gunung Rabong (Kelantan). Also on the Eastern Range. Fairly common at the hill-stations. **Habitat:** Forests, the forest edge. **Habits:** Frequents the canopy, feeding in clusters of fruiting mistle-toes and also figs. Will also descend low to feed in fruiting *Melastoma* shrubs in cleared overgrown areas.

628. Scarlet-backed Flowerpecker, *Dicaeum cruentatum* Plate 69

Sepah Bunga Belakang Merah

Description: 9 cm (3.5 in). **Male:** Broad red band running from crown to rump diagnostic. Sides of head, wings and tail black; underparts white with greyish tinge on flanks. **Female:** From other female and nondescript flowerpeckers by red rump. **Juvenile (not illustrated):** Like female but lacks red in plumage; base of bill orange-red. **Voice:** A sharp metallic *tik*, while foraging. In flight, this note is repeated, *tik-tik-tik-tik*… Song is a thin *tissit-tissit*… repeated persistently; distinct from Orange-bellied Flowerpecker (625). **Range:** The Himalayas and southern China through South-East Asia to Sumatra and Borneo. **Status and distribution:** Common and widespread resident from low elevations up to 240 m, south to Singapore. Also on Pulau Pinang, Pulau Langkawi (Kedah), Pulau Pangkor (Perak) and Pulau Tioman (Pahang). Recorded as high as 820 m at The Gap (Selangor) and at 870 m on the slopes of Gunung Raya (Pulau Langkawi). A common bird in wooded suburban gardens. **Habitat:** Forests, the forest edge, mangroves, scrub, plantations, gardens. **Habits:** Frequents the canopy of primary forests, feeding in clusters of fruiting mistle-toes. At the forest edge, frequently feeds in fruiting *Melastoma* shrubs. Also enters gardens especially with fruiting *Muntingia* trees.

SUNBIRDS AND SPIDERHUNTERS

ORDER *Passeriformes* **FAMILY** *Nectariniidae*

World: *131 species* **Peninsular Malaysia and Singapore:** *18 species*

Small to large size with long down-curved bills. Active and arboreal, feeding largely on nectar and small insects. Members of this family are very active and will constantly flit among flowering trees and shrubs, probing and piercing blossoms to feed on nectar. Will also glean small invertebrates from twigs and foliage. Flight is typically darting and is over short distances. Will also hover in front of blossoms while taking nectar. Usually solitary, but may gather in larger groups in flowering forest trees. Represented by the following groups and genera: **Sunbirds:** Small to medium size. Sexually dimorphic, males have bright iridescent plumage. Females and juveniles are drab coloured and are difficult to identify; association with the unmistakable males is often one of the clues of making positive identification. In some species, males assume a duller eclipse plumage after breeding. Calls sharp metallic *chit* notes, songs high pitched twittering notes. Nests are hanging purse-shaped structures. Genera *Chalcoparia* (629), *Hypogrammica* (630) and *Anthreptes* (631–633), bills short and straight. Genera *Leptocoma* (634, 635) and *Cinnyris* (636), bills longer and down-curved. Many display bright orange or yellow pectoral tufts on the underwing, visible only during display. Genus *Aethopyga* (637–639). Bills strongly curved. Males have elongated tail feathers and display prominent yellow rump patches. Females have graduated or rounded tails. **Spiderhunters,** genus *Arachnothera* (640–646). Larger and more robust than sunbirds. Bills extremely long and down-curved. Plumage mainly olive and yellow. Sexes virtually alike, males of some species have orange or yellow pectoral tufts concealed in the underwing. Feed largely on insects on the inflorescences of flowers. Calls largely high-pitched explosive chattering notes. Flight swift and undulating. Nests are cup-shaped structures, usually sewn to the underside of large leaves, using cobwebs.

629. Ruby-cheeked Sunbird, *Chalcoparia singalensis* Plate 70
Kelicap Pipi Merah

Description: 11 cm (4.5 in). From male Brown-throated (632) and Red-throated Sunbirds (633) by smaller size and straighter bill. **Male:** Pale rufous throat and iridescent coppery red ear coverts diagnostic. Upperparts iridescent dark green; underparts yellow. **Female:** Pale rufous throat together with dull olive green upperparts diagnostic. **Juvenile (not illustrated):** Like female but lacks pale rufous throat. **Voice:** A disyllabic *whee-eet,* with rising inflection. Also shrill trills. **Range:** North-East India and Myanmar through South-East Asia to the Greater Sundas. **Status and distribution:** Fairly common resident from low elevations up to 900 m, south to Johor. Also on Pulau Langkawi (Kedah) and Pulau Pinang. **Habitat:** Forests (including peatswamp), the forest edge, the mangrove edge, plantations, scrub, wooded gardens. **Habits:** Frequents the canopy and middle storey. Active and noisy, flitting through the foliage amongst clumps of flowers. Congregates in flowering trees and shrubs with other sunbirds. Will also glean insects from twigs and foliage. Displaying male often jerks tail vigourously.

630. Purple-naped Sunbird, *Hypogramma hypogrammicum* Plate 70
Kelicap Rimba

Description: 15 cm (6 in). From all sunbirds by larger size and distinctive yellowish-white underparts with dark streaks. Upperparts dark olive. **Male:** Has iridescent purple nape and rump. **Female:** Lacks purple nape and rump. **Voice:** A sharp *chewp*. **Range:** Myanmar through South-East Asia to Sumatra and Borneo. **Status and distribution:** Fairly common resident from low elevations up to 1200 m, south to Johor. Locally extinct in Singapore since 1950 (L and G). **Habitat:** Forests, the forest edge. **Habits:** Frequents the middle and lower storeys. Often feeds in low flowering shrubs in the forest undergrowth. Pairs have been observed indulging in courtship rituals on the ground.

631. Plain Sunbird, *Anthreptes simplex* Plate 70
Kelicap Kelabu

Description: 13 cm (5 in). Relatively straight bill distinctive. **Male:** Upperparts bright olive green; underparts dull greyish olive; forehead dark iridescent green (often difficult to discern in field). **Female:** Like male but lacks green forehead. From other female sunbirds by the lack of yellow on underparts. **Juvenile (not illustrated):** Bill and legs orange-yellow. **Voice:** Unknown from review area. **Range:** Tenasserim (Myanmar) through the Malay Peninsula to Sumatra and Borneo. **Status and distribution:** Fairly common resident from low elevations up to 900 m, south to Singapore, where with unconfirmed sightings, its status is still indeterminate (W and H). **Habitat:** Forests, the forest edge. **Habits:** Frequents the middle storey, gathering to feed in flowering trees and shrubs.

632. Brown-throated Sunbird, *Anthreptes malacensis* Plate 70
Kelicap Mayang Kelapa

Description: 14 cm (5.5 in). **Male:** Pale brown throat and sides of head diagnostic. Upperparts dark iridescent green with rufescent olive tinge on wings; patch on shoulder, rump and uppertail coverts iridescent purple; underparts bright yellow. **Female:** Olive upperparts and bright yellowish underparts distinctive. Has indistinct narrow yellowish patch above and below eye. From female Olive-backed Sunbird (636) by larger size, straighter bill, and the lack of white in tail **Voice:** A high pitched chatter, *chip-chip-chiap*. Territorial call of male is *chip-chew*, persistently uttered. **Range:** Myanmar to southern Indo-China through the Malay Peninsula south to the Greater Sundas, Bali and the Philippines. **Status and distribution:** Common and widespread resident at low elevations, south to Singapore. Also on Pulau Langkawi (Kedah), Pulau Pinang, Pulau Tioman (Pahang), Pulau Kapas (Terengganu), Pulau Pangkor (Perak), Pulau Pemanggil and Pulau Aur (Johor). A conspicuous garden bird. **Habitat:** Mangroves, beach scrub, riverine vegetation, the forest edge, plantations, gardens. **Habits:** Frequents trees as well as low shrubs. Very active and noisy, flitting about in search of nectar and small insects. Often feeds high on flowering coconut palms. A known brood host of the Violet Cuckoo (255).

633. Red-throated Sunbird, *Anthreptes rhodolaemus* Plate 70
Kelicap Pinang

Description: 13 cm (5 in). **Male:** From male Brown-throated Sunbird (632) by light brick red throat and dark maroon-red sides of head and upperwing coverts. Iridescent purple shoulder patch much reduced. Underparts dull yellowish. **Female:** Like female Brown-throated Sunbird but underparts dull yellowish olive with denser yellowish tinge down centre. **Voice:** A rising high-pitched *ua'is* or *tsuu'i* (Robson, 2008). Other vocalizations inadequately described. **Range:** Tenasserim (Myanmar) through the Malay Peninsula to Sumatra and Borneo. This forest-associated species is assumed to have undergone a moderately rapid population decline owing to extensive clearance of lowland rainforest but is not considered more threatened because of its ability to utilize secondary habitats. **NT. Status and distribution:** Uncommon and localized resident at low elevations usually up to 400 m, south to Johor. Has been also recorded up to 900 m at Cameron Highlands (Pahang) (W2). Also on Pulau Pinang. Locally extinct in Singapore since 1879 (L and G). **Habitat:** Forests, the forest edge. **Habits:** Frequents the canopy and crowns of trees. Frequently gathers to feed in flowering forest trees. **Note:** Does not usually overlap with Brown-throated Sunbird due to different habitat preferences. However at the parkland trees at the Taman Negara Headquarters at Kuala Tahan (Pahang), both may be seen together, the Red-throated Sunbird usually feeding higher.

634. Purple-throated Sunbird, *Leptocoma sperata* Plate 70
Kelicap Nibong

Description: 10 cm (4 in). Tiny size. **Male:** Iridescent green cap, iridescent purple throat together with dark blue mantle and dull maroon breast and belly diagnostic. Appears all black in poor light. **Female:** Dark olive above, whitish yellow below. Throat paler than that of female Olive-backed Sunbird (636). Lacks pectoral tufts. **Voice:** A thin *sisisi* and intermittent *wheep* (A.T. Edgar in W2). Other vocalizations inadequately described. **Range:** North-East India and Myanmar to southern Vietnam through the Malay Peninsula to the Greater Sundas, Bali and the Philippines. **Status and distribution:** Locally common resident at low elevations along the coastal plains, south to Singapore. Inland, local at low elevations and hills up to 1217 m on Gunung Jerai (Kedah). Also on Bukit Bendara (Penang Hill), Gunung Kledang (Perak), Pulau Langkawi (Kedah) and Pulau Tioman (Pahang). **Habitat:** Forests, the forest edge, mangroves (on Pulau Tioman only), beach scrub, secondary growth, plantations, wooded gardens. **Habits:** Frequents the middle and lower storeys, gathering in flowering trees and shrubs with other species. In primary forest, will also feed in the canopy. Singing male holds body vertical, with bill up (W2). A known brood host of the Violet Cuckoo (255).

635. Copper-throated Sunbird, *Leptocoma calcostetha* Plate 70
Kelicap Bakau

Description: 14 cm (5.5 in). Graduated tail distinctive. **Male:** From male Purple-throated Sunbird (634) by larger size and iridescent coppery throat. Cap and shoulders bright iridescent green; breast and belly iridescent violet purple. Appears all black in poor light. Orange and yellow pectoral tufts usually concealed in underwing. **Female:** Head greyish; throat white; upperparts dull olive; breast and belly yellow; tail black with bold whitish tips beneath. **Voice:** Not well documented in review area. **Range:** Myanmar and the Malay Peninsula and southern Vietnam to the Greater Sundas and the Philippines. **Status and distribution:** Locally common resident along the coastal plains, south to Singapore, where it is uncommon. Also on Pulau Pinang and Pulau Langkawi (Kedah). **Habitat:** Mangroves, beach scrub, coconut groves. **Habits:** Keeps mainly to the upper and middle storeys. Frequents flowering trees and shrubs, usually with other species.

636. Olive-backed Sunbird, *Cinnyris jugularis* Plate 70
Kelicap Tekak Biru

Description: 11 cm (4.5 in). **Male:** Dark iridescent blue face, throat and breast (appears black in poor light) contrasting with yellow belly and undertail coverts diagnostic. Upperparts dark olive. Yellow to orange pectoral tufts usually concealed in underwing. **Male eclipse (not illustrated):** Dark iridescent blue parts reduced to broad stripe along centre of throat. **Female:** Lacks dark iridescent blue parts. From other female sunbirds by bright yellow underparts. From female Brown-throated Sunbird (632) by broad white tips to underside of tail. **Voice:** A loud, rising, *sweet* while foraging. Also a sharp *tik –tik-tik* in flight. Song consists of wheezy or reeling trills. **Range:**

Myanmar and southern China through South-East Asia to northern Australia and the Solomons. **Status and distribution:** Common and widespread resident at low elevations, confined mainly to the coastal plains, south to Singapore. Also occurs inland near coastal districts but is absent from interior districts. Also on Pulau Pinang, Pulau Tioman (Pahang) and Pulau Langkawi (Kedah) where it occurs up to 820 m. A conspicuous garden bird. **Habitat:** Mangroves, beach scrub, secondary growth, plantations, gardens. **Habits:** Frequents the middle and lower storeys. Often enters gardens to feed in flower beds and potted plants. Displaying males while perched often rapidly flap wings and expose pectoral tufts. Will also nest in the verandahs and porches of houses in cities. Tame.

637. Black-throated Sunbird, *Aethopyga saturata* Plate 71
Kelicap Tekak Hitam

Description: 11 cm (4.5 in). Elongate tail of male extends a further 3 cm (1 in). **Male:** Maroon mantle together with iridescent violet crown, rump and tail diagnostic. Throat and breast black with a violet tinge; belly pale greyish yellow. In poor forest light, colourful iridescent parts appear black. **Female:** From other female sunbirds by yellow rump. Upperparts olive grey; underparts paler. **Voice:** Rapid high pitched twittering notes, *tsi-tsi-tsi-tsi*... **Range:** The Himalayas and southern China through South-East Asia to the Malay Peninsula. **Status and distribution:** Fairly common resident above 800 m in the Larut Hills (Perak) and the Titiwangsa Range south to Gunung Telapa Buruk (Negeri Sembilan). Also on Gunung Benom (Pahang), Gunung Rabong (Kelantan) and in the Eastern Range on Gunung Tapis (Pahang), and Gunung Padang (Terengganu). Very common at the hill-stations. **Habitat:** Forests, the forest edge. **Habits:** Frequents the middle and lower storeys. Often hovers over flowering plants before settling on a stem to feed. Enters hill-station gardens frequently, flitting amongst the flower beds and potted plants. Tame and confiding.

638. Crimson Sunbird, *Aethopyga siparaja* Plate 71
Kelicap Sepah Raja

Description: 11 cm (4.5 in). Elongate tail of male extends a further 3 cm (1 in). **Male:** Bright crimson head, mantle, throat and breast together with iridescent violet tail diagnostic. Forehead dark iridescent green; narrow malar stripe dark iridescent violet. **Male eclipse (not illustrated):** Lacks elongate tail; dark olive plumage with diffused red throat and breast. **Female:** Upperparts dark olive green; underparts pale yellow. **Voice:** A very rapid and high pitched *chit-chit-chew* (L and G). **Range:** India and southern China through South-East Asia to the Greater Sundas, Sulawesi, and the Philippines. **Status and distribution:** Locally common resident at low elevations, mostly along coastal plains, south to Singapore. Also on Pulau Pinang and Pulau Langkawi (Kedah) where it occurs at all altitudes. **Habitat:** Forests, the forest edge, mangroves, beach scrub, plantations, wooded gardens. **Habits:** Frequents the middle and lower storeys. Gathers to feed in low flowering shrubs. Where common, will enter gardens frequently. Tame and confiding. **Note:** Does not usually overlap with the Scarlet Sunbird (639), which is more common inland.

639. Scarlet Sunbird, *Aethopyga temminckii* Plate 71
Kelicap Merah

Description: 11 cm (4.5 in). Elongate tail of male extends a further 3 cm (1 in). **Male:** From male Crimson Sunbird (638) by scarlet sides of head, breast, mantle, and tail. Malar stripe and line around head from forehead through nape iridescent purple. Rump yellow. Belly grey; undertail coverts grey with yellow tinge. **Female:** From female Crimson Sunbird by greyish head and scarlet tinge on wings and tail. Underparts yellowish olive, denser on centre of belly and undertail coverts. **Voice:** No information from the review area. **Range:** The Malay Peninsula, Sumatra, and Borneo. **Status and distribution:** Uncommon and localized resident from low elevations up to 1500 m, south to Johor. **Habitat:** Forests. **Habits:** Usually frequents the middle and lower storeys. General habits similar to Crimson Sunbird. Sometimes frequents the forest edge at the hill-stations.

640. Little Spiderhunter, *Arachnothera longirostra* Plate 71
Kelicap-Sabit Kecil

Description: 16 cm (6.25 in). Small size. Whitish throat merging into yellowish olive breast and yellow belly and undertail coverts diagnostic. Upperparts dull olive; narrow supercilium white; thin malar stripe black; outer tail feathers tipped white. Both sexes have yellow and orange pectoral tufts. **Voice:** A sharp and high pitched *chit-chit-chit*... uttered while perched and in flight. Also a quick, repeated, *chip-chee*. Song is a four note phrase, *chip-chip-chip-chiap*, uttered repeatedly. **Range:** India and southern China through South-East Asia to the Greater Sundas and the Philippines. **Status and distribution:** Fairly common resident from low elevations up to 1670 m, south to Singapore, where numbers are declining (W and H). **Habitat:** Forests, the forest edge, plantations, wooded gardens. **Habits:** Frequents the lower storey. Has a partiality for wild bananas and ginger. Frequently enters plantations and gardens especially where bananas are cultivated. Hovers efficiently. Will also pick off spiders from webs. Flies low at great speed with ease and dexterity. A known brood host of the Violet Cuckoo (255).

641. Thick-billed Spiderhunter, *Arachnothera crassirostris* Plate 71
Kelicap-Sabit Paruh Tebal

Description: 17 cm (6.5 in). From Little Spiderhunter (640) by thicker bill and lack of white in plumage. Throat and breast greyish olive, merging into yellow belly and undertail coverts. Narrow black line through eye. Indistinct yellow patches above and below eye. Male has orange pectoral tufts. **Voice:** A series of strident chatters, *chit-chit-chit*... **Range:** The Malay Peninsula to Sumatra and Borneo. **Status and distribution:** Uncommon and localized resident from low elevations up to 1200 m, south to Singapore, where it is rare and confined to the central catchment forest (W and H). **Habitat:** Forests, the forest edge. **Habits:** Forages in the canopy and mid storey but nests in the lower storey. Like other spiderhunters, will visit flowering shrubs. Also forages in the canopy for insects. Other habits little known.

642. Long-billed Spiderhunter, *Arachnothera robusta* Plate 71
Kelicap-Sabit Paruh Panjang

Description: 22 cm (8.5 in). Long and thick bill diagnostic. Sides of head dark olive green; throat and breast yellowish green with faint streaks, merging into pale yellow belly and undertail coverts. Male has orange-yellow pectoral tufts. **Voice:** A loud *chup-chup-chup...* in flight. Utters a single *chup* while perched. **Range:** The Malay Peninsula to the Greater Sundas. **Status and distribution:** Locally common resident from low elevations up to 1520m m, south to Johor. **Habitat:** Forests (including peat-swamp), the forest edge. **Habits:** Frequents the canopy, often in flowering trees. Bold and aggressive and will chase away other nectarines from a food source. At forest edge settlements, especially at the Taman Negara headquarters at Kuala Tahan (Pahang), will sometimes descend to feed in flower beds and flowering shrubs. Will sometimes perch on bare branches.

643. Spectacled Spiderhunter, *Arachnothera flavigaster* Plate 71
Kelicap-Sabit Besar

Description: 22 cm (8.5 in). Large and robust appearance distinctive. Thick, broad based bill, broad yellow eyering and yellow ear coverts diagnostic (see Yellow-eared Spiderhunter (644)). **Voice:** A loud and powerful disyllabic *chak-chak*, uttered while perched or in flight. **Range:** The Malay Peninsula to Sumatra and Borneo. **Status and distribution:** Locally common resident from low elevations up to 300 m, south to Johor. Locally extinct in Singapore since 1920 (L and G) **Habitat:** Forests, the forest edge, plantations, wooded gardens. **Habits:** Frequents the canopy, gathering in flowering trees. Will sometimes hover while probing blossoms. Bold and aggressive and will sometimes displace other nectarines from a food source.

644. Yellow-eared Spiderhunter, *Arachnothera chrysogenys* Plate 71
Kelicap-Sabit Telinga Kuning

Description: 18 cm (7 in). Like Spectacled Spiderhunter (643) but smaller with proportionately longer and thinner bill, narrower and less distinct yellow eyering and more conspicuous yellow ear coverts. **Voice:** A high pitched *chik*. **Range:** Tenasserim (Myanmar) and the Malay Peninsula to the Greater Sundas. **Status and distribution:** Locally common resident from low elevations up to 2010 m, south to Singapore where it is rare (W and H). Also on Pulau Pinang. **Habitat:** Forests, the forest edge, plantations, wooded gardens. **Habits:** Frequents the canopy and the crowns of smaller trees. Gathers to feed in flowering trees with other species. Will descend to feed in flowering shrubs in suburban gardens.

645. Grey-breasted Spiderhunter, *Arachnothera affinis* Plate 71
Kelicap-Sabit Dada Kelabu

Description: 18 cm (7 in). Bright olive green upperparts together with pale greyish underparts with slight streaking on throat and upper breast diagnostic. **Voice:**

Strident chattering notes, uttered while perched and in flight. **Range:** Tenasserim (Myanmar) through the Malay Peninsula to the Greater Sundas and Bali. **Status and distribution:** Locally common resident from low elevations up to 1120 m, south to Johor. Locally extinct in Singapore since 1950 (L and G). **Habitat:** Forests, the forest edge. **Habits:** Frequents the canopy as well as the middle and lower storeys. Gathers in flowering trees. Has a partiality for wild bananas and ginger. Flies fast with ease and dexterity.

646. Streaked Spiderhunter, *Arachnothera magna* Plate 71
Kelicap-Sabit Berjalur

Description: 19 cm (7.5 in). Bold blackish streaks on bright olive upperparts and whitish underparts together with bright orange legs diagnostic. **Voice:** A high pitched metallic *chisikik* while foraging. Also a sharp *ka-chik* in flight. **Range:** The Himalayas and southern China discontinuously through South-East Asia to the Malay Peninsula. **Status and distribution:** Fairly common resident from 820–1900 m in the Larut Hills (Perak) and the Titiwangsa Range, south to Gunung Telapa Buruk (Negeri Sembilan). Also on Gunung Benom and Gunung Tahan (Pahang) and on the Eastern Range. Very common and conspicuous at the hill-stations. **Habitat:** Forests, the forest edge. **Habits:** Frequents the canopy, middle and lower storeys. Has a great partiality for wild bananas. Perches on service wires at the hill-stations. Often visits flower beds and potted plants in hill-station gardens. Very vocal. Tame and confiding. Flight strongly undulating.

OLD WORLD SPARROWS AND ALLIES

ORDER *Passeriformes* **FAMILY** *Passeridae*

World: *45 species* **Peninsular Malaysia and Singapore:** *3 species*

Small to medium sized with short conical bills and notched tails. Most are clad in combinations of browns, chestnut, and greys, and are sexually dimorphic in many species, the males are usually more attractive. Sparrows are found in most open habitats from desert, scrub, and farmland to towns and cities where they are gregarious, perching on trees but foraging entirely on the ground on seeds and grain. They are sedentary and many species have become commensals of man, utilizing occupied buildings as convenient nesting sites. Calls are largely harsh chips and chirps.

647. Plain-backed Sparrow, *Passer flaveolus* Plate 74
Ciak Sawah

Description: 15 cm (6 in). From Eurasian Tree Sparrow (649) by unstreaked upperparts. **Male:** Crown, back of neck and rump grey green; mantle, scapulars and sides of head behind eye chestnut; lores and throat black; cheek patch; forecrown and underparts pale yellow. **Female:** Upperparts dull light brown with pale yellow or buff supercilium. Underparts dull greenish yellow. **Voice:** A loud and clear disyllabic *filip*; less harsh but richer than Eurasian Tree Sparrow with a more metallic quality.

Range: Except for northern Myanmar and northern Vietnam, continental South-East Asia, to the Malay Peninsula. **Status and distribution:** Uncommon and localized resident at low elevations, confined to the coastal plains, south to Kuala Selangor (Selangor) in the west and Dungun (Terengganu) in the east. Also on Pulau Pinang and Pulau Langkawi (Kedah). Previously ranged south to Kuantan (Pahang) in the east and Tanjung Tuan (Melaka) in the west but are no longer seen at these locations today. **Habitat:** Beach scrub, coconut groves, ricefields, gardens. **Habits:** Feeds in small groups in open areas. Less a commensal of man than Eurasian Tree Sparrow, keeping more to ricefields and villages with rice granaries. Sometimes feeds and roosts with the former.

648. House Sparrow, *Passer domesticus* Plate 72
Ciak Rumah

Description: 15 cm (6 in). **Breeding male:** Differs from Eurasian Tree Sparrow (649) by grey crown, chestnut band from back of eye curving downwards towards sides of nape, more extensive black on throat and upper breast (appearing like a broad bib) and the lack of black patch on ear coverts. Bill black. Mantle dull chestnut with blackish and pale brownish streaks, with underparts from lower breast and below, pale greyish-white. **Non-breeding male (not illustrated):** Crown tinged brown, mantle greyer, black bib and chestnut sides of nape appear mottled with pale feather tips. Bill pale horn. **Female:** Nondescript brownish with pale supercilium and underparts and blackish-brown streaks on mantle and scapulars. Bill pale horn. **Voice:** Unknown from review area. **Range:** North Africa and the islands of the north Atlantic, Europe and Asia east to Siberia, South-West Asia, Tibet, the Indian subcontinent, Sri Lanka, Myanmar, Thailand, Laos, and Cambodia. Widely introduced to the Americas, southern Africa, New Zealand, Eastern Australia, and also islands in the Indian and Pacific oceans. **Status and distribution:** Introduced. First observed in January 1995, a feral population of about 100 birds was seen in an open market area at Pasir Panjang (Singapore) in early September 1997 (W and H), with many nesting in warehouse-like buildings. The origin of this population is unknown. With no other records from elsewhere, the population is assumed to be probable shipborne arrivals as the site is by the habour. Also on Jurong Island. **Habitat:** Open urban habitat. **Habits:** Like Eurasian Tree Sparrow. **Note:** W and H have included this species on the Singapore list.

649. Eurasian Tree Sparrow, *Passer montanus* Plate 74
Ciak Pokok

Description: 15 cm (6 in). Dark chestnut cap, white cheek patch, black throat and patch on ear coverts diagnostic. Brownish upperparts and chestnut wings boldly streaked with black. Underparts pale greyish, merging into whitish on belly. **Voice:** A repertoire of harsh chirping notes. **Range:** Europe, temperate Asia and the Himalayas through South-East Asia to Sulawesi and the Lesser Sundas. In Borneo, feral populations have been established from shipborne specimens. Widely introduced in North America, Australia, New Zealand, and the western Pacific. **Status and distribution:** Abundant and widespread resident principally at low elevations, south to Singapore. Also recorded above 1500 m in the settled areas of Cameron Highlands and Fraser's Hill (Pahang). **Habitat:** Towns, rural settlements, gardens, scrub. **Habits:** Strictly a commensal of man, frequenting buildings and settlements in small groups. Nests under the eaves of buildings. Feeds largely on grass seeds but will also take scraps of leftovers from roadside food stalls and restaurants. Will also forage at rubbish dumps. Tame and confiding.

WEAVERS AND ALLIES

ORDER *Passeriformes* **FAMILY** *Ploceidae*

World: *109 species* **Peninsular Malaysia and Singapore:** *2 species*

(This cosmopolitan family, which has its centre of distribution in sub-Saharan Africa, also includes the malimbes, queleas, fodies, bishops, and widowbirds which do not occur in Asia and hence the review area. For this purpose only the true weavers are described below.)

Medium sized birds with short and powerful conical bills, specially adapted for feeding on seeds and grain. Highly gregarious, nesting in colonies. Nests elaborately woven structures with lateral openings, usually suspended from branches in most species. Sexually dimorphic in most species, the males have a distinct breeding plumage; otherwise like females.

650. Baya Weaver, *Ploceus philippinus* Plate 74

Burung Tempua

Description: 15 cm (6 in). **Breeding male:** Forehead, crown and nape bright yellow; bill and facial mask black; throat brownish. **Non-breeding male and female:** Underparts pale tawny brown; upperparts (including crown) dark with buff streaks; long supercilium pale buff brown; throat pale; bill horn coloured with pinkish base. The characteristic nest is a neatly woven flask shaped structure with a long downward tubular entrance, suspended from branches or coconut palm fronds. **Voice:** A repertoire of harsh chirpings, uttered rapidly. Song, a series of chatters and dry wheezy rattles. **Range:** The Indian subcontinent through South-East Asia and the Malay Peninsula to Sumatra and Java. **Status and distribution:** Fairly common resident at low elevations, south to Singapore, where it is uncommon, due to loss of long grass areas (W and H). **Habitat:** Ricefields, reedbeds and adjacent scrub, wet grassland, plantations, cultivated land. **Habits:** Associates in compact flocks of up to 40. Flight swift, direct and co-ordinated with the whole flock turning in unison. Feeds largely on seeds and ripe paddy, often with munias (654–659). Nests in small colonies in trees usually in association with nesting red ants and hornets. Outside breeding, roosts communally in reedbeds and other tall dense vegetation.

651. Streaked Weaver, *Ploceus manyar* Plate 72

Burung Tempua Lorek

Description: 13.5 cm (5.3 in). **Breeding male:** Similar to breeding male Baya Weaver (650) but throat all black and breast and flanks boldly streaked blackish. **Non-breed-**

ing male and female: Have brown ear coverts, bordered above by prominent whitish supercilium, joining yellowish buff band below which terminates at base of mandible, creating a typical bunting-type face pattern Differs from buntings (671–673) by much thicker bill and shorter and rounded tail which lack the white sides. The nest is a globular structure with a short entrance tube and is usually placed in low vegetation or reedbeds. **Voice:** Song is a soft continuous trill, *see-see-see-see*, terminating with *o-chee*. Contact calls are loud *chirt* notes (Robson, 2008). **Range:** The Indian subcontinent and South-West China through much of continental South-East Asia, discontinuously to Java and Bali. **Status and distribution:** Introduced. Rare feral resident in Singapore, with the first confirmed sightings at Lorong Halus on 12–13 April 1997, with further sightings at Marina East in 2003 and again at Lorong Halus in 2004 (W and H). **Habitat:** In review area, long grass and rural habitats **Habits:** Associates in very small groups, exclusively on its own. Other habits similar to Baya Weaver. **Note:** W and H have included this species on the Singapore list.

PARROTFINCHES AND MUNIAS

ORDER *Passeriformes* **FAMILY** *Estrildidae*

World: *138 species* **Peninsular Malaysia and Singapore:** *9 species*

Small birds with thick conical bills, specially adapted for eating seeds. Resemble true finches externally but differ in internal anatomical structure. Plumage striking and colourful in parrotfinches, genus *Erythrura* (652–653). Munias, genus *Lonchura* (654–659) are clad in varying shades of brown patterned with varying amounts of black or white and have distinct piping flocking calls. Java Sparrow, genus *Padda* (660), is larger than munias with a thicker bill. Tails pointed or rounded. Flight swift and undulating. Diet consists almost exclusively of seeds and most species are to a great extent nomadic, frequently moving over large areas to abundant food sources. Highly gregarious, flying in compact flocks. Nests are neat globular structures with lateral openings, usually placed in low vegetation or branches. Old nests are often used as roosts in many species. Sexes alike in most species.

652. Tawny-breasted Parrotfinch, *Erythrura hyperythra* Plate 72
Ciak-Hijau Dada Perang

Description: 10 cm (4 in). Like Pin-tailed Parrotfinch (653) but tail all green and rounded. *Erythura hyperythra malayana*: **Female:** Entire upperparts dull grass green with slightly bluish crown. Face, sides of head and ear coverts chestnut, deeper on lores. Underparts tawny chestnut, more intense on throat with flanks tinged greenish. Legs greenish yellow. The male of this subspecies has not yet been described. In nominate *E. h. hyperythra*, the male is like female but is brighter green above with black patch on extreme forehead and dull dark blue forecrown. **Voice:** A very soft munia like *seet-seet-seet*...uttered by a flock (Wilbur Goh). **Range:** The Malay Peninsula to Java, Borneo, Sulawesi, the Lesser Sundas and the Philippines **Status and distribution:** Very rare resident, known only from mist-netted specimens at altitudes between 1000 and 1800 m in the Cameron Highlands (Pahang), and more recently an individual sighted at Bukit Larut (Perak) in June 2002 (W2) and a party of four near the summit of Gunung Batu Brinchang, Cameron Highlands (Pahang) on 12 June 2006 (Wilbur Goh and Yang Chong, BiW). **Habitat:** Forests. **Habits:** Frequents the lower storeys, especially amongst bamboo thickets, solitarily or in small parties. Other habits little known.

653. Pin-tailed Parrotfinch, *Erythrura prasina* Plate 72
Ciak-Hijau Ekor Runcing

Description: 15 cm (6 in) including 2 cm (1 in) of tail of male. Bright grass green upperparts and red pointed tail diagnostic. **Male:** Elongate red pointed tail distinctive. Throat and face blue; lores black; underparts rich tawny buff, sometimes tinged reddish on breast and upper belly; rump red. **Female:** Tail shorter. Throat and upper breast pale greyish blue; underparts pale tawny buff; rump green, concolorous with upperparts. Legs pinkish in both sexes. **Voice:** High pitched *zit-zit* calls, uttered while foraging. **Range:** Indo-China through Thailand and the Malay Peninsula to the Greater Sundas. **Status and distribution:** Uncommon and localized resident principally in hilly country from low elevations up to 1500 m, south to Selangor and central Pahang. **Habitat:** Forests with bamboo thickets. **Habits:** Frequents the lower storeys in small loose parties. Movements rather nomadic and is to a great extent influenced by the availability of seeding bamboo.

654. White-bellied Munia, *Lonchura leucogastra* Plate 72
Pipit Padi

Description: 11 cm (4.5 in). From White-rumped Munia (655) by dark rump which is concolorous with mantle. Tail is shorter without notch and is marked pale yellowish on upperside. Flanks, undertail coverts and thighs blackish brown with white restricted to lower breast and upper belly. **Voice:** Flocking call is a piping *prrrt,* without the trilling quality of White-rumped Munia. **Range:** Tenasserim (Myanmar) through the Malay Peninsula to Sumatra, Borneo and the Philippines. **Status and distribution:** Uncommon and localized resident from low elevations up to 700 m, south to Johor. **Habitat:** Forests, the forest edge, clearings in forests and along riverbanks. **Habits:** Generally in pairs or in small loose flocks, skulking in the underbrush of forests and forest edges. Sometimes feeds in the open. Does not associate with other munias in typical open country habitats.

655. White-rumped Munia, *Lonchura striata* Plate 72
Pipit Tuli

Description: 11 cm (4.5 in). Blackish brown plumage and whitish lower breast and lower belly together with long pointed and notched tail diagnostic. Undertail coverts pale brownish. Dark plumage has faint buff scales and streaks on mantle. In flight, white rump is conspicuous. **Juvenile (not illustrated):** Like adult but rump buff white. **Voice:** Flocking call is a trilling *prrrit* **Range:** The Indian subcontinent through southern China to South-East Asia and Sumatra. **Status and distribution:** Fairly common resident from low elevations up to 2000 m, south to Singapore where it is rare. Also on Pulau Pinang, Pulau Langkawi (Kedah), Pulau Pangkor (Perak), Pulau Redang (Terengganu) and Pulau Tioman (Pahang). **Habitat:** The forest edge, scrub, plantations, ricefields, clearings beside roads and riverbanks. The only munia that occurs at the hill-stations. **Habits:** Usually in small noisy groups, feeding in the scrub and secondary growth at the forest edge as well as at grassy verges of roads in wooded country. Sometimes associates with other munias in smaller numbers especially in ricefields.

656. Javan Munia, *Lonchura leucogastroides* Plate 72
Pipit Jawa

Description: 11 cm (4.5 in). From White-rumped Munia (655) by shorter unnotched tail and whiter breast and belly. Face and upper breast black with blackish brown undertail coverts. **Voice:** Flocking call is a shrill *pi-i*. Also utters soft cheeping notes. **Range:** Southern Sumatra, Java, Bali and Lombok. **Status and distribution:** Introduced. Feral populations were well established in Singapore since 1922 but have declined and become less common due to development of rural areas. **Habitat:** Scrub, clearings in wooded areas, fringes of reservoirs, wooded gardens. **Habits:** Gathers in small flocks, feeding largely on grass seeds in grassy verges of roadsides and on lawns.

657. Scaly-breasted Munia, *Lonchura punctulata* Plate 72
Pipit Pinang

Description: 11 cm (4.5 in). Plumage pattern distinctive with brown upperparts contrasting with brownish scales on breast and flanks with whitish centre of belly and vent. **Juvenile:** Brown upperparts, paler below. Lacks scales. From juveniles of Chestnut (658) and White-headed Munias (659) by greyish black bill. **Voice:** Flocking call is a piping *ki-dee*. **Range:** The Indian subcontinent and southern China discontinuously through South-East Asia to the Malay Peninsula, Sumatra, Java, Sulawesi, the Lesser Sundas, and the Philippines. Widely introduced in other parts of the world. **Status and distribution:** Common and widespread resident at low elevations, south to Singapore. Uncommon in open country habitats in the late nineteenth and early twentieth centuries, it is probable that it is not indigenous to the Malay Peninsula but might be feral, originating from introduced stock (M and W). **Habitat:** Open country (including wet and dry grassland), ricefields, plantations, scrub, gardens. **Habits:** Gathers in large groups, often in association with other munias in open country habitats. Typically seen in flight, carrying long blades of grass to construct nests. Often nests in low bushes and trees in rural and urban gardens.

658. Chestnut Munia, *Lonchura malacca* Plate 72
Pipit Rawa

Description: 11 cm (4.5 in). Black head and upper breast, bluish grey bill together with rich chestnut plumage diagnostic. **Juvenile:** Plumage rich buff brown, paler and more rufous below. **Voice:** Flocking call is a thin *pwi,* with a reedy quality. **Range:** The Indian subcontinent and southern China discontinuously east through South-East Asia, Sumatra, Borneo, Sulawesi, and the Philippines. **Status and distribution:** Fairly common resident principally at low elevations, south to Singapore. Individuals with buff-white crown, face, and nape, which are consistent with the Javan subspecies *L. m. ferruginosa* seen since late 1986 are presumably escaped captives (W and H). **Habitat:** Wet grasslands, ricefields, swampy vegetation, reedbeds. **Habits:** Gathers in large flocks especially in ricefields, often in association with Scaly-breasted (656) and White-rumped Munias (654). Generally prefers wetter habitats than other munias. Forms large communal roosts in reedbeds.

659. White-headed Munia, *Lonchura maja* — Plate 72

Pipit Uban

Description: 11 cm (4.5 in). White head and upper breast contrasting sharply with brown iris, bluish grey bill and dark chestnut plumage diagnostic. **Juvenile:** From juvenile Chestnut Munia (658) by whitish buff crescentic shape on ear coverts and underparts. **Voice:** Flocking call is a thin piping *puip*. **Range:** The Malay Peninsula to Sumatra, Java and Bali. **Status and distribution:** Fairly common resident from low elevations up to 300 m, south to Singapore, where it is uncommon. **Habitat:** Open grassland, ricefields, scrub, gardens. **Habits:** Like other munias, gathers in large flocks (usually in much less numbers than Chestnut Munia). Associates with other munias, foraging in open country habitats, especially ricefields.

660. Java Sparrow, *Padda oryzivora* — Plate 72

Ciak Jawa

Description: 16 cm (6.25 in). Larger and more robust with more thickset bill than munias. Pale grey plumage, red bill and legs together with black head with broad white cheek patches diagnostic. Lower belly pale pinkish; undertail coverts white and tail black. **Juvenile (not illustrated):** Like adult but tinged brown on upperparts and buff on underparts. **Voice:** A quiet *tup,* uttered in flight. Also low churring calls. Also a soft chattering and rather unmusical sequence ending in a whining *ti-tiu*. **Range:** Java and Bali. Widely introduced elsewhere. Its popularity as a cage bird has resulted in intense trapping for the trade, resulting in population declines. **VU. Status and distribution:** Introduced. Feral populations, derived from largely escaped caged specimens are established in small numbers in a number of towns in Peninsular Malaysia, notably Kangar (Perlis), Alor Star (Kedah), Pulau Pinang, Ipoh (Perak), Kuala Lumpur, Petaling Jaya (Selangor), Seremban (Negeri Sembilan) and Melaka. In Singapore, feral populations have disappeared but occasional escapees still occur. **Habitat:** Towns, cities, gardens, open country, ricefields. **Habits:** Usually in small loose parties. In ricefields, feeds on ripening paddy. In towns and cities, sometimes associates with Eurasian Tree Sparrow (649), feeding on food scraps on the ground.

WAGTAILS AND PIPITS

ORDER *Passeriformes* **FAMILY** *Motacillidae*

World: *67 species* **Peninsular Malaysia and Singapore:** *9 species*

Small to medium-sized slender built birds with rather long legs and tails which have white outer tail feathers. Inhabit primarily open country. Largely terrestrial, feeding on insects caught on the ground. Fly with low dipping flight. Gregarious. Sexes nearly alike in most species. Most species occurring here are migratory. Represented by the following groups and genera:

Wagtails, genera *Dendronanthus* (661) and *Motacilla* (662–665). Have bright plumage with bold markings. Long tails usually wagged up and down in the latter; the former sways body sideways. **Pipits,** genus *Anthus* (666–669). Plumage largely dull brown with dark streaks. Tails proportionately shorter than wagtails. Often difficult to identify; their distinct call notes are often the best field mark.

661. Forest Wagtail, *Dendronanthus indicus* Plate 73

Kedidi-Air Hutan

Description: 18 cm (7 in). Broad black crescentic breast patch separated from broken black crescentic patch by creamy white breast band together with whitish wing-bars diagnostic. Upperparts olive brown with thin whitish supercilium. Underparts creamy white. **Voice:** A sharp and distinct metallic *pink*. **Range:** Breeding in far eastern Russia, Korea and northern China; wintering south to the Indian subcontinent and South-East Asia to the Greater Sundas. **Status and distribution:** Locally common passage migrant and winter visitor principally at low elevations, south to Singapore, where it is uncommon (W and H). **Habitat:** Forests, the forest edge, mangroves, plantations, scrub. **Habits:** Usually solitary or in pairs, foraging on the ground. Frequently sways body sideways. Will perch on trees. Sometimes tame. Has been observed roosting communally in large numbers in scrubland and reedbeds.

662. White Wagtail, *Motacilla alba* Plate 73

Kedidi-Air Hitam-Putih

Description: 19 cm (7.5 in). From Ashy Minivet (407) by terrestrial habits. Pied plumage distinctive and variable according to subspecies. *M. a. leucopsis*: **Breeding male:** Black throat, breast, crown, nape and mantle. **Non-breeding male:** Lacks black throat. **Female:** Like non-breeding male but mantle grey. From *M. a. ocularis* by lack of black eyeline. *M. a. ocularis*: Shows narrow black eyeline and pale grey mantle in all plumages. Crown, nape and tail black. Forehead, head and entire underparts white. **Breeding (not illustrated):** Black throat and breast. **Non-breeding:** Lacks black throat. **Voice:** A hard and sharp *chissik*. **Range:** Breeding throughout Europe and temperate Asia, South and South-East Asia to northern Indo-China; northern populations are migratory, wintering south to Africa, South and South-East Asia to the Malay Peninsula, Borneo, and the Philippines. **Status and distribution:** Rare and localized winter visitor principally at low elevations, recorded from Bagan Belat, Seberang Prai (Pulau Pinang), south to Singapore. Has also been recorded at about 1500 m at Cameron Highlands (Pahang). A roosting party of seven at Sungai Balang, Muar, (Johor) from December 2002–February 2003 comprised at least four *M. a. leucopsis* and two *M. a. occularis* (W2). **Habitat:** Open country, reedbeds, mangroves (as roosting sites only). **Habits:** Usually solitary, feeding along streams, roadsides, and open areas. Shy.

663. Citrine Wagtail, *Motacilla citreola* — Plate 73
Kedidi-Air Kepala Kuning

Description: 19 cm (7.5 in). **Breeding male:** Bright yellow head and underparts contrasting with either grey or black mantle with broad white wingbar diagnostic. **Female and non-breeding male:** From female Yellow Wagtail (664) by yellow lores and yellow supercilium which curves around margin of dark ear coverts. In flight shows white wingbars. **Juvenile:** Buff forehead, grey crown and nape separate it from Yellow Wagtail. Has pale supercilium. Buff breast has faint gorget of black spots. Underparts brownish white. **Voice:** A wheezy *zeeep*, similar to Yellow Wagtail. **Range:** Breeding throughout the Middle East, the Indian subcontinent and China; wintering south to Myanmar with stragglers reaching Thailand and the Malay Peninsula. **Status and distribution:** Very rare winter visitor, recorded twice at low elevations in Singapore; an adult male in breeding plumage at Punggol on 11 March 1989, a juvenile at Tuas between 12-18 December 1994, and two juveniles at Serangoon on 4 October 1998 (W and H). **Habitat:** Open country. **Habits:** Like other *Motacilla* wagtails.

664. Yellow Wagtail, *Motacilla flava* — Plate 73
Kedidi-Air Kuning

Description: 18 cm (7 in). From Grey Wagtail (665) by proportionately shorter tail, olive-green upperparts and the lack of white wingbars in flight. Breeding birds have bright yellow underparts. Plumage variable according to subspecies. *M. f. simillima.* **Breeding male:** Pale blue-grey head and white supercilium distinctive. **Non-breeding and female:** Whitish underparts. *M. f. taivana*. Has yellow supercilium in both breeding and non-breeding plumages. **Voice:** A distinct wheezy *shreeep*, in flight. **Range:** Breeding throughout Europe, temperate Asia and western North America; wintering south to Africa, South and South-East Asia to Australia. **Status and distribution:** Fairly common passage migrant and winter visitor from low elevations up to 1200 m at Fraser's Hill (Pahang), south to Singapore. More common than Grey Wagtail in the flat lowlands. **Habitat:** Open grasslands, airfields, disused tin mines, sludge beds, margins of freshwater swamps, mudflats (on passage only), reedbeds. **Habits:** Feeds in small to large groups in open grassy areas, often together with Pacific Golden Plover (142). Roosts communally in reedbeds.

665. Grey Wagtail, *Motacilla cinerea* — Plate 73
Kedidi-Air Kelabu

Description: 19 cm (7.5 in). Grey head and mantle with narrow white supercilium together with yellow rump and undertail coverts diagnostic. Underparts vary from whitish to bright yellow. In flight, shows narrow white wingbars **Breeding male:** Black throat patch. **Female and non-breeding male:** Lacks black throat patch. **Voice:** A sharp *chisik*, with a metallic quality. **Range:** Breeding throughout Europe and temperate Asia; wintering south to Africa, South and South-East Asia to New Guinea. **Status and distribution:** Fairly common winter visitor from low elevations up to 2000 m, south to Singapore where it is uncommon (W and H). The most common wagtail in hilly country throughout Peninsular Malaysia. **Habitat:** Grassy verges of hilly roads, margins of forested rivers and streams, margins of freshwater swamps, disused tin mines. **Habits:** Usually solitary, feeding along grassy verges of hilly roads and along wet gravel banks of hill streams. Constantly wags tail. Shy.

666. Paddyfield Pipit, *Anthus rufulus* Plate 73

Pipit-Tanah Padang

Description: 16 cm (6.25 in). Long-tailed and long-legged pipit with distinctive upright carriage. Tawny brown plumage (paler below) with lightly streaked upperparts (excluding rump and uppertail coverts). Streaking on underparts confined to breast. Shows pale buff supercilium, indistinct thin blackish malar stripe and dark lores. Juveniles are more heavily streaked and have rufous wash on flanks. Legs pink with long hindclaw. **Voice:** A single loud *chup* uttered in flight, repeated with each dip. Flight song is a rapid *chik-a-chik-a-chik*, repeated in undulating flight. **Range:** The Indian subcontinent and Sri Lanka east to southern China and South-East Asia to the Malay Peninsula, Sumatra, the Riau archipelago, the Greater Sundas and the Philippines east to Sulawesi and the Lesser Sundas. **Status and distribution:** Common and widespread resident from low elevations up to 1500 m, south to Singapore. Frequently seen on playing fields and golf courses. **Habitat:** Open country, harvested ricefields, airfields, gardens. **Habits:** Usually forages solitarily or in pairs; larger groups tend to congregate outside breeding. Moves around in short brief runs before pausing. Has spectacular courtship flight; rising briefly before descending with rapidly flapping outstretched wings, uttering flight song (see voice). **Note:** The migratory Richard's Pipit *A. richardi* of northern temperate Asia which winters south to North-East Borneo could turn up in the review area. It is larger (20 cm (8 in)), has pale lores and heavier streaking on breast. The call is a loud and distinct *schreep*.

667. Red-throated Pipit, *Anthus cervinus* Plate 73

Pipit-Tanah Leher Merah

Description: 16 cm (6.5 in). From Paddyfield Pipit (666) by shorter tail, more compact appearance and bold blackish streaks on brownish upperparts especially on rump and uppertail coverts. Shows two pale lines on mantle. Only female has thin black malar stripe. **Breeding:** Throat, upper breast and supercilium pinkish with belly and undertail coverts creamy buff marked with bold blackish streaks on sides of breast and flanks. Females are more heavily streaked and have less pink on breast. **Non-breeding:** Throat and supercilium whitish, ear coverts brown. Creamy buff underparts boldly streaked on breast and flanks. Shows some pink on throat. **Voice:** A thin *zi*, uttered in flight. **Range:** Breeding throughout northern Europe and northern temperate Asia; wintering south through Africa, northern India, South-East Asia and the Malay

Peninsula to Borneo and Sulawesi. **Status and distribution:** Uncommon and localized winter visitor in small numbers at low elevations, south to Singapore. **Habitat:** Harvested ricefields, open country, disused tin mines, margins of freshwater swamps. **Habits:** Usually solitary, sometimes in small groups; occasionally in large numbers during passage. Prefers wetter habitats than Paddyfield Pipit.

668. Olive Backed-pipit, *Anthus hodgsoni* — Plate 73
Pipit-Tanah Pokok

Description: 16 cm (6.25 in). Small and comparatively short-tailed pipit with diagnostic white supercilium above black hind margin of ear coverts. Throat and underparts whitish with narrow blackish malar stripe joining bold dark streaks on breast with finer streaks on flanks. Entire upperparts olive-greenish with faint or no streaking. **Voice:** A faint *si-si-si-si* while perched. Flight call is a rather distinct and coarse *teez*. **Range:** Breeding in the Himalayas, eastern Russia, China, Korea and Japan; wintering south to South and South-East Asia to the Malay Peninsula, Borneo and the Philippines. **Status and distribution:** Locally common winter visitor in small numbers principally in hilly country from 450–2000 m, south to central Selangor. **Habitat:** The forest edge, scrub, tea plantations (at Cameron Highlands). **Habits:** Usually in small groups, feeding largely on the ground. Most arboreal pipit, alighting in trees when disturbed. Will also enter hill-station telecommunication stations in the morning to feed on moths and other insects stranded there by lights the previous night.

669. Blyth's Pipit, *Anthus godlewskii* — Plate 73
Pipit-Tanah Paruh Pendek

Description: 17 cm (6.5 in). At rest, has a similar upright carriage like Paddyfield Pipit (666) but has markedly shorter legs, shorter and more pointed bill, shorter tail, shorter hind claw and is slightly more pot-bellied in appearance. Upperparts more heavily streaked. Legs flesh-coloured, lacking the pinkish or reddish tinge of Paddyfield Pipit. **Voice:** A soft *chep* in flight (Bakewell, 2010). Other calls unknown from review area. **Range:** Breeding in Siberia, Mongolia, and northern China, wintering south to the Indian subcontinent to Myanmar, straggling to north-east Thailand and the Malay Peninsula. **Status and distribution:** Vagrant, an individual at Chuping (Perlis) on 7 January 2010 (Bakewell, 2010). **Habitat:** In review area, sandy track running through young rubber land. **Habits:** Generally similar to Paddyfield Pipit. Other habits not adequately observed.

FINCHES

ORDER *Passeriformes* **FAMILY** *Fringillidae*

World: *179 species* **Peninsular Malaysia:** *1 species*

(This cosmopolitan family also includes the Chaffinches, Serins, Seedeaters, Grosbeaks, Siskins, Rosefinches and Crossbills which do not occur in the review area. For this purpose only the Bullfinches, genus *Pyrrhula* will be described below.)

Small to medium sized birds with stubby conical bills and notched tails. Feed largely on seeds and small insects.

670. Brown Bullfinch, *Pyrrhula nipalensis* Plate 74
Pipit Gunung

Description: 17 cm (6.5 in). Short and stubby greyish bill, long and moderately notched black tail and white cheeks diagnostic. White rump conspicuous in flight. Plumage ashy-brown; centre of belly and undertail coverts white; wings black with purplish sheen; lores and band around base of bill brown. Male has dark red streak on outer web of innermost secondary, in female yellow (usually inconspicuous in the field). **Voice:** A pleasing disyllabic piping *pi-piu*, uttered both while perched and in flight. **Range:** The Himalayas and southern China discontinuously through South-East Asia to the Malay Peninsula. **Status and distribution:** Rare and localized resident normally above 1500 m in the Titiwangsa Range, south to Gunung Ulu Kali (Pahang/Selangor). Also on Gunung Tahan and Gunung Benom (Pahang). Fairly regular on Gunung Batu Brinchang in the Cameron Highlands (Pahang). Used to be fairly regular since pre World War 2, in the grounds of a bungalow at 1400 m at Fraser's Hill (Pahang), where stands of exotic conifers had been planted. Not seen there since 1992 **Habitat:** Ericaceous forests, with stands of *Dacrydium* conifers. **Habits:** An upper montane resident, usually in pairs or small loose groups. Flight swift and undulating. Keeps to the tree tops but sometimes descends to forage in lower branches and on the ground. Feeds largely on seeds and cones of native conifers in which it nests.

BUNTINGS AND ALLIES

ORDER *Passeriformes* **FAMILY** *Emberizidae*

World: *270 species* **Peninsular Malaysia and Singapore:** *3 species*

(This cosmopolitan family also includes the New World Sparrows, Brush Finches and Cardinals, Nearctic, and Neotropical species which do not occur in the review area. For this purpose only the Buntings will be described below.)

Finch-like birds with conical bills. Differ from finches by longer tails and bill structure; the cutting edge of the maxilla being wavy-edged. Sexually dimorphic, males are brighter coloured than females especially during breeding. Feed mostly on seeds on the ground. Flight undulating. Gregarious outside breeding.

671. Chestnut-eared Bunting, *Emberiza fucata* Plate 74
Pipit Pipi Coklat

Description: 16 cm (6.25 in). Chestnut ear coverts (with white border on lower edge) and shoulder patch together with thin black malar stripe running down into narrow blackish upper breast band distinctive. Edge of outer tail feathers white. **Male breeding:** Crown, nape and sides of neck grey with black streaks on crown; upperparts rufous brown with bold black streaks; rump rufous buff; white throat and breast separated from buff underparts by chestnut lower breast band. **Male non-breeding (not illustrated):** Head tinged buff; streaks on underparts faint. **Female:** Like non-breeding male but lacks chestnut lower breast band. Ear coverts duller chestnut; underparts more uniform buff, especially on breast and lower belly with pale centre of belly. **Voice:** Unknown from review area. **Range:** Breeds in the Himalayas, eastern Russia, China, Korea and Japan. Winters in southern China and eastern India through South-East Asia. **Status and distribution:** Vagrant, a single sight record from Batu Berendam (Melaka) on 28 February 1968. **Habitat:** In review area, reedbeds and marsh vegetation adjacent to ricefields. **Habits:** Similar to Yellow-breasted Bunting (673).

672. Chestnut Bunting, *Emberiza rutila* Plate 74

Pipit Coklat

Description: 14 cm (5.5 in). **Male breeding:** Bright chestnut head and upperparts contrasting with bright yellow breast and underparts diagnostic. **Male non-breeding:** Duller than breeding with pale fringes to chestnut upperparts. **Female:** Very similar to female Yellow-breasted Bunting (673) with pale unbordered ear coverts and rufous-chestnut rump. **Voice:** Unknown from review area. **Range:** Breeding in the Eastern Palaearctic; its known wintering range is North-East India and southern China, through continental South-East Asia, except the Malay Peninsula. **Status and distribution:** A male at the Kuala Selangor Nature Park (Selangor) on 30 October 2004 (Barnard Lau, 2004, RCS). **Habitat:** In review area, the landward edge of mangrove forest. **Habits:** The individual was briefly observed perched on a tree. Other habits were not noted.

673. Yellow-breasted Bunting, *Emberiza aureola* Plate 74

Pipit Dada Kuning

Description: 15 cm (6 in). **Male breeding:** Black face and throat, chestnut crown, nape, back and breast band together with white shoulder patch (conspicuous in flight) diagnostic. Underparts bright yellow. **Male non-breeding:** Chestnut breast band broken; black on head restricted to narrow border along buff ear coverts; coronal stripe blackish brown; supercilium and median coronal stripe light buff. **Female:** Like non-breeding male but lacks broken chestnut breast band and white shoulder patch. Underparts paler yellow with dark streaks on flanks. Upperparts boldly streaked black and buff brown. **Voice:** A short monosyllabic *zip*. **Range:** Breeds throughout north-eastern Europe and temperate Asia. Winters in eastern India and South-East Asia to the Malay Peninsula. Although it is abundant locally, anecdotal evidence suggests that overall it has undergone a very rapid population decline owing mainly to trapping on its wintering grounds. **VU. Status and distribution:** Rare and localized winter visitor at low elevations, south to Singapore. **Habitat:** Reedbeds, ricefields, scrub. **Habits:** Moves about in small parties, perching on reeds and bushes. Often comes to feed on the ground. Roosts in small groups in reedbeds. **Note:** Birds seen in Peninsular Malaysia and Singapore are mostly in non-breeding plumage.

Appendix 1 Table of night bird calls

	Species	Typical call	Other calls
267	Barn Owl *Tyto alba*	A loud screech	Rasping notes and hissing notes
268	Oriental Bay-owl *Phodilus badius*	A musical series of mournful whistles	Unknown
269	White-fronted Scops-owl *Otus sagittatus*	A hollow *hoo* like 270 with abrupt start and end.	Soft moans
270	Reddish Scops-owl *Otus rufescens*	A hollow high-pitched *hoo* without any inflection	Not known
271	Mountain Scops-owl *Otus spilocephalus.*	A soft and fluty *who-who,* with a one second interval between notes.	Not known
272	Collared Scops-owl *Otus bakkamoena*	A soft *bo-ou*, with both notes running together	Not known
273	Oriental Scops-owl *Otus sunia*	A purring *toik-o-toik*	Not known
274	Barred Eagle-owl *Bubo sumatranus*	A deep *hu-hu*	Harsh gecko-like *gagagagogo*. A hooting scream
275	Dusky Eagle-owl *Bubo coromandus*	A deep and gradually accelerating *kok, ko-ko-kokokoko.*	A loud scream and a wheezy throaty *ee-ee*
276	Brown Fish-owl *Ketupa zeylonensis*	Laughing hoots in a definite rhythm	A low hoarse scream
277	Buffy Fish-owl *Ketupa ketupu*	A series of long-drawn out shrill and wheezy notes	A rapid *pup-pup-pup-pup-pup-pup*
278	Spotted Wood-owl *Strix seluputo*	A rising mellow *hoo,* uttered every five seconds	A series of rapid *hoo* notes
279	Brown Wood-owl *Strix leptogrammica*	A deep *hoo-hoo*	A rapid *go-ko-kaloo*
280	Collared Owlet *Glaucidium brodiei*	A series of long monotonous *ponk, ponk-ponk-ponk*, with a brief pause after the initial note.	Not known
281	Brown Boobook *Ninox scutulata*	A distinct *hoo-wup*, the second note briefer and rising	Not known
283	Large Frogmouth *Batrachostomus auritus*	A loud, deep, and hollow tremulous rattle, *kworrrrr*	Not known
284	Gould's Frogmouth *Batrachostomus stellatus*	A distinct *whoo-wheeow*, the second syllable longer and falling	Not known
285	Javan Frogmouth *Batrachostomus javensis*	An ascending *wheeow*	A shrill descending *kerr-ker-ker*. Falling *gwa* notes.

286	Malaysian Eared-nightjar *Eurostopodus temminckii*	A rich *tip-ti- tiew*	Not known
287	Great Eared-nightjar *Eurostopodus macrotis*	A loud *pit, whee-wheeow*	Not known
288	Grey Nightjar *Caprimulgus indicus*	A rapid *chook-chook-chook-chook*	Not known
289	Large-tailed Nightjar *Caprimulgus macrurus*	A loud *chonk,* uttered every second	A rapid purring *kru-kru-kru-kru*
290	Savanna Nightjar *Caprimulgus affinis*	A harsh rasping *chweeez*	Not known

Appendix 2 Table of barbet calls

	Species	Typical call	Other calls
340	Fire-tufted Barbet *Psilopogon pyrrolophus*	A series of shrill buzzing trills	A sharp *kik*
341	Linetated Barbet *Megalaima lineata*	*Po-prrp,* with the second syllable stressed	A rapid rolling trill
342	Gold-whiskered Barbet *Megalaima chrysopogon*	A low pitched *tehup*, uttered rapidly	A loud and slow rolling trill
343	Red-crowned Barbet *Megalaima rafflesii*	*tuk, tuk-tuk-tuk-tuk-tuk*	Not known
344	Red-throated Barbet *Megalaima mystacophanos*	*tok, tok-tok-tok*	A rapid rolling trill
345	Golden-throated Barbet *Megalaima franklinii*	*pi-priu*	A rapid rolling trill
346	Black-browed Barbet *Megalaima oorti*	*to-ka-krrr*	Not known
347	Yellow-crowned Barbet *Megalaima henricii*	*trrr, tok-tok-tok-tok-trrr..*	Not known
348	Blue-eared Barbet *Megalaima australis*	A rapid and repeated *krik-krik..*	A repeated *priu*
349	Coppersmith Barbet *Megalaima haemacephala*	A series of repeated *wout* notes	Not known
350	Brown Barbet *Calorhamphus fuliginosus*	High-pitched thin squeeks, *pseuoo*	A repeated *see-se-see*

Appendix 3 Hills, peaks, hill-stations, and mountain passes cited in the text

Peninsular Malaysia

Bukit Larut (Perak): Hill-station in the Larut Hills with highest point at 1200m, 4°52′ N: 100° 48′ E.

Cameron Highlands (Pahang): Hill-station and surrounding peaks in the Titiwangsa Range with the highest point at 2031 m, 4°28′ N: 101°23′ E.

Fraser's Hill (Pahang/Selangor): Hill-station in the Titiwangsa Range with highest point at 1310 m, 3°42′ N: 101° 45′ E.

Genting Highlands (Pahang): Hill-station in the Titiwangsa Range with highest point at 1774 m, 3°26′ N: 101° 48′ E.

Genting Peres (Negeri Sembilan): Pass through the Titiwangsa Range at 2° 51′ N: 102° 00′ E.

Genting Semangkok/The Gap (Selangor): Pass through the Titiwangsa Range with maximum altitude of 822 m, 3° 42′ N: 101° 45′ E.

Genting Sempah (Pahang/Selangor): Pass through the Titiwangsa Range with maximum altitude of 634 m, 3° 26′ N: 101° 48′ E.

Gunung Benom (Pahang): Isolated mountain at 2107 m, 3° 50′ N: 102° 05′ E.

Gunung Bunga Buah (Selangor): Peak in the Titiwangsa Range at 1430 m, 3° 22′ N: 101° 45′ E.

Gunung Batu Brinchang (Pahang): Highest peak in the Cameron Highlands at 2031 m, 3° 50′ N: 102° 05′ E.

Gunung Gagau (Terengganu): Isolated mountain within Taman Negara at 1377 m, 4° 47′ N: 102° 39′ E.

Gunung Gelemat (Terengganu): Peak in the Eastern Range at 1480m, 4°.36′N: 102° 52′ E.

Gunung Inas (Perak): Highest peak in the Larut Hills at 1801 m, 5° 16′ N: 100° 56′ E.

Gunung Jerai (Kedah): Isolated mountain at 2107 m, 5° 47′ N: 100° 25′ E.

Gunung Kledang (Perak): Peak in the Kledang Range at 808 m, 4° 38′ N: 101° 01′ E.

Gunung Korbu (Perak/Pahang): Highest peak in the Titiwangsa Range at 2183 m, 4° 41′ N: 101° 18′ E.

Gunung Lawit (Terengganu): Peak in the Eastern Range at 1519m, 5° 25′ N: 102° 34′ E

Gunung Ledang (Johor): Isolated mountain at 1276 m, 2° 22′ N: 102° 36′ E.

Gunung Mandi Angin (Terengganu): Isolated mountain within Taman Negara at 1460 m, 4° 40′ N: 102° 50′ E.

Gunung Menuang Gasing (Selangor): Peak in the Titiwangsa Range at 1493 m, 3° 16′ N: 101° 54′ E.

Gunung Padang (Terengganu): Peak in the Eastern Range at 1314 m, 4° 53′ N: 102° 49′ E.

Gunung Perlis (Pahang): Isolated mountain within Taman Negara at 1284 m, 4° 43′ N: 102° 21′ E.

Gunung Rabong (Kelantan): Isolated mountain at 1538 m, 4° 50′ N: 102° 04′ E.

Gunung Semanggol (Perak): Isolated hill in the Kerian District at 390 m, 4° 56 N: 100° 40′ E.

Gunung Tahan (Pahang): Large isolated mountain within Taman Negara. The highest peak in Peninsular Malaysia at 2187 m, 4° 38′ N: 102° 14′ E.

Gunung Tampin (Negeri Sembilan): Peak in the Titiwangsa Range at 764 m, 2° 31′ N: 102° 13′ E.

Gunung Tapis (Pahang): Peak in the Eastern Range at 1511 m, 4° 01′ N: 102° 54′ E.

Gunung Telapa Buruk (Negeri Sembilan) Peak in the Titiwangsa Range at 1193 m, 2° 50′ N: 102° 04′ E.

Gunung Ulu Kali (Pahang/Selangor): Highest peak in the Genting Highlands at 1774 m, 3° 26′ N: 101° 46′ E.

Singapore

Bukit Timah: Isolated hill and the highest peak on Singapore island at 163 m, 1° 22′ N: 103° 44′ E.

Mount Faber (Singapore): Ridge in southern Singapore at 106 m, 1° 17′ N: 103° 49′ E.

Appendix 4 Islands and islets cited in the text

Peninsular Malaysia

Pulau Aur (Johor): Small island 67 km offshore in the South China Sea, at 2° 27′ N: 104° 32′ E.

Pulau Babi Besar (Johor): Small island 17 km offshore in the South China Sea, at 2° 27′ N: 103° 59′ E.

Pulau Jarak (Perak): Small island 62 km offshore, at 3° 59′ N: 100° 00′ E.

Pulau Kapas (Terengganu): Small island 5 km offshore, at 5° 13′ N: 103° 16′ E.

Pulau Ketam (Selangor): Large mangrove island and fishing settlement 10 km offshore, at 3° 02′ N: 101° 15′ E.

Pulau Langkawi (Kedah): The largest island in the Langkawi group, 25 km offshore, at 6° 22′ N: 99° 49′ E. The highest point is Gunung Raya at 881 m.

Pulau Pangkor (Perak): Small island 2 km offshore, at 4° 14′ N: 103° 34′ E.

Pulau Pemanggil (Johor): Small island 52 km offshore in the South China Sea, at 2° 35′ N: 104° 20′ E.

Pulau Perak (Kedah): Large rock 142 km offshore north-west of Pulau Pinang, at 5° 43′ N: 99° 00′ E.

Pulau Perhentian Besar (Terengganu): Small island 17.5 km offshore, at 5° 54′ N: 102° 44′ E.

Pulau Perhentian Kecil (Terengganu): Small island 17 km offshore, at 5° 54′ N: 102° 44′ E.

Pulau Pinang (Penang Island): Island state of Peninsular Malaysia together with Seberang Prai, east on the adjacent mainland. 5 km offshore, at 5° 20′ N: 100° 10′ E. The highest point is Bukit Bendara (Penang Hill) at 771 m.

Pulau Pisang (Johor): Small island 13 km offshore in the Straits of Melaka, at 1° 28′ N: 103° 16′ E.

Pulau Pintu Gedung (Selangor): Large estuarine mangrove islet at 2° 25′ N: 101° 15′ E.

Pulau Redang (Terengganu): The largest island of the Redang group, 26 km offshore at 5° 47′ N: 103° 01′ E.

Pulau Rembia (Perak): Small island in the Sembilan group, 18 km offshore, at 4° 02′ N: 100° 33′ E.

Pulau Rengis (Pahang): Small islet off the west coast of Pulau Tioman at 2° 49′ N: 104° 08′ E.

Pulau Sembilan (Perak): A group of islands 17 km offshore between 4° 00′ N– 4° 08′ N: 100 30′ E –100 36′ E.

Pulau Serenggeh (Terengganu): Small island 18 km offshore, at 5° 57′ N: 102° 40′ E.

Pulau Tengah (Selangor): Small mangrove island 10 km offshore, at 3° 00′ N: 101° 15′ E.

Pulau Tenggol (Terengganu): Small island 28 km offshore, at 4° 48′ N: 103° 41′ E.

Pulau Tinggi (Johor): Small island 14 km offshore in the South China Sea at 2° 18′ N: 104° 07′ E.

Pulau Tioman (Pahang): The largest island of the Tioman group 39 km offshore, at 2° 47′ N: 104° 10′ E. The highest point is Gunung Kajang at 939 m.
Pulau Yu (Johor): Small island 22.5 km offshore in the South China Sea at 2° 07′ N: 104° 15′ E.

Singapore

Pulau Bukum: Small island 5 km south of the main island at 1° 14′ N: 103° 46′ E.
Pulau Busing: Small island 7 km south of the main island at 1° 14′ N: 103° 45′ E.
Pulau Hantu: Small island 7 km south of the main island at 1° 13′ N: 103° 45′ E.
Pulau Salu: Small island 10 km south of the main island at 1° 13′ N: 103° 43′ E.
Pulau Semakau: Small island 7.5 km south of the main island at 1° 13′ N: 103° 46′ E.
Pulau Sudong: Small island 10 km south of the main island at 1° 12′ N: 103° 44′ E.
Pulau Tekong Besar: Large island 4 km north-east of the main island at 1° 25′ N: 104° 03′ E.
Pulau Tekong Kecil: Small island 4 km north-east of the main island at 1° 25′ N: 104° 01′ E.
Pulau Ubin: Large island 2 km north-east of the main island at 1° 25′ N: 103° 58′ E.
Sentosa Island: Small island 2 km south of the main island at 1° 15′ N: 103° 50′ E.
St John's Island: Small island 4 km south-east of Pulau Sentosa at 1° 14′ N: 103° 51′ E.

Appendix 5 Conservation bodies, ornithological clubs, publications, and museums

Department of Wildlife and National Parks, Peninsular Malaysia

The law enforcement agency of wildlife laws in Peninsular Malaysia. Carries out bird-ringing projects and captive breeding of pheasants. Also conducts educational programmes on wildlife awareness and conservation. Maintains a bird skin collection in its museum. Publishes a quarterly bulletin *Berita Perhilitan* and a journal, *The Journal of Wildlife and Parks*.

Address: Km 10, Jalan Cheras, 56100 Kuala Lumpur, Malaysia.
Website: http://www.wildlife.gov.my Email: pakp@wildlife.gov.my.

National Parks Board, Singapore

The National Parks Board (NParks) is primarily responsible for parks, roadside planting and horticultural maintenance in Singapore, as well as maintaining all nature reserves and protected areas and safeguarding the flora and fauna within them.

Address: Singapore Botanic Gardens, 1 Cluny Road, Singapore 259569.
Website: http://www.nparks.gov.sg Email: nparks_mailbox@nparks.gov.sg.

The Malaysian Nature Society (MNS)

The oldest non-governmental organization in Malaysia. Established in 1940, it promotes nature conservation as well as conservation awareness. Organizes scientific expeditions, field trips, workshops, and conferences. Also helps to set up nature clubs in schools throughout the country. MNS publishes a quarterly journal, *The Malayan Nature Journal* and a magazine, *The Malaysian Naturalist*. Has a Bird Conservation Council (MNS-BCC) which focuses on conservation of birds and their habitats. The MNS-BCC publishes a quarterly bulletin, *Suara Enggang*. MNS is the BirdLife International Partner in Malaysia.

Address: P.O. Box 10750, 50724 Kuala Lumpur, Malaysia.
Website: http://www.mns.org.my Email: mns@mns.org.my

The Nature Society (Singapore)

The main non-governmental organization in Singapore. Promotes nature conservation as well as nature education. Has an active Bird Group which compiles and updates data on all bird species occurring in Singapore as well as the neighbouring Malaysian state of Johor. Also organizes field trips. Publishes a quarterly bulletin, *Singapore Avifauna* and a journal *Iora*. The Nature Society (Singapore) is the BirdLife International Partner in Singapore.

Address: 510, Geylang Road, #02–05, The Sunflower, Singapore 259569.
Website: http://www.nss.org.sg Email: contact@nss.org.sg.

World Wide Fund for Nature Malaysia (WWFM)

Promotes and provides funds for nature conservation in Malaysia. Conducts surveys related to conservation. Also promotes nature education in schools. Publishes a quarterly magazine *Green Heart*, a magazine on conservation as well as an annual report which gives an overview of all activities carried out within a calendar year.

Address: 49, Jalan SS23/15, 47301 Petaling Jaya, Selangor, Malaysia.
Website: www.wwf.org.my Email: wwf@org.my

Wetlands International

An international body concerned largely with sustaining and restoring wetlands, their resources and biodiversity for future generations. Regularly conducts surveys and monitors wetlands. Conducts counts on migratory shorebirds as well as other wetland birds. Publishes a newsletter *Wetlands*.

Address: 3A31, Block A, Kelana Centre Point, Kelana Jaya, Jalan SS7/19, 47301 Petaling Jaya, Selangor, Malaysia.
Website: www.wetlands.org Email: malaysia@wetlands.org.my

BirdLife International

International ornithological body which studies bird populations as well as habitats throughout the world with special emphasis on conservation. It acts as the Bird Specialist Group of the IUCN (World Conservation Union) and is instrumental in preparing the International Red Data Book of endangered birds. Publishes a number of books, periodicals and lists which includes *Birds to Watch*, a checklist of globally endangered species of birds. Coordinates the compilation, collation and publication of Important Bird Areas (IBA) directories throughout the world.

Address: Wellbrook Court, Girton Road, Cambridge CB3 0NA, United Kingdom. Website: www.birdlife.org Email: birdlife@birdlife.org

Oriental Bird Club

Founded in 1985, the Oriental Bird Club promotes interest in birds of the Oriental region as well as their conservation. Publishes an annual journal, *Forktail* and a bulletin *Birding ASIA*. Has representatives in both Malaysia and Singapore.

Address: P.O. Box 324, Bedford, MK42 0WG. UK.
Website: www.orientalbirdclub.org. Email: mail@orientalbirdclub.org.

Museums and collections

Raffles Museum of Biodiversity Research, National University of Singapore

Has an excellent collection and maintains the entire Raffles collection of bird skins from South-East Asia. Publishes the *Raffles Bulletin of Zoology*.

Address: Department of Biological Sciences, Faculty of Science, Science Building (S6) Level 3, Science Drive 2, Singapore 117546. Website: http://rmbr.nus.edu.sg

The *Raffles Bulletin of Zoology* is available online at http://rmbr.nus.edu.sg/rbz/. For Zoological Reference Collection research visits, contact Mr Lim Kok Peng, Kelvin at kelvinlim@nus.edu.sg

Institute of Biological Sciences, Faculty of Science, University of Malaya

Maintains a limited collection of bird skins in its museum.

Address: 50603 Kuala Lumpur, Malaysia. Website: http://biology.um.edu.my

Appendix 6 A checklist of the birds of Peninsular Malaysia and Singapore

The categories below follow *A Checklist of the Birds of Peninsular Malaysia* (MNS-Bird Conservation Council, 2010) with adaptations. This is the source for Peninsular Malaysia. The source for Singapore is from Wang and Hails (2007).

No.	Family	Common name	Scientific name	Status			Notes
				PM	SIN	GCS	
1	**Phasianidae**	Long-billed Partridge	*Rhizothera longirostris*	R, 2	–	NT	
2		Black Partridge	*Melanoperdix niger*	R, 1	L. EXT	VU	
3		Blue-breasted Quail	*Coturnix chinensis*	R, 3	R, 2		
4		Malaysian Partridge	*Arborophila campbelli*	R, 3	–		
5		Chestnut-necklaced Partridge	*Arborophila charltonii*	R, 2	–	NT	
6		Ferruginous Partridge	*Caloperdix oculeus*	R, 2	–	NT	
7		Crested Partridge	*Rollulus rouloul*	R, 4	–	NT	
8		Red Junglefowl	*Gallus gallus*	R, 5	R, 4		
9		Crestless Fireback	*Lophura erythrophthalma*	R, 2	L. EXT	VU	
10		Crested Fireback	*Lophura ignita*	R, 4	–	NT	
11		Mountain Peacock-pheasant	*Polyplectron inopinatum*	R, 3	–	VU	
12		Malaysian Peacock-pheasant	*Polyplectron malacense*	R, 3	L. EXT	VU	
13		Crested Argus	*Rheinardia ocellata*	R, 1	–	NT	
14		Great Argus	*Argusianus argus*	R, 4	L. EXT	NT	
15		Green Peafowl	*Pavo muticus*	L. EXT	–	VU	
16	**Anatidae**	Wandering Whistling-duck	*Dendrocygna arcuata*	?	F, 2		
17		Lesser Whistling-duck	*Dendrocygna javanica*	R, 4	R, 3		
18		White-winged Duck	*Cairina scutulata*	V, 1	–	EN	
19		Cotton Pygmy-goose	*Nettapus coromandelianus*	R, 3	D, 2		
20		Gadwall	*Anas strepera*	–	V, 1		

No.	Family	Common name	Scientific name	Status			Notes
				PM	SIN	GCS	
21		Eurasian Wigeon	*Anas penelope*	V, 1	V, 1		
22		Northern Shoveler	*Anas clypeata*	M, 2	M, 2		
23		Northern Pintail	*Anas acuta*	M, 2	V, 1		
24		Garganey	*Anas querquedula*	M, 3	M, 3		
25		Eurasian Teal	*Anas crecca*	V, 1	V, 1		
26		Tufted Duck	*Aythya fuligula*	V, 1	V, 1		
27	**Procellariidae**	Streaked Shearwater	*Calonectris leucomelas*	M, 1	–		
28		Wedge-tailed Shearwater	*Puffinus pacificus*	M, 1	V, 1		
29		Short-tailed Shearwater	*Puffinus tenuirostris*	M, 1	–		
30		Bulwer's Petrel	*Bulweria bulwerii*	M, 1	–		
31	**Hydrobatidae**	Wilson's Storm-petrel	*Oceanites oceanicus*	M, 1	–		
32		Swinhoe's Storm-petrel	*Oceanodroma monorhis*	M, 2	M, 2		
33	**Podicipedidae**	Little Grebe	*Tachybaptus ruficollis*	R, M, 4	R, 2		
34	**Ciconiidae**	Milky Stork	*Mycteria cinerea*	R, 2	–	**VU**	
35		Painted Stork	*Mycteria leucocephala*	F, 3	–	**NT**	
36		Asian Openbill	*Anastomus oscitans*	V, 1	–		
37		Woolly-necked Stork	*Ciconia episcopus*	V, 1	–		
38		Sunda Stork	*Ciconia stormi*	R, ?	–	**EN**	
39		Lesser Adjutant	*Leptoptilos javanicus*	R, 4	D, 1	**VU**	
40	**Threskiornithidae**	Black-headed Ibis	*Threskiornis melanocephalus*	V, 1		**NT**	
41		Glossy Ibis	*Plegadis falcinellus*	–	V, 1		
42	**Ardeidae**	Great Bittern	*Botaurus stellaris*	M, 1	M, 1		
43		Yellow Bittern	*Ixobrychus sinensis*	R, M, 4	R, M, 3		

No.	Family	Common name	Scientific name	Status			Notes
				PM	SIN	GCS	
44		Chestnut Bittern	*Ixobrychus eurhythmus*	M, 3	M, 3		
45		Cinnamon Bittern	*Ixobrychus cinnamomeus*	R, M, 4	R, 4, M, 2		
46		Black Bittern	*Dupetor flavicollis*	M, 3	M, 3		
47		Rufous-necked Night-heron	*Gorsachius melanolophus*	M, 2	M, 2		
48		Black-crowned Night-heron	*Nycticorax nycticorax*	R, 5	R, 3		
49		Little Heron	*Butorides striata*	R, M, 5	R, 4, M, 2		
50		Indian Pond- heron	*Ardeola grayii*	M, 3	–		
51		Chinese Pond- heron	*Ardeola bacchus*	M, 6	M, 3		
52		Javan Pond -heron	*Ardeola speciosa*	M, 4	–		
53		Cattle Egret	*Bubulcus ibis*	R, M, 5	R, M, 4		
54		Grey Heron	*Ardea cinerea*	R, 5	R, 4		
55		Great-billed Heron	*Ardea sumatrana*	R, 2	R, 2		
56		Purple Heron	*Ardea purpurea*	R, M, 5	R, 4		
57		Great Egret	*Casmerodius albus*	R, 1 M, 5	M, 4		
58		Intermediate Egret	*Mesophoyx intermedia*	M, 4	M, 3		
59		Little Egret	*Egretta garzetta*	R, 3, M, 6	M, 5		
60		Pacific Reef-egret	*Egretta sacra*	R, 4	R, 3		
61		Chinese Egret	*Egretta eulophotes*	M, 3	M, 3	**VU**	
62	**Fregatidae**	Christmas Island Frigatebird	*Fregata andrewsi*	M, 2	M, 2	**CR**	
63		Great Frigatebird	*Fregata minor*	V, 1	–		
64		Lesser Frigatebird	*Fregata ariel*	M, 4	M, 2		
65	**Pelecanidae**	Great White Pelican	*Pelecanus onocrotalus*	V, 1	?		

No.	Family	Common name	Scientific name	Status			Notes
				PM	SIN	GCS	
66		Spot-billed Pelican	*Pelecanus philippensis*	V, 1	–		
67	**Sulidae**	Masked Booby	*Sula dactylatra*	V, 1	–		
68		Red-footed Booby	*Sula sula*	V, 1	–		
69		Brown Booby	*Sula leucogaster*	R, M, 2	D, 1		
70	**Phalacrocoracidae**	Little Cormorant	*Phalacrocorax niger*	M, 2	–		
71		Great Cormorant	*Phalacrocorax carbo*	M, 2	–		
72	**Anhingidae**	Oriental Darter	*Anhinga melanogaster*	V, 1	–		
73	**Falconidae**	Black-thighed Falconet	*Microhierax fringillarius*	R, 4	D, 2		
74		Lesser Kestrel	*Falco naumanni*	–	V, 1		
75		Common Kestrel	*Falco tinnunculus*	M, 2	M, 2		
76		Amur Falcon	*Falco amurensis*	V, 1	V, 1		
77		Eurasian Hobby	*Falco subbuteo*	V, 1	–		
78		Oriental Hobby	*Falco severus*	–	V, 1		
79		Peregrine Falcon	*Falco peregrinus*	R, M, 4	M, 3		
80	**Accipitridae**	Osprey	*Pandion haliaetus*	M, 1	M, 1		
81		Jerdon's Baza	*Aviceda jerdoni*	R, 2	D, 1		
82		Black Baza	*Aviceda leuphotes*	M, 4	M, 4		
83		Oriental Honey buzzard	*Pernis ptilorhyncus*	R, M, 5	M, 5		
84		Bat Hawk	*Macheiramphus alcinus*	R, 3	D, 2		
85		Black-winged Kite	*Elanus caeruleus*	R, 5	R, 4		
86		Black Kite	*Milvus migrans*	M, 4	M, 3		
87		Brahminy Kite	*Haliastur indus*	R, 6	R, 5		
88		White-bellied Sea-eagle	*Haliaeetus leucogaster*	R, 5	R, 4		

No.	Family	Common name	Scientific name	Status			Notes
				PM	SIN	GCS	
89		Lesser Fish-eagle	*Ichthyophaga humilis*	R, 4	_	**NT**	
90		Grey-headed Fish-eagle	*Ichthyophaga ichthyaetus*	R, 3	R, 2	**NT**	
91		White-rumped Vulture	*Gyps bengalensis*	L. EXT	_	**CR**	
92		Indian Vulture	*Gyps indicus*	V, 1	_	**CR**	
93		Himalayan Griffon	*Gyps himalayensis*	V, 1	V, 1		
94		Red-headed Vulture	*Sarcogyps calvus*	L. EXT	_	**CR**	
95		Cinereous Vulture	*Aegypius monachus*	V, 1	_	**NT**	
96		Short-toed Snake-eagle	*Circaetus gallicus*	M, 2	M, 2		
97		Crested Serpent-eagle	*Spilornis cheela*	R, 6	R, 2		
98		Eastern Marsh-harrier	*Circus spilonotus*	M, 4	M, 3		
99		Hen Harrier	*Circus cyaneus*	M, 1	M, 1		
100		Pied Harrier	*Circus melanoleucos*	M, 3	M, 2		
101		Crested Goshawk	*Accipiter trivirgatus*	R, 5	R, 2		
102		Shikra	*Accipiter badius*	M, 1	_		
103		Chinese Sparrowhawk	*Accipiter soloensis*	M, 4	M, 3		
104		Japanese Sparrowhawk	*Accipiter gularis*	M, 5	M, 4		
105		Besra	*Accipiter virgatus*		V, 1		
106		Eurasian Sparrowhawk	*Accipiter nisus*	V, 1	_		
107		Grey-faced Buzzard	*Butastur indicus*	M, 4	M, 3		
108		Common Buzzard	*Buteo buteo*	M, 3	M, 3		
109		Black Eagle	*Ictinaetus malayensis*	R, 4	_		
110		Greater Spotted Eagle	*Aquila clanga*	M, 3	M, 2	**VU**	
111		Steppe Eagle	*Aquila nipalensis*	M, 2	M, 2		

No.	Family	Common name	Scientific name	Status			Notes
				PM	SIN	GCS	
112		Eastern Imperial Eagle	*Aquila heliaca*	M, 2	M, 2	**VU**	
113		Booted Eagle	*Hieraaetus pennatus*	M, 2	M, 2		
114		Rufous-bellied Eagle	*Lophptrorchis kienerii*	R, M, 4	M, 2		
115		Changeable Hawk-eagle	*Nisaetus cirrhatus*	R, 5	R, 3		
116		Mountain Hawk-eagle	*Nisaetus nipalensis*	R, 2	–		
117		Blyth's Hawk-eagle	*Nisaetus alboniger*	R, 4	D, 2		
118		Wallace's Hawk-eagle	*Nisaetus nanus*	R, 3	–		
119	**Rallidae**	Red-legged Crake	*Rallina fasciata*	R, M, 3	R, M, 3		
120		Slaty-legged Crake	*Rallina eurizonoides*	M, 2	M, 2		
121		Slaty-breasted Rail	*Gallirallus striatus*	R, 4	R, 4		
122		White-breasted Waterhen	*Amaurornis phoenicurus*	R, M, 6	R, M, 6		
123		Baillon's Crake	*Porzana pusilla*	M, 3	M, 2		
124		Ruddy-breasted Crake	*Porzana fusca*	R, M, 4	R, M, 3		
125		Band-bellied Crake	*Porzana paykullii*	M, 1	–	**NT**	
126		White browed Crake	*Porzana cinerea*	R, 5	R, 5		
127		Watercock	*Gallicrex cinerea*	R, M, 4	R, M, 3		
128		Purple Swamphen	*Porphyrio porphyrio*	R, 5	R, 2		
129		Common Moorhen	*Gallinula chloropus*	R, 5	R, 4		
130		Common Coot	*Fulica atra*	V, 1	V, 1		
131	**Heliornithidae**	Masked Finfoot	*Heliopais personatus*	M, 2	V, 1	**EN**	
132	**Gruidae**	Sarus Crane	*Grus antigone*	L. EXT	–	**VU**	
133	**Turnicidae**	Small Buttonquail	*Turnix sylvaticus*	R, 2	–		
134		Barred Buttonquail	*Turnix suscitator*	R, 5	R, 4		

No.	Family	Common name	Scientific name	Status			Notes
				PM	SIN	GCS	
135	**Burhinidae**	Beach Thick-knee	*Esacus neglectus*	?	R, 1	**NT**	
136	**Haematopodidae**	Eurasian Oystercatcher	*Haematopus ostralegus*	V, 1	–		
137	**Dromadidae**	Crab Plover	*Dromas ardeola*	V, 1	–		
138	**Recurvirostridae**	Black-winged Stilt	*Himantopus himantopus*	R. M, 5	M, 4		
139	**Charadriidae**	Yellow-wattled Lapwing	*Vanellus malabaricus*	V, 1	–		
140		Grey-headed Lapwing	*Vanellus cinereus*	M, 3	–		
141		Red-wattled Lapwing	*Vanellus indicus*	R, 4	R, 2		
142		Pacific Golden Plover	*Pluvialis fulva*	M, 6	M, 5		
143		Grey Plover	*Pluvialis squatarola*	M, 4	M, 4		
144		Common Ringed Plover	*Charadrius hiaticula*	V, 1	V, 1		
145		Long-billed Plover	*Charadrius placidus*	V, 1	–		
146		Little Ringed Plover	*Charadrius dubius*	M, 5	M, 5		
147		Kentish Plover	*Charadrius alexandrinus*	M, 4	M, 4		
148		Malaysian Plover	*Charadrius peronii*	R, 3	R, 2	**NT**	
149		Lesser Sand Plover	*Charadrius mongolus*	M, 6	M, 5		
150		Greater Sand Plover	*Charadrius leschenaultii*	M, 5	M, 4		
151		Oriental Plover	*Charadrius veredus*	M, 1	M, 1		
152	**Rostratulidae**	Greater Paintedsnipe	*Rostratula benghalensis*	R, 4	R, 4		
153	**Jacanidae**	Pheasant-tailed Jacana	*Hydrophasianus chirurgus*	M, 3	M, 3		
154		Bronze-winged Jacana	*Metopidius indicus*	V, 1	–		
155	**Scolopacidae**	Eurasian Woodcock	*Scolopax rusticola*	M, 1	V, 1		
156		Pintail Snipe	*Gallinago stenura*	M, 4	M, 4		
157		Swinhoe's Snipe	*Gallinago megala*	M, 3	M, 3		

No.	Family	Common name	Scientific name	Status			Notes
				PM	SIN	GCS	
158		Common Snipe	*Gallinago gallinago*	M, 3	M, 3		
159		Asian Dowitcher	*Limnodromus semipalmatus*	M, 3	M, 3	**NT**	
160		Black-tailed Godwit	*Limosa limosa*	M, 4	M, 4	**NT**	
161		Bar-tailed Godwit	*Limosa lapponica*	M, 5	M, 5		
162		Little Curlew	*Numenius minutus*	V, 1	V, 1		
163		Whimbrel	*Numenius phaeopus*	M, 5	M, 5		
164		Eurasian Curlew	*Numenius arquata*	M, 5	M, 5		
165		Far Eastern Curlew	*Numenius madagascariensis*	M, 4	M, 3	**NT**	
166		Spotted Redshank	*Tringa erythropus*	M, 1	M, 1		
167		Common Redshank	*Tringa totanus*	M, 6	M, 6		
168		Marsh Sandpiper	*Tringa stagnatilis*	M, 5	M, 5		
169		Common Greenshank	*Tringa nebularia*	M, 5	M, 5		
170		Spotted Greenshank	*Tringa guttifer*	M, 3	M, 2	**EN**	
171		Green Sandpiper	*Tringa ochropus*	M, 2	M, 2		
172		Wood Sandpiper	*Tringa glareola*	M, 5	M, 4		
173		Grey-tailed Tattler	*Tringa brevipes*	M, 3	M, 3		
174		Terek Sandpiper	*Xenus cinereus*	M, 5	M, 5		
175		Common Sandpiper	*Actitis hypoleucos*	M, 7	M, 6		
176		Ruddy Turnstone	*Arenaria interpres*	M, 5	M, 5		
177		Great Knot	*Calidris tenuirostris*	M, 5	M, 5		
178		Red Knot	*Calidris canutus*	M, 3	M, 3		
179		Sanderling	*Calidris alba*	M, 3	M, 3		
180		Red-neck Stint	*Calidris ruficollis*	M, 5	M, 5		

No.	Family	Common name	Scientific name	Status			Notes
				PM	SIN	GCS	
181		Little Stint	*Calidris minuta*	M, 3	–		
182		Temminck's Stint	*Calidris temminckii*	M, 3	M, 3		
183		Long-toed Stint	*Calidris subminuta*	M, 4	M, 4		
184		Pectoral Sandpiper	*Calidris melanotos*	V, 1	V, 1		
185		Sharp-tailed Sandpiper	*Calidris acuminata*	V, 1	V, 1		
186		Curlew Sandpiper	*Calidris ferruginea*	M, 5	M, 4		
187		Dunlin	*Calidris alpina*	M, 2	M, 2		
188		Spoon-billed Sandpiper	*Eurynorhynchus pymeus*	V, 1	V, 1	**CR**	
189		Broad-billed Sandpiper	*Limicola falcinellus*	M, 4	M, 4		
190		Ruff	*Philomachus pugnax*	M, 3	M, 3		
191		Red-necked Phalarope	*Phalaropus lobatus*	M, 2	V, 1		
192	**Glareolidae**	Oriental Pratincole	*Glareola maldivarum*	R, M, 4	M, 4		
193		Small Pratincole	*Glareola lactea*	–	V, 1		
194	**Laridae**	Heuglin's Gull	*Larus heuglini*	V, 1	–		
195		Brown-headed Gull	*Larus brunnicephalus*	M, 3	V, 1		
196		Common Black-headed Gull	*Larus ridibundus*	M, 2	M, 2		
197		Laughing Gull	*Larus atricilla*	V, 1	–		
198		Gull-billed Tern	*Gelochelidon nilotica*	M, 4	M, 2		
199		Caspian Tern	*Hydroprogne caspia*	M, 4	M, 2		
200		Lesser Crested Tern	*Thalasseus bengalensis*	M, 3	M, 3		
201		Great Crested Tern	*Thalasseus bergii*	M, 4	M, 4		
202		Roseate Tern	*Sterna dougallii*	R, 3	D, 2		
203		Black-naped Tern	*Sterna sumatrana*	R, 4	R, 3		

No.	Family	Common name	Scientific name	Status			Notes
				PM	SIN	GCS	
204		Common Tern	*Sterna hirundo*	M, 4	M, 3		
205		Little Tern	*Sterna albifrons*	R, M, 5	R, M, 4		
206		Aleutian Tern	*Sterna aleutica*	M, 2	M, 2		
207		Bridled Tern	*Sterna anaethetus*	R, 4	D, 3		
208		Sooty Tern	*Sterna fuscata*	V, 1	–		
209		Whiskered Tern	*Chlidonias hybrida*	M, 5	M, 3		
210		White-winged Tern	*Chlidonias leucopterus*	M, 5	M, 4		
211		Brown Noddy	*Anous stolidus*	R, 2	–		
212	**Stercorariidae**	Pomarine Jaeger	*Stercorarius pomarinus*	M, 2	–		
213		Parasitic Jaeger	*Stercorarius parasiticus*	V, 1	–		
214		Long-tailed Jaeger	*Stercorarius longicaudus*	M, 2	–		
215	**Columbidae**	Rock Pigeon	*Columba livia*	F, 7	F, 6		
216		Red Collared-dove	*Streptopelia tranquebarica*	R, 3	R, 4		
217		Spotted Dove	*Streptopelia chinensis*	R, 7	R, 6		
218		Barred Cuckoo-dove	*Macropygia unchall*	R, 3	–		
219		Little Cuckoo-dove	*Macropygia ruficeps*	R, 5	–		
220		Emerald Dove	*Chalcophaps indica*	R, 6	R, 3		
221		Zebra Dove	*Geopelia striata*	R, 7	R, 6		
222		Nicobar Pigeon	*Caloenas nicobarica*	R, 2	–	NT	
223		Cinnamon-headed Green-pigeon	*Treron fulvicollis*	R, 3	D, 2	NT	
224		Little Green-pigeon	*Treron olax*	R, 4	R, 2		
225		Pink-necked Green-pigeon	*Treron vernans*	R, 6	R, 5		
226		Orange-breasted Green-pigeon	*Treron bicinctus*	R, 3	–		

No.	Family	Common name	Scientific name	Status			Notes
				PM	SIN	GCS	
227		Thick-billed Green-pigeon	*Treron curvirostra*	R, 5	R, 3		
228		Large Green-pigeon	*Treron capellei*	R, 4	_	**VU**	
229		Yellow-vented Green-pigeon	*Treron seimundi*	R, 3	_		
230		Wedge-tailed Green-pigeon	*Treron sphenurus*	R, 2	_		
231		Jambu Fruit-dove	*Ptilinopus jambu*	R, 4	D, 3	**NT**	
232		Green Imperial-pigeon	*Ducula aenea*	R, 3	D, 2		
233		Mountain Imperial-pigeon	*Ducula badia*	R, 5	_		
234		Pied Imperial-pigeon	*Ducula bicolor*	R, 4	D, 2		
235	**Psittacidae**	Blue-crowned Hanging-parrot	*Loriculus galgulus*	R, 6	R, 3		
236		Tanimbar Corella	*Cacatua goffini*	_	F, 4	**NT**	
237		Yellow- crested Cockatoo	*Cacatua sulphurea*	_	F, 2		
238		Rose-ringed Parakeet	*Psittacula krameri*	F, 1	F, 1		
239		Red-breasted Parakeet	*Psittacula alexandrii*	_	F, 4		
240		Reddish-cheeked Parakeet	*Psittacula longicauda*	R, 5	R, 4	**NT**	
241		Blue-rumped Parrot	*Psittinus cyanurus*	R, 4	R, 2	**NT**	
242	**Cuculidae**	Pied Cuckoo	*Clamator jacobinus*	V, 1	_		
243		Chestnut-winged Cuckoo	*Clamator coromandus*	M, 3	M, 3		
244		Large Hawk-cuckoo	*Hierococcyx sparveriodes*	R, M, 4	M, 3		
245		Moustached Hawk-cuckoo	*Hierococcyx vagans*	R, 2	_	**NT**	
246		Hodgson's Hawk-cuckoo	*Hierococcyx fugax*	R, M, 3	D1, M, 3		
247		Indian Cuckoo	*Cuculus micropterus*	R, M, 5	M, 4		
248		Oriental Cuckoo	*Cuculus saturatus*	R, M, 3	V, 1		
249		Banded Bay Cuckoo	*Cacomantis sonnerattii*	R, 4	R, 3		

No.	Family	Common name	Scientific name	Status			Notes
				PM	SIN	GCS	
250		Plaintive Cuckoo	*Cacomantis merulinus*	R, 5	R, 3		
251		Rusty-breasted Cuckoo	*Cacomantis sepulcralis*	R, 4	R, 3		
252		Little Bronze-cuckoo	*Chrysococcyx minutillus*	R, 4	R, 4		
253		Horsfield's Bronze-cuckoo	*Chrysococcyx basalis*	V, 1	M, 2		
254		Asian Emerald Cuckoo	*Chrysococcyx maculatus*	M, 3	–		
255		Violet Cuckoo	*Chrysococcyx xanthorhynchus*	R, M, 4	R, M, 3		
256		Drongo Cuckoo	*Surniculus lugubris*	R, M, 6	R, M, 3		
257		Asian Koel	*Eudynamys scolopaceus*	R, M, 7	R, M, 4		
258		Black-bellied Malkoha	*Rhopodytes diardii*	R, 4	L. EXT	**NT**	
259		Chestnut-bellied Malkoha	*Rhopodytes sumatranus*	R, 3	R, 2	**NT**	
260		Green-billed Malkoha	*Rhopodytes tristis*	R, 5	–		
261		Raffles's Malkoha	*Rhinortha chlorophaea*	R, 5	L. EXT		
262		Red-billed Malkoha	*Zanclostomus javanicus*	R, 4	L. EXT		
263		Chestnut-breasted Malkoha	*Zanclostomus curvirostris*	R, 5			
264		Short-toed Coucal	*Centropus rectunguis*	R, 2	–	**VU**	
265		Greater Coucal	*Centropus sinensis*	R, 6	R, 3		
266		Lesser Coucal	*Centropus bengalensis*	R, 5	R, 4		
267	**Tytonidae**	Barn Owl	*Tyto alba*	R, 5	R, 3		
268		Oriental Bay-owl	*Phodilus badius*	R, 3	L. EXT		
269	**Strigidae**	White-fronted Scops-owl	*Otus sagittatus*	R, 2	–	**VU**	
270		Reddish Scops-owl	*Otus rufescens*	R, 3	–	**NT**	
271		Mountain Scops-owl	*Otus spilocephalus*	R, 4	–		

No.	Family	Common name	Scientific name	Status			Notes
				PM	SIN	GCS	
272		Collared Scops-owl	*Otus bakkamoena*	R, 5	R, 5		
273		Oriental Scops-owl	*Otus sunia*	M, 4	M, 2		
274		Barred Eagle-owl	*Bubo sumatranus*	R, 4	D, 1		
275		Dusky Eagle-owl	*Bubo coromandus*	R, 2	_		
276		Brown Fish-owl	*Ketupa zeylonensis*	R, 2	_		
277		Buffy Fish-owl	*Ketupa ketupu*	R, 6	R, 2		
278		Spotted Wood-owl	*Strix seloputo*	R, 6	R, 5		
279		Brown Wood-owl	*Strix leptogrammica*	R, 4	_		
280		Collared Owlet	*Glaucidium brodiei*	R, 4	_		
281		Brown Boobook	*Ninox scutulata*	R, M, 5	R, M, 3		
282		Short-eared Owl	*Asio flammeus*	V, 1	M, 1		
283	**Podargidae**	Large Frogmouth	*Batrachostomus auritus*	R, 3	_	**NT**	
284		Gould's Frogmouth	*Batrachostomus stellatus*	R, 4	L.EXT	**NT**	
285		Javan Frogmouth	*Batrachostomus javensis*	R, 4	_		
286	**Caprimulgidae**	Malaysian Eared-nightjar	*Eurostopodus temminckii*	R, 5	R, 3		
287		Great Eared-nightjar	*Eurostopodus macrotis*	R, 4	_		
288		Grey Nightjar	*Caprimulgus indicus*	M, 4	M, 2		
289		Large-tailed Nightjar	*Caprimulgus macrurus*	R, 7	R, 6		
290		Savanna Nightjar	*Caprimulgus affinis*	R, 5	R, 5		
291	**Hemiprocnidae**	Grey-rumped Treeswift	*Hemiprocne longipennis*	R, 5	R, 4		
292		Whiskered Treeswift	*Hemiprocne comata*	R, 4	D, 1		
293	**Apodidae**	Waterfall Swift	*Hydrochous gigas*	R, 1	_	**NT**	
294		Glossy Swiftlet	*Collocalia esculenta*	R, 5	R, 2		

No.	Family	Common name	Scientific name	Status			Notes
				PM	SIN	GCS	
295		Himalayan Swiftlet	*Aerodramus brevirostris*	M, 3	?		
296		Black-nest Swiftlet	*Aerodramus maximus*	R, 5	R, 4		
297		Edible-nest Swiftlet	*Aerodramus fuciphaga*	R, 5	R, 4		
298		Silver-rumped Spinetail	*Rhaphidura leucopygialis*	R, 5	?		
299		White-throated Needletail	*Hirundapus caudacutus*	M, 2	M, 1		
300		Silver-backed Needletail	*Hirundapus cochinchinensis*	M, 3	M, 2		
301		Brown–backed Needletail	*Hirundapus giganteus*	R, M, 5	R, 4		
302		Asian Palm-swift	*Cypsiurus balasiensis*	R, 5	R, 5		
303		Fork-tailed Swift	*Apus pacificus*	M, 4	M, 4		
304		House Swift	*Apus affinis*	R, 7	R, 6		
305	**Trogonidae**	Red-naped Trogon	*Harpactes kasumba*	R, 4	L. EXT	NT	
306		Diard's Trogon	*Harpactes diardii*	R, 4	L. EXT	NT	
307		Cinnamon-rumped Trogon	*Harpactes orrhophaeus*	R, 3	_	NT	
308		Scarlet-rumped Trogon	*Harpactes duvaucelii*	R, 5	_	NT	
309		Orange-breasted Trogon	*Harpactes oreskios*	R, 4	_		
310		Red-headed Trogon	*Harpactes erythrocephalus*	R, 4	_		
311	**Coraciidae**	Indian Roller	*Coracias benghalensis*	R, 4	_		
312		Dollarbird	*Eurystomus orientalis*	R, M, 5	R, 1, M, 1		
313	**Alcedinidae**	Rufous-collared Kingfisher	*Actenoides concretus*	R, 3	L. EXT	NT	
314		Banded Kingfisher	*Lacedo pulchella*	R, 5	L. EXT		
315		Brown-winged Storkbill	*Pelargopsis amauroptera*	R, 3	_	NT	
316		Common Storkbill	*Pelargopsis capensis*	R, 5	R, 3		
317		Ruddy Kingfisher	*Halcyon coromanda*	R, M, 3	R, M, 2		

No.	Family	Common name	Scientific name	Status			Notes
				PM	SIN	GCS	
318		White-throated Kingfisher	*Halcyon smyrnensis*	R, 7	R, 6		
319		Black-capped Kingfisher	*Halcyon pileata*	M, 5	M, 3		
320		Collared Kingfisher	*Todiramphus chloris*	R, M, 6	R, M, 7		
321		Oriental Dwarf Kingfisher	*Ceyx erithaca*	R, M, 4	M, 3		
322		Blue-banded Kingfisher	*Alcedo euryzona*	R, 3	_	VU	
323		Blue-eared Kingfisher	*Alcedo meninting*	R, 4	R, 3		
324		Common Kingfisher	*Alcedo atthis*	R, M, 5	M, 5		
325	**Meropidae**	Red-bearded Bee-eater	*Nyctyornis amictus*	R, 4	_		
326		Blue-tailed Bee-eater	*Merops philippinus*	R, M, 6	R, M, 5		
327		Blue-throated Bee-eater	*Merops viridis*	R, M, 5	R, M, 5		
328		Chestnut-headed Bee-eater	*Merops leschenaulti*	R, 4	_		
329	**Upupidae**	Eurasian Hoopoe	*Upupa epops*	M, 2	M, 2		
330	**Bucerotidae**	Bushy-crested Hornbill	*Anorrhinus galeritus*	R, 4	_		
331		Oriental Pied Hornbill	*Anthracoceros albirostris*	R, 5	R2, D2		
332		Black Hornbill	*Anthracoceros malayanus*	R, 4	_	NT	
333		Great Hornbill	*Buceros bicornis*	R, 4	_	NT	
334		Rhinoceros Hornbill	*Buceros rhinoceros*	R, 4	L. EXT	NT	
335		Helmeted Hornbill	*Rhinoplax vigil*	R, 3	L. EXT	NT	
336		White-crowned Hornbill	*Berenicornis comatus*	R, 3	_	NT	
337		Wrinkled Hornbill	*Aceros corrugatus*	R, 3	_	NT	
338		Plain-pouched Hornbill	*Rhyticeros subruficollis*	R, 3	_	VU	
339		Wreathed Hornbill	*Rhyticeros undulatus*	R, 4	_		
340	**Ramphastidae**	Fire-tufted Barbet	*Psilopogon pyrrolophus*	R, 5	_		

No.	Family	Common name	Scientific name	Status			Notes
				PM	SIN	GCS	
341		Lineated Barbet	*Megalaima lineata*	R, 5	F, 2		
342		Gold-whiskered Barbet	*Megalaima chrysopogon*	R, 5			
343		Red-crowned Barbet	*Megalaima rafflesii*	R, 4	R, 3	**NT**	
344		Red-throated Barbet	*Megalaima mystacophanos*	R, 4	L. EXT	**NT**	
345		Golden-throated Barbet	*Megalaima franklinii*	R, 4	–		
346		Black-browed Barbet	*Megalaima oorti*	R, 5	–		
347		Yellow-crowned Barbet	*Megalaima henricii*	R, 4	L. EXT	**NT**	
348		Blue-eared Barbet	*Megalaima australis*	R, 5	L. EXT		
349		Coppersmith Barbet	*Megalaima haemacephala*	R, 6	R, 5		
350		Brown Barbet	*Calorhamphus fuliginosus*	R, 5	L. EXT		
351	**Indicatoridae**	Malaysian Honeyguide	*Indicator archipelagicus*	R, 2	–	**NT**	
352	**Picidae**	Eurasian Wryneck	*Jynx torquilla*	V, 1	–		
353		Speckled Piculet	*Picumnus innominatus*	R, 4	–		
354		Rufous Piculet	*Sasia abnormis*	R, 5	–		
355		Sunda Woodpecker	*Dendrocopos moluccensis*	R, 5	R, 5		
356		Grey-capped Woodpecker	*Dendrocopos canicapillus*	R, 4	L. EXT		
357		Rufous Woodpecker	*Celeus brachyurus*	R, 5	R, 3		
358		White-bellied Woodpecker	*Dryocopus javensis*	R, 4	R, 2		
359		Banded Yellownape	*Picus miniaceus*	R, 5	R, 4		
360		Crimson-winged Yellownape	*Picus puniceus*	R, 5	L. EXT		
361		Lesser Yellownape	*Picus chlorolophus*	R, 4	–		
362		Chequer-throated Yellownape	*Picus mentalis*	R, 5	L. EXT		
363		Greater Yellownape	*Picus flavinucha*	R, 4	–		

No.	Family	Common name	Scientific name	Status			Notes
				PM	SIN	GCS	
364		Laced Woodpecker	*Picus vittatus*	R, 4	R, 3		
365		Streak-breasted Woodpecker	*Picus viridanus*	R, 3	–		
366		Grey-faced Woodpecker	*Picus canus*	R, 1	–		
367		Olive-backed Woodpecker	*Dinopium rafflesii*	R, 3	L. EXT	**NT**	
368		Common Flameback	*Dinopium javanense*	R, 5	R, 4		
369		Greater Flameback	*Chrysocolaptes lucidus*	R, 3	L. EXT		
370		Bamboo Woodpecker	*Gecinulus viridis*	R, 3	–		
371		Maroon Woodpecker	*Blythipicus rubiginosus*	R, 5	L. EXT		
372		Bay Woodpecker	*Blythipicus pyrrhotis*	R, 3	–		
373		Orange-backed Woodpecker	*Reinwardtipicus validus*	R, 4	L. EXT		
374		Buff-rumped Woodpecker	*Meiglyptes tristis*	R, 5	L. EXT		
375		Buff-necked Woodpecker	*Meiglyptes tukki*	R, 4	L. EXT	**NT**	
376		Grey-and-Buff Woodpecker	*Hemicircus concretus*	R, 4	L. EXT		
377		Great Slaty Woodpecker	*Mulleripicus pulverulentus*	R, 4	L. EXT		
378	**Eurylaimidae**	Little Green-broadbill	*Calyptomena viridis*	R, 4	L. EXT	**NT**	
379		Black-and-Red Broadbill	*Cymbirhynchus macrorhynchus*	R, 4	L. EXT		
380		Long-tailed Broadbill	*Psarisomus dalhousiae*	R, 4	–		
381		Silver-breasted Broadbill	*Serilophus lunatus*	R, 4	–		
382		Banded Broadbill	*Eurylaimus javanicus*	R, 4	L. EXT		
383		Black-and-Yellow Broadbill	*Eurylaimus ochromalus*	R, 4	L. EXT	**NT**	
384		Dusky Broadbill	*Corydon sumatranus*	R, 3	L. EXT		
385	**Pittidae**	Rusty-naped Pitta	*Pitta oatesi*	R, 3	–		

No.	Family	Common name	Scientific name	Status			Notes
				PM	SIN	GCS	
386		Giant Pitta	*Pitta caerulea*	R, 2	–	**NT**	
387		Banded Pitta	*Pitta guajana*	R, 4	–		
388		Garnet Pitta	*Pitta granatina*	R, 4	L. EXT	**NT**	
389		Hooded Pitta	*Pitta sordida*	R, M, 3	M, 2		
390		Blue-winged Pitta	*Pitta moluccensis*	R, M, 4	M, 4		
391		Mangrove Pitta	*Pitta megarhyncha*	R, 3	R, 2	**NT**	
392	**Acanthizidae**	Golden-bellied Gerygone	*Gerygone sulphurea*	R, 5	R, 5		
393	**Eupetidae**	Rail-babbler	*Eupetes macrocerus*	R, 3	–	**NT**	
394	**Incertae sedis**	Large Woodshrike	*Tephrodornis virgatus*	R, 5	L. EXT		
395		Rufous-winged Philentoma	*Philentoma pyrhoptera*	R, 5	L. EXT		
396		Maroon-breasted Philentoma	*Philentoma velata*	R, 4	L. EXT	**NT**	
397	**Artamidae**	Ashy Woodswallow	*Artamus fuscus*	V, 1	–		
398		White-breasted Woodswallow	*Artamus leucorhynchus*	R, 4	–		
399	**Aegithinidae**	Common Iora	*Aegithina tiphia*	R, 6	R, 5		
400		Green Iora	*Aegithina viridissima*	R, 5	L. EXT	**NT**	
401		Great Iora	*Aegithina lafresnayei*	R, 3	–		
402	**Campephagidae**	Javan Cuckooshrike	*Coracina javensis*	R, 4	–		
403		Bar-bellied Cuckooshrike	*Coracina striata*	R, 3	L. EXT		
404		Lesser Cuckooshrike	*Coracina fimbriata*	R, 4	R, 2		
405		Pied Triller	*Lalage nigra*	R, 5	R, 5		
406		Rosy Minivet	*Pericrocotus roseus*	V, 1	–		
407		Ashy Minivet	*Pericrocotus divaricatus*	M, 4	M, 3		
408		Fiery Minivet	*Pericrocotus igneus*	R, 4	L. EXT	**NT**	

No.	Family	Common name	Scientific name	Status			Notes
				PM	SIN	GCS	
409		Grey-chinned Minivet	*Pericrocotus solaris*	R, 5	_		
410		Scarlet Minivet	*Pericrocotus flammeus*	R, 5	R, 2		
411		Bar-winged Flycatcher -shrike	*Hemipus picatus*	R, 5	_		
412		Black-winged Flycatcher- shrike	*Hemipus hirundinaceus*	R, 4	_		
413	**Pachycephalidae**	Mangrove Whistler	*Pachycephala cinerea*	R, 4	R, 3		
414	**Laniidae**	Tiger Shrike	*Lanius tigrinus*	M, 5	M, 4		
415		Brown Shrike	*Lanius cristatus*	M, 6	M, 5		
416		Long-tailed Shrike	*Lanius schach*	R, 4	R, 4		
417	**Vireonidae**	White-browed Shrike -babbler	*Pteruthius flaviscapis*	R, 4	_		
418		Black-eared Shrike- babbler	*Pteruthius melanotis*	R, 4	_		
419		White-bellied Erpornis	*Erpornis zantholeuca*	R, 5	_		
420	**Oriolidae**	Dark-throated Oriole	*Oriolus xanthonotus*	R, 4	L. EXT	NT	
421		Eurasian Golden Oriole	*Oriolus oriolus*	V, 1	_		
422		Black-naped Oriole	*Oriolus chinensis*	R, M, 7	R, 6, M, 2		
423		Black-hooded Oriole	*Oriolus xanthornus*	R, M, 3	_		
424		Black-and-crimson Oriole	*Oriolus cruentus*	R, 4	_		
425	**Dicruridae**	Black Drongo	*Dicrurus macrocercus*	M, 5	M, 3		
426		Ashy Drongo	*Dicrurus leucophaeus*	R, M, 5	D, 2		
427		Crow-billed Drongo	*Dicrurus annectans*	M, 3	M, 3		
428		Bronzed Drongo	*Dicrurus aeneus*	R, 4	L. EXT		
429		Lesser Racquet-tailed Drongo	*Dicrurus remifer*	R, 4	_		
430		Greater Racquet-tailed Drongo	*Dicrurus paradiseus*	R, 5	R, 4		
431	**Rhipiduridae**	White-throated Fantail	*Rhipidura albicollis*	R, 4	_		

No.	Family	Common name	Scientific name	Status			Notes
				PM	SIN	GCS	
432		Pied Fantail	*Rhipidura javanica*	R, 5	R, 5		
433		Spotted Fantail	*Rhipidura perlata*	R, 3	–		
434	**Monarchidae**	Black-naped Monarch	*Hypothymis azurea*	R, 4	R, 2		
435		Asian Paradise-flycatcher	*Terpsiphone paradisi*	R, M, 4	R, 4, M, 2		
436		Japanese Paradise-flycatcher	*Terpsiphone atrocaudata*	M, 3	M, 2	**NT**	
437	**Corvidae**	Crested Jay	*Platylophus galericulatus*	R, 3	–	**NT**	
438		Black Magpie	*Platysmurus leucopterus*	R, 4	–	**NT**	
439		Common Green Magpie	*Cissa chinensis*	R, 4	–		
440		Racquet-tailed Treepie	*Crypsirina temia*	R, 2	–		
441		House Crow	*Corvus splendens*	F, 6	F, 4		
442		Slender-billed Crow	*Corvus enca*	R, 3	–		
443		Large-billed Crow	*Corvus macrorhynchos*	R, 5	R, 4		
444	**Paridae**	Great Tit	*Parus major*	R, 4	–		
445		Sultan Tit	*Melanochlora sultanea*	R, 4	–		
446	**Hirundinidae**	Collared Sand Martin	*Riparia riparia*	M, 3	M, 3		
447		Dusky Crag-martin	*Ptyonoprogne concolor*	R, 3	–		
448		Barn Swallow	*Hirundo rustica*	M, 7	M, 6		
449		Pacific Swallow	*Hirundo tahitica*	R, 7	R, 5		
450		Red-rumped Swallow	*Cecropis daurica*	M, 4	M, 3		
451		Striated Swallow	*Cecropis striolata*	R, 4	–		
452		Asian House-martin	*Delichon dasypus*	M, 3	M, 3		
453	**Pycnonotidae**	Straw-headed Bulbul	*Pycnonotus zeylanicus*	R, 3	R, 3	**VU**	
454		Black-and-white Bulbul	*Pycnonotus melanoleucos*	R, 3	L. EXT	**NT**	

No.	Family	Common name	Scientific name	Status			Notes
				PM	SIN	GCS	
455		Black-headed Bulbul	*Pycnonotus atriceps*	R, 4	R, 2		
456		Black-crested Bulbul	*Pycnonotus melanicterus*	R, 4	_		
457		Scaly-breasted Bulbul	*Pycnonotus squamatus*	R, 3	_	**NT**	
458		Grey-bellied Bulbul	*Pycnonotus cyaniventris*	R, 4	L. EXT	**NT**	
459		Red-whiskered Bulbul	*Pycnonotus jocosus*	R, 3	F, 3		
460		Sooty-headed Bulbul	*Pycnonotus aurigaster*	_	F, 2		
461		Puff-backed Bulbul	*Pycnonotus eutilotus*	R, 3	L. EXT	**NT**	
462		Yellow-vented Bulbul	*Pycnonyus goiavier*	R, 7	R, 6		
463		Stripe-throated Bulbul	*Pycnonotus finlaysoni*	R, 6	_		
464		Olive-winged Bulbul	*Pycnonotus plumosus*	R, 5	R, 4		
465		Streak-eared Bulbul	*Pycnonotus blanfordi*	R, 4	_		
466		Cream-vented Bulbul	*Pycnonotus simplex*	R, 4	R, 3		
467		Red-eyed Bulbul	*Pycnonotus brunneus*	R, 5	R, 2		
468		Spectacled Bulbul	*Pycnonotus erythrophthalmos*	R, 5	L. EXT		
469		Sulphur-throated Bulbul	*Alophoixus finschii*	R, 4	_	**NT**	
470		Ochraceous Bulbul	*Alophoixus ochraceus*	R, 5	_		
471		Grey-cheeked Bulbul	*Alophoixus bres*	R, 4	_		
472		Yellow-bellied Bulbul	*Alophoixus phaeocephalus*	R, 4	L. EXT		
473		Hairy-backed Bulbul	*Tricholestes criniger*	R, 5	_		
474		Buff-vented Bulbul	*Iole olivacea*	R, 5	R, 2	**NT**	
475		Streaked Bulbul	*Ixos malaccensis*	R, 4	D, 2	**NT**	
476		Mountain Bulbul	*Hypsipetes mcclellandii*	R, 5	_		
477		Ashy Bulbul	*Hemixos flavala*	R, 4	D, 2		

No.	Family	Common name	Scientific name	Status			Notes
				PM	SIN	GCS	
478	**Sylviidae**	Lanceolated Warbler	*Locustella lanceolata*	M, 3	M, 2		
479		Rusty-rumped Warbler	*Loscustella certhiola*	M, 3	M, 3		
480		Dusky Warbler	*Phylloscopus fuscatus*	M, 2	M, 1		
481		Radde's Warbler	*Phylloscopus schwarzi*	V, 1	–		
482		Yellow-browed Warbler	*Phylloscopus inornatus*	M, 4	M, 2		
483		Arctic Warbler	*Phylloscopus borealis*	M, 6	M, 5		
484		Two-barred Warbler	*Phylloscopus plumbeitarsus*	V, 1	–		
485		Pale-legged Leaf-warbler	*Phylloscopus tenellipes*	M, 2	–		
486		Eastern-crowned Warbler	*Phylloscopus coronatus*	M, 4	M, 3		
487		Mountain Leaf -warbler	*Phylloscopus trivirgatus*	R, 5	–		
488		Plain-tailed Warbler	*Seicercus soror*	V, 1	–		
489		Chestnut-crowned Warbler	*Seicercus castaniceps*	R, 4	–		
490		Yellow-breasted Warbler	*Seicercus montis*	R, 4	–		
491		Yellow-bellied Warbler	*Abroscopus superciliaris*	R, 5	–		
492		Black-browed Reed-warbler	*Acrocephalus bistrigiceps*	M, 4	M, 3		
493		Oriental Reed-warbler	*Acrocephalus orientalis*	M, 5	M, 5		
494		Thick-billed Warbler	*Acrocephalus aedon*	V, 1	–		
495	Timaliidae	Oriental White-eye	*Zosterops palpebrosus*	R, 5	L. EXT		
496		Everett's White-eye	*Zosterops everetti*	R, 4	–		
497		Rufous-winged Fulvetta	*Pseudominla castaneceps*	R, 4	–		
498		Brown Fulvetta	*Alcippe brunneicauda*	R, 4	–	**NT**	
499		Mountain Fulvetta	*Alcippe peracensis*	R, 5	–		
500		Large Scimitar-babbler	*Pomatorhinus hypoleucos*	R, 3	–		

No.	Family	Common name	Scientific name	Status			Notes
				PM	SIN	GCS	
501		Chestnut-backed Scimitar-babbler	*Pomatorhinus montanus*	R, 4	_		
502		Rufous-fronted Babbler	*Stachyris rufifrons*	R, 4	_		
503		Golden Babbler	*Stachyris chrysaea*	R, 5	_		
504		Chestnut-winged Babbler	*Stachyris erythroptera*	R, 5	R, 3		
505		Grey-throated Babbler	*Stachyris nigriceps*	R, 5	_		
506		Grey-headed Babbler	*Stachyris poliocephala*	R, 4	_		
507		White-necked Babbler	*Stachyris leucotis*	R, 2	_	NT	
508		Black-throated Babbler	*Stachyris nigricollis*	R, 3	L. EXT	NT	
509		Chestnut-rumped Babbler	*Stachyris maculata*	R, 5	L. EXT	NT	
510		Striped Tit-babbler	*Macronous gularis*	R, 6	R, 5		
511		Fluffy-backed Tit-babbler	*Macronous ptilosus*	R, 5	L. EXT	NT	
512		Puff-throated Babbler	*Pellorneum ruficeps*	R, 5	_		
513		Buff-breasted Babbler	*Pellorneum tickelli*	R, 4	_		
514		Black-capped Babbler	*Pellorneum capistratum*	R, 4	L. EXT		
515		Moustached Babbler	*Malacopteron magnirostre*	R, 5	R, 1		
516		Sooty-capped Babbler	*Malacopteron affine*	R, 5	_	NT	
517		Scaly-crowned Babbler	*Malacopteron cinereum*	R, 5	_		
518		Rufous-crowned Babbler	*Malacopteron magnum*	R, 5	_	NT	
519		Grey-breasted Babbler	*Ophrydornis albogulare*	R, 2	L. EXT	NT	
520		White-chested Babbler	*Trichastoma rostratum*	R, 4	R, 2	NT	
521		Ferruginous Babbler	*Trichastoma bicolor*	R, 4	_		
522		Abbott's Babbler	*Malacocincla abbotti*	R, 5	R, 4		

No.	Family	Common name	Scientific name	Status			Notes
				PM	SIN	GCS	
523		Horsfield's Babbler	*Malacocincla sepiarium*	R, 4	_		
524		Short-tailed Babbler	*Malacocincla malaccense*	R, 4	R, 3	**NT**	
525		Striped Wren-babbler	*Kenopia striata*	R, 3	L. EXT	**NT**	
526		Large Wren-babbler	*Napothera macrodactyla*	R, 3	L. EXT	**NT**	
527		Marbled Wren-babbler	*Napothera marmorata*	R, 3	_		
528		Streaked Wren-babbler	*Napothera brevicaudata*	R, 4	_		
529		Eye-browed Wren-babbler	*Napothera epilepidota*	R, 3	_		
530		Pygmy Wren-babbler	*Pnoepyga pusilla*	R, 4	_		
531		Collared Babbler	*Gampsorhynchus torquatus*	R, 3	_		
532		Cutia	*Cutia nipalensis*	R, 3	_		
533		White-crested Laughingthrush	*Garrulax leucolophus*	F, 2	F, 4		
534		Black Laughingthrush	*Garrulax lugubris*	R, 4	_		
535		Spectacled Laughingthrush	*Garrulax mitratus*	R, 5	_		
536		Hwamei	*Garrulax canorus*	_	F, 3		
537		Chestnut crowned Laughingthrush	*Garrulax erythrocephalus*	R, 4	_		
538		Silver-eared Mesia	*Leiothrix argentauris*	R, 5	_		
539		Long-tailed Sibia	*Heterophasia picaoides*	R, 5	_		
540		Blue-winged Minla	*Minla cyanouroptera*	R, 5	_		
541		Chestnut-tailed Minla	*Minla strigula*	R, 4	_		
542	**Cisticolidae**	Zitting Cisticola	*Cisticola juncidis*	R, 4	R, 4		
543		Hill Prinia	*Prinia atrogularis*	R, 2	_		
544		Rufescent Prinia	*Prinia rufescens*	R, 4	_		

No.	Family	Common name	Scientific name	Status			Notes
				PM	SIN	GCS	
545		Yellow-bellied Prinia	*Prinia flaviventris*	R, 5	R, 4		
546		Mountain Tailorbird	*Phyllergates cuculatus*	R, 4	–		
547		Common Tailorbird	*Orthotomus sutorius*	R, 6	R, 5		
548		Dark-necked Tailorbird	*Orthotomus atrogularis*	R, 5	R, 4		
549		Rufous-tailed Tailorbird	*Orthotomus sericeus*	R, 4	R, 3		
550		Ashy Tailorbird	*Orthotomus ruficeps*	R, 5	R, 4		
551	**Sittidae**	Velvet-fronted Nuthatch	*Sitta frontalis*	R, 4	–		
552		Blue Nuthatch	*Sitta azurea*	R, 4	–		
553	**Sturnidae**	Asian Glossy Starling	*Aplonis panayensis*	R, 6	R, 5		
554		Purple-backed Starling	*Sturnus sturninus*	M, 4	M, 4		
555		Chestnut-cheeked Starling	*Sturnus philippensis*	V, 2	V, 1		
556		White-shouldered Starling	*Sturnus sinensis*	M, 3	M, 3		
557		Rosy Starling	*Sturnus roseus*	V, 1	V, 1		
558		Black-winged Myna	*Acridotheres melanopterus*	–	F, 1	**CR**	
559		Common Myna	*Acridotheres tristis*	R, 7	R, 6		
560		Jungle Myna	*Acridotheres fuscus*	R, 4	–		
561		Javan Myna	*Acridotheres javanicus*	F, 6	F, 7		
562		White-vented Myna	*Acridotheres grandis*	F, 4	–		
563		Crested Myna	*Acridotheres cristatellus*	F, 4	F, 2		
564		Golden-crested Myna	*Ampeliceps coronatus*	V, 1	–		
565		Hill Myna	*Gracula religiosa*	R, 5	R, 3		
566	**Turdidae**	White-throated Rock-thrush	*Monticola gularis*	M, 2	M, 2		
567		Blue Rock-thrush	*Monticola solitarius*	R, M, 4	M, 2		

No.	Family	Common name	Scientific name	Status			Notes
				PM	SIN	GCS	
568		Malaysian Whistling-thrush	*Myophonus robinsoni*	R, 3	–		
569		Blue Whistling-thrush	*Myophonus caeruleus*	R, 4, M?	–		
570		Chestnut-capped Thrush	*Zoothera interpres*	R, 2	–	**NT**	
571		Orange-headed Thrush	*Zoothera citrina*	M, 4	M, 2		
572		Siberian Thrush	*Zoothera sibirica*	M, 4	M, 3		
573		Eurasian Scaly Thrush	*Zoothera dauma*	V, 1	–		
574		Eyebrowed Thrush	*Turdus obscurus*	M, 4	M, 3		
575		Dark-throated Thrush	*Turdus ruficollis*	V, 1	–		
576		Lesser Shortwing	*Brachypteryx leucophrys*	R, 4	–		
577		Common Stonechat	*Saxicola torquatus*	M, 3	M, 2		
578		Siberian Rubythroat	*Luscinia calliope*	V, 1	–		
579		Rufous-headed Robin	*Luscinia ruficeps*	V, 1	–	**VU**	
580		Siberian Blue Robin	*Luscinia cyane*	M, 5	M, 4		
581		Oriental Magpie Robin	*Copsychus saularis*	R, 6	R, 3		
582		White-rumped Shama	*Copsychus malabaricus*	R, 5	R, 2		
583		Rufous-tailed Shama	*Trichixos pyrropygus*	R, 3	–	**NT**	
584		Chestnut-naped Forktail	*Enicurus ruficapillus*	R, 4	–	**NT**	
585		Slaty-backed Forktail	*Enicurus schistaceus*	R, 4	–		
586		White-crowned Forktail	*Enicurus leschenaulti*	R, 3	–		
587		White-tailed Robin	*Myiomela leucura*	R, 4	–		
588	**Muscicapidae**	Taiga Flycatcher	*Ficedula albicilla*	M, 2	–		
589		Rufous-browed Flycatcher	*Ficedula solitaris*	R, 5	–		
590		Snowy-browed Flycatcher	*Ficedula hyperythra*	R, 4	–		

No.	Family	Common name	Scientific name	Status			Notes
				PM	SIN	GCS	
591		Rufous-chested Flycatcher	*Ficedula dumetoria*	R, 3	–	NT	
592		Little Pied Flycatcher	*Ficedula westermanni*	R, 5	–		
593		Yellow-rumped Flycatcher	*Ficedula zanthopygia*	M, 5	M, 4		
594		Narcissus Flycatcher	*Ficedula narcissina*	M, 3	M, 2		
595		Mugimaki Flycatcher	*Ficedula mugimaki*	M, 4	M, 3		
596		Blue-and-White Flycatcher	*Cyanoptila cyanomelana*	M, 3	M, 2		
597		Large Niltava	*Niltava grandis*	R, 4	–		
598		Rufous-vented Niltava	*Niltava sumatrana*	R, 3	–		
599		Pygmy Blue-flycatcher	*Muscicapella hodgsoni*	R, 3	–		
600		White-tailed Flycatcher	*Cyornis cancretus*	R, 2	–		
601		Pale Blue- flycatcher	*Cyornis unicolor*	R, 4	–		
602		Blue-throated Flycatcher	*Cyornis rubeculoides*	M, 3	M, 1		
603		Hill Blue-flycatcher	*Cyornis banyumas*	R, M, 4	–		
604		Tickell's Blue-flycatcher	*Cyornis tickelliae*	R, M, 5	–		
605		Mangrove Blue-flycatcher	*Cyornis rufigastra*	R, 4	R, 2		
606		Malaysian Blue-flycatcher	*Cyornis turcosus*	R, 3	–	NT	
607		Brown-chested Jungle-flycatcher	*Rhinomyias brunneatus*	M, 2	M, 2	VU	
608		Grey-chested Jungle-flycatcher	*Rhinomyias umbratilis*	R, 3	–	NT	
609		Verditer Flycatcher	*Eumyias thalassinus*	R, 5	–		
610		Dark-sided Flycatcher	*Muscicapa sibirica*	M, 4	M, 3		
611		Ferruginous Flycatcher	*Muscicapa ferruginea*	M, 4	M, 3		
612		Asian Brown Flycatcher	*Muscicapa dauurica*	R, 3, M, 6	M, 4		

No.	Family	Common name	Scientific name	Status			Notes
				PM	SIN	GCS	
613		Grey-headed Canary -flycatcher	*Culicicapa ceylonensis*	R, 5	_		
614	**Irenidae**	Asian Fairy-bluebird	*Irena puella*	R, 5	R, 4		
615	**Chloropseidae**	Greater Green Leafbird	*Chloropsis sonnerati*	R, 5	R, 3		
616		Lesser Green Leafbird	*Chloropsis cyanopogon*	R, 4	R, 3	**NT**	
617		Blue-winged Leafbird	*Chloropsis cochinchinensis*	R, 5	R, 4		
618		Orange-bellied Leafbird	*Chloropsis hardwickii*	R, 4	_		
619	**Dicaeidae**	Yellow-breasted Flowerpecker	*Prionochilus maculatus*	R, 5	L. EXT		
620		Crimson-breasted Flowerpecker	*Prionochilus percussus*	R, 4	_		
621		Scarlet-breasted Flowerpecker	*Prionochilus thoracicus*	R, 3	_	**NT**	
622		Thick-billed Flowerpecker	*Dicaeum agile*	R, 3	D, 2		
623		Brown-backed Flowerpecker	*Dicaeum everetti*	R, 1	_	**NT**	
624		Yellow-vented Flowerpecker	*Dicaeum chrysorrheum*	R, 4	R, 3		
625		Orange-bellied Flowerpecker	*Dicaeum trigonostigma*	R, 5	R, 4		
626		Plain Flowerpecker	*Dicaeum concolor*	R, 4	L. EXT		
627		Fire-breasted Flowerpecker	*Dicaeum ignipectus*	R, 5			
628		Scarlet-backed Flowerpecker	*Dicaeum cruentatum*	R, 6	R, 5		
629	**Nectariinidae**	Ruby-cheeked Sunbird	*Chalcoparia singalensis*	R, 5	_		
630		Purple-naped Sunbird	*Hypogramma hypogrammicum*	R, 5	L. EXT		
631		Plain Sunbird	*Anthreptes simplex*	R, 5	?		
632		Brown-throated Sunbird	*Anthreptes malacensis*	R, 6	R, 5		
633		Red-throated Sunbird	*Anthreptes rhodolaemus*	R, 3	L. EXT	**NT**	
634		Purple-throated Sunbird	*Leptocoma sperata*	R, 4	R, 3		

No.	Family	Common name	Scientific name	Status			Notes
				PM	SIN	GCS	
635		Copper-throated Sunbird	*Leptocoma calcostetha*	R, 4	R, 3		
636		Olive-backed Sunbird	*Cinnyris jugularis*	R, 6	R, 5		
637		Black-throated Sunbird	*Aethopyga saturata*	R, 5	–		
638		Crimson Sunbird	*Aethopyia siparaja*	R, 4	R, 4		
639		Scarlet Sunbird	*Aethopyga temminckii*	R, 3	–		
640		Little Spiderhunter	*Arachnothera longirostra*	R, 5	R, 4		
641		Thick-billed Spiderhunter	*Arachnothera crassirostris*	R, 3	R, 2		
642		Long-billed Spiderhunter	*Arachnothera robusta*	R, 4	–		
643		Spectacled Spiderhunter	*Arachnothera flavigaster*	R, 4	L. EXT		
644		Yellow-eared Spiderhunter	*Arachnothera chrysogenys*	R, 4	R, 2		
645		Grey-breasted Spiderhunter	*Arachnothera affinis*	R, 4	L. EXT		
646		Streaked Spiderhunter	*Arachnothera magna*	R, 5	–		
647	**Passeridae**	Plain-backed Sparrow	*Passer flaveolus*	R, 3	–		
648		House Sparrow	*Passer domesticus*	–	F, 2		
649		Eurasian Tree Sparrow	*Passer montanus*	R, 7	R, 6		
650	**Ploceidae**	Baya Weaver	*Ploceus philippinus*	R, 5	R, 3		
651		Streaked Weaver	*Ploceus manyar*	–	F, 2		
652	**Estrildidae**	Tawny-breasted Parrotfinch	*Erythrura hyperythra*	R, 2	–		
653		Pin-tailed Parrotfinch	*Erythrura prasina*	R, 3	–		
654		White-bellied Munia	*Lonchura leucogastra*	R, 3	–		
655		White-rumped Munia	*Lonchura striata*	R, 5	R, 2		
656		Javan Munia	*Lonchura leucogastroides*	–	F, 2		
657		Scaly-breasted Munia	*Lonchura punctulata*	R, 6	R, 5		

No.	Family	Common name	Scientific name	Status			Notes
				PM	SIN	GCS	
658		Chestnut Munia	*Lonchura malacca*	R, 5	R, 4		
659		White-headed Munia	*Lonchura maja*	R, 5	R, 3		
660		Java Sparrow	*Padda oryzivora*	F, 2	F, 1	**VU**	
661	**Motacillidae**	Forest Wagtail	*Dendronanthus indicus*	M, 4	M, 3		
662		White Wagtail	*Motacilla alba*	M, 2	M, 2		
663		Citrine Wagtail	*Motacilla citreola*	–	M, 1		
664		Yellow Wagtail	*Motacilla flava*	M, 5	M, 4		
665		Grey Wagtail	*Motacilla cinerea*	M, 5	M, 3		
666		Paddyfield Pipit	*Anthus rufulus*	R, 6	R, 5		
667		Red-throated Pipit	*Anthus cervinus*	M, 3	M, 3		
668		Olive-backed Pipit	*Anthus hodgsoni*	M, 4	–		
669		Blyth's Pipit	*Anthus godlewskii*	V, 1	–		
670	**Fringillidae**	Brown Bullfinch	*Pyrrhula nipalensis*	R, 2	–		
671	**Emberizidae**	Chestnut-eared Bunting	*Emberiza fucata*	V, 1	–		
672		Chestnut Bunting	*Emberiza rutila*	V, 1	–		
673		Yellow-breasted Bunting	*Emberiza aureola*	M, 1	M, 1	**VU**	

Key to the checklist

Abbreviations

PM Peninsular Malaysia
SIN Singapore
GCS Global Conservation Status according to IUCN Red List 8 (IUCN, 2009).

Categories

(a) Status of birds

R (Resident)	A species is considered resident if it is known to breed within the review area.
M (Migrant)	A species is considered a migrant if it uses the review area as a staging point (passage migrant) or as a wintering ground during both northern and southern hemisphere winters (winter visitor).
V (Vagrant)	Migratory species which end up in areas outside their normal wintering range, usually with only one or two records.
F (Feral)	Non indigenous species which have established breeding populations in both Peninsular Malaysia and Singapore.
D (Dispersals)	Refers to resident populations from Peninsular Malaysia which disperse over a wider area, including Singapore, outside the breeding season.
L. EXT (Locally extinct)	A former breeding or presumed breeding species which no longer occurs in the review area. In the context of Singapore, refers to resident species which have died out due to changes in habitat on the island since 1819.
? (Status uncertain)	Very little or no information on the actual status of the bird.

(b) Abundance

1	Very rare	About one or two reports per year.
2	Rare and localized	One or two sites where there is a slim chance of recording it.
3	Uncommon and localized	Some sites where there is a moderate chance of recording it.
4	Locally common	A good chance of recording the species in the right habitat.
5	Fairly common	Fairly commonly encountered in the right habitat.
6	Common and widespread	Common in a variety of habitats.
7	Abundant and widespread	Easily seen or heard in a variety of habitats.

(c) Global conservation status

NT. (Near Threatened)	A species currently not classified under any of the endangered categories but is close to qualifying for them in the near future in the lack of conservation efforts.
VU. (Vulnerable)	A species facing a high risk of extinction in the wild.
EN. (Endangered)	A species facing a very high risk of extinction in the wild.
CR. (Critically Endangered)	A species considered to be facing an extremely high risk of extinction in the wild.

Bibliography

Allen, F.G.H. (1955). The nesting of the Bluewinged Pitta (*Pitta megarhyncha* Schleg). *Malay Nat J* **10**: 67–8.

Alström, P. (1997). Field identification of Asian *Gyps* vultures. *Bull Oriental Bird Club* **25**: 32–49.

Alström, P. and Olsson, U. (1999). The Golden-spectacled Warbler: a complex of sibling species, including a previously undescribed species. *Ibis* **141**: 545–68.

Alström, P. and Olsson, U. (2000). Golden-spectacled Warbler systematics. *Ibis* **142**: 495–500.

Bakewell, D.N. (1988). Little Stint (*Calidris minuta)* at Kuala Selangor, Sept 30–Oct 1, 1987. *Enggang* **1**(10): 9.

Bakewell, D.N. (1989). Nordmann's Greenshanks *Tringa guttifer* in winter. *Enggang* **2**(1): 10–11.

Bakewell, D.N. (1989a). Sekinchan Harrier Roost. *Enggang* **2**(1): 1.

Bakewell, D.N. (1989b). Common swifts and swallows of the Klang Valley. *Enggang* **2**(7): 7–11.

Bakewell, D.N. (2006). A record of Small Buttonquail *Turnix sylvaticus* in Kedah, Malaysia, 10 July 2006. *Suara Enggang* **14**(4): 9–11.

Bakewell, D.N. (2007). First documented record of Eurasian Wryneck *Jynx torquilla* in Malaysia, at Sungai Dua rice fields, Penang. *Suara Enggang* **15**(2): 9–11.

Bakewell, D.N. (2007a). A record of Pectoral Sandpiper *Calidris melanotos* at Pulau Burung landfill site, Pulau Pinang, 12 September 2007. *Suara Enggang* **15**(5): 11–13.

Bakewell, D.N. (complier) (2009). Kapar Power Station waterbirds report 2008: MNS-Bird Conservation Council (MNS-BCC Waterbirds Group Report No.1). Kuala Lumpur.

Bakewell, D.N. (2010). Blyth's Pipit *Anthus godlewskii* Chuping Sugarcane Plantation, Perlis, 7 January 2010. *Suara Enggang* **18**(1): 7–10.

Bakewell, D.N. (2010a). New records for Malaysia – The work of the Malaysian Nature Society Records Committee (MNS-BCC RC) and updates to the status of birds in Malaysia. *BirdingASIA* **13**: 22–9.

Bakewell, D.N. and Eng, T.C. (2007). A new bird for Malaysia: Rosy Minivet *Pericrocotus roseus* at Bintang Hijau Forest Reserve, Perak. *Suara Enggang* **15**(2): 5–8.

Bakewell, D.N., Lim K.C., bin Mohd Sah, S.A., and Muin, M.A. (2010). Malaysian Whistling-thrush *Myophonus robinsoni* – field identification, distribution and conservation concerns. *Birding ASIA* **14**: 23–9.

Bakewell, D.N., Lim K.C., and bin Mohd Sah, S.A. (2010a). *A preliminary assessment of the field identification, status, distribution and population of regionally endemic whistling-thrush taxa in Peninsular Malaysia*. Report on a survey funded by the Tan Kean Cheong Bird Conservation Memorial Fund, Perak, Malaysia.

Batchelor, D.M. (1958). A checklist of the birds seen at Asahan, Malacca Part 1. *Malay Nat J* **12**: 164–82.

BirdLife International (2004). *Important Bird in Asia: key sites for conservation*. Cambridge, UK: BirdLife International. (BirdLife Conservation Series No. 13).

BirdLife International (2010). *Data Zone (Search for Species)* (http://www.birdlife.org/datazone/speciesfactsheet.php)

Brisbin, I.L., Jr, Townsend P.A., Okimoto, R., and Amjato, G. (2002). Characterizations of the genetic status of populations of Red Junglefowl. *JBHNS*, **99**(2).

Bromley, E.H. (1952). A note on birds seen at Maxwell's Hill, Perak, April 1950 and February 1951. *Bull Raffles's Museum* **24**: 199–218.

Bucknill, J.A.S. and Chasen, F.N. (1927). *The Birds of Singapore Island*. Government Printers, Singapore.

Chan, S. (2001). Repeated Sightings of the Dusky Eagle Owl (*Bubo coromandus*), at Bukit Merah, Kerian District, Perak. *Suara Enggang* **9**: 12–16.

Chasen, F.N. (1939). *The Birds of the Malay Peninsula Vol IV (The birds of the low country jungle and scrub)*. H.F. and G. Witherby Ltd, London.

Chiu S.C. (2005). Observations on the breeding ecology of Oriental Honey-buzzard *Pernis ptilorhyncus torquatus* in Malaysia. Paper presented at the ARRCN 4th Symposium on Asian Raptors.

Chiu S.C. (2009). Oriental Darter *Anhinga melanogaster* at Kinta Nature Park, Perak, 13 December 2009, *Suara Enggang* **17**(6): 27.

Chong, M.H.N. (1994). Masked Finfoot *Heliopais personata* in Peninsular Malaysia. *Bull Oriental Bird Club* **20**: 28–31.

Chong, M.H.N. (1996). To the Editor: Occurrence of Himalayan Griffon Vulture *Gyps himalayensis* in Malaysia. *Bull Oriental Bird Club* **24**: 42–3.

Chong, M.H.N and Sutari, S. (2010). *Khazanah Burung Bukit Fraser – Panduan Bergambar*. Tinta Publishers, Kuala Lumpur.

Choo, T.C. (2009). Pied Cuckoo *Clamator jacobinus* at Penanti, Pulau Pinang, 6 December 2009, *Suara Enggang* **17**(6): 26.

Choudhury, A. (1991). New for Assam: the White-vented Myna. *Bull Oriental Bird Club* **13**: 20–1.

Clark, W.S. and Schmitt, N.J. (1993). Field Identification of the Rufous-bellied Eagle *Hieraaetus kienerii*. *Forktail* **8**: 7–9.

Coates, B.J. and Bishop, K.D. (1997). *A Guide to The Birds of Wallacea*. Dove Publications, Alderney, Queensland.

Collar, N.J., Crosby, M.J., and Stattersfield, A.J. (1994). *Birds to watch 2, the world list of threatened birds*. BirdLife International, Cambridge.

Collar, N.J. and Robson, C. (2007). Family Timaliidae (Babblers). In del Hoyo, J., Elliot, A., and Christie, D.A (eds) *Handbook of the Birds of the World*, 12, pp. 70–291. Lynx Edicions, Barcelona.

Davison, G.W.H. (1982). Systematics within the genus *Arborophila* Hodgson. *Federation Museums Journal* **27**: 125–34.

Davison, G.W.H. (ed) (1988). *Endau Rompin, A Malaysian Heritage*. Malayan Nature Society, Kuala Lumpur.

Davison, G.W.H., Phillipps, K., and dan Alias, K. (1989). *Pengenalan Burung-Burung Malaysia*. WWF Malaysia, Kuala Lumpur.

Declacour, J. (1977). *The pheasants of the world* (2nd edn). Spur Publications, Surrey.

Dickinson, E.C. (ed) (2003). *The Howard and Moore Complete Checklist of the Birds of the World* (3rd edn). Christopher Helm, London.

Dickinson, E.C. and Dekker, R.W.R.J. (2001). Systematic notes on Asian birds. 13. A preliminary review of the Hirundinidae. *Zool Verh Leiden* **335**: 127–44.

Dobby, E.H.G. (1973). *Southeast Asia*. University of London Press, Ltd, London.

Duckett, J.E and Duckett, N.E. (1982). Notes on the Large-tailed Nightjar *(Caprimulgus macrurus* Horsfield). *Malay Nat J* **35**: 299–313.

Duckett, J.E. (1988). The Brown Shrike (*Lanius cristatus). Malayan Nat* **41**(3 and 4): 6–8.

Enticott, J. and Tipling, D. (1997). *Seabirds of The World*. Stackpole Books, Pennsylvania, PA.

Fry, C.H., Fry, K., and Harris, A. (1992). *Kingfishers, bee-eaters and rollers*. Christopher Helm, London.

Gill, F. and Wright, M. (2006). *Birds of the World – Recommended English Names*. Christopher Helm, London.

Glenister, A.G. (1971). *The Birds of the Malay Peninsula, Singapore & Penang*. Oxford University Press, Kuala Lumpur.

Gooders, J. (1975). *An illustrated survey of the bird families of the world*. Hamlyn, London.

Gosler, A. (ed) (1991). *The Hamlyn Photographic Guide to the Birds of the World*. Hamlyn, London.

Green, A.J. (1991). Focus on the White-winged Wood Duck. *Bull Oriental Club* **14**: 25–7.

Green, A.J. (1993). The biology of the White-winged Duck *Cairina scutulata. Forktail* **8**: 65–82.

Green, P. (1990). First international recovery of a Blue-throated Bee-eater *Merops viridis. Malayan Nat* **43**(3): 33.

Gregory-Smith, R. (1993). Golden-spectacled Warbler *Seicercus burkii*: a new species for Malaysia. *Forktail* **9**: 153.

Hails, C. (1982). Of swifts and swallows. *Malayan Nat* **35**(4): 25–8.

Hails, C. and Jarvis, F. (1987). *Birds of Singapore*. Times Edition, Singapore.

Harrison, P. (1987). *Seabirds of the world: A photographic guide*. Stephen Green Press, Lexington, MA.

Ho, H.C. and Sutari, S. (1997). Spectacular movements of hornbills, possibly Plain-Pouched Hornbills *Aceros subruficollis*, in Peninsular Malaysia. *Bull Oriental Bird Club* **25**: 59–61.

Howes, J.R. (1988). Focus on Egrets. *Enggang* **1**(12): 5–8.

Hum K.C. (2008). A first record in Southeast Asia, Long-legged Buzzard *Buteo rufinus* X Upland Buzzard *Buteo hemilasius SuaraEnggang* **16**(2): 17–18.

Indra, S. and Irshad, M. (2009). A brief observation of predation of Great Hornbill *Buceros bicornis* by Mountain Hawk-eagles *Nisaetus nipalensis* in Pulau Langkawi, Kedah. *Suara Enggang* **17**(6): 11.

Ismail, A.R. (1981). *Senarai nama-nama burung Semenanjung Malaysia dan Asia Tenggara*. Perhilitan, Kuala Lumpur.

Ismail, H.M. and Mohd, N.Y. (1998). The status and current distribution of Crested Argus *Rheinardia ocellata* in Peninsular Malaysia. *Bird Conservation International* **9**: 325–30.

IUCN (2009). *IUCN Red List of Threatened Species* (http://www.iucnredlist.org/).

Jeyarajasingam, A. (1983). The White Egrets of Kuala Gula. *Nat Malaysiana* **8**(2): 26–31.

Jeyarajasingam, A. (1983). Observations on the nest of the Silver-breasted Broadbill (Aves, Eurylaimidae). *Malayan Nat* **37**(2): 14–16.

Jeyarajasingam, A. (1984). The Black-shouldered Kite. *Nat Malaysiana* **9**(3): 12–15.

Jeyarajasingam, A. (1990). A note on birds observed and heard at Bukit Larut (Maxwell's Hill), Perak, in June 1988. *Malayan Nat* **43**(3): 29–33.

Jeyarajasingam, A. (1994). Wedge-tailed Pigeon *Treron sphenura* in Malaysia *Bull Oriental Bird Club* **20**: 47–9.

Jeyarajasingam, A. (1995). Recent records. *Malayan Nat* **49**(1): 32–5.

Jeyarajasingam, A (in prep.). Recommended changes to the English names of some birds of the Oriental Region.

Jeyarajasingam, A (in prep. a). The broadbills of the genus *Calyptomena*. A revision of the English names.

Jeyarajasingam, A (in prep. b). The kingfishers of the genus *Pelargopsis*. A revision of the English names.

Jonsson, L. (1992). *Birds of Europe with North Africa and the Middle East*. Christopher Helm, London.

Kennerley, P. (1989). Does the Rufous-headed Robin spend the winter in Malaysia? *Enggang* **2**(12): 11–12.

Kennerley, P., Bakewell, D.N., and Round, P.D. (2008). Rediscovery of a long lost *Charadrius* plover from South-East Asia. *Forktail* **24**: 6379.

Khoo S.S. (2001). The Green Iora as a brood host of the Banded Bay Cuckoo. *Suara Enggang* 5: 8–9.

King, B.F., Woodcock, M.W., and Dickinson, E.C. (1975). *A Field Guide to the Birds of South-East Asia*. Collins, London.

King, B. F. (1978). A new race of *Pitta oatesi* from Peninsular Malaysia. *Bull Brit Orn Cl* **98**: 109–13.

King, B.F. (2002). The *Hierococcyx fugax,* Hodgson's Hawk-cuckoo complex, *Bull Brit Orn Cl* **122**: 74–80.

King, B.F. (2002a). Species limits in the Brown Boobook *Ninox scutulata* complex. *Bull Brit Orn Cl* **122**: 250–7

King, B.F. (2004). The taxonomic status of the three subspecies of *Cuculus saturatus. Bull Brit Orn Cl* **125**: 48–55

Lambert, F. (1989). Large Green Pigeon. *Bull Oriental Bird Club* **10**: 22–4.

Lambert, F. and Woodcock, M. (1996). *Pittas, Broadbills and Asities*. Pica Press, London.

Lambert, F. (1996). Identification of pittas in the *brachyura* complex in Asia. *Bull Oriental Bird Club* **23**: 31–5.

Law, H.D. (1997). *Malaysian Nature Conservation and Environment Policy: Malaysia Today*. Asia Strategy and Leadership Institute, Kuala Lumpur: pp. 37–40.

Lee, P. (2009). Two Rail-babblers *Eupetes macrocerus* competing for territory. *Suara Enggang* **17**(3): 26–7.

Lekagul, B. and Round, P.D. (1991). *A Guide to the Birds of Thailand*. Saha Karn Bhaet Co. Ltd, Bangkok.

Li, Z.W.D., Siti H.Y., Howes, J., and Ilias, R. (2006). *Status Overview and Recommendations for the Conservation of Milky Stork* Mycteria cinerea *in Malaysia: Final Report of the 2004/2006 Milky Stork Field Surveys in the Matang Mangrove Forest, Perak, Malaysia*. Wetlands International and the Department of Wildlife and National Parks, Peninsular Malaysia, Kuala Lumpur, Malaysia.

Lim, K.Y. (1997). Sustainable Forest Management in Malaysia: Malaysia Today. Asian Strategy and Leadership Institute, Kuala Lumpur: pp. 27–36.

Lim, K.C. (2007). Eurasian Wryneck – a retrospective sighting on 17 October 1990 at Malim Nawar, Perak. *Suara Enggang* **15**(2): 11–12.

Lim, K.C. (2008). A record of Asian Openbill *Anastomus oscitans* in Perlis, North Peninsular Malaysia. *Suara Enggang* **16**(2): 22–3.

Lim, K.C. and Lim, S.Y. (2006). A Jaeger rescued from Temengor Lake, Perak. *Suara Enggang* **14**(4): 5–7.

Lim, K.S. (1990). Checklist of the Birds of Singapore. *SINAV* Vol 4. Special Supplement.

Lim, K.S. (1991). *Pocket Checklist of the Birds of Singapore*. The Nature Society (Singapore), Singapore.

Lim, K.S. (1992). *Vanishing Birds of Singapore*. The Nature Society (Singapore).

Lim, K.S. (1994). Panti Forest Reserve, Johor, Malaysia. *Bull Oriental Bird Club* **20**: 35–6.

Lim, K.S. and Gardner, D. (1997). *Birds: An Illustrated Field Guide to the Birds of Singapore*. Sun Tree Publishing Limited, Singapore.

Lim, K.S. (1997). Interesting Birds: Notes on Malaysian Blue Flycatcher at Sungei Sedili Besar, Johor. *Singapore Avifauna* **11**(2): 26–7.

Lim K.S. (in prep.). Short notes on a new brood host of Asian Drongo- cuckoo *Surniculus lugubris* in the Thai-Malay Peninsula.

MacKinnon, J. and Phillipps, K. (1993). *A Field Guide to the Birds of Borneo, Sumatra, Java and Bali*. Oxford University Press, London.

Madge, S.G. (1969). Notes on the breeding of the Bushy-crested Hornbill *Annorhinus galeritus*. *Malay Nat J* 23: 1–16.

Madge, S. and Burn, H. (1987). *Wildfowl: An identification guide to the ducks, geese and swans of the world*. Christopher Helm, London

Madoc, G.C. (1956). *An Introduction to Malayan Birds*. Malayan Nature Society, Kuala Lumpur.

Madoc, G.C. (1956a). A further note on the Lesser Cuckoo. *Malay Nat J* **10**: 97–103.

Madoc, G.C. and Allen, F.G.H. (1956). Postscript to Pulau Perak. *Malay Nat J* **10**: 131–3.

Malaysian Nature Society (2011). *Malaysia Bird Report 2008*. (Compiled by David Bakewell). Malaysian Nature Society (MNS Conservation Publication No 11), Kuala Lumpur.

McClure, H.E. (1963). Is this one of the rarest birds in the world? *Malay Nat J* **17**: 185–7.

McClure, H.E. (1963). Northern visitors to Malaya. *Malay Nat J* **17**: 232–4.

McClure, H.E. (1964). Know your barbets. *Malay Nat J* **18**: 45–9.

McGowan, P.I.K. (1991). On the trail of the Malaysian Peacock-Pheasant. *Bull Oriental Bird Club* **14**: 16–20.

Medway, Lord and Wells, D.R. (1976). *The Birds of the Malay Peninsula Vol V (Conclusion and survey of every species)*. H.F. and G. Witherby and Penerbit Universiti Malaya, London & Kuala Lumpur.

Mees, G.F. (1996). Geographical variation in birds of Java. *Publications of the Nuttall Ornithological Club* **26**: 13–20.

MNS-Bird Conservation Council (2010). *A checklist of the Birds of Peninsular Malaysia*. Malaysian Nature Society (MNS Conservation Publication No.10): Kuala Lumpur.

Myers, S. (2009). *A Field Guide to The Birds of Borneo*. New Holland Publishers Ltd, London.

Nelson, J.B. (1980). *Seabirds: their biology and ecology*. Hamlyn, London.

Newmark, G.H. (1955). Birds seen at Maxwell's Hill, April, 1955. *Malay Nat J* **10**: 9–12.

Noramly, G. (1984). A tale of two woodpeckers and some dollarbirds. *Malayan Nat* **38**(1): 15–20.

Noramly, G. (1987). Bird Watching for Beginners: Myna Matters. *Malayan Nat* **41**(1): 34–5.

Noramly, G. and Jeyarajasingam, A. (1985). Bird Notes 1984 from the Selangor Branch Bird Group. *Malayan Nat* **39**(1&2): 47–50.

Noramly, M. and Noramly, G. (1985). Layang-layang, island paradise for seabirds. *Nat Malaysiana* **10**(1): 14–23.

Noske, R.A. (1991). Field identification and ecology of the Greater Goldenback *Chrysocolaptes lucidus* in Malaysia. *Forktail* **6**: 72–4.

Onley D. and Scofield, P. (2007). *Albatrosses, Petrels, and Shearwaters of the World*. Oxford University Press, Oxford.

Ooi, B.Y. (1998). Ulu Dedap in March. *Suara Enggang – May–June 1998*: 5–6.

Ooi, B.Y. and Chiu, S.C. (2008). Sighting of Ashy Woodswallow *Artamus fuscus* in Ulu Dedap, Kampung Gajah, Perak. *Suara Enggang* **16** (2): 18–19.

Payne, R.B. and Klitz, K. (2005). *The Cuckoos*. Oxford University Press, Oxford.

Perennou, C. and Mundkur, T. (1991). *Asian Waterfowl Census 1991*. IWRB, Slimbridge.

Peterson, A.T. and Brisbin Jr, I.L. (1998). Genetic endangerment of wild Red Junglefowl *Gallus gallus*? *Bird Conservation International* **8**: 387–94.

Pfander, P. and Schmigalew, S. (2001). Extensive hybridization of Long-legged *Buteo rufinus* and Upland Buzzard *B. hemilasius*. *Ornithologische Mitteilung* **53**: 344–9.

Pizzey, G. and Knight, F. (1997). *The Graham Pizzey and Frank Knight Field Guide to the Birds of Australia*. HarperCollins Pty Limited, Sydney.

Prentice, R.C. (1988). Observations of the nuptial behaviour of the Black-and- Yellow Broadbill *Eurylaimus ochromalus*. *Enggang* **1**(4): 8.

Rasmussen, P.C. and Anderton, J.C. (2005). *Birds of South Asia: The Ripley Guide, 1–2*. Smithsonian Institution and Lynx Edicions, Washington D.C. and Barcelona.

Ratnam, L. (1976). Night Heron. *Nat Malaysiana* **1**(2): 22–7.

Ratnam, L. (1977). Storks and Kuala Gula. *Nat Malaysiana* **2**(4): 38–45.

Reville, B.J. (1989). *A Visitor's Guide to the Birds of Christmas Island, Indian Ocean*. Christmas Island Natural History Association, Christmas Island.

Robinson, H.C. (1927). *The Birds of the Malay Peninsula Vol I (The commoner birds)*. H.F. & G. Witherby Ltd, London.

Robinson, H.C. (1928). *The Birds of the Malay Peninsula Vol II (The birds of the hill stations)*. H.F. and G. Witherby Ltd, London.

Robinson, H.C. and Chasen, F.N. (1936). *The Birds of the Malay Peninsula Vol III (Sporting birds; birds of the shore and estuaries)*. H.F. and G. Witherby Ltd, London.

Robson, C. (2008). *A Field Guide to The Birds of South-East Asia*. New Holland Publishers Ltd, London.

Round, P.D., Treesucon, U., and Eames, J.C. (1989). A breeding record of the Giant Pitta *Pitta caerulea* from Thailand. *Forktail* **5**: 35–47.

Sebastian, A.C. (1994). Kuala Selangor Nature Park – rich and accessible. *Bull Oriental Bird Club* **20**: 33–4.

Sibley, G. and Monroe Jr, B. (1990). *Distribution and taxonomy of birds of the world*. Yale University Press, CT.

Siti, H.Y. (1988). *Kuang Gunung*. *Perhilitan* **8**(1): 2–5.

Slater, P., Slater, P., and Slater, R. (1986). *The Slater Field Guide to Australian Birds*. Lansdowne, Sydney.

Smythies, B.E. (1999). *The Birds of Borneo – 4th Edition*. Natural History Publications, Kota Kinabalu.

Soepadmo, E. (1980). Mangroves. *Nat Malaysiana* **5**(1): 14 –23.

Sonobe, K. and Robinson, J.W. (eds) (1982). *A Field Guide to the Birds of Japan*. Wild Bird Society of Japan, Tokyo.

Sonobe, K. and Usui, S. (eds) (1993). *A Field Guide to the Waterbirds of Asia*. Wild Bird Society of Japan, Tokyo.

Speight, G.J. (2007). A record of Pectoral Sandpiper on Pulau Pinang Island, Sunday 25 March 1984. *Suara Enggang* **15**(5): 13–14.

Strange, M. and Jeyarajasingam, A. (1993). *Birds: A Photographic Guide to The Birds of Peninsular Malaysia and Singapore*. Sun Tree Publishing Limited, Singapore.

Tan, C.E. (2006). A Hoopoe Sighting on Mainland Penang. *Suara Enggang* **14**(5): 4–6.

Tebb, G., Vernon, P.K., and Craig, M. (2003). Laughing Gull *Larus atricilla* in Malaysia: the first record for Asia. *Forktail* **19**: 131–2.

Tweedie, M.W.F. (1970). *Common Birds of the Malay Peninsula – 2nd Edition*. Longmans, Singapore.

Vaurie, C. (1959). *The Birds of the Palearctic fauna. Vol 1: Passeriformes*. Witherby, London.

Wang, L.K. and Hails, C.J. (2007). An annotated checklist of the birds of Singapore. *Raffles Bull Zool* Suppl No. **15**: 1–179.

Wells, D.R. (1982). Bird Report: 1974–1975. *Malay Nat J* **36**(2): 61–85.

Wells, D.R. (1983). Bird Report: 1976–1977. *Malay Nat J* **36**(3): 197–218.

Wells, D.R. (1984). Bird Report: 1978–1979. *Malay Nat J* **38**: 113–50.

Wells, D.R. (1986). Bird Report: 1980–1981. *Malay Nat J* **39**: 279–98.

Wells, D.R. (1990). Bird Report: 1982–1987. *Malay Nat J* **43**: 116–210.

Wells, D.R. (1992). Night migration at Fraser's Hill, Peninsular Malaysia. *Bull Oriental Bird Club* **16**: 21–23.

Wells, D.R. (1999). *The Birds of the Thai-Malay Peninsula – Vol 1 (non-passerines)*. Academic Press, London.

Wells, D.R. (2007). *The Birds of the Thai-Malay Peninsula – Vol 2 (passerines)*. Christopher Helm, London.

White, T. (1984). *A Field Guide to the Bird Songs of Southeast Asia*. The British Library National Sound Archive, London.

Whitmore, T.C. (1984). *Tropical Rain Forests of the Far East*. English Language Book Society/Oxford University Press, London.

Yap, K.T. (1995). Observation Notes: Sighting of Great Hornbill at Gunung Telapa Buruk, Negeri Sembilan. *Malayan Nat* **49**(1): 35.

Yeap, C.A. (2005). Report on Birds of Langkawi Archipelago. *Malay Nat J* **57**(1): 91–144.

Yong, D. (1990). Records of the Streaked Spiderhunter from Gunung Tahan. *Enggang* **3**(3): 26

Yong, D.L. (2010). Commentary: Red Junglefowl making a comeback in Singapore? *Singapore Avifauna* **24**(3): 31–2.

Index

Plate numbers are shown in **bold**.

N

Q

R

S

U

V

W

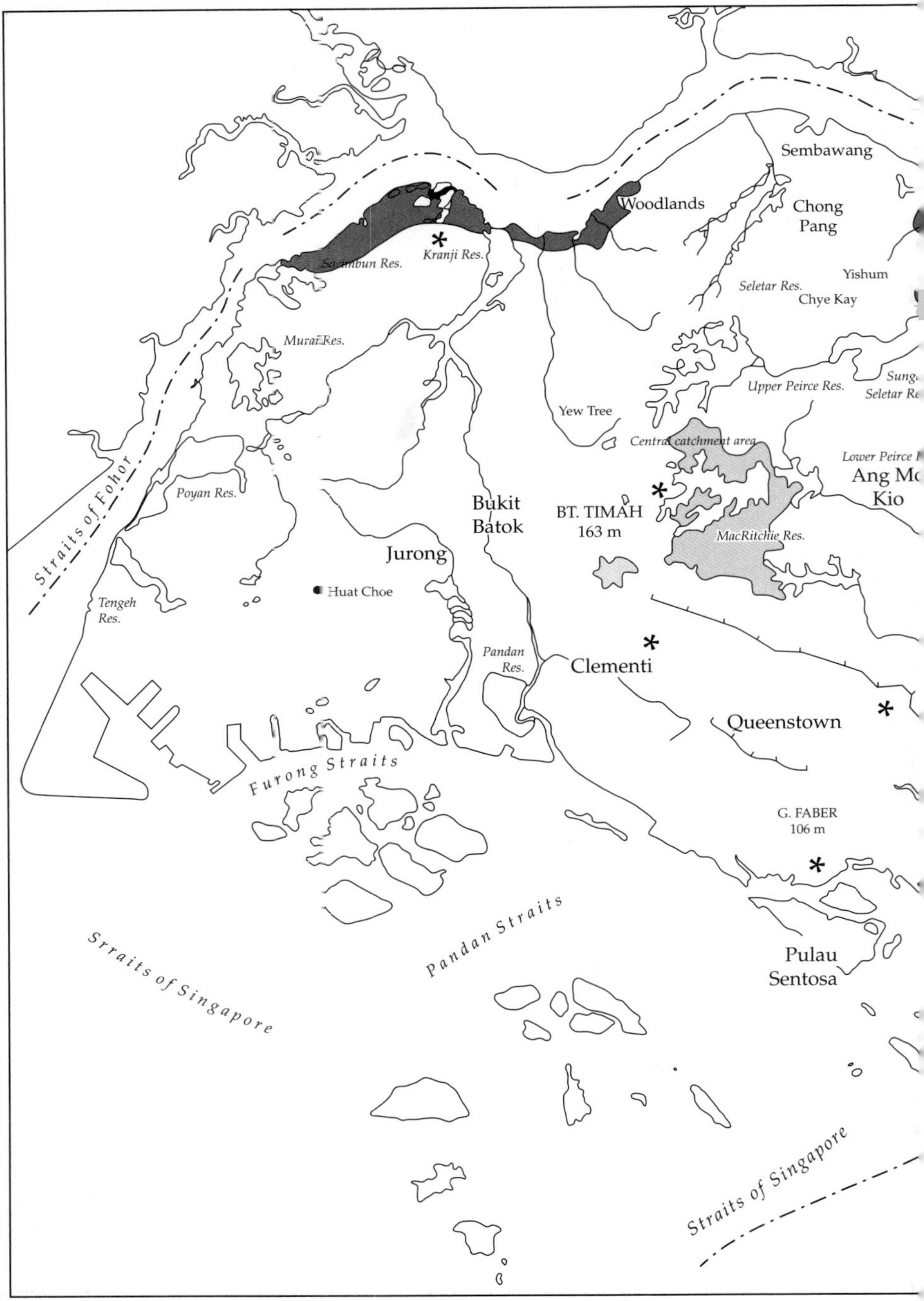

Sembawang
Woodlands
Chong Pang
Kranji Res.
Yishum
Seletar Res.
Chye Kay
Murai Res.
Upper Peirce Res.
Yew Tree
Central catchment area
Straits of Fohor
Poyan Res.
Bukit Batok
BT. TIMAH 163 m
MacRitchie Res.
Jurong
Huat Choe
Tengeh Res.
Pandan Res.
Clementi
Queenstown
Furong Straits
G. FABER 106 m
Pandan Straits
Srraits of Singapore
Pulau Sentosa
Straits of Singapore